Handbook of Item Response Theory

VOLUME TWO

Statistical Tools

Handbook of Item Response Theory, Three-Volume Set

Handbook of Item Response Theory, Volume One: Models

Handbook of Item Response Theory, Volume Two: Statistical Tools

Handbook of Item Response Theory, Volume Three: Applications

Chapman & Hall/CRC

Statistics in the Social and Behavioral Sciences Series

Aims and scope

Large and complex datasets are becoming prevalent in the social and behavioral sciences and statistical methods are crucial for the analysis and interpretation of such data. This series aims to capture new developments in statistical methodology with particular relevance to applications in the social and behavioral sciences. It seeks to promote appropriate use of statistical, econometric and psychometric methods in these applied sciences by publishing a broad range of reference works, textbooks and handbooks.

The scope of the series is wide, including applications of statistical methodology in sociology, psychology, economics, education, marketing research, political science, criminology, public policy, demography, survey methodology and official statistics. The titles included in the series are designed to appeal to applied statisticians, as well as students, researchers and practitioners from the above disciplines. The inclusion of real examples and case studies is therefore essential.

Published Titles

Analyzing Spatial Models of Choice and Judgment with R
David A. Armstrong II, Ryan Bakker, Royce Carroll, Christopher Hare, Keith T. Poole, and Howard Rosenthal

Analysis of Multivariate Social Science Data, Second Edition
David J. Bartholomew, Fiona Steele, Irini Moustaki, and Jane I. Galbraith

Latent Markov Models for Longitudinal Data
Francesco Bartolucci, Alessio Farcomeni, and Fulvia Pennoni

Statistical Test Theory for the Behavioral Sciences
Dato N. M. de Gruijter and Leo J. Th. van der Kamp

Multivariable Modeling and Multivariate Analysis for the Behavioral Sciences
Brian S. Everitt

Multilevel Modeling Using R
W. Holmes Finch, Jocelyn E. Bolin, and Ken Kelley

Bayesian Methods: A Social and Behavioral Sciences Approach, Third Edition
Jeff Gill

Multiple Correspondence Analysis and Related Methods
Michael Greenacre and Jorg Blasius

Applied Survey Data Analysis
Steven G. Heeringa, Brady T. West, and Patricia A. Berglund

Informative Hypotheses: Theory and Practice for Behavioral and Social Scientists
Herbert Hoijtink

Generalized Structured Component Analysis: A Component-Based Approach to Structural Equation Modeling
Heungsun Hwang and Yoshio Takane

Statistical Studies of Income, Poverty and Inequality in Europe: Computing and Graphics in R Using EU-SILC
Nicholas T. Longford

Foundations of Factor Analysis, Second Edition
Stanley A. Mulaik

Linear Causal Modeling with Structural Equations
Stanley A. Mulaik

Age–Period–Cohort Models: Approaches and Analyses with Aggregate Data
Robert M. O'Brien

Handbook of International Large-Scale Assessment: Background, Technical Issues, and Methods of Data Analysis
Leslie Rutkowski, Matthias von Davier, and David Rutkowski

Generalized Linear Models for Categorical and Continuous Limited Dependent Variables
Michael Smithson and Edgar C. Merkle

Incomplete Categorical Data Design: Non-Randomized Response Techniques for Sensitive Questions in Surveys
Guo-Liang Tian and Man-Lai Tang

Handbook of Item Response Theory, Volume 1: Models
Wim J. van der Linden

Handbook of Item Response Theory, Volume 2: Statistical Tools
Wim J. van der Linden

Handbook of Item Response Theory, Volume 3: Applications
Wim J. van der Linden

Computerized Multistage Testing: Theory and Applications
Duanli Yan, Alina A. von Davier, and Charles Lewis

Chapman & Hall/CRC
Statistics in the Social and Behavioral Sciences Series

Handbook of Item Response Theory

VOLUME TWO

Statistical Tools

Edited by
Wim J. van der Linden
Pacific Metrics
Monterey, California

CRC Press
Taylor & Francis Group
Boca Raton London New York

CRC Press is an imprint of the
Taylor & Francis Group, an **informa** business
A CHAPMAN & HALL BOOK

CRC Press
Taylor & Francis Group
6000 Broken Sound Parkway NW, Suite 300
Boca Raton, FL 33487-2742

Printed on acid-free paper
Version Date: 20151130

International Standard Book Number-13: 978-1-4665-1432-4 (Hardback)

Library of Congress Cataloging-in-Publication Data

Names: Linden, Wim J. van der, editor.
Title: Handbook of item response theory / Wim J. van der Linden, [editor].
Description: Boca Raton, FL : CRC Press, 2015- | Series: Chapman & Hall/CRC Statistics in the Social and Behavioral Sciences. | Includes bibliographical references and index.
Identifiers: LCCN 2015034163 | ISBN 9781466514324 (alk. paper : vol. 2)
Subjects: LCSH: Item response theory. | Psychometrics. | Psychology--Mathematical models.
Classification: LCC BF39.2.I84 H35 2015 | DDC 150.28/7--dc23
LC record available at http://lccn.loc.gov/2015034163

Visit the Taylor & Francis Web site at
http://www.taylorandfrancis.com

and the CRC Press Web site at
http://www.crcpress.com

Contents

Section III Parameter Estimation

Section IV Model Fit and Comparison

Contents for *Models*

Contents for *Applications*

Preface

Item response theory (IRT) has its origins in pioneering work by Louis Thurstone in the 1920s, a handful of authors such as Lawley, Mosier, and Richardson in the 1940s, and more decisive work by Alan Birnbaum, Frederic Lord, and George Rasch in the 1950s and 1960s. The major breakthrough it presents is the solution to one of the fundamental flaws inherent in classical test theory—its systematic confounding of what we measure with the properties of the test items used to measure it.

Test administrations are observational studies, in which test takers receive a set of items and we observe their responses. The responses are the joint effects of both the properties of the items and abilities of the test takers. As in any other observational study, it would be a methodological error to attribute the effects to only one of these underlying causal factors. Nevertheless, it seems we need to be able to do so. If new items are field tested, the interest is exclusively in their properties, and any confounding with the abilities of the largely arbitrary selection of test takers used in the study would bias our inferences about them. Likewise, if examinees are tested, the interest is only in their abilities and we do not want their scores to be biased by the incidental properties of the items. Classical test theory does create such biases. For instance, it treats the p-values of the items as their difficult parameters, but these values equally depend on the abilities of the sample of test takers used in the field test. In spite of the terminology, the same holds for its item-discrimination parameters and the definition of test reliability. On the other hand, the number-correct scores classical test theory typically is used for are scores equally indicative of the difficulty of the test as the abilities of test takers. In fact, the tradition of indexing such parameters and scores by the items or test takers only systematically hides this confounding.

IRT solves the problem by recognizing each response as the outcome of a distinct probability experiment that has to be modeled with separate parameters for the item and test-taker effects. Consequently, its item parameters allow us to correct for item effects when we estimate the abilities. Likewise, the presence of the ability parameters allows us to correct for their effects when estimating the item parameter. One of the best introductions to this change of the paradigm is Rasch (1960, Chapter 1), which is mandatory reading for anyone with an interest in the subject. The chapter places the new paradigm in the wider context of the research tradition still found in the behavioral and social sciences with its persistent interest in vaguely defined "populations" of subjects, who, except for some random noise, are treated as exchangeable, as well as its use of statistical techniques as correlation coefficients, analysis of variance, and hypothesis testing that assume "random sampling" from them.

The developments since the original conceptualization of IRT have remained rapid. When Ron Hambleton and I edited an earlier handbook of IRT (van der Linden and Hambleton, 1997), we had the impression that its 28 chapters pretty much summarized what could be said about the subject. But now, nearly two decades later, three volumes with roughly the same number of chapters each appear to be necessary. And I still feel I have to apologize to all the researchers and practitioners whose original contributions to the vast literature on IRT are not included in this new handbook. Not only have the original models for dichotomous responses been supplemented with numerous models for different response formats or response processes, it is now clear, for instance, that models for response times on test items require the same type of parameterization to account both for the item and test-taker effects. Another major development has been the recognition of the need of deeper parameterization due to a multilevel or hierarchical structure of the response data. This development has led to the possibility to introduce explanatory covariates, group structures with an impact on the item or ability parameters, mixtures of response processes, higher-level relationships between responses and response times, or special structures of the item domain, for instance, due to the use of rule-based item generation. Meanwhile, it has also become clear how to embed IRT in the wider development of generalized latent variable modeling. And as a result of all these extensions and new insights, we are now keener in our choice of treating the model parameter as fixed or random. Volume One of this handbook covers most of these developments. Each of its chapters basically reviews one model. However, all chapters have the common format of an introductory section with some history of the model and a motivation of its relevance, and then continue with sections that present the model more formally, treat the estimation of its parameters, show how to evaluate its fit to empirical data, and illustrate the use of the model through an empirical example. The last section discusses further applications and remaining research issues.

As any other type of probabilistic modeling, IRT heavily depends on the use of statistical tools for the treatment of its models and their applications. Nevertheless, systematic introductions and review with an emphasis on their relevance to IRT are hardly found in the statistical literature. Volume Two is to fill this void. Its chapters are on topics such as common probability distributions, the issue of models with both intentional and nuisance parameters, the use of information criteria, methods for dealing with missing data, model identification issues, and several topics in parameter estimation and model fit and comparison. It is especially in these last two areas that recent developments have been overwhelming. For instance, when the previous handbook of IRT was produced, Bayesian approaches had already gained some ground but were certainly not common. But thanks to the computational success of Markov chain Monte Carlo (MCMC) methods, these approaches have now become standard, especially for the more complex models in the second half of Volume One.

The chapters of Volume Three review several applications of IRT to the daily practice of testing. Although each of the chosen topics in the areas of item calibration and analysis, person fit and scoring, and test design have ample resources in the larger literature on test theory, the current chapters exclusively highlight the contributions IRT has brought to them. This volume also offers chapters with reviews of how IRT has advanced areas such as large-scale educational assessments, psychological testing, cognitive diagnosis, health measurement, marketing research, or the more general area of measurement of change. The volume concludes with an extensive review of computer software programs available for running any of the models and applications in Volumes One and Three.

I expect this *Handbook of Item Response Theory* to serve as a daily resource of information to researchers and practitioners in the field of IRT as well as a textbook to novices. To serve them better, all chapters are self-contained. But their common core of notation and extensive cross-referencing allow readers of one of the chapters to consult others for background information without much interruption.

I am grateful to all my authors for their belief in this project and the time they have spent on their chapters. It has been a true privilege to work with each of them. The same holds for Ron Hambleton who was willing to serve as my sparring partner during the conception of the plan for this handbook. John Kimmel, executive editor, statistics,

Chapman & Hall/CRC has been a permanent source of helpful information during the production of this book. I thank him for his support as well.

Wim J. van der Linden
Monterey, California

References

Rasch, G. 1960. *Probabilistic Models for Some Intelligence and Attainment Tests.* Copenhagen: Danish Institute for Educational Research.

van der Linden, W. J. and Hambleton, R. K. (eds.) 1997. *Handbook of Modern Item Response Theory.* New York: Springer.

Contributors

Murray Aitkin is an honorary professorial fellow in mathematics and statistics at the University of Melbourne. His interests are in general statistical modeling, likelihood and Bayesian inference, and statistical computing. He has held teaching, research, and consulting positions in Australia, the United Kingdom, the United States, and Israel.

James H. Albert earned his PhD from Purdue University in 1979 with a specialty in Bayesian modeling. He is currently a professor in mathematics and statistics at Bowling Green State University where he has been on the faculty since 1979. Dr. Albert is a fellow of the American Statistical Association and has been an editor of the *American Statistician* and the *Journal of Quantitative Analysis of Sports*. His research interests are in Bayesian modeling of categorical data, the analysis of sports data, and statistical education.

Jodi M. Casabianca is an assistant professor of quantitative methods in educational psychology in the College of Education at the University of Texas at Austin. Dr. Casabianca completed a postdoctoral fellowship in the Carnegie Mellon and RAND Traineeships (CMART) in Methodology and Interdisciplinary Research program which was funded by the Institute of Education Sciences. She earned her PhD and MA in psychometrics from Fordham University, and her MS in applied and mathematical statistics and BA in statistics and psychology from Rutgers University. Her research interests are in psychometrics and educational measurement, specifically measurement models for rated data and computational methodology for large-scale testing and assessment.

Hua-Hua Chang is a professor of educational psychology, psychology, and statistics at the University of Illinois at Urbana-Champaign (UIUC). He earned his PhD in statistics from UIUC in 1992. His research interests include computerized testing, statistically detecting biased items, cognitive diagnosis, and asymptotic properties in IRT. Dr. Chang is the editor-in-chief of *Applied Psychological Measurement*, past president of the Psychometric Society (2012–2013), and a fellow of American Educational Research Association (AERA).

Sun-Joo Cho is an assistant professor at Peabody College of Vanderbilt University. Her research topics include generalized latent variable models and their parameter estimation, with a focus on item response models.

Allan S. Cohen earned his PhD from the University of Iowa in 1972, with specialties in measurement and applied statistics. He holds the title of Aderhold professor of research methods at the University of Georgia, where he has been on the faculty since 2003. Dr. Cohen received a career achievements award from National Council on Measurement in Education (NCME) in 2012. His research interests are in the areas of IRT and test construction.

Cees A. W. Glas is chair of the Department of Research Methodology, Measurement, and Data Analysis, faculty of behavioural, management, and social sciences, at the University of Twente in the Netherlands. He has participated in numerous research projects including projects of the Law School Admission Council and the OECD International Educational Survey PISA. He has published articles, book chapters, and supervised doctoral theses on topics such as testing the fit of IRT models, Bayesian estimation of multidimensional and multilevel IRT models using MCMC, modeling with nonignorable missing data, concurrent modeling of item responses and textual input, and the application of computerized adaptive testing in health assessment and organizational psychology.

Shelby J. Haberman earned his PhD in statistics from the University of Chicago in 1970. He is a distinguished presidential appointee at the Educational Testing Service, where he has served since 2002. Previously, he served as a faculty member at the University of Chicago, Hebrew University, and Northwestern University. Dr. Haberman is a fellow of the Institute of Mathematical Statistics, the American Statistical Association, and the American Association for the Advancement of Science. His research has emphasized the analysis of qualitative data.

Ronald K. Hambleton earned his PhD from the University of Toronto in Canada in 1969, with specialties in psychometric methods and applied statistics. He holds the title of a Distinguished University Professor at the University of Massachusetts where he has been on the faculty since 1969. Dr. Hambleton received career achievement awards from NCME in 1993, from AERA in 2005, and from the American Psychological Association (APA) in 2006. His research interests are in the areas of score reporting, standard setting, applications of IRT, and test adaptation methodology.

Heinz Holling earned his PhD from the Free University of Berlin in Germany in 1980, with specialities in psychometrics. Currently, he is a professor of statistics and quantitative methods at the University of Münster. His research focuses on optimal design, meta-analysis, and rule-based item generation. Dr. Holling has published numerous articles in international journals such as *Psychometrika*, *Biometrika*, and *Journal of Statistical Planning and Inference*. He is the author of several books on statistics, and his research has been continuously funded by the German Research Foundation and Federal Ministry of Education and Research.

Minjeong Jeon earned her PhD from the University of California, Berkeley in quantitative methods and evaluation in 2012. Dr. Jeon is currently a postdoctoral scholar at the Graduate School of Education, University of California, Berkeley. Her research interests include IRT models, multilevel models, and research methods.

Matthew S. Johnson is an associate professor of statistics and education and chair of the Department of Human Development at Teachers College of Columbia University. He earned his PhD in statistics from Carnegie Mellon University in 2001. Prior to joining Teachers College, Dr. Johnson was an associate professor at Baruch College of the City University of New York, and was an associate research scientist in the Center for Large-Scale Assessment of Educational Testing Service. He is a past editor of the *Journal of Educational and Behavioral Statistics*, and on the editorial board of *Psychometrika*. Dr. Johnson's research interests broadly focus on statistical models used in psychological and educational measurement.

Brian W. Junker is a professor of statistics and associate dean for academic affairs in the Dietrich College of Humanities and Social Sciences at Carnegie Mellon University. He has broad interests in psychometrics, education research, and applied statistics, ranging from nonparametric and Bayesian IRT to MCMC and other computing and estimation methods, and rating protocols for teacher quality, educational data mining, social network analysis, and mixed membership modeling. He earned a BA in mathematics from the University of Minnesota, and an MS in mathematics and PhD in statistics from the University of Illinois.

Robert J. Mislevy earned his PhD from the University of Chicago in 1981. Dr. Mislevy is the Frederic M. Lord Chair in Measurement and Statistics at Educational Testing Service, and professor emeritus at the University of Maryland. His specialties are psychometrics, Bayesian statistics, and assessment design. His contributions include the multiple-imputation approach for latent variables in the National Assessment of Educational Progress, an evidence-centered framework for assessment design and with Cisco Systems, design and analysis methods for simulation-based assessments.

Tim Moses earned his PhD from the University of Washington in 2003, with specialties in measurement, statistics, and research design. Dr. Moses is currently a senior psychometrician at the College Board. His research interests are statistical applications that inform investigations in large-scale testing and social science.

Richard J. Patz is the chief measurement officer at ACT, with responsibilities for research and development. His research interests include statistical methods, assessment design, and management of judgmental processes in education and assessment. He served as president of the NCME in 2015–2016. Dr. Patz earned a BA in mathematics from Grinnell College, and a MS and PhD in statistics from Carnegie Mellon University.

Sophia Rabe-Hesketh is a professor of education and biostatistics at the University of California at Berkeley, and professor of social statistics at the Institute of Education, University of London. Her research on multilevel, latent variable, and longitudinal modeling has been published in *Psychometrika*, *Biometrics*, and *Journal of Econometrics*, among others. Her six authored books include *Generalized Latent Variable Modeling* and *Multilevel and Longitudinal Modeling Using Stata* (both with Anders Skrondal).

Frank Rijmen earned his PhD from the University of Leuven in Belgium in 2002, with specialties in psychometric methods and quantitative psychology. He holds the title of principal research scientist and director of psychometric services and research at McGraw-Hill Education CTB where he has been working since 2013. Dr. Rijmen received the 2003 Psychometric Society Dissertation Award. His research interests are in the areas of IRT, latent class models, generalized linear mixed models, and graphical models.

Ernesto San Martín earned his PhD in statistics from the Université Catholique de Louvain, Belgium, in 2000. He has been a visiting professor at the faculty of psychology, KU Leuven, Belgium, and at the Center for Econometric and Operational Research CORE. Dr. San Martin is one of the professors who founded the Measurement Center MIDE UC, Chile, where applied research and educational services are developed. Currently, he is associate professor at the faculty of mathematics and the faculty of education, Pontificia Universidad Católica de Chile, Chile. His research interests are identification problems in models involving latent variables, school effectiveness and value-added models, mathematical statistics, and the history of probability and statistics.

Rainer Schwabe earned his PhD in 1985 and his habilitation in 1993, both from the Free University of Berlin in Germany, with specialties in stochastic approximation and optimal design of experiments. Dr. Schwabe was a deputy head of the Institute of Medical Biometry at the University of Tübingen and is now a university professor of mathematical statistics and its applications at the University of Magdeburg, where he has been on the faculty since 2002. His research interests are in the area of optimal design of experiments in nonstandard settings with applications in biosciences and psychometrics.

Sandip Sinharay is the chief statistician at Pacific Metrics Corporation in Monterey, California. He earned his PhD from the Department of Statistics at Iowa State University in 2001. Dr. Sinharay received the Award for Technical or Scientific Contribution to the Field of Educational Measurement from NCME in 2009 and 2015. His research interests include Bayesian statistics, reporting of diagnostic scores, model checking and model selection methods, IRT, and application of statistical methods to education.

Wim J. van der Linden is distinguished scientist and director of Research Innovation, Pacific Metrics Corporation, Monterey, California and professor emeritus of measurement and data analysis, University of Twente. Dr. van der Linden earned his PhD in psychometrics from the University of Amsterdam in 1981. His research interests include test theory, computerized adaptive testing, optimal test assembly, parameter linking, test equating, and response-time modeling, as well as decision theory and its application to problems of educational decision making. He is a past president of the Psychometric Society and the NCME and has received career achievement awards from NCME, Association of Test Publishers (ATP), and AERA.

Nathan M. VanHoudnos is currently a postdoctoral fellow at the Institute for Policy Research at Northwestern University. Dr. VanHoudnos earned his PhD in statistics and public policy from Carnegie Mellon University. Prior to Carnegie Mellon, he earned a BS in physics at the University of Illinois, Champaign-Urbana, and worked in computing services as a research programmer. His interest in education research stems from his experience as a Teach for America corps member in the Mississippi Delta. His research agenda is broadly organized around serving the learning sciences in the same way that a biostatistician serves the medical sciences. His current interests include generalized linear mixed models and their applications to effect size estimation, meta-analysis, missing data problems, and statistical disclosure control.

Chung Wang earned her PhD in quantitative psychology from the University of Illinois at Urbana-Champaign (UIUC) in 2012. She is currently an assistant professor of quantitative/psychometric methods in the Department of Psychology at the University of Minnesota. Her research interests include multidimensional and multilevel IRT, computerized adaptive testing, cognitive diagnosis modeling, and semiparametric modeling for item response-time analysis. Dr. Wang received the NCME Alicia Cascallar Award (2013), the NCME Jason Millman Promising Measurement Scholar Award (2014), and the AERA Early Career Award (2015).

Craig S. Wells is an associate professor at the University of Massachusetts Amherst in the research in evaluation measurement and psychometrics concentration and also serves as an associate director in the Center for Educational Assessment. Dr. Wells earned his PhD from the University of Wisconsin at Madison in 2004 in educational psychology with a specialization in quantitative methods. His research interests include the study of item parameter drift, differential item functioning, and assessment of IRT model fit. He also has a keen interest in the philosophy of science and its applications to social science research.

Zhiliang Ying earned his PhD in statistics from Columbia University in 1987. He is currently a professor of statistics in the Department of Statistics at Columbia University. Dr. Ying is an elected fellow of both the Institute of Mathematical Statistics and the American Statistical Association. In 2004, he was awarded the Morningside Gold Medal of Applied Mathematics. His research interests include educational and psychological measurement, survival analysis, sequential analysis, longitudinal data analysis, stochastic processes, semiparametric inference, biostatistics, and educational statistics.

Section I

Basic Tools

1

Logit, Probit, and Other Response Functions

James H. Albert

CONTENTS

1.1 Introduction

A general statistical problem is to model a binomial probability in terms of a vector of explanatory variables. We observe independent responses $y_1, \ldots, y_N$, where y_i is assumed binomial with sample size n_i and probability of success p_i. (In the special case where the sample sizes $n_i = \cdots = n_N = 1$, we have Bernoulli observations.) Corresponding to the ith observation, we also observe k covariates $x_{i1}, \ldots, x_{ik}$; we let $\mathbf{x_i} = (x_{i1}, \ldots, x_{ik})$ denote the vector of covariates. We relate the binomial probabilities with the covariates by means of the equation

$$g(p_i) = \mathbf{x_i}\beta, \tag{1.1}$$

where $\beta = (\beta_1, \ldots, \beta_k)$ is a vector of unknown regression coefficients. We call the function g the link function or response function. It is the function that links or connects the binomial probability, a value between 0 and 1, with the real-valued linear function of the covariates, often called the linear predictor. The purpose of this chapter is to provide a general overview on the choice for the response function g.

Cox and Snell (1989) describe two general strategies on what they call "dependency relations" between the probability and the linear predictor. The first approach expresses p in terms of x in a flexible way that provides a clear interpretation and leads to a relatively easy statistical analysis. The second approach relates the binary situation to a regression problem involving a latent continuous response variable.

In this chapter, we provide a general discussion on response functions for probabilities. After some initial considerations about new scales for proportions are discussed, we provide some graphical comparisons of four common symmetric response functions and one asymmetric function. Section 1.4 describes the attractive aspects of the logit response function from the viewpoint of interpretation and Section 1.5 illustrates the interpretation for a simple logistic model and a 2 by 2 contingency table. Section 1.6 provides an introduction to the probit link function (used, for instance, in bioassay problems), and Section 1.7 shows how a probit regression model can be viewed as a normal regression problem with missing data. Section 1.8 gives an overview of other response functions including the popular angular transformation and the use of family of link functions. We conclude in Section 1.9 with a historical perspective on the response functions focusing on the relative use of the logit and probit links and provide some comments on the future use of response functions.

1.2 Some Initial Considerations

Tukey (1977, chap. 15) talks about choosing a new scale for fraction data f that falls between 0 and 1. The problem of choosing a new scale for a fraction f is similar to the problem of choosing a response function for a probability p, so Tukey's comments are relevant to our problem. Two initial criteria used for choosing a new scale are (1) the new scale should have a value of 0 when $f = 0.5$ and (2) the fractions f and $1 - f$ should be treated in a symmetric fashion where swapping f with $1 - f$ will change the sign but not the size of the reexpressed fraction. In the response function setting, since it is often arbitrary to define a success as $y = 1$ or $y = 0$, this second criterion would lead to the choice of a symmetric response function.

Another issue with fractions or proportion is that the variability of the values is not constant across 0 to 1. The variance of a fraction is equal to $f(1 - f)$ and fractions close to the boundary values will have small variation. Tukey introduced the *froot* and *flog*, reexpressions defined respectively by $\sqrt{f} - \sqrt{1 - f}$ and $\log f - \log(1 - f)$, and the purpose of these reexpressions was to stretch the ends of the scale of the fraction f. As one moves from a folded fraction $f - (1 - f)$ to a froot to a flog, the tails are stretched more towards negative and positive infinity.

1.3 Comparison of Response Functions

1.3.1 Symmetric Functions

Cox and Snell (1989) provide a helpful comparison of four response functions for probabilities. Suppose we have a single continuous covariate x and we wish to represent a probability p in terms of x. We first consider symmetric response functions of the general form $p = F(x)$ that satisfies the property $1 - p = F(-x)$. Without loss of generality, we assume that the value $x = 0$ maps to a probability of $p = 0.5$. One can write the logistic curve with scale parameter β as follows:

$$p_{LT}(x) = \frac{e^{\beta x}}{1 + e^{\beta x}}. \tag{1.2}$$

Alternatively, we can suppose that p is a linear function of x of the form

$$p_{LN}(x) = \begin{cases} 1 & x > \frac{1}{2}\gamma^{-1} \\ \frac{1}{2} + \gamma x & |x| \leq \frac{1}{2}\gamma^{-1} \\ 0 & x < -\frac{1}{2}\gamma^{-1}. \end{cases} \tag{1.3}$$

In the probit transformation, we assume that the probability p is the cumulative distribution of a normal with mean 0 and standard deviation $1/\xi$:

$$p_N(x) = \Phi(\xi x). \tag{1.4}$$

Last, using the angular transform, the probability of success is given by

$$p_A(x) = \begin{cases} 1 & x > \frac{1}{4}\pi/\eta \\ \sin^2\left(\eta x + \frac{1}{4}\pi\right) & |x| \leq \frac{1}{4}\pi/\eta \\ 0 & x < -\frac{1}{4}\pi/\eta. \end{cases} \tag{1.5}$$

To compare the four response functions, Cox and Snell set the logistic scale $\beta = 1$ and then chose values of the linear, probit, and angular constants γ, ξ, η so that the four curves agreed approximately when $p = 0.8$. The corresponding matching values are $\gamma = 0.216$, $\xi = 0.607$, and $\eta = 0.232$.

Figure 1.1 displays the logit, linear, probit, and angular response curves for values of the covariate x between 0 and 5. Since all four curves are symmetric about 0, one would observe the same comparison for negatives values of the covariate. We see that the curves

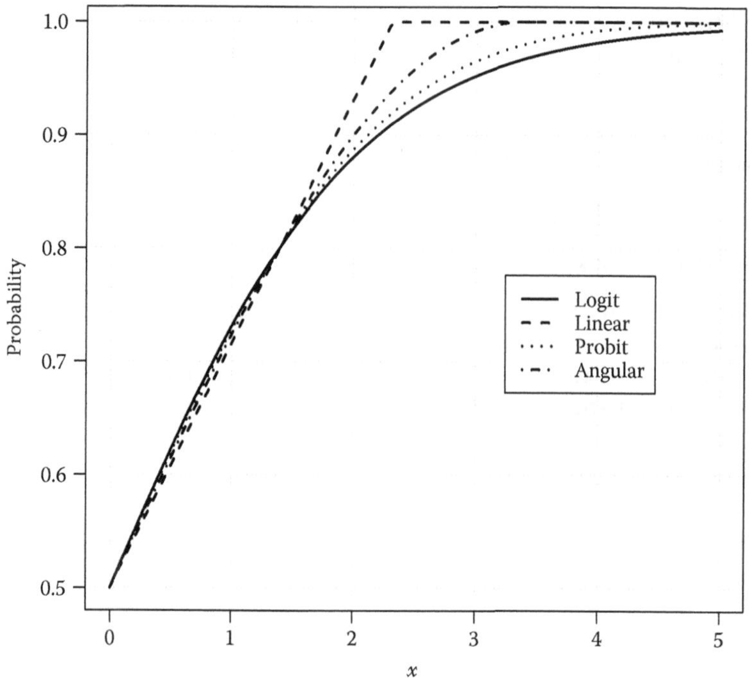

FIGURE 1.1
Comparison of logit, linear, probit, and angular response curves as a function of the covariate variable x. Scaling constants of the four curves are chosen so the curves agree when the probability p is equal to 0.8.

are very similar for probabilities between 0.5 and 0.8 (values of x between 0 and 1.5). For larger values of the covariate, the linear and angular curves increase more rapidly toward $p = 1$. Generally, the logit and probit response curves agree closely for all probability values. As Cox and Snell state, one would expect the two curves to differ substantially only in cases where the probability of success is either very large or very small.

1.3.2 Asymmetric Functions

In some situations, it may not be desirable to use a symmetric response function. Suppose we assume that the population of tolerances (Section 1.6 below) satisfies the extreme value distribution of the form

$$f(x) = \beta_2 \exp\left[(\beta_1 + \beta_2 x) - \exp(\beta_1 + \beta_2 x)\right]. \quad (1.6)$$

This choice of tolerance distribution leads to the response curve

$$p = 1 - \exp[-\exp(\beta_1 + \beta_2 x)], \quad (1.7)$$

leading to the complementary log–log representation

$$\log[-\log(1 - p)] = \beta_1 + \beta_2 x. \quad (1.8)$$

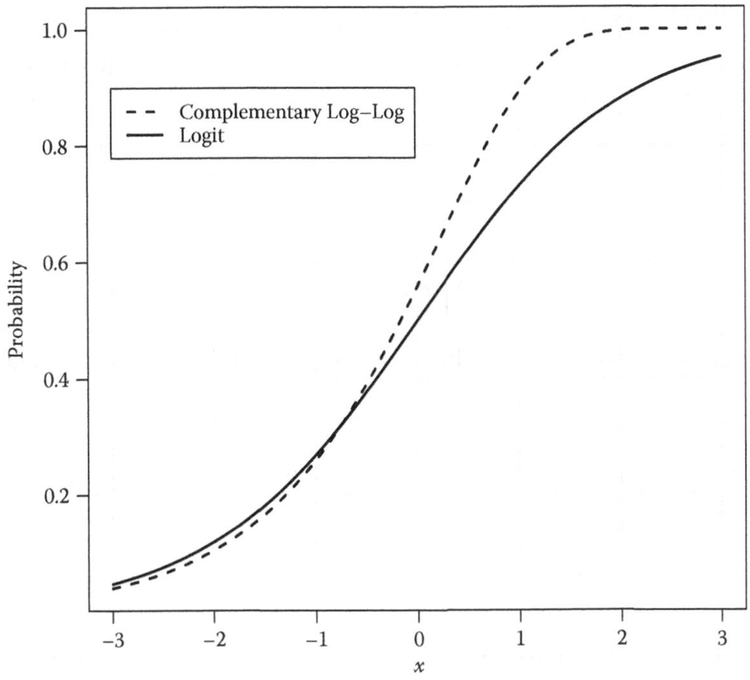

FIGURE 1.2
Comparison of complementary log–log (with $\beta_1 = -0.2$, $\beta_2 = 1$) and logit response functions.

Figure 1.2 compares the response curves of the complementary log–log function with $\beta_1 = -0.2$ and $\beta_2 = 1$ with the logit. Both curves are similar over the range of probabilities between 0.1 and 0.4, but differ for probabilities larger than 0.6.

1.4 Logit Response Function

Arguably the most popular response function is the logit where the response probability p_i is related to the vector of covariates $\mathbf{x}_i$ by means of the relationship

$$\text{logit}(p_i) = \log\left(\frac{p_i}{1-p_i}\right) = \mathbf{x}_i\beta. \tag{1.9}$$

Since one is typically interested in the probability p_i, it is perhaps more useful to express this relationship as a function of p_i, the inverse-logit form:

$$p_i = \text{inv logit}(\beta) = \frac{\exp(\mathbf{x}_i\beta)}{1+\exp(\mathbf{x}_i\beta)}. \tag{1.10}$$

Cox and Snell (1989, p. 19) describe the logit as the "most useful analogue of binary response data of the linear model for normally distributed data" for a number of reasons.

1.4.1 Connection with Exponential Family

Binary response modeling with a logit link fits "naturally" within the class of generalized linear models. For a binomial response y with sample size n, one can write the sampling density as

$$f(y|p) = \binom{n}{y} p^y (1-p)^{n-y}$$
$$= \binom{n}{y} \exp\left[y \log\left(\frac{p}{1-p}\right) + n \log(1-p) \right]. \quad (1.11)$$

The binomial can be seen as a member of the exponential family of the general form

$$f(y|\theta) = h(y) \exp[\eta(\theta) T(y) - A(\theta)], \quad (1.12)$$

where $\eta(\theta)$ is the so-called natural parameter (Haberman, Chapter 4). Here the natural parameter is the logit of the probability

$$\eta(\theta) = \log\left(\frac{p}{1-p}\right), \quad (1.13)$$

and the binomial variable is the sufficient statistic $T(y) = y$. This exponential family representation of binomial modeling with a logistic link leads to convenient expressions of the likelihood function and the associated inference procedures.

1.4.2 Interpretation in Terms of Probabilities

For a logistic function with a single continuous covariate, it is straightforward to compute the derivative of the inverse-logit function with respect to the covariate and this leads to a simple interpretation of the regression slope. If one computes the derivative of the probability

$$p = \frac{\exp(\beta_0 + \beta_1 x)}{1 + \exp(\beta_0 + \beta_1 x)} \quad (1.14)$$

with respect to x, one obtains the expression

$$\frac{dp}{dx} = p(1-p)\beta_1. \quad (1.15)$$

When the fitted probability p is close to 0.5, then

$$\frac{dp}{dx} \approx \frac{\beta_1}{4}. \quad (1.16)$$

So for fitted probabilities close to 0.5, $\beta_1/4$ represents the change in the probability of success for a unit increase in the covariate x. Gelman and Hill (2007) call this the "divide by four" interpretation of a logistic regression coefficient.

This interpretation is also useful for a logistic multiple regression model of the form

$$p = \frac{\exp(\beta_0 + \beta_1 x_1 + \cdots + \beta_k x_k)}{1 + \exp(\beta_0 + \beta_1 x_1 + \cdots + \beta_k x_k)}. \quad (1.17)$$

For fitted probabilities in a neighborhood of 0.5, $\beta_j/4$ represents the change in the probability for a unit increase in the covariate x_j assuming that the other covariates $\{x_i, i \neq j\}$ are held fixed.

1.4.3 Interpretation in Terms of Odds

One can also interpret the slope parameter β_1 in terms of odds. We express the logistic model in the "odds form" as

$$odds = \frac{p}{1-p} = \exp(\beta_0 + \beta_1 x). \tag{1.18}$$

As one increases the covariate x by one unit, the ratio of the odds of success at $x+1$ to the odds of success at x is given by

$$\frac{odds(x+1)}{odds(x)} = \frac{\exp(\beta_0 + \beta_1(x+1))}{\exp(\beta_0 + \beta_1 x)} = \exp(\beta_1). \tag{1.19}$$

In the multiple regression setting, $\exp(\beta_j)$ is the ratio of the odds of the success at $x_j + 1$ to the odds of success at x_j when the other covariates are held fixed.

1.5 Examples of Logistic Modeling

1.5.1 Simple Logistic Modeling

As an example, we consider a sample of 358 college students who are taking a business calculus course. Suppose we say that a student is successful if he or she obtains a grade of B or higher and we believe that the chance of success is related to the student's high school-grade point average (HSGPA). We define the binary response $y = 1$ and $y = 0$ if the student is successful or not, and we let x denote the standardized HSGPA

$$x = \frac{HSGPA - mean(HSGPA)}{sd(HSGPA)}. \tag{1.20}$$

We model the probability the student is successful, $p = Prob(y = 1)$, by means of the simple logistic model

$$\log\left(\frac{p}{1-p}\right) = \beta_0 + \beta_1 x. \tag{1.21}$$

Since the probability of success is the quantity of interest, it is perhaps more helpful to write the model in terms of the probability:

$$p = \frac{\exp(\beta_0 + \beta_1 x)}{1 + \exp(\beta_0 + \beta_1 x)}.$$

Figure 1.3 displays a scatterplot of the standardized HSGPAs and jittered values of the binary response y. The fitted logistic model is given by

$$\log\left(\frac{p}{1-p}\right) = 0.008 + 0.908x, \tag{1.22}$$

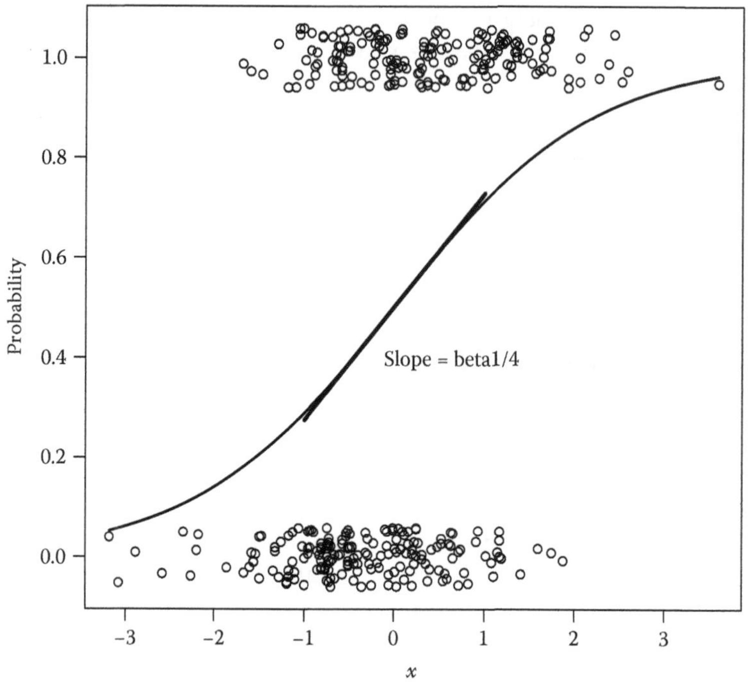

FIGURE 1.3
Scatterplot of the standardized HSGPA and jittered binary responses for the placement test example. The fitted logistic curve is placed on top of the scatterplot. In addition, the thick line represents a linear approximation to the logistic curve at the point (0, 0.5).

which leads to the fitted probability

$$p = \frac{\exp(0.008 + 0.908x)}{1 + \exp(0.008 + 0.908x)}. \tag{1.23}$$

The fitted probability curve is drawn on top of the scatterplot in Figure 1.3.

We give different interpretations of the parameters of this logistic model:

1. *Interpreting the intercept.* The intercept β_0 is the fitted logit when the predictor x is equal to zero. When the student has an average HSGPA value, the standardized predictor $x = 0$ and the fitted logit of success is 0.008 corresponding to a fitted probability of success of $p = \exp(0.008) \approx 0.5$. A student with average high school grades has a 50% chance of succeeding in the course.
2. *Interpreting the slope by looking at difference in probabilities.* The slope parameter β_1 tells us how the probability of success changes as a function of the covariate x. A simple way to interpret this parameter is to compute the difference in probabilities for two covariate values of interest. For a student of average ability ($x = 0$), one predicts the probability of success to be $\exp(0.008)/(1 + \exp(0.008)) = 0.502$, and for a student with a HSGPA one standard deviation above the mean, one predicts the probability to be $\exp(0.008 + 0.908)/(1 + \exp(0.008 + 0.908)) = 0.714$. An increase in one standard deviation in HSGPA results in a change of $0.714 - 0.502 = 0.212$ in the probability of success.

3. *Interpreting the slope using the derivative.* Recall that the derivative of the probability p with respect to x was equal to $p(1-p)\beta_1$. In this example, $\beta_1 = 0.908$ and $\beta_1/4 = 0.908/4 = 0.227$. For success probabilities close to one half, one predicts the probability of success to increase by 23% as the HSGPA increases by one standard deviation. Note that this derivative approximation gives a similar answer to the exact difference in probabilities between $x = 0$ and $x = 1$.
4. *Interpreting the slope using odds.* One can also interpret the slope parameter β_1 as terms of a multiplicative effect using odds. As one increases the covariate x by one unit, the ratio of the odds of success at $x + 1$ to the odds of success at x is given by $\exp(\beta_1)$. In our example, $\exp(\beta_1) = \exp(0.908) = 2.48$. If one student has a HSGPA one standard deviation higher than another student, one would predict the odds of success of the first student to be 2.48 times higher than the odds of success of the second student.

1.5.2 Interpretation for a Two-Way Contingency Table

Logistic modeling provides a convenient interpretation of parameters when the input variable is categorical. Following the discussion in Fleiss (1981), suppose we are interested in studying the relationship between smoking and lung cancer in a community. Let x denote the amount of a specified pollutant in the atmosphere in this particular community. One can represent the mortality rate from cigarette smokers by means of the logistic model

$$p_S = \frac{\exp(ax + b_S)}{1 + \exp(ax + b_S)}. \tag{1.24}$$

Likewise, assume that the mortality rate for nonsmokers has the logistic representation

$$p_N = \frac{\exp(ax + b_N)}{1 + \exp(ax + b_N)}. \tag{1.25}$$

Here, a represents the dependence of mortality on the air pollutant; since the same parameter a is used for both smokers and nonsmokers, we are assuming no synergistic effect of smoking and air pollution on mortality.

The odds that a smoker dies of lung cancer is then given by

$$\begin{aligned} \Omega_S &= \frac{p_S}{1 - p_S} \\ &= \exp(ax + b_S), \end{aligned} \tag{1.26}$$

and likewise the odds of a nonsmoker dying of lung cancer is

$$\begin{aligned} \Omega_N &= \frac{p_N}{1 - p_N} \\ &= \exp(ax + b_N). \end{aligned} \tag{1.27}$$

So under this logistic model, the ratio of the odds of a smoker dying of lung cancer to the odds of a nonsmoker dying of lung cancer, the odds ratio, is given by

$$\begin{aligned} \alpha &= \frac{\Omega_S}{\Omega_N} \\ &= \frac{\exp(ax + b_S)}{\exp(ax + b_N)} \\ &= \exp(b_S - b_N). \end{aligned} \tag{1.28}$$

The logarithm of the odds ratio is simply the difference in the logistic regression parameters:

$$\log \alpha = b_S - b_N. \tag{1.29}$$

The odds ratio α is a popular measure of association in a 2 by 2 contingency table. Here, the table would categorize residents in the community by smoking behavior and cause of mortality, where $a, b, c,$ and d are the observed counts of residents in the four categories.

	No Lung Cancer	Lung Cancer
Smoker	a	b
Nonsmoker	c	d

Bishop et al. (1975) and Fleiss (1981) describe many useful properties of the odds ratio for measuring association in a two-way table.

1.6 Probit Link

1.6.1 Probit Response Function in Biological Assay

Finney (2009) gives an extensive discussion of the use of the probit link in biological assay or bioassay. In this type of scientific experiment, a stimulus (e.g., a vitamin, a drug, or a mental test) is applied to a subject (e.g., an animal, a plant, or a piece of tissue) at an intensity of the stimulus, and a response is produced by the subject. The observed response is of the quantal, or all-or-nothing, type. Examples of quantal responses are death, a failure of a spore to germinate, or a cure of an unhealthy condition.

The occurrence of a quantal response depends on the intensity or concentration of the stimulus. For each subject, there will be a particular level of intensity below which the response does not occur and above which the response does occur—this level is called the subject's tolerance. It is natural to assume that the tolerances vary among the respondents in population.

For the analysis of quantal response data, one needs to make an assumption about the distribution of the tolerances over the population of interest. Finney describes situations such as insects responding to concentrations of an insecticide where the tolerances may have a right-skewed population distribution. Although there are situations where one would not expect the tolerances to be normally distributed, it will usually be possible in general to transform the tolerances, say by a power transformation, so that the tolerances are approximately normal.

For different concentrations of dose of an insecticide, one can graph the percentage of insects responding against the dose. When the stimulus is measured in terms of log dose ("dosage unit" in Finney), the plot of percentage responding against log dose takes on the form of the distribution function of the normal.

1.6.2 Estimation of the Tolerance Distribution

Finney (2009) described a typical test where successive batches of insects are exposed to different concentrations of the insecticide, and after a suitable time interval, the counts of insects dead and alive are recorded. Table 1.1 gives results of an insecticidal test of Martin (1942), who sprayed batches of 50 Macrosiphoniella sanborni with different concentrations of the rotenone. The objective is to use these data to estimate the mean μ and the variance σ^2 of the normal tolerance distribution.

Gaddum (1933) proposed to transform the probability of a response, p, by the normal equivalent deviate (NED) y, defined by

$$p = \int_{-\infty}^{y} \frac{1}{\sqrt{2\pi}} \exp\left\{-\frac{1}{2}z^2\right\} dz. \tag{1.30}$$

In modern language, NED is the value of the inverse cumulative distribution function of the standard normal evaluated at the probability p. In general, Finney notes that if p is the probability of response at a log dose x from a general $N(\mu, \sigma^2)$ tolerance distribution where

$$p = \int_{-\infty}^{x} \frac{1}{\sigma\sqrt{2\pi}} \exp\left\{-\frac{1}{2\sigma^2}(x-\mu)^2\right\} dx, \tag{1.31}$$

then

$$y = (x - \mu)/\sigma. \tag{1.32}$$

The relation between the log dose and the NED of the probability of response is a straight line.

TABLE 1.1

Results of an Insecticidal Test of Martin (1942)

	Dose of Rotenone (mg/L)	No. of Insects (n)	No. Affected (r)	% Kill (p)
1	10.20	50	44	88
2	7.70	49	42	86
3	5.10	46	24	52
4	3.80	48	16	33
5	2.60	50	6	12
6	0.00	49	0	0

TABLE 1.2

Calculations to Estimate the Parameters of a Normal Tolerance Distribution from Finney (2009)

	Log. Dose	N	p	Em.probit
1	1.01	50	88	6.17
2	0.89	49	86	6.08
3	0.71	46	52	5.05
4	0.58	48	33	4.56
5	0.41	50	12	3.83

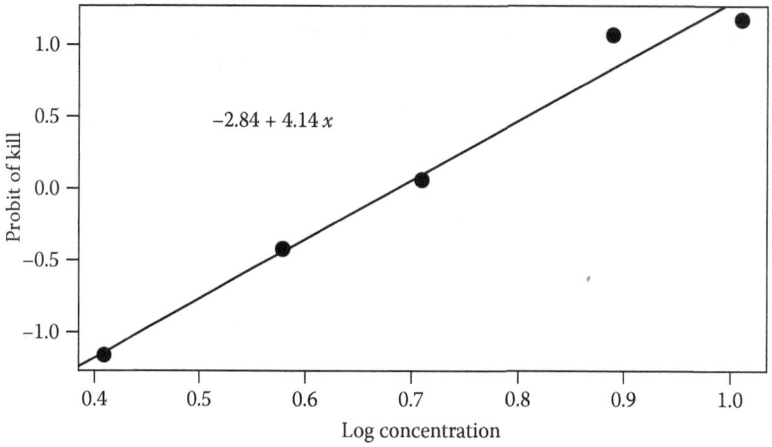

FIGURE 1.4
Scatterplot of empirical probits and log dose levels. The probit regression line and its equation are displayed.

Finney actually defines the probit of the proportion p as the abscissa y that corresponds to a probability p in a normal distribution with mean 5 and variance 1. Bliss (1943) increased the corresponding NED by 5 to avoid negative values. (In the days where biologists did not have access to calculating machines, it was attractive to avoid negative quantities.) In current usage, the probit is defined as the NED. In item response theory, response functions based on the probit are typically referred to as a normal-ogive functions.

A simple graphical method is proposed to estimate the parameters of the normal tolerance distribution. The calculations are illustrated in Table 1.2 for the insect dataset. For each of the log dose values with observed kill proportion p, one computes the empirical probit $\Phi^{-1}(p)$. One plots the log dose (horizontal) against the empirical probit (vertical) and fits a line (see Figure 1.4). The log LD 50 is the value of the log dose where the kill proportion is equal to 0.50. The slope of the line is an estimate at $1/\sigma$, the reciprocal of the standard deviation of the tolerance distribution. For this data, a best-line fit is given by

$$y = 4.14(x - 0.686). \tag{1.33}$$

So an estimate at log LD 50 is 0.686 and an estimate at σ is given by $1/4.14 = 0.242$.

1.6.3 Application to Item Response Models

In the early days of item response modeling (before 1960), the probit link was used in the development of the item characteristic curve. If $P_i(\theta)$ represents the probability an examinee with ability θ answers item i correctly, then Lord (1952, 1953) proposed the model

$$P_i(\theta) = \int_{\infty}^{a_i(\theta - b_i)} \phi(z) dz,$$

where $\phi()$ is the standard normal density function and a_i and b_i are parameters characterizing item i. Lord and Novick (1968) review this normal-ogive item response model in Chapter 16. In this same text, Birnbaum (1968) in Chapter 17 introduces the logistic item response model which has advantages of "mathematical convenience." This model replaces the normal cumulative distribution function with a logistic cumulative distribution function. After this date, the logistic was generally preferred to the probit for item response modeling.

1.7 Data Augmentation and the Probit Link

1.7.1 A Missing-Data Perspective

Albert and Chib (1993) showed that a probit regression model can be viewed as a missing data problem. In a medical example, suppose a binary response y_i is an indicator of survival, where $y_i = 1$ indicates that the ith person survived and $y_i = 0$ indicates the person did not survive. Suppose there exists a continuous measurement of care Z_i of health such that if Z_i is positive, then the person survives; otherwise ($Z_i \leq 0$) the person does not survive. Moreover, the health measurement is related to a set of covariates by the normal regression model

$$Z_i = \mathbf{x}_i \beta + \epsilon_i, \tag{1.34}$$

where $\epsilon_1, \ldots, \epsilon_n$ are a random sample from a standard normal distribution. It is straightforward to show that

$$P(y_i = 1) = P(Z_i > 0) = \Phi(\mathbf{x}_i \beta). \tag{1.35}$$

So the probit model can be seen as a missing-data problem where one has a normal regression model on latent or unobserved data $Z_1, \ldots, Z_n$ and the observed responses are missing or incomplete in that we only observe them if $Z_i > 0$ ($y_i = 1$) or $Z_i \leq 0$ ($y_i = 0$).

1.7.2 Advantages of This Perspective

This missing data perspective shows that probit regression is essentially normal regression on underlying continuous response data. One attractive aspect of this perspective is that it motivates a straightforward use of Gibbs sampling (Gelfand and Smith, 1990) to fit a probit model from a Bayesian perspective.

Also, new regression models for binary and ordinal response data can be found by use of models built on the underlying continuous latent response data. For example, Albert and Chib (1993) generalized the probit response model by representing the underlying tolerance distribution as a mixture of normal distributions. The use of a single cutpoint on normal latent response data motivates a binary probit regression model. If one generalizes by using several cutpoints on the latent data, one develops the popular cumulative probit model for ordinal response data.

1.7.3 Application to Item Response Models

Albert (1992) demonstrated that this missing data approach leads to an "automatic" approach for fitting a two-parameter item response model with a probit link. Suppose the latent observation Z_{ij} for the ith examinee to the jth item is distributed

$$Z_{ij} \sim N(a_j\theta_i - b_j, 1),$$

where θ_i is the examinee's ability and a_j and b_j are characteristics of the jth item. If one define the binary observation y_{ij} equal to one if Z_{ij} is positive, and $y_{ij} = 0$ otherwise, then this model can be shown equivalent to the two-parameter normal ogive model.

Albert (1992) shows the missing data representation leads to a Gibbs sampling approach (Gelfand and Smith (1990)) for sampling from the posterior distribution of the ability and item parameters. From the viewpoint of model fitting, the use of the probit link is natural since the Gibbs sampling algorithm is based on sampling from familiar normal distributions. The ease of fitting this high-dimensional posterior distribution has increased the popularity of the probit link for item response models. Patz and Junker (1999) describe similar type of Gibbs sampling algorithms for item response models with a logistic link, although these algorithms are not as automatic since they require the input of some tuning parameters.

1.8 Other Link Functions

1.8.1 Angular Transformation

As mentioned in Section 1.1, one difficulty of binomial data y is that the variance of y is not constant across values of the probability p. In the case, where the variability of a response is not constant, one general approach is to reexpress the response by a nonlinear transformation $T(y)$ so that the variance of T is approximately constant (not depending on the proportion p). Then one can apply linear regression models that assume a constant variance with the transformed response data. Such a reexpression is called a *variance-stabilizing transformation.*

A popular variance-stabilizing transformation for proportion data is the angular reexpression. If $\hat{p}$ is the sample proportion, then the angular transformation is defined by

$$T(\hat{p}) = \sin^{-1}\left(\sqrt{\hat{p}}\right). \tag{1.36}$$

Fisher (1974) gave a general discussion of the angular transformation and attributes Bartlett (1945) and Anscombe (1948) with early an discussion of this proposal.

1.8.2 Family of Link Functions

In the situation where there is no reason a priori to choose a particular response function, a general approach (e.g., Czado and Santner, 1992) is to consider the use of a general family of response functions and then estimate the unknown parameters of the family from the data.

Guerrero and Johnson (1982) considered the use of the response function $p = F_\lambda(z)$, where F_λ is the Box–Cox family of distributions:

$$F_\lambda(z) = \begin{cases} 0 & \text{if } z < -1/\lambda, \lambda > 0 \\ \dfrac{(1+\lambda z)^{1/\lambda}}{(1+\lambda z)^{1/\lambda}+1} & \text{if } 1+\lambda z > 0, \lambda \neq 0 \\ 1 & \text{if } z > -1/\lambda, \lambda < 0. \end{cases} \tag{1.37}$$

Every $F_\lambda(z)$, except for $F_0(z)$, has one-sided bounded support; $F_\lambda(z)$ is negatively skewed for $\lambda < 0$ and positively skewed for $\lambda > 0$.

Another choice of response functions is the Burr family (modification of scale-shaped family proposed by Burr (1942))

$$F_\psi(z) = 1 - (1 - \exp(z))^{-\psi}, \tag{1.38}$$

for $\psi > 0$ and z real. The distribution is positively skewed if $\psi < 1$ and negatively skewed if $\psi > 1$ but also exhibits kurtosis. One attractive feature of this family is that it includes the logit ($\psi = 1$) and the extreme minimum value distribution ($\psi \to \infty$).

Another possible family of symmetric links used by Aranda-Ordaz (1981) and Albert and Chib (1997) is defined as a function of the probability p as

$$g_\rho(p) = \frac{2}{\rho} \times \frac{p^\rho - (1-p)^\rho}{p^p + (1-p)^\rho}. \tag{1.39}$$

The values of $\rho = 0.0$, 0.4, and 1.0 correspond respectively to the logit, (approximately) probit, and linear links (see Figure 1.5).

1.9 History of the Logit and Probit Response Curves

Cramer (2011) wrote an extensive discussion of the origins and the development of the logit and probit response curves. In this section, we summarize some of the key contributors described in Cramer's book.

1.9.1 History of the Logistic Function

The logistic function was initially proposed in the nineteenth century to model population growth. One starts with considering the growth of a quantity $W(t)$ as a function of time t. If $dW(t)/dt$ represents the derivative (the growth rate), a basic model assumption is that the growth is proportional to the quantity's size:

$$dW(t)/dt = \beta W(t). \tag{1.40}$$

This assumption corresponds to exponential growth:

$$W(t) = A\exp(\beta t). \tag{1.41}$$

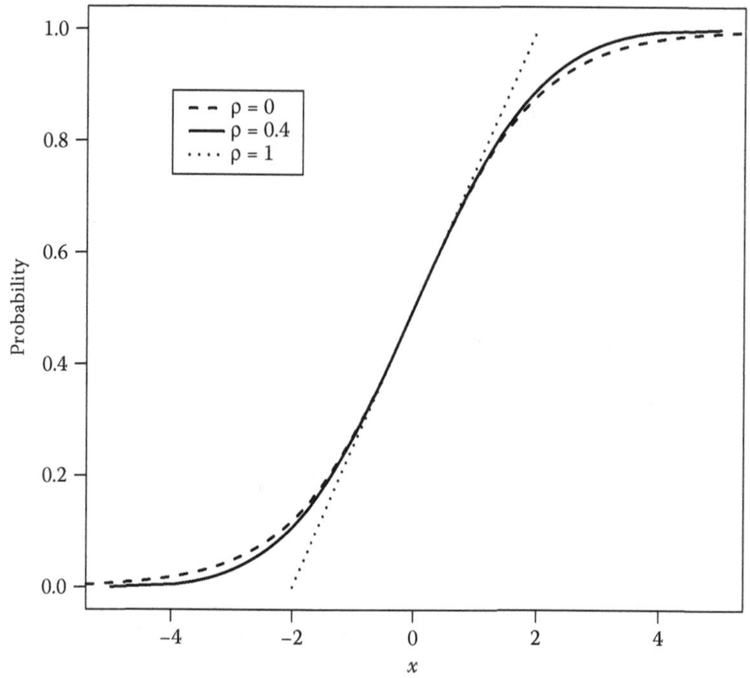

FIGURE 1.5
Comparison of Aranda–Ordaz family of response functions for $\rho = 0, 0.4$, and 1.0.

This model is useful for representing the growth of a developing country, but does not provide a good description of the growth of a mature country. The French mathematician Pierre–Francois Verhulst modified the differential equation by adding an extra term representing resistance to further growth:

$$dW(t)/dt = \beta W(t) - \phi(W(t)). \tag{1.42}$$

If ϕ is chosen to be a quadratic function, then we can rewrite the differential equation as

$$dW(t)/dt = \beta W(t)(\Omega - W(t)), \tag{1.43}$$

where Ω represents the upper limit of the growth of W as time approaches infinity. Growth is now proportional both to the current population and the population remaining for expansion. If one reexpresses $W(t)$ as the proportion $P(t) = W(t)/\Omega$, we can rewrite the differential equation as

$$dP(t)/dt = \beta P(t)\{1 - P(t)\}, \tag{1.44}$$

and the solution is the logistic function

$$P(t) = \frac{\exp(\alpha + \beta t)}{1 + \exp(\alpha + \beta t)}. \tag{1.45}$$

The population at time t is given by

$$W(t) = \Omega \frac{\exp(\alpha + \beta t)}{1 + \exp(\alpha + \beta t)}. \tag{1.46}$$

This contribution including the term logistic was published by Verhulst (1838, 1845, 1847). It is interesting that Verhulst's teacher Alphonse Quetelet did not appear to be excited about his student's contribution and said little about the logistic curve in his writings.

The logistic curve was discovered again as a model for population should be growth by Pearl and Reed (1920), who were unaware of Verhulst's contribution. These researchers evidently thought the logistic curve was a very useful growth model and applied it to a wide variety of populations including fruit flies, the population of the North Africa French colonies and the growth of cantaloupes. Later, Pearl and Reed (1922) rediscovered Verhulst's contribution but didn't give Verhulst much credit. Yule (1925), in his Presidential Address in the Royal Statistical Society in 1925, devoted an appendix to Verhulst's contribution and revives the use of the name logistic.

Cramer describes a second root of the logistic function in chemistry where it was used to model autocatalytic or chain reactions. In these processes, the product itself acts as a catalyst for the process and the raw material supply is fixed. The logistic function is derived as a solution to a similar type of differential equation for the process. Reed and Berkson (1929) described work of the chemist Wilhelm Ostwald in 1883. Cramer noticed that the logistic model "simple and effective" and comments that the logistic curve is still used to model market penetration of new products and technologies.

1.9.2 History of the Probit Function

Cramer (2011) explains that the genesis of the probit model and the transformation of frequencies to normal deviates comes from nineteenth century work by the German scholar Fechner in psychophysics. In experimental tests of the ability of humans to distinguish differences in weight, Fechner (1860) discovered that responses to a fixed stimulus were not uniform and that led to the use of normal deviates. In 70 years between Fechner (1860) and the 1930s, many researchers rediscovered this result, leading to the influential papers of Gaddum (1933) and Bliss (1934a,b). These papers began the standard bioassay methodology and Bliss introduced the term probit in brief notes in *Science*.

In the standard bioassay methodology described by Bliss and Gaddum (see Section 1.6), the stimulus is fixed and the responses are random due to the variability of the tolerances of the individuals. A probit of a relative frequency f was defined to the equivalent normal deviate z such that the cumulative normal distribution at z is equal to f. In the probit methodology, one used a line to relate the relative frequencies with the logarithm of the stimulus. Cramer believes that the "full flowering" of the probit methodology coincided with the first edition of Finney's (1947) book . At this time, probit analyses become more popular in other disciplines where one wished to model a binary response in terms of one or more covariates. For example, Adam (1958) used lognormal curves to learn how willingness to buy cigarette lighters depended on price.

1.9.3 Logit or Probit?

Cramer believed the use of the logistic response curve as an alternative to the probit curve was first discussed by Berkson in a large number of papers (see e.g., Berkson, 1951, 1980). Although the logistic and normal distributions are very similar in shape and the inverse of the logistic is much simpler than the definition of the probit, there was much controversy surrounding Berkson's writings, especially among the biometric establishment.

Why wasn't the logit model better received during this particular period? First, the logit was considered an inferior method for bioassay since it was not related to an underlying normal distribution for the tolerance levels. Second, Berkson's argument for the logit was not helped by his simultaneous attack on maximum likelihood estimation and his preference of the minimum chi-squared.

One big advantage of the logit over the probit was in the ease in computation, even in finding maximum likelihood estimates. During the 1950s, all numerical statistical work would be done by hand, and logit and probit models were often fit using special graph paper where best fits corresponded to straight lines (see Figure 1.4). Logit modeling was more popular in "workfloor practice" due to the relative simplicity of calculation.

Although there was significant ideological conflict in the bioassay environment on the choice of logit or probit, this conflict eventually abated with other developments in statistics and biometrics. The logit model began to be more widely used in a variety of disciplines, forgetting the original population growth motivation for the model.

The analytical advantages of the logit response function for binary responses began to be more universally recognized. Cox wrote a series of influential papers that cumulated with a textbook in 1969. The use of the logit model for bioassay was generalized to logistic regression with multiple covariates, and later associations of the logistic model with discriminant analysis and loglinear models were developed. The logit transformation or log odds arise naturally in case-control studies in epidemiology.

1.9.4 The Present and Future

Cramer (2011) documented the rise of the logit in the statistical literature by use of a table that gives the number of articles in journals of the Royal Statistical Society and the American Statistical Association containing the word probit or logit in 5-year periods between 1935 and 1994. In the period 1950–1954, there were 50 probit papers and 15 logit papers; in contrast, in the period 1990–1994, there were 127 probit papers and 311 logit papers. Cramer comments that the rise of the logit appears to be primarily due to its much wider use in statistical theory and applications.

To check the relative popularity of the probit and logit response curves, the *Current Index to Statistics* was to used to count the number of published statistics articles containing probit or logit in the keywords or title during different periods. Table 1.3 gives the counts of probit and logit papers and the percentage of logit papers for each period. It is interesting to observe that more than 60% of the papers in the period from 1971 to 1990 used logits, but the percentage of logit decreased to 52%–55% from 1991 to 2011. This table seems to indicate that both response functions are still very popular among researchers in statistics.

TABLE 1.3

Count of Number of Published Papers with Either Probit or Logit in the Title or Keywords Listed in the *Current Index of Statistics* during Different Time Periods

Date	Probit	Logit	% Logit
1971–1980	38	64	63
1981–1990	114	171	60
1991–2000	203	222	52
2001–2011	208	253	55

References

Adam, D. 1958. *Les reactions du consommateur devant les prix*. Paris: Sedes.
Albert, J. H. 1992. Bayesian estimation of normal ogive item response curves using Gibbs sampling. *Journal of Educational Statistics, 17*, 251–269.
Albert, J. H. and Chib, S. 1993. Bayesian analysis of binary and polychotomous response data. *Journal of the American Statistical Association, 88*, 669–679.
Albert, J. and Chib, S. 1997. Bayesian tests and model diagnostics in conditionally independent hierarchical models. *Journal of the American Statistical Association, 92*, 916–925.
Anscombe, F. J. 1948. The transformation of poisson, binomial and negative-binomial data. *Biometrika, 35*, 246–254.
Aranda-Ordaz, F. J. 1981. On two families of transformations to additivity for binary response data (Corr: V70 pp. 303). *Biometrika, 68*, 357–363.
Bartlett, M. S. 1945. The use of transformations. *Biometrics, 3*, 39–52.
Berkson, J. 1951. Why I prefer logits to probits. *Biometrics, 9*, 327–339.
Berkson, J. 1980. Minimum chi-square, not maximum likelihood! *The Annals of Statistics, 8*, 457–487.
Birnbaum, A. 1968. Some latent trait models and their use in inferring an examinee's ability. In F. M. Lord and M. R. Novick, *Statistical Theories of Mental Test Scores*. Reading MA: Addison-Wesley.
Bishop, Y. M. M., Fienberg, S. E. and Holland, P. W. 1975. *Discrete Multivariate Analyses: Theory and Practice*. Cambridge, MA: MIT Press.
Bliss, C. I. 1934a. The method of probits. *Science, 79*, 38–39.
Bliss, C. I. 1934b. The method of probits – A correction. *Science, 79*, 409–410.
Burr, I. W. 1942. Cumulative frequency functions. *The Annals of Mathematical Statistics, 13*, 215–232.
Cox, D. R. and Snell, E. J. 1989. *Analysis of Binary Data*. London, UK: Chapman & Hall Ltd.
Cramer, J. S. 2011. *Logit Models from Economics and Other Fields*. Cambridge, UK: Cambridge University Press.
Czado, C. and Santner, T. J. 1992. The effect of link misspecification on binary regression inference. *Journal of Statistical Planning and Inference, 33*, 213–231.
Fechner, G. T. 1860. *Elemente der Psychophysik*. Leipzig: Breitkopf und Hartel.
Finney, D. 2009. *Probit Analysis*. Cambridge University Press.
Fisher, R. 1974. The analysis of variance with various binomial transformations. In *Collected Papers*, R. A. Fisher, Volume V 1948–62, pp. 295–304. University of Adelaide.
Fleiss, J. L. 1981. *Statistical Methods for Rates and Proportions*. New York: John Wiley & Sons.
Gaddum, J. H. 1933. *Report on Biological Standards III: Methods of Biological Assay Depending on Quantal Response*. London: Medical Research Council.
Gelfand, A. E. and Smith, A. F. M. 1990. Sampling-based approaches to calculating marginal densities. *Journal of the American Statistical Association, 85*, 398–409.
Gelman, A. and Hill, J. 2007. *Data Analysis Using Regression and Multilevel/Hierarchical Models*. Cambridge, UK: Cambridge University Press.
Guerrero, V. M. and Johnson, R. A. 1982. Use of the Box-Cox transformation with binary response models. *Biometrika, 69*, 309–314.
Lord, F. M. 1952. A theory of test scores. *Psychometric Monograph, 7*.
Lord, F. M. 1953. The relation of test score to the trait underlying the test. *Educational and Psychological Measurement, 28*, 989–1020.
Lord, F. M. and Novick, M. R. 1968. *Statistical Theories of Mental Test Scores*. Reading, Mass: Addison-Wesley.
Martin, J. T. 1942. The problem of the evaluation of rotenone-containing plants. vi. the toxicity of L-elliptone and of poisons applied jointly, with further observations on the rotenone equivalent method of assessing the toxicity of derris root. *Annals of Applied Biology, 29*, 69–81.
Patz, R. J. and Junker, B. W. 1999. A straightforward approach to Markov chain Monte Carlo methods for item response models. *Journal of Educational and Behavioral Statistics, 24*, 146–178.

Pearl, R. and Reed, L. J. 1920. On the rate of growth of the population of the United States and its mathematical representation. *Proceedings of the National Academy of Sciences, 6,* 275-288.

Pearl, R. and Reed, L. J. 1922. A further note on the mathematical theory of population growth. *Proceedings of the National Academy of Sciences, 8,* 365–368.

Rasch, G. 1966. *Probabilistic Models for Some Intelligence and Attainment Tests.* Copenhagen: Danish Institute for Educational Research.

Reed, L. J. and Berkson, J. 1929. The application of logistic function to experimental data. *The Journal of Physical Chemistry, 33,* 760–779.

Tukey, J. W. 1977. *Exploratory Data Analysis.* New-York: Addison-Wesley.

Verhulst, P. F. 1838. Notice sur la loi que la population suit dans son accroissement [Instructions on the law that the population follows in its growth]. *Correspondance mathematique et Physique, publiee par A. Quetelet, 10,* 113–120.

Verhulst, P. F. 1845. Recherches mathematiques sur la loi d'accroissement de la population [Mathematics research on the law of population growth]. *Nouveauz Memoires de l'Academie Royale des Sciences, des Lettres et des Beaux-Arts de Belgique, 18,* 1–38.

Verhulst, P. F. 1847. Deuxieme memoire sur la loi d'accroissement de la population [Second memorandum on the law of population growth]. *Nouveauz Memoires de l'Academie Royale des Sciences, des Lettres et des Beaux-Arts de Belgique, 20,* 1–32.

Yule, G. U. 1925. The growth of population and the factors which control it. *Journal of the Royal Statistical Society, 38,* 1–59.

2

Discrete Distributions

Jodi M. Casabianca and Brian W. Junker

CONTENTS

2.1 Introduction

In this chapter, we review the elementary features of several basic parametric distributions for discrete random variables that form the building blocks of item response theory (IRT) and related psychometric models. Many IRT models are defined directly in terms of these distributions or their generalizations. For each distribution, we will review the form of the probability mass function (pmf), that is, the function $f(u) = PU = u$ for a response variable U with that distribution, as well as some basic features of the distribution. Most of the distributions reviewed here are members of the exponential family of distributions. The textbook of DeGroot and Schervish (2001) is the primary source for the general properties of these distributions. Other useful sources include Johnson and Kotz (1969) and Novick and Jackson (1974).

A random variable U with an exponential-family distribution has a density or pmf if U is discrete of the following form:

$$f(u;\eta) = h(u)\exp(\eta^T \mathbf{T}(u) - A(\eta)) \tag{2.1}$$

where u is the value of U, η is the (possibly multivariate) parameter, called the natural parameter(s), $h(u)$ is base measure or an arbitrary function of u that does not depend on η, $\mathbf{T}(u)$ is a vector of sufficient statistics for each coordinate of η, and $A(\eta)$ is the normalizing constant chosen so that the area under $f(u;\eta)$ is equal one. This definition also applies if U is multivariate. For an in-depth treatment of exponential-family distributions, see Volume Two, Chapter 4.

We will also indicate the *conjugate prior distribution* for likelihoods with these random variables. The *likelihood* for (possibly multidimensional) data U is (any function proportional to) the conditional density or pmf of U given (possibly multidimensional) parameter η, $f(u|\eta)$; the *prior distribution* can be represented by a marginal density or pmf $f(\eta;\tau)$ with (possibly multidimensional) hyperparameter τ; and the *posterior distribution* is represented by the posterior density $f(\eta|u;\tau) = C \cdot (u|\eta)f(\eta;\tau)$, where C is the normalizing constant that does not depend on η, chosen so that $f(\eta|u;\tau)$ has total mass one. We say that the prior $f(\eta;\tau)$ is *conjugate* to the likelihood $f(u|\eta)$ if the posterior $f(\eta|u;\tau)$ can be rewritten as a member $f(\eta|\tau^*)$ of the same parametric family as the prior $f(\eta;\tau)$, where τ^* may be any function of τ and the data u. If the likelihood is a member of the exponential family, or nearly so, then it is often easy to "read off" a conjugate prior distribution, by interchanging the roles of data U and parameter η.

Finally, we will indicate the *generalized linear model* (glm) corresponding to many of these discrete distributions, since glms play a central role in building IRT and other psychometric models from these distributions. It is useful to think of glms as exponential-family models in which the natural parameters, η, are replaced with a linear function of covariates or predictors of interest. The monographs of McCullagh and Nelder (1989) and Dobson (2011) include more information about glms.

2.2 Distributions

2.2.1 Bernoulli Distribution

The Bernoulli distribution is a basic building block for IRT models, and indeed, for most discrete-response psychometric models. We say that the random variable U has *Bernoulli distribution* with parameter p, if

$$U = \begin{cases} 1, & \textit{with probability } p \\ 0, & \textit{with probability } 1-p \end{cases} \tag{2.2}$$

For example, if a multiple-choice test item has $A = 4$ four choices $a = 1, 2, 3, 4$, and option 3 is correct, then we can record the item response as $U = 1$ if option 3 is chosen, and $U = 0$ otherwise; p is then the probability that the item is correct, that is, $P\{U = 1\} = p$. (Note that p may or may not be equal to 1/4 in this case, depending on the characteristics of the test item and of the person responding to that item.) Any response that can be scored in two categories—a short answer response with no partial credit, a preference for strategy I versus strategy II, etc.—can be coded as Bernoulli's random variable U. The Bernoulli distribution is the basic model for wrong/right item scoring in dichotomous item response models, as will be seen throughout this handbook.

Theorem 2.1

The pmf for Bernoulli's random variable U can be written as follows:

$$Ber(u;p) = P\{U = u; p\} = \begin{cases} p^u(1-p)^{1-u} & \text{for } u = 0, 1, \\ 0 & \text{otherwise} \end{cases} \tag{2.3}$$

Theorem 2.2

The mean and variance of U are as follows:

$$E(U) = p \tag{2.4}$$

$$\text{Var}(U) = p(1-p) \tag{2.5}$$

If $\mathbf{U} = (U_1, U_2, \ldots, U_n)$ is a random sample (independent, identically distributed, or *iid*) from Bernoulli distribution, the likelihood is given as follows:

$$f(u_1, \ldots, u_n; p) = \prod_{i=1}^{n} p^{u_i}(1-p)^{1-u_i} = p^x(1-p)^{n-x} \tag{2.6}$$

where $x = \sum_{i=1}^{n} u_i$ is the number of successes. Interchanging the role of data and parameter, this has the form of a *Beta* distribution. Indeed, for a Beta prior density,

$$Beta(p; \alpha, \beta) = \frac{\Gamma(\alpha+\beta)}{\Gamma(\alpha)\Gamma(\beta)} p^{\alpha-1}(1-p)^{\beta-1}, \quad \alpha > 0, \ \beta > 0 \tag{2.7}$$

we can easily verify that the posterior density for p is $Beta(p|\alpha + x, \beta + n - x)$, so that the conjugate prior distribution for Bernoulli is the Beta distribution.

Here, $\Gamma(z) = \int_0^{\infty} e^{-t} t^{z-1}\, dt$ is the *Gamma function*, defined for any positive continuous z as an improper integral. It is the unique logarithmically convex extension of the factorial function: If k is a positive integer, then $\Gamma(k) = (k-1)! = (k-1)(k-2)\cdots 1$, the number of ways of arranging $k-1$ distinct objects.

It is interesting to note that when $\alpha = \beta = 1$, $Beta(p; \alpha, \beta)$ reduces to a uniform distribution on the unit interval. $Beta(p; 1, 1)$ is often used as an "uninformative" or "objective" prior distribution, because it weights all $p \in [0, 1]$ equally—it is a "flat" prior distribution.

2.2.2 Binomial Distribution

A *binomial* random variable X simply counts the number of successes in a random sample of n Bernoulli random variables, $X = \sum_{i=1}^{n} U_i$. For example, X might be the total score on a test of dichotomously scored items with equal difficulties, that is, equal probabilities $P\{U_i = 1\} = p$ for all $i = 1, \ldots, n$ (for total-score distributions, see Volume Two, Chapter 6). Since the total X could be achieved by getting the items in any of the $\binom{n}{x} = \frac{n!}{x!(n-x)!}$ possible subsets of $(U_1, \ldots, U_n)$ right ($U_i = 1$), and the rest of the items wrong ($U_i = 0$), we have

Theorem 2.3

A random variable X has a binomial distribution with parameters n and p, if X follows the pmf

$$Bin(x; n, p) = \begin{cases} \binom{n}{x} p^x(1-p)^{n-x} & \text{for } x = 0, 1, 2, \ldots, n, \\ 0 & \text{otherwise} \end{cases} \tag{2.8}$$

Here, n is the number of trials ($n > 0$), and similar to the notation for the Bernoulli distribution, p is the probability of success ($0 \leq p \leq 1$), and x is the total number of successes.

Theorem 2.4

The mean and variance of a binomial random variable are as follows:

$$E(X) = np \tag{2.9}$$

$$\mathrm{Var}(X) = np(1 - p) \tag{2.10}$$

It is easy to see from the binomial pmf that the conjugate prior distribution is again $Beta(p; \alpha, \beta)$. The posterior distribution will be $Beta(p|\alpha + x, \beta + n - x)$.

It is interesting to compute the distribution of the number-correct score for n independent Bernoulli test items $U_1, \ldots, U_n$ with identical success probability p, using a $Beta(\alpha, \beta)$ prior density for p. The distribution of X in this case is the Beta-Binomial distribution

$$BetaBin(x; \alpha, \beta, n) = \int_0^1 Bin(x|n, p)\, Beta(p; \alpha, \beta)\, dp \tag{2.11}$$

If we model variation in p among a population of examinees with a $Beta(\alpha, \beta)$ population distribution, then the distribution of total scores on this test will follow the $BetaBin(x; \alpha, \beta)$ pmf (Lord and Novick, 1968; van der Linden, see Chapter 6).

2.2.3 Negative Binomial Distribution

The binomial distribution models the number of successes in a finite number n of independent, identical Bernoulli trials with success probability p. If, instead of a fixed number of trials, we keep repeating independent Bernoulli trials until the rth success, then the number of failures X in the sequence has a *negative binomial* distribution. One might encounter the negative binomial distribution as the number of trials required by a student in a mastery testing situation in which r correct answers are required to declare mastery (Wilcox, 1981).

Theorem 2.5

A random variable X has the negative binomial distribution with parameters r and p, if X follows the pmf

$$NegBin(x; r, p) = \begin{cases} \binom{r + x - 1}{x} p^r (1 - p)^x & \text{for } x = 0, 1, 2, \ldots, \\ 0 & \text{otherwise} \end{cases} \tag{2.12}$$

Theorem 2.6

The mean and variance of the negative binomial distribution are as follows:

$$E(X) = \frac{r(1-p)}{p} \tag{2.13}$$

$$\text{Var}(X) = \frac{r(1-p)}{p^2} \tag{2.14}$$

It is noted that the domain of X is unbounded: X can be any nonnegative integer. A special case of the negative binomial distribution occurs when $r = 1$. In this case, the negative binomial distribution is called a *geometric* distribution with parameter p; here, we model the number of failures X that occur before the first success occurs. For example, the geometric distribution is a simple model for the number of trials required by an examinee to complete an "answer until correct" item format. It is easy to verify that the geometric distribution has the *memoryless* property: For any $t \geq s \geq 0$, $P\{X = t|X \geq s\} = P\{X = t - s\}$, that is, if we have been waiting for $s - 1$ trials without a success, the distribution of the number of additional trials needed before the first success does not depend on how long we have already been waiting—a long sequence of failures in a set of trials has no effect on the future outcome of the sampling process.

The conjugate prior distribution for a negative binomial random variable X with parameters r and p is again $Beta(p; \alpha, \beta)$. The posterior distribution will be $Beta(p|\alpha + r, \beta + x)$.

It is worth noting that each of the Bernoulli, binomial, and negative binomial distributions have the same conjugate prior distribution, that is, as a function of the parameter p, the pmfs are proportional to one another. Indeed, any two likelihoods proportional to each other will have the same conjugate prior distribution.

2.2.4 Multinomial Distribution

The *multinomial* distribution is a generalization of the binomial distribution involving $k \geq 2$ categories instead of just two. In n independent trials, let $X_1, \ldots, X_K$ be the number of trials resulting in outcome k, $k = 1, \ldots, K$. When $n = 1$, the multinomial distribution reduces to a general discrete distribution on K categories and is the basis for most polytomous item response models, as will be discussed throughout this handbook. In other settings, the multinomial distribution might be used to model the number of students, out of n students, at various achievement levels (e.g., below basic, basic, proficient, and advanced).

Theorem 2.7

Suppose that the components X_k of the random vector $\mathbf{X} = (X_1, X_2, \ldots, X_K)$ count the number of K mutually exclusive events that occur in n trials, each with $P(X_k = x_k) = p_k$. The pmf for the random vector $\mathbf{X}$ is as follows:

$$Multi\,(\mathbf{x}; n, \boldsymbol{p}) = \begin{cases} \binom{n}{x_1, \ldots, x_K} p_1^{x_1} \cdots p_K^{x_K} & \text{for positive } x_1, \ldots, x_K \\ & \text{such that } x_1 + \cdots + x_K = n \\ 0 & \text{otherwise} \end{cases} \tag{2.15}$$

where $\mathbf{x} = (x_1, \ldots, x_K)$, $\boldsymbol{p} = (p_1, \ldots, p_k)$, $0 \leq p_k \leq 1$ and $\sum_{k=1}^{K} p_k = 1$.

Theorem 2.8

The mean and variance of X_k are as follows:

$$E(X_k) = np_k \tag{2.16}$$

$$\text{Var}(X_k) = np_k(1 - p_k) \tag{2.17}$$

Theorem 2.9

The covariance of X_k and X_j is as follows:

$$Cov(X_k, X_j) = -np_k p_j \tag{2.18}$$

The individual components X_k of a multinomial random vector are binomial with parameters n and p_k. While the n trials are independent, groups of responses are not; in fact, X_k and X_j will be negatively correlated. The relationship between the multinomial and the Poisson distributions is discussed in Section 2.2.6.

The conjugate prior distribution for the multinomial distribution is the *Dirichlet distribution* for the probability vector $\boldsymbol{p} = (p_1, \ldots, p_K)$ with parameters $\boldsymbol{\alpha} = (\alpha_1, \ldots, \alpha_K)$, $\alpha_k > 0$, $k = 1, \ldots, K$, and density

$$Dir(\boldsymbol{p}; \boldsymbol{\alpha}) = \frac{\Gamma\left(\sum_{k=1}^{K} \alpha_k\right)}{\prod_{k=1}^{K} \Gamma(\alpha_k)} \prod_{k=1}^{K} p_k^{\alpha_{k-1}} \tag{2.19}$$

Note that the Dirichlet density is defined to be nonzero only on the simplex of $\boldsymbol{p}$'s satisfying $p_k \geq 0$ for all k, and $\sum_{k=1}^{K} p_k = 1$. Similar to the multinomial distribution generalizes the binomial to more than two categories, the Dirichlet distribution generalizes the Beta distribution to more than two probabilities. If we take all $\alpha_k = 1$, we get a flat or uninformative prior distribution (uniform on the probability simplex). The posterior distribution for the multinomial distribution with $Dir(\boldsymbol{p}; \boldsymbol{\alpha})$ prior will be $Dir(\boldsymbol{p}|\boldsymbol{\alpha} + \mathbf{x})$ where $\boldsymbol{\alpha} + \mathbf{x}$ is the vector with coordinates $\alpha_k + x_k, k = 1, \ldots, K$.

2.2.5 Hypergeometric Distribution

The *hypergeometric* distribution describes Bernoulli sampling without replacement from a finite population—it is the analog of the binomial distribution for typical survey-sampling applications. Consider a finite population of size N in which there are M "target" or "success" cases. If we sample n units without replacement from that population, the number X of successes in the sample follows a hypergeometric distribution.

Theorem 2.10

The random variable X has a hypergeometric distribution with parameter M, if it follows the pmf

$$Hyper(x; N, M, n) = \begin{cases} \dfrac{\binom{M}{x}\binom{N-M}{n-x}}{\binom{N}{n}} & \text{for } \max\{0, n+M-N\} \leq x \leq \min\{M, n\}, \\ 0 & \text{otherwise} \end{cases} \tag{2.20}$$

Theorem 2.11

The mean and variance of the hypergeometric distribution are as follows:

$$E(X) = n\frac{M}{N} \tag{2.21}$$

$$\text{Var}(X) = n \cdot \frac{M}{N} \cdot \frac{(N-M)}{N} \cdot \frac{N-n}{N-1} = np(1-p)\frac{N-n}{N-1} \tag{2.22}$$

where $p = M/N$.

The hypergeometric distribution is not an exponential-family distribution, but it still has a fairly simple conjugate prior distribution. The conjugate prior pmf for M is the $BetaBin(M; \alpha, \beta, N)$ defined previously. The posterior distribution is again $BetaBin(M|\alpha + x, \beta + n - x, N)$ (Dyer and Pierce, 1993).

2.2.6 Poisson Distribution

The Poisson distribution is an alternative to the negative binomial distribution, for non-negative integers, which are often counts of events in finite time or space. The *multiplicative Poisson model* or the *Rasch Poisson counts model* (Volume One, Chapter 15; Jansen and Van Duijn, 1992; Lord and Novick, 1968; Rasch, 1960) is an IRT-like model based on the Poisson distribution, modeling the number of errors examinees make in a fixed quantity of work.

Theorem 2.12

The random variable X has a Poisson distribution with parameter λ, if X follows the pmf

$$Poisson(x; \lambda) = \begin{cases} \frac{e^{-\lambda}\lambda^x}{x!} & \text{for } x = 0, 1, 2, \ldots \\ 0 & \text{otherwise} \end{cases} \tag{2.23}$$

The parameter λ is always positive ($\lambda > 0$).

Theorem 2.13

The mean and variance of X are as follows:

$$E(X) = \lambda \tag{2.24}$$

$$\mathrm{Var}(X) = \lambda \tag{2.25}$$

Theorem 2.14

If the random variables $X_1, \ldots, X_n$ are independent and if X_i has a Poisson distribution with mean λ_i, $i = 1, \ldots, n$, then the sum $X_1 + \cdots + X_n$ has a Poisson distribution with mean $\lambda_1 + \cdots + \lambda_n$.

The Poisson distribution approximates the binomial distribution when n is large and p is small.

Theorem 2.15

Suppose that $X \sim B(n, p)$ and let $n \to \infty$ and $p \to 0$ in such a way that $np \to \lambda$, where λ is a constant. Then

$$\lim_{\substack{n\to\infty,\\ np\to\lambda}} \binom{n}{x} p^x (1-p)^{n-x} = \frac{e^{-\lambda}\lambda^x}{x!} \tag{2.26}$$

Thus, in the limit, a binomial random variable X will have a Poisson distribution with parameter λ.

The conjugate prior distribution for the λ parameter of the Poisson distribution is the Gamma distribution with parameters α and $\beta (\alpha > 0, \beta > 0)$, with density

$$\Gamma(\lambda; \alpha, \beta) = \frac{\beta^\alpha}{\Gamma(\alpha)} \lambda^{\alpha-1} e^{-\beta\lambda}, \quad \lambda \geq 0 \tag{2.27}$$

The posterior distribution of λ is again a Gamma distribution with density $Gamma(\lambda|\alpha + x, \beta + 1)$.

The Poisson distribution plays a key role in understanding classical discrete multivariate statistics, that is, the statistics of contingency tables. Suppose $X_1, \ldots, X_k$ are independent counts, distributed as $Poisson(x_i; \lambda_i)$, $i = 1, \ldots, k$, and let $T = X_1 + \cdots + X_k$. Then, we can rewrite the pmf for $X_1, \ldots, X_k$ given $T = t$ as a conditional multinomial pmf times a single Poisson for the total:

$$f(x_1, \ldots, x_k|t) = Multi(x_1, \ldots x_k|t; p_1, \ldots, p_k) \cdot Poisson(t; \lambda_+) \tag{2.28}$$

where $p_i = \lambda_i/(\lambda_1 + \cdots + \lambda_k)$, $i = 1, \ldots, k$ and $\lambda_+ = \lambda_1 + \cdots + \lambda_k$. This fact is intimately connected with the notion that maximum likelihood estimates for log-linear models of contingency tables have the same properties under Poisson or multinomial sampling models (Birch, 1963; Lang, 1996). On the other hand, if $X_1 \sim Bin(n_1, p_1)$ and $X_2 \sim Bin(n_2, p_2)$ are independent, then it is easy to see that the pmf of X_1 given $X_1 + X_2 = t$ will be hypergeometric

$$f(x_1|t) = Hyper(x_1|t; n_1) \tag{2.29}$$

and this is the basis of Fisher's exact test of independence for 2 × 2 tables under multinomial or Poisson sampling (Andersen, 1980; Fisher, 1935) and its generalizations.

2.3 Generalized Linear Model

In this section, we change the notation slightly, to conform to standard notation for regression models. Let $\mathbf{Y} = (Y_1, \ldots, Y_k)^T$ be a vector of response variables, and let $\mathbf{X}$ be a matrix of predictors. The family of *glms* generalizes the familiar Normal linear regression model to a family of nonlinear regression models in which the data follow an exponential-family distribution, conditional on a linear predictor. A glm is composed of (i) a probability distribution from the exponential family for the dependent variables Y, (ii) a linear predictor function $\eta = \mathbf{X}\beta$, and (iii) a link function $g(\mu)$ applied to each component of $\eta = (\eta_1, \ldots, \eta_k)$ that defines the relationship between the linear predictor $\mathbf{X}\beta$ and the mean of the distribution function for a response variable, $E(\mathbf{Y}) = g^{-1}(\eta) = g^{-1}(\mathbf{X}\beta)$.

Many IRT and related models can be constructed as glms or generalizations of glms for multiway data. For example, we may write the Bernoulli pmf as an exponential-family model

$$Ber(y; p) = p^y(1-p)^{1-y} = (1-p)\exp\left[y\ln\frac{p}{1-p}\right] = h(y)\exp(\eta T(y) - A(\eta)) \tag{2.30}$$

with $h(y) \equiv 1$, $A(\eta) = -\ln(1-p) = \ln(1+e^{\eta})$, $\eta = g(p) = \ln p/(1-p) \equiv \text{logit}\, p$, and $T(y) = y$. If we let $\eta = \mathbf{X}\beta$ for a matrix of predictors $\mathbf{X}$ and vector of regression coefficients β, we obtain the usual logistic regression model

$$\text{logit}\, \boldsymbol{p} = \mathbf{X}\beta \tag{2.31}$$

Given a two-way data matrix $[Y_{pi}]$ with persons in rows and items in columns, we may arrange the linear model $\mathbf{X}\beta$ so that $\eta_{pi} = \theta_p - b_i$ and observe that $p_{pi} = e^{\eta}_{pi}/(1 + e^{\eta}_{pi})$, so that the model for the (p, i)th element of the data matrix is

$$Ber(y_{pi}; \theta_p, b_i) = \left[\frac{e^{\theta_p - b_i}}{1 + e^{\theta_p - b_i}}\right]^{y_{pi}} \left[\frac{1}{1 + e^{\theta_p - b_i}}\right]^{1-y_{pi}} \tag{2.32}$$

This is, of course, the familiar Rasch (1960) model for dichotomous response data. The familiar two-parameter logistic (2PL) IRT model replaces η_{pi} with $\eta^{2PL}_{pi} = a_i(\theta_p - b_i)$, and the three-parameter logistic (3PL) IRT model replaces $p^{2PL}_{pi} = \frac{e^{a_i(\theta_p - b_i)}}{1+e^{a_i(\theta_p - b_i)}}$ with the mixture $p^{3PL}_{pi} = c_i + (1 - c_i)p^{2PL}_{pi}$ (for these models, see Volume One, Chapters 2 and 3). Note that the 2PL and 3PL models are no longer strictly glms, since η_{pi} can no longer be expressed as a linear function of the parameters a_i, b_i, and θ_p in either model, and in the 3PL model, the item success probability is a mixture model; however, both models are "close" to glms and so intuitions and computational strategies for glms may still be useful.

TABLE 2.1
Exponential-Family Form for Discrete Distributions Using Standard glm Notation

Distribution	Exponential Form	Natural Link Function	Conjugate Posterior
Bernoulli	$f(y;\eta) = (1+e^{\eta})^{-1}\exp(\eta y)$	$\eta = \ln\left(\frac{p}{1-p}\right)$	$Beta(p\|\alpha + y, \beta + n - y)$
Binomial (with known n)	$f(y;\eta) = \binom{n}{y}\exp(y\eta + n\ln(1+e^{\eta}))$	$\eta = \ln\left(\frac{p}{1-p}\right)$	$Beta(p\|\alpha + y, \beta + n - y)$
Negative binomial (with known r)	$f(y;\eta) = \binom{y+r-1}{y}\exp(y\eta + r\ln(1-e^{\eta}))$	$\eta = \ln p$	$Beta(p\|\alpha + r, \beta + y)$
Multinomial (with known n)	$f(\boldsymbol{y};\boldsymbol{\eta}) = \frac{n!}{\prod_{i=1}^{k} y_i}\exp(\boldsymbol{\eta}^T\boldsymbol{y})$	$\boldsymbol{\eta} = \ln\boldsymbol{p}$	$\mathrm{Dir}(\boldsymbol{p}\|\boldsymbol{\alpha} + \boldsymbol{y})$
Poisson	$p(y;\eta) = \frac{1}{y!}\exp(y\eta - \exp(\eta))$	$\eta = \ln\lambda$	$Gamma(\lambda\|\alpha + y, \beta + 1)$

Another example of how an IRT-like model can be constructed as a glm begins with the Poisson pmf, written as an exponential-family model

$$Poisson(y;\lambda) = \frac{e^{-\lambda}\lambda^y}{y!} = \frac{1}{y!}\exp(y \cdot \ln\lambda - \exp(\ln\lambda)) = h(y)\exp(\eta T(y) - A(\eta)) \tag{2.33}$$

with $h(y) \equiv 1/y!$, $A(\eta) = \exp(\eta)$, $\eta = \ln\lambda$, and $T(y) = y$. Much like in the previous example, if we let $\boldsymbol{\eta} = \ln(\mathbf{X}\boldsymbol{\beta})$ for a matrix of predictors $\mathbf{X}$ and regression coefficients β, we obtain what looks like a Poisson regression model

$$\ln\boldsymbol{\lambda} = \mathbf{X}\boldsymbol{\beta} \quad \text{or reduced to: } \boldsymbol{\lambda} = e^{\mathbf{X}\boldsymbol{\beta}} \tag{2.34}$$

Given a two-way data matrix or scores $[Y_{pt}]$ with persons in rows and tests in columns, we may arrange the linear model, so that $\eta_{pt} = \ln(b_t\theta_p) = \ln b_t + \ln\theta_p$, where b_t is the test difficulty parameter for test t and θ_p is the trait parameter for person p. Note that $\lambda_{pt} = e^{\eta_{pt}} = e^{\ln b_t\theta_p} = b_t\theta_p$, so that the model for the (p,t)th element of the two-way data matrix is

$$Poisson(y_{pt}; b_t, \theta_p) = \frac{e^{-b_t\theta_p}(b_t\theta_p)^{y_{pt}}}{y_{pt}!} \tag{2.35}$$

This gives the probability of test taker p obtaining score the y_{pt} on test t under the Rasch Poisson counts model (Rasch, 1960; Volume One, Chapter 15).

We can generalize this procedure to obtain a general approach for building IRT and IRT-like models using glms:

1. Identify the probability function for the scored item response variable Y and express this model in an exponential-family form, $f(y|\eta)$ using Table 2.1
2. Identify the link function
3. Write the linear predictor in terms of person and item effects and rewrite $f(y;\eta)$ in terms of person and item effects

Table 2.1 gives the exponential-family form and link function for the discrete distributions mentioned in this chapter,* and other IRT and IRT-like models can be derived analogously to the IRT and Rasch Poisson counts models above, using this approach.

References

Andersen, E. B. 1980. *Discrete Statistical Models with Social Science Applications*. New York: North-Holland.

Birch, M. W. 1963. Maximum likelihood in three-way contingency tables. *Journal of the Royal Statistical Society, B*, 25, 220–233.

DeGroot, M. H. and Schervish, M. J. 2001. *Probability and Statistics*. Reading, MA: Addison-Wesley.

Dobson, A. 2011. *An Introduction to Generalized Linear Models*. New York: Chapman-Hall/CRC.

* The hypergeometric distribution is not part of the exponential family of distributions, and therefore does not appear in Table 2.1.

Dyer, D. and Pierce, R. L. 1993. On the choice of the prior distribution in hypergeometric sampling. *Communications in Statistics—Theory and Methods*, 22, 2125–2146.
Fisher, R. A. 1935. The logic of inductive inference (with discussion). *Journal of the Royal Statistical Society*, 98, 39–82.
Jansen, M. G. H. and van Duijn, M. A. J. 1992. Extensions of Rasch's multiplicative Poisson model. *Psychometrika*, 57, 405–414.
Johnson, N. and Kotz, S. 1969. *Discrete Distributions*. New York: John Wiley and Sons.
Lang, J. B. 1996. On the comparison of multinomial and Poisson loglinear models. *Journal of the Royal Statistical Society, B*, 58, 253–266.
Lord, F. M. and Novick, M. R. 1968. *Statistical Theories of Mental Test Scores*. Reading, MA: Addison-Wesley.
McCullagh, P. and Nelder, J. A. 1989. *Generalized Linear Models*. New York: Chapman-Hall/CRC.
Novick, M. R. and Jackson, P. H. 1974. *Statistical Methods for Educational and Psychological Research*. New York: McGraw-Hill.
Rasch, G. 1960. *Probabilistic Models for Some Intelligence and Attainment Tests*. Copenhagen: Danish Institute for Educational Research. Expanded edition, University of Chicago Press, 1980.
Wilcox, R. R. 1981. A cautionary note on estimating the reliability of a mastery test with the beta-binomial. *Applied Psychological Measurement*, 5, 531–537.

3

Multivariate Normal Distribution

Jodi M. Casabianca and Brian W. Junker

CONTENTS

3.1 Introduction

In this chapter, we review several basic features of the multivariate normal distribution. Section 3.2 considers general properties of the multivariate normal density and Section 3.3 considers the sampling distribution of the maximum likelihood estimators (MLEs) of the mean vector and variance–covariance matrix based on iid (independent, identically distributed) random sampling of a multivariate normal distribution. Section 3.4 reviews the standard conjugate distributions for Bayesian inference with the multivariate normal distribution and Section 3.5 considers various generalizations and robustifications of the normal model. The properties of the multivariate normal distribution are well known and available in many places; our primary sources are the texts by Johnson and Wichern (1998) and Morrison (2005). A classic and comprehensive treatment is given by Anderson's (2003) text.

In item response theory (IRT), the multivariate normal and its generalizations are most often used as the underlying variables distribution for a data-augmentation version of the normal-ogive model (Bartholomew and Knott, 1999; Fox, 2010), as a population distribution for the proficiency parameters θ_p, for persons $p = 1, \ldots, P$, and as a prior distribution for other model parameters (e.g., difficulty parameters b_i, log-discrimination parameters $\log a_i$, etc.) whose domain is the entire real line or Euclidean space. Especially, in the latter two cases, the multivariate normal distribution serves to link standard IRT modeling with hierarchical linear models (HLMs) and other structures that incorporate various group structures and other dependence on covariates, to better model item responses in terms of the contexts in which they are situated (e.g., Fox, 2003, 2005a,b, 2010).

Because of the wide variety of applications of the normal distribution in IRT, we will depart somewhat from the notation used in the rest of the book in this chapter. We use $\mathbf{X} = (X_1, X_2, \ldots, X_K)^T$ to represent a generic K-dimensional random vector and X to represent a generic random variable. Their observed values will be denoted $\mathbf{x}$ and x, respectively. Also, because very many densities will be discussed, we will use the generic notation $f()$ to denote a density. The particular role or nature of $f()$ will be clear from context.

3.2 Multivariate Normal Density

The univariate normal density with mean μ and variance σ^2 for a random variable X is as follows:

$$f(x; \mu, \sigma^2) = \frac{1}{\sqrt{2\pi\sigma^2}} e^{-\frac{1}{2}(x-\mu/\sigma)^2} \quad -\infty < x < \infty \tag{3.1}$$

As it is well known, $E[X] = \mu$, and $\mathrm{Var}(X) = E[(X - \mu)^2] = \sigma^2$. We often write $X \sim N(\mu, \sigma^2)$ to convey that the random variable X has this density. The quantity in the exponent of the normal density thus measures the square of the distance between realizations of the random variable, X, and the mean μ, scaled in units of the standard deviation σ.

The multivariate normal density, for $\mathbf{X} \in \Re^K$, is as follows:

$$f(\mathbf{x}; \boldsymbol{\mu}, \boldsymbol{\Sigma}) = \frac{1}{(2\pi)^{K/2}|\boldsymbol{\Sigma}|^{1/2}} e^{-(1/2)(\mathbf{x}-\boldsymbol{\mu})^T \boldsymbol{\Sigma}^{-1}(\mathbf{x}-\boldsymbol{\mu})} \tag{3.2}$$

where again $E[\mathbf{X}] = \boldsymbol{\mu} = (\mu_1, \ldots, \mu_K)^T$ is the mean vector and $\mathrm{Var}(\mathbf{X}) = E[(\mathbf{X} - \boldsymbol{\mu})(\mathbf{X} - \boldsymbol{\mu})^T] = \boldsymbol{\Sigma}$ is the $K \times K$ symmetric nonnegative-definite variance–covariance matrix. We often write $\mathbf{X} \sim N_K(\boldsymbol{\mu}, \boldsymbol{\Sigma})$, or just $\mathbf{X} \sim N(\boldsymbol{\mu}, \boldsymbol{\Sigma})$ if the dimension K is clear from context, to indicate that $\mathbf{X}$ follows the multivariate normal density. The quantity $(\mathbf{x} - \boldsymbol{\mu})^T \boldsymbol{\Sigma}^{-1}(\mathbf{x} - \boldsymbol{\mu})$ in the exponent is the squared distance from $\mathbf{x}$ to $\boldsymbol{\mu}$, again scaled by the variance–covariance matrix $\boldsymbol{\Sigma}$. In other contexts, this is called the *Mahalanobis distance* between $\mathbf{x}$ and $\boldsymbol{\mu}$ (Morrison, 2005).

The following theorem gives some properties of the multivariate normal random variables.

Theorem 3.1

If $\mathbf{X} \sim N_K(\boldsymbol{\mu}, \boldsymbol{\Sigma})$, then

1. $E[\mathbf{X}] = \boldsymbol{\mu}$, and $\mathrm{Var}(\mathbf{X}) = E[(\mathbf{X} - \boldsymbol{\mu})(\mathbf{X} - \boldsymbol{\mu})^T] = \boldsymbol{\Sigma}$ (Johnson and Wichern, 1998).
2. If $\mathbf{W} = \mathbf{A}\mathbf{X} + \mathbf{b}$ for a constant matrix $\mathbf{A}$ and a constant vector $\mathbf{b}$, then $\mathbf{W} \sim N_K(\mathbf{A}\boldsymbol{\mu} + \mathbf{b}, \mathbf{A}\boldsymbol{\Sigma}\mathbf{A}^T)$ (Johnson and Wichern, 1998).
3. *Cholesky decomposition*: There exists a lower-triangular matrix $\mathbf{L}$ with nonnegative diagonal entries, such that $\boldsymbol{\Sigma} = \mathbf{L}\mathbf{L}^T$ (Rencher, 2002). If $\mathbf{X} \sim N_K(\mathbf{0}, \mathbf{I}_{K \times K})$ then $\mathbf{W} = \mathbf{L}\mathbf{X} + \boldsymbol{\mu} \sim N_K(\boldsymbol{\mu}, \boldsymbol{\Sigma})$.
4. $\boldsymbol{\Sigma}$ is diagonal if and only if the components of $\mathbf{X}$ are mutually independent (Johnson and Wichern, 1998).

5. If $\mathbf{X}$ is partitioned into disjoint subvectors $\mathbf{X}_1$ and $\mathbf{X}_2$, and we write the following equation:

$$\begin{pmatrix}\mathbf{X}_1\\ \mathbf{X}_2\end{pmatrix} \sim N\left[\begin{pmatrix}\boldsymbol{\mu}_1\\ \boldsymbol{\mu}_2\end{pmatrix}, \begin{pmatrix}\boldsymbol{\Sigma}_{11} & \boldsymbol{\Sigma}_{12}\\ \boldsymbol{\Sigma}_{21} & \boldsymbol{\Sigma}_{22}\end{pmatrix}\right] \tag{3.3}$$

then the conditional distribution of $\mathbf{X}_1$, given $\mathbf{X}_2 = \mathbf{x}_2$, is also multivariate normal, with mean $\boldsymbol{\mu}_{1|2}$ and variance–covariance matrix $\boldsymbol{\Sigma}_{11|2}$

$$\boldsymbol{\mu}_{1|2} = \boldsymbol{\mu}_1 + \boldsymbol{\Sigma}_{12}\boldsymbol{\Sigma}_{22}^{-1}(\mathbf{x}_2 - \boldsymbol{\mu}_2) \tag{3.4}$$

$$\boldsymbol{\Sigma}_{11|2} = \boldsymbol{\Sigma}_{11} - \boldsymbol{\Sigma}_{12}\boldsymbol{\Sigma}_{22}^{-1}\boldsymbol{\Sigma}_{12} \tag{3.5}$$

(Johnson and Wichern, 1998).

3.2.1 Geometry of the Multivariate Normal Density

A useful geometric property of the multivariate normal distribution is that it is log quadratic: $\log f(\mathbf{x})$ is a simple linear function of the symmetric nonnegative definite quadratic form $(\mathbf{x} - \boldsymbol{\mu})^T \boldsymbol{\Sigma}^{-1}(\mathbf{x} - \boldsymbol{\mu})$. This means that the level sets $\{\mathbf{x} : f(\mathbf{x}) = K\}$, or equivalently (after taking logs and omitting irrelevant additive constants) the level sets $\{\mathbf{x} : (\mathbf{x} - \boldsymbol{\mu})^T \boldsymbol{\Sigma}^{-1}(\mathbf{x} - \boldsymbol{\mu}) = c^2\}$, will be ellipsoids. Figure 3.1 depicts bivariate normal density plots and the same densities using contour plots for two sets of variables with equal variance; variables in subplot (a) are uncorrelated and variables in subplot (b) are highly correlated. The contours are exactly the level sets $\{\mathbf{x} : (\mathbf{x} - \boldsymbol{\mu})^T \boldsymbol{\Sigma}^{-1}(\mathbf{x} - \boldsymbol{\mu}) = c^2\}$ for various values of c^2.

Finding the principal axes of the ellipsoids is straightforward with Lagrange multipliers. For example, to find the first principal axis, we want to find the point $\mathbf{x}$ that is a maximum distance from $\boldsymbol{\mu}$ (maximize the squared distance $(\mathbf{x} - \boldsymbol{\mu})^T (\mathbf{x} - \boldsymbol{\mu})$) that is still on the contour (satisfies the constraint $(\mathbf{x} - \boldsymbol{\mu})^T \boldsymbol{\Sigma}^{-1}(\mathbf{x} - \boldsymbol{\mu}) = c^2$). Differentiating the Lagrange-multiplier objective function

$$g_\lambda(\mathbf{x}) = (\mathbf{x} - \boldsymbol{\mu})^T (\mathbf{x} - \boldsymbol{\mu}) - \lambda\left[(\mathbf{x} - \boldsymbol{\mu})^T \boldsymbol{\Sigma}^{-1}(\mathbf{x} - \boldsymbol{\mu}) - c^2\right] \tag{3.6}$$

with respect to $\mathbf{x}$ and setting these derivatives equal to zero leads to the eigenvalue/eigenvector problem

$$(\boldsymbol{\Sigma} - \lambda I)(\mathbf{x} - \boldsymbol{\mu}) = 0 \tag{3.7}$$

It is now a matter of calculation to observe that the eigenvalues can be ordered so that

$$\lambda_1 \geq \lambda_2 \geq \cdots \geq \lambda_K \tag{3.8}$$

with corresponding mutually orthogonal eigenvectors $\mathbf{y}_k = (\mathbf{x}_k - \boldsymbol{\mu})$ lying along the principal axes of the ellipsoid with half-lengths $c\sqrt{\lambda_k}$.

Let $\mathbf{A}$ be the $K \times K$ matrix with columns $\mathbf{a}_k = \mathbf{y}_k / \|\mathbf{y}_k\|$. Then, $\mathbf{A}^T\mathbf{A} = I$, the identity matrix, and $\mathbf{A}^T \boldsymbol{\Sigma} \mathbf{A} = \text{diag}(\lambda_1, \ldots, \lambda_K)$, the $K \times K$ diagonal matrix with diagonal elements $\lambda_1, \ldots, \lambda_K$. Now, consider the random vector $\mathbf{W} = \mathbf{A}^T\ (\mathbf{X} - \boldsymbol{\mu})$. It is easy to verify that $\mathbf{W} \sim N\ (\mathbf{0}, \text{diag}(\lambda_1, \ldots, \lambda_K))$. The components W_k of $\mathbf{W}$ are the (population) *principal components* of $\mathbf{X}$.

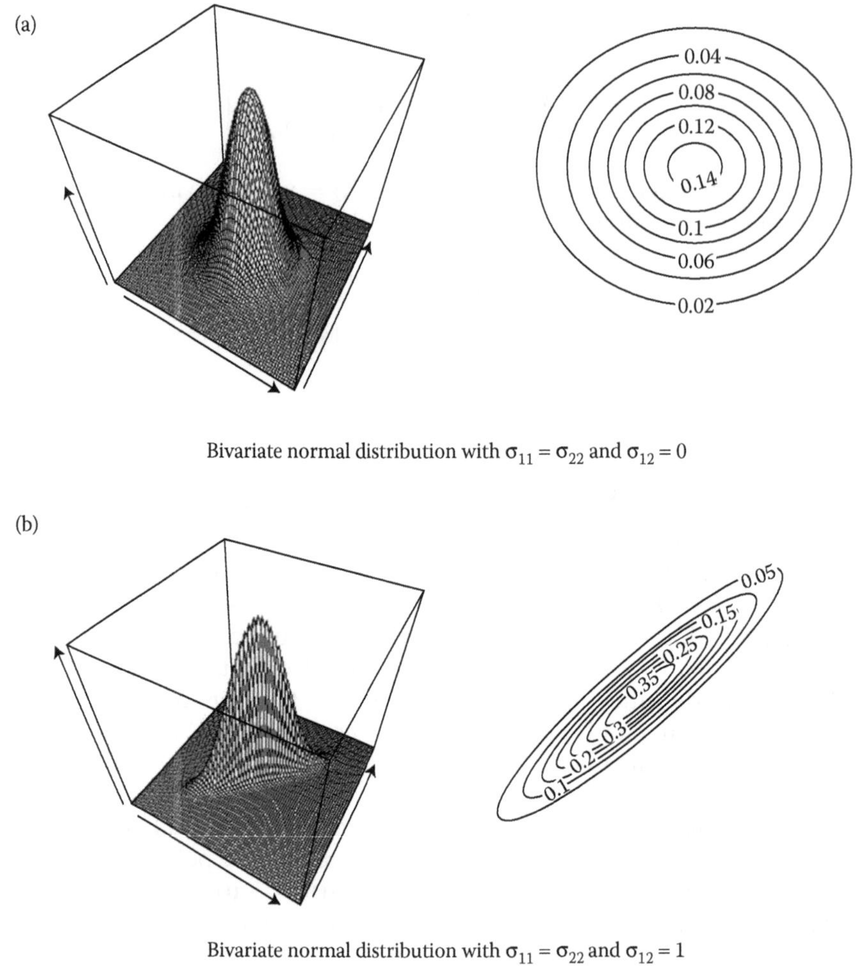

FIGURE 3.1
Bivariate density plots. Plots in (a) are the bivariate density and contour plots for two uncorrelated variables $\sigma_{12} = 0$ and plots in (b) are the bivariate density and contour plots for two perfectly correlated variables $\sigma_{12} = 1$.

3.3 Sampling from a Multivariate Normal Distribution

3.3.1 Multivariate Normal Likelihood

Let a set of N $K \times 1$ vectors $\mathbf{X}_1, \mathbf{X}_2, \ldots, \mathbf{X}_N$ represent an iid random sample from a multivariate normal population with mean vector $\boldsymbol{\mu}$ and covariance matrix $\boldsymbol{\Sigma}$. Since densities of independent random variables multiply, it is easy to observe that the joint density will be as follows:

$$f(\mathbf{x}_1, \ldots, \mathbf{x}_N | \boldsymbol{\mu}, \boldsymbol{\Sigma}) = \frac{1}{(2\pi)^{NK/2} |\boldsymbol{\Sigma}|^{N/2}} \exp\left[-\frac{1}{2} \sum_{n=1}^{N} (\mathbf{x}_n - \boldsymbol{\mu})^T \boldsymbol{\Sigma}^{-1} (\mathbf{x}_n - \boldsymbol{\mu}) \right] \tag{3.9}$$

The log of this density can be written (apart from some additive constants) as follows:

$$L(\mu, \Sigma) = -\frac{N}{2}\log|\Sigma| - \frac{1}{2}\sum_{n=1}^{N}(\mathbf{x}_n - \mu)^T\Sigma^{-1}(\mathbf{x}_n - \mu)$$

$$= -\frac{N}{2}\log|\Sigma| - \frac{1}{2}\sum_{n=1}^{N}(\mathbf{x}_n - \bar{\mathbf{x}})^T\Sigma^{-1}(\mathbf{x}_n - \bar{\mathbf{x}}) - \frac{1}{2}\sum_{n=1}^{N}(\bar{\mathbf{x}} - \mu)^T\Sigma^{-1}(\bar{\mathbf{x}} - \mu) \quad (3.10)$$

The value of μ that maximizes $L(\mu, \Sigma)$ for any Σ is $\hat{\mu} = \bar{\mathbf{x}}$, since that makes the third term in this loglikelihood equal to zero, so that $\hat{\mu} = \bar{\mathbf{x}}$ is the MLE. Furthermore, the first two terms of $L(\mu, \Sigma)$ may be rewritten as $(N/2)\log|\Sigma^{-1}| - (1/2)\text{tr}\mathbf{A}\Sigma^{-1}$ where $\mathbf{A} = \sum_{n=1}^{N}(\mathbf{x}_n - \bar{\mathbf{x}})(\mathbf{x}_n - \bar{\mathbf{x}})^T$, and from this and a little calculus, the MLE for Σ may be deduced. Summarizing,

Theorem 3.2

If $\mathbf{X}_1, \ldots, \mathbf{X}_N \sim N_K(\mu, \Sigma)$, then the MLE for μ is as follows:

$$\hat{\mu} = \bar{\mathbf{x}} = \frac{1}{N}\sum_{n=1}^{N}\mathbf{x}_n \quad (3.11)$$

The MLE for Σ is

$$\widehat{\Sigma} = \frac{1}{N}\sum_{n=1}^{N}(\mathbf{x}_n - \bar{\mathbf{x}})(\mathbf{x}_n - \bar{\mathbf{x}})^T \quad (3.12)$$

Note that $\bar{\mathbf{X}}$ and $\widehat{\Sigma}$ are sufficient statistics; they contain all of the information about μ and Σ in the data matrix $\mathbf{X}$. Furthermore, note that $\widehat{\Sigma}$ is a biased estimate of Σ; the unbiased estimator $\mathbf{S} = (1/N - 1)\sum_{n=1}^{N}(\mathbf{x}_n - \bar{\mathbf{x}})(\mathbf{x}_n - \bar{\mathbf{x}})^T$ is often used instead.

3.3.2 Sampling Distribution of $\bar{\mathbf{X}}$ and S

The sampling distributions of $\bar{\mathbf{X}}$ and $\mathbf{S}$ are easily generalized from the univariate case. In the univariate case, $\bar{X}$ and S are independent, $\bar{X} \sim N(\mu, \sigma^2/N)$, and $(N-1) \cdot S/\sigma^2 \sim \chi^2_{N-1}$, a χ-squared distribution with $N-1$ degrees of freedom. Recall that $\sum_{n=1}^{N}(X_n - \mu)^2/\sigma^2 = Z_1^2 + Z_2^2 + \cdots Z_N^2 \sim \chi^2_N$ by definition, because it is a sum of squares of independent standard normals; intuitively, we lose one degree of freedom for $(N-1) \cdot S/\sigma^2 = \sum_{n=1}^{N}(X_n - \bar{X})^2/\sigma^2$ because we are estimating the mean μ with $\bar{X}$ in calculating S.

In the multivariate case where we have the random sample $\mathbf{X}_1, \mathbf{X}_2, \ldots, \mathbf{X}_N$, the following three theorems apply:

Theorem 3.3

If $\mathbf{X}_1, \ldots, \mathbf{X}_N \sim N_K(\mu, \Sigma)$ are iid, then

1. $\bar{\mathbf{X}}$ and $\mathbf{S}$ are independent.

2. $\bar{\mathbf{X}}$ is distributed as a K-variate normal with parameters $\boldsymbol{\mu}$ and $\boldsymbol{\Sigma}/N$, $\bar{\mathbf{X}} \sim N(\boldsymbol{\mu}, \boldsymbol{\Sigma}/N)$.
3. $(N-1)\cdot\mathbf{S}$ is distributed as a Wishart random variable with parameter $\boldsymbol{\Sigma}$ and $N-1$ degrees of freedom, $(N-1)\cdot\mathbf{S} \sim W_{N-1}(\boldsymbol{\Sigma})$.

By definition, $\sum_{n=1}^{N}(\mathbf{x}_n - \boldsymbol{\mu})(\mathbf{x}_n - \boldsymbol{\mu})^T = \mathbf{Z}_1\mathbf{Z}_1^T + \mathbf{Z}_2\mathbf{Z}_2^T + \cdots + \mathbf{Z}_N\mathbf{Z}_N^T \sim W_N(\boldsymbol{\Sigma})$, since it is the sum of outer products of N independent $N(\mathbf{0}, \boldsymbol{\Sigma})$ random vectors. Once again, we lose one degree of freedom for $(N-1)\cdot\mathbf{S} = \sum_{n=1}^{N}(\mathbf{x}_n - \bar{\mathbf{x}})(\mathbf{x}_n - \bar{\mathbf{x}})^T$ since we are estimating the mean $\boldsymbol{\mu}$ with $\bar{\mathbf{X}}$. The Wishart density for a positive nonnegative definite random $K \times K$ matrix $\mathbf{A}$ with $D > K$ degrees of freedom and parameter $\boldsymbol{\Sigma}$ is as follows:

$$\omega_{D-1}(\mathbf{A}|\boldsymbol{\Sigma}) = \frac{|\mathbf{A}|^{(D-K-2)/2}e^{-tr[\mathbf{A}\boldsymbol{\Sigma}^{-1}]/2}}{2^{K(D-1)/2}\pi^{K(K-1)/4}|\boldsymbol{\Sigma}|^{(D-1)/2}\prod_{k=1}^{K}\Gamma((1/2)(D-k))} \tag{3.13}$$

3.4 Conjugate Families

Recall that if the *likelihood* for data $\mathbf{X}$ is (any function proportional to) the conditional density of $\mathbf{X}$ given (possibly multidimensional) parameter $\boldsymbol{\eta}$, $f(\mathbf{x}|\boldsymbol{\eta})$, then a *conjugate family of prior distributions* for $f(\mathbf{x}|\boldsymbol{\eta})$ is a parametric family of densities $f(\boldsymbol{\eta};\boldsymbol{\tau})$ for $\boldsymbol{\eta}$ with (possibly multidimensional) hyperparameter $\boldsymbol{\tau}$, such that for any member of the conjugate family, the *posterior distribution* $f(\boldsymbol{\eta}|\mathbf{x};\boldsymbol{\tau}) = f(\mathbf{x}|\boldsymbol{\eta})f(\boldsymbol{\eta};\boldsymbol{\tau})/\int_{\mathbf{h}} f(\mathbf{x}|\mathbf{h})f(\mathbf{h};\boldsymbol{\tau})\,d\mathbf{h}$ can be rewritten as $f(\boldsymbol{\eta}|\boldsymbol{\tau}^*)$, a member of the same parametric family as $f(\boldsymbol{\eta}|\boldsymbol{\tau})$, where $\boldsymbol{\tau}^* = \boldsymbol{\tau}^*(\mathbf{x},\boldsymbol{\tau})$ is some function of $\mathbf{x}$ and $\boldsymbol{\tau}$. Since only the form of $f(\boldsymbol{\eta}|\boldsymbol{\tau}^*)$ as a function of $\boldsymbol{\eta}$ matters, in verifying that $f(\boldsymbol{\eta}|\boldsymbol{\tau}^*)$ and $f(\boldsymbol{\eta};\boldsymbol{\tau})$ belong to the same parametric family, it is usual to ignore multiplicative constants that do not depend on $\boldsymbol{\eta}$.

For example, if $X_1,\ldots,X_N$ are iid $N(\mu,\sigma^2)$, then the likelihood is

$$f(x_1,\ldots,x_N|\mu,\sigma^2) = \prod_{n=1}^{N}\frac{1}{\sqrt{2\pi}\sigma}e^{-(1/2\sigma^2)(x_n-\mu)^2} \propto \frac{1}{\sqrt{2\pi\sigma^2/n}}e^{-(1/(2\sigma^2/n))(\bar{x}-\mu)^2} \tag{3.14}$$

as a function of μ, as would be expected, since $\bar{x}$ is sufficient for μ.

If we assume σ^2 is known and place a normal $f(\mu;\mu_0,\tau_0^2) = (1/\sqrt{2\pi}\tau_0)e^{-(1/2\tau_0^2)(\mu-\mu_0)^2}$ prior on μ, then the posterior density for μ will be

$$\begin{aligned} f(\mu|x_1,\ldots,x_N;\mu_N,\tau_N^2) &\propto f(x_1,\ldots,x_N|\mu,\sigma^2)f(\mu|\mu_0,\tau_0^2) \\ &= \frac{1}{\sqrt{2\pi\sigma^2/n}}e^{-(1/(2\sigma^2/n))(\bar{x}-\mu)^2}\frac{1}{\sqrt{2\pi}\tau_0}e^{-(1/2\tau_0^2)(\mu-\mu_0)^2} \\ &\propto \frac{1}{\sqrt{2\pi}\tau_N}e^{-(1/2\tau_N^2)(\mu-\mu_N)^2} \end{aligned} \tag{3.15}$$

after completing the square, collecting terms, and identifying the normalizing constant, where

$$\tau_N^2 = \frac{1}{1/(\sigma^2/N) + 1/\tau_0^2} \tag{3.16}$$

$$\mu_N = \left(\frac{\tau_0^2}{\tau_0^2 + \sigma^2/N}\right)\bar{x} + \left(\frac{\sigma^2/N}{\tau_0^2 + \sigma^2/N}\right)\mu_0 \tag{3.17}$$

In this case, the posterior mean is $\mu_N = \rho_N \bar{x} + (1 - \rho_N)\mu_0$ where $\rho_N = \tau_0^2/(\tau_0^2 + \sigma^2/n)$ is the classical reliability coefficient. Thus, if the prior distribution is $\mu \sim N(\mu_0, \tau_0)$, then the posterior distribution will be $\mu|x_1, \ldots, x_N \sim N(\mu_N, \tau_N)$; this shows that the normal distribution is the conjugate prior for a normal mean μ, when the variance σ^2 is known.

Further calculation (as shown in Gelman et al., 2004) shows that when *both* the mean μ and the variance σ^2 are unknown, and the data $X_1, \ldots, X_N$ are an iid sample from $N(\mu, \sigma^2)$, the joint distribution

$$\mu|\sigma^2 \sim N(\mu_0, \sigma^2/\kappa_0) \tag{3.18}$$

$$\sigma^2 \sim \text{Inv-}\chi^2(\nu_0, \sigma_0^2) \tag{3.19}$$

is the conjugate prior, where the notation "$\sigma^2 \sim \text{Inv} - \chi^2(\nu_0, \sigma_0^2)$" means that $\sigma_0^2/\sigma^2 \sim \chi^2_{\nu 0}$. Here, κ_0 and ν_0 are hyperparmeters that function as "prior sample sizes"—the larger κ_0 and ν_0, the greater the influence of the prior on the posterior. In this case, the joint posterior distribution for μ, σ^2 is

$$\mu|\sigma^2, x_1, x_2, \ldots, x_N \sim N(\mu_N, \sigma_N^2/\kappa_N) \tag{3.20}$$

$$\sigma^2|x_1, x_2, \ldots, x_N \sim \text{Inv} - \chi^2(\nu_N, \sigma_N^2) \tag{3.21}$$

where

$$\kappa_N = \kappa_0 + N$$

$$\nu_N = \nu_0 + N$$

$$\nu_N \sigma_N^2 = \nu_0 \sigma_0^2 + (N-1)S^2 + \frac{\kappa_0 N}{\kappa_0 + N}(x - \mu_0)^2$$

$$\mu_N = \left(\frac{\sigma^2/\kappa_0}{\sigma^2/\kappa_0 + \sigma^2/N}\right)\bar{x} + \left(\frac{\sigma^2/N}{\sigma^2/\kappa_0 + \sigma^2/N}\right)\mu_0 = \left(\frac{N}{\kappa_0 + N}\right)\bar{x} + \left(\frac{\kappa_0}{\kappa_0 + N}\right)\mu_0$$

with $S^2 = (1/N - 1)\sum_{n=1}^{N}(x_n - \bar{x})^2$.

Although this is the conjugate family when μ and σ^2 are both unknown, the forced dependence between μ and σ^2 in the prior is often awkward in applications. Common alternatives are as follows:

1. To replace σ^2/κ_0 with an arbitrary τ^2 in the conditional prior for $\mu|\sigma^2$, forcing independence. In this case, the conditional posterior for $\mu|\sigma^2$ is as in the "σ^2 known" case above, but the marginal posterior for σ^2 is neither conjugate nor in closed form (it is not difficult to calculate however) or

2. To make the conditional prior for $\mu|\sigma^2$ very flat/uninformative by letting $\kappa_0 \to 0$. In this case, the posterior distributions for $\mu|\sigma^2$ and for σ^2 mimic the sampling distributions of the MLEs

If one wishes for a noninformative prior for σ^2 (analogous to the flat prior choice for μ in (b) above), a choice that preserves conjugacy would be to take the degrees of freedom $\nu_0 = 0$ in the $\text{Inv} - \chi^2(\nu_0, \sigma_0^2)$ prior for σ^2. This leads to a prior $f(\sigma^2) \propto 1/\sigma^2$. Another noninformative choice is the Jeffreys prior for σ^2 (proportional to the square root of the Fisher information for the parameter), which in this case leads to the prior $f(\sigma) \propto 1/\sigma$. The Jeffreys prior for $\log \sigma^2$ (or equivalently $\log \sigma$) is simply $f(\sigma^2) \propto 1$. All of these choices are improper priors and care must be taken that the posterior turns out to be proper. One common way to avoid this issue is to force the prior to be proper, say, by taking $\sigma^2 \sim \text{Unif}(0, M)$ for some suitably large number M.

The conjugate prior distribution for a multivariate normal distribution with parameters $\boldsymbol{\mu}$ and $\boldsymbol{\Sigma}$ has a form similar to that of the univariate case, but with multivariate normal and inverse-Wishart densities replacing the univariate normal and inverse-χ-squared densities. In particular, assuming sampling $\mathbf{x}_1, \ldots, \mathbf{x}_N$ iid from $N(\boldsymbol{\mu}, \boldsymbol{\Sigma})$, the joint prior for $\boldsymbol{\mu}$ and $\boldsymbol{\Sigma}$ is of the following form:

$$\boldsymbol{\mu}|\boldsymbol{\Sigma} \sim N(\boldsymbol{\mu}_0, \boldsymbol{\Sigma}/\kappa_0) \tag{3.22}$$

$$\boldsymbol{\Sigma} \sim \text{Inv-Wishart}(\nu_0, \boldsymbol{\Sigma}_0^{-1}) \tag{3.23}$$

where the notation "$\Sigma \sim \text{Inv-Wishart}(\nu_0, \boldsymbol{\Sigma}_0^{-1})$" means that $\boldsymbol{\Sigma}_0\boldsymbol{\Sigma}^{-1} \sim \omega_{\nu_0}(\boldsymbol{\Sigma}_0)$, the usual Wishart distribution with ν_0 degrees of freedom and parameter matrix $\boldsymbol{\Sigma}_0$. Again, κ_0 and ν_0 function as prior sample sizes. Then, the joint posterior distribution will be as follows:

$$\boldsymbol{\mu}|\boldsymbol{\Sigma}, x_1, x_2, \ldots, x_N \sim N(\boldsymbol{\mu}_N, \boldsymbol{\Sigma}/\kappa_N) \tag{3.24}$$

$$\boldsymbol{\Sigma}|x_1, x_2, \ldots, x_N \sim \text{Inv-Wishart}(\nu_N, \boldsymbol{\Sigma}_N^{-1}) \tag{3.25}$$

where

$$\kappa_N = \kappa_0 + N$$

$$\nu_N = \nu_0 + N$$

$$\nu_N\boldsymbol{\Sigma}_N = \nu_0\boldsymbol{\Sigma}_0 + (N-1)\mathbf{S} + \frac{\kappa_0 N}{\kappa_0 + N}(\bar{\mathbf{x}} - \boldsymbol{\mu}_0)(\bar{\mathbf{x}} - \boldsymbol{\mu}_0)^T$$

$$\boldsymbol{\mu}_N = \left(\frac{N}{\kappa_0 + N}\right)\bar{\mathbf{x}} + \left(\frac{\kappa_0}{\kappa_0 + N}\right)\boldsymbol{\mu}_0$$

with $\mathbf{S} = (1/N-1)\sum_{n=1}^{N}(\mathbf{x}_n - \bar{\mathbf{x}})(\mathbf{x}_n - \bar{\mathbf{x}})^T$ (Gelman et al., 2004). Once again the joint conjugate prior forces some prior dependence between $\boldsymbol{\mu}$ and $\boldsymbol{\Sigma}$ that may be awkward in practice. And once again the common fixes are as follows:

1. Replace $\boldsymbol{\Sigma}/\kappa_0$ with an arbitrary $\boldsymbol{\Lambda}_0$ in the conditional prior for $\boldsymbol{\mu}|\boldsymbol{\Sigma}$, forcing independence or
2. Make the conditional prior for $\boldsymbol{\mu}|\boldsymbol{\Sigma}$ very flat/uninformative by letting $\kappa_0 \to 0$

The details, generally analogous to the univariate normal case, are worked out in many places; in particular, see Gelman et al. (2004). Gelman et al. (2004) suggest another way to reduce the prior dependence between μ and Σ. Sun and Berger (2007) provide an extensive discussion of several "default" objective/noninformative prior choices for the multivariate normal.

3.5 Generalizations of the Multivariate Normal Distribution

Assessing multivariate normality in higher dimensions is challenging. A well-known theorem (Johnson and Wichern, 1998; see also Anderson, 2003) states that $\mathbf{X}$ is a multivariate normal vector if and only if each linear combination $\mathbf{a}^T\mathbf{X}$ of its components is univariate normal, but this is seldom feasible to show in practice. Instead one often only checks the one- and two-dimensional margins of $\mathbf{X}$; for example, examining a Q—Q plot for each of the K components of $\mathbf{X}$, and in addition, examining bivariate scatterplots of each possible pair of variables to determine whether the data points yield an elliptical appearance. (See Johnson and Wichern (1998) for more information on techniques for assessing multivariate normality.)

There is no doubt that true multivariate normality is a rare property for a multivariate dataset. Although the latent ability variable in IRT is often assumed to be normally distributed, the data may not conform well to this assumption at all (Casabianca, 2011; Casabianca et al., 2010; Moran and Dresher, 2007; Woods and Lin, 2009; Woods and Thissen, 2006). In these and other cases, robust alternatives to the multivariate normal distribution can be considered.

An important class of generalizations of the multivariate normal distribution is the family of *multivariate elliptical distributions*, so named because each of its level sets defines an ellipsoid (Fang and Zhang, 1990; Branco and Dey, 2002). The K-dimensional random variable $\mathbf{X}$ has an elliptical distribution if and only if its characteristic function (Billingsley, 1995; Lukacs, 1970) is of the form

$$\Psi(\mathbf{t}) = E[\exp(i\mathbf{t}^T\mathbf{X})] = -\exp(i\mathbf{t}^T\mu)\psi(\mathbf{t}^T\Sigma\mathbf{t}) \tag{3.26}$$

where as usual μ is a K-dimensional vector and Σ is a symmetric nonnegative definite $K \times K$ matrix, and $\mathbf{t} \in \Re^K$. When a density $f(\mathbf{x})$ exists for $\mathbf{X}$, it has the form

$$f_g(\mathbf{x}; \mu, \Sigma) = \frac{1}{|\Sigma|^{1/2}} g[(\mathbf{x}-\mu)^T\Sigma^{-1}(\mathbf{x}-\mu)] \tag{3.27}$$

where $g()$ is itself a univariate density; $g()$ is called the generator density for $f()$. The density defines a location/scale family with location parameter μ and scale parameter Σ. The parameter μ is the median of $f()$ in all cases, and if $E[\mathbf{X}]$ exists, $E[\mathbf{X}] = \mu$. If $\text{Var}(\mathbf{X})$ exists, $\text{Var}(\mathbf{X}) = -(\partial\psi(\mathbf{0})/\partial\mathbf{t}) \cdot \Sigma$. A little calculation shows that if $\mathbf{W} = \mathbf{AX} + \mathbf{b}$ is elliptically distributed with location μ and scale Σ, then $\mathbf{W} = \mathbf{AX} + \mathbf{b}$ is again elliptically distributed, with location $\mathbf{A}\mu + \mathbf{b}$ and scale $\mathbf{A}\Sigma\mathbf{A}^T$.

Elliptical distributions are used as a tool for generalizing normal-theory structural equations modeling (e.g., Shapiro and Browne, 1987; Schumacker and Cheevatanarak, 2000). The special case of the K-dimensional multivariate-t distribution on ν degrees of freedom with density

$$f(\mathbf{x};\boldsymbol{\mu},\boldsymbol{\Sigma}) = \frac{\Gamma((v+K)/2)}{\Gamma(v/2)v^{K/2}\pi^{K/2}}|\boldsymbol{\Sigma}|^{-1/2} \times \left(1+\frac{1}{v}(\mathbf{x}-\boldsymbol{\mu})^T\boldsymbol{\Sigma}^{-1}(\mathbf{x}-\boldsymbol{\mu})\right)^{-(v+K)/2} \tag{3.28}$$

has been used as the error distribution in robust Bayesian and non-Bayesian linear regression modeling since at least Zellner (1976); more recently, robust regression with general elliptical error distributions and the particular case of scale mixtures of normals (of which the univariate-t and multivariate-t are also examples) has also been studied (e.g., Fernandez and Steel, 2000).

A further generalization is the family of *skew-elliptical distributions* (e.g., Branco and Dey, 2001, 2002). When it exists, the density of a skew-elliptical distribution is of the form

$$f_{g_1,g_2}(\mathbf{x}|\boldsymbol{\mu},\boldsymbol{\Sigma},\boldsymbol{\lambda}) = 2f_{g_1}(\mathbf{x}|\boldsymbol{\mu},\boldsymbol{\Sigma})F_{g_2}(\boldsymbol{\lambda}^T(\mathbf{x}-\boldsymbol{\mu})) \tag{3.29}$$

where $f_{g1}()$ is the density of a multivariate elliptical distribution, and $F_{g2}()$ is the cumulative distribution function of a (possibly different) univariate elliptical distribution with location parameter 0 and scale parameter 1. The vector parameter $\boldsymbol{\lambda}$ is a skewness parameter; when $\boldsymbol{\lambda} = \mathbf{0}$, the skew-elliptical density reduces to a symmetric elliptical density.

When the generator densities $g_1(x) = g_2(x) = \phi(x)$, the standard normal density, we obtain the special case of the *skew-normal distributions*. It has been observed (Moran and Dresher, 2007) that the empirical distribution of the latent proficiency variable in IRT models applied to large-scale educational surveys sometimes exhibits some nontrivial skewing, which if unmodeled can cause bias in estimating item parameters and features of the proficiency distribution. Skew-normal and related distributions have been proposed (Xu and Jia, 2011) as a way of accounting for this skewing in the modeling of such data, with as few extra parameters as possible.

Two additional classes of transformed normal distributions used in IRT are the *lognormal* and *logit-normal distributions*. A random variable X has a lognormal distribution when its logarithm is normally distributed. The density of the lognormal distribution is of the following form:

$$f(x;\mu,\sigma^2) = \frac{1}{x\sqrt{2\pi\sigma^2}}e^{-(\ln x-\mu)^2/2\sigma^2} \quad x > 0 \tag{3.30}$$

where μ and σ^2 are the mean and variance of the variable's natural log. This distribution is an alternative to Gamma and Weibull distributions for nonnegative continuous random variables and is used in psychometrics to model examinee response times (e.g., van der Linden, 2006) and latent processes in decision making (e.g., Rouder et al., 2014). It is also used as a prior distribution in Bayesian estimation of nonnegative parameters, for example the discrimination parameter in two- and three-parameter logistic IRT models (van der Linden, 2006).

Similarly, a random variable X has a logit-normal distribution when its logit is normally distributed. The density of the logit-normal distribution is of the following form:

$$f(x;\mu,\sigma^2) = \frac{1}{\sigma\sqrt{2\pi}}e^{-(\text{logit}(x)-\mu)^2/2\sigma^2}\frac{1}{x(1-x)} \quad 0 < x < 1 \tag{3.31}$$

Here, μ and σ^2 are the mean and variance of the variable's logit, and x is a proportion, bounded by 0 and 1. A multivariate generalization of the logit-normal distribution (Aitchison, 1985) has been used in latent Dirichlet allocation models for text classification (Blei and

Lafferty, 2007) and in mixed membership models for strategy choice in cognitive diagnosis (Galyardt, 2012).

Acknowledgment

This work was supported by a postdoctoral fellowship at Carnegie Mellon University and Rand Corporation, through Grant #R305B1000012 from the Institute of Education Sciences, U.S. Department of Education.

References

Aitchison, J. 1985. A general class of distributions on the simplex. *Journal of the Royal Statistical Society. Series B (Methodological)*, 47,136–146.

Anderson, T. W. 2003. *Introduction to Multivariate Statistical Analysis*. New York: John Wiley and Sons.

Bartholomew, D. J. and Knott, M. 1999. *Latent Variable Models and Factor Analysis*. London: Arnold. (Kendall's Library of Statistics 7).

Billingsley, P. 1995. *Probability and Measure* (3rd ed.). New York: John Wiley & Sons.

Blei, D. and Lafferty, J. 2007. A correlated topic model of science. *Annals of Applied Statistics*, 1, 17–35.

Branco, M. and Dey, D. K. 2001. A general class of multivariate skew-elliptical distributions. *Journal of Multivariate Analysis*, 79, 99–113.

Branco, M. and Dey, D. K. 2002. Regression model under skew-elliptical error distribution. *The Journal of Mathematical Sciences, Delhi, New Series*, 1, 151–169.

Casabianca, J. M. 2011. *Loglinear Smoothing for the Latent Trait Distribution: A Two-Tiered Evaluation*. (Doctoral dissertation), ProQuest dissertations and theses. (Accession Order No. AAT 3474125.)

Casabianca, J. M., Xu, X., Jia, Y., and Lewis, C. 2010. Estimation of item parameters when the underlying latent trait distribution of test takers is nonnormal. *Paper Presented at the Meeting of the National Council for Measurement in Education*, Denver, Colorado.

Fang, K. T. and Zhang, Y. T. 1990. *Generalized Multivariate Analysis*. New York: Springer.

Fernandez, C. and Steel, M. F. J. 2000. Bayesian regression analysis with scale mixtures of normals. *Econometric Theory*, 16, 80–101.

Fox, J. P. 2003. Stochastic EM for estimating the parameters of a multilevel IRT model. *British Journal of Mathematical and Stat Psychology*, 56, 65–81.

Fox, J. P. 2005a. Multilevel IRT model assessment. In van der Ark, L. A., Croon, M. A., and Sijtsma, K. (Eds.), *New Developments in Categorical Data Analysis for the Social and Behavioral Sciences* (pp. 227–252). Mahwah, NJ: Lawrence Erlbaum.

Fox, J. P. 2005b. Multilevel IRT using dichotomous and polytomous response data. *British Journal of Mathematical and Statistical Psychology*, 58, 145–172.

Fox, J. P. 2010. *Bayesian Item Response Modeling*. New York: Springer.

Gelman, A., Carlin, J. B., Stern, H. A., and Rubin, D. B. 2004. *Bayesian Data Analysis*. New York: John Wiley and Sons.

Galyardt, A. 2012. *Mixed Membership Distributions with Applications to Modeling Multiple Strategy Usage*. PhD dissertation, Department of Statistics, Carnegie Mellon University, Pittsburgh, PA.

Johnson, R. A. and Wichern, D. W. 1998. *Applied Multivariate Statistical Analysis*. Upper Saddle River, NJ: Prentice-Hall.

Lukacs, E. 1970. *Characteristic Functions*. London: Griffin.

Moran, R. and Dresher, A. 2007. Results from NAEP marginal estimation research on multivariate scales. *Paper Presented at the Meeting of the National Council for Measurement in Education*, Chicago, IL.

Morrison, D. F. 2005. *Multivariate Statistical Methods*. Belmont, CA: Thomson Brooks Cole.

Rencher, A. C. 2002. *Methods of Multivariate Analysis*. New York: John Wiley and Sons.

Rouder, J. N., Province, J. M., Morey, R. D., and Heathcote, A. 2014. The lognormal race: A cognitive-process model of choice and latency with desirable psychometric properties. *Psychometrika*, 1–23.

Schumacker, R. E. and Cheevatanarak, S. 2000. A comparison of normal and elliptical estimation methods in structural equations models. *Paper Presented at the Meeting of the American Educational Research Association*, New Orleans, LA (ERIC Document Reproduction Service No. ED441872). Retrieved April 21, 2012, from http://www.eric.ed.gov/PDFS/ED441872.pdf.

Shapiro, A. and Browne, M. W. 1987. Analysis of covariance structures under elliptical distributions. *Journal of the American Statistical Association*, 82, 1092–1097.

Sun, D. and Berger, J. O. 2007. Objective Bayesian analysis for the multivariate normal model. *Bayesian Statistics*, 8, 525–562.

van der Linden, W. J. 2006. A lognormal model for response times on test items. *Journal of Educational and Behavioral Statistics*, 31, 181–204.

Woods, C. M. and Lin, N. 2009. IRT with estimation of the latent density using Davidian curves. *Applied Psychological Measurement*, 33, 102–117.

Woods, C. M. and Thissen, D. 2006. IRT with estimation of the latent population distribution using spline-based densities. *Psychometrika*, 71, 281–301.

Xu, X. and Jia, Y. 2011. *The Sensitivity of Parameter Estimates to the Latent Ability Distribution* (Research Report 11-40). Princeton, NJ: Educational Testing Service.

Zellner, A. 1976. Bayesian and non-Bayesian analysis of the regression model with multivariate Student-t error terms. *Journal of the American Statistical Association*, 71, 400–405.

4

Exponential Family Distributions Relevant to IRT

Shelby J. Haberman

CONTENTS

4.1 Introduction

Exponential families are families of probability distributions characterized by a sufficient statistic. In item response model, their treatment emphasizes loglinear models for polytomous random variables and normal probability models for continuous random variables. Generally, exponential families are widely applied in statistical theory, for they include beta distributions, Dirichlet distributions, exponential distributions, gamma distributions, normal distributions, negative binomial distributions, Poisson distributions, and Weibull distributions (Morris, 1982; McCullagh and Nelder, 1989; Kotz et al., 2000; see Chapters 2 and 3). In this chapter, Section 4.2 provides a general definition of exponential families and applies the definition to polytomous and continuous random vectors. Section 4.3 applies exponential families to IRT. In Section 4.4, special features of exponential families are applied to estimation problems encountered in IRT. Some concluding remarks concerning the role of exponential families are provided in Section 4.5.

4.2 Definition of Exponential Families

An exponential family is a collection of probability distributions. In this chapter, these probability distributions are used to describe finite-dimensional random vectors. The description of an exponential family involves four steps. A nonempty set S of vectors is specified such that all distributions in the exponential family are defined on S. A vector-valued function $\mathbf{T}$ is then specified on the set S. Given $\mathbf{T}$ and S, a parameter space is then specified for the exponential family. Finally, the exponential family is specified by use of a density kernel that involves an exponential function of a linear combination of the vector $\mathbf{T}$.

For a positive integer n, the probability distributions are all defined on a nonempty set S of n-dimensional vectors. For a positive integer q, the exponential family may be characterized by a random vector $\mathbf{X}$ with range S and by a q-dimensional vector-valued function $\mathbf{T}$ on S (Lehmann, 1959; Berk, 1972; Barndorff–Nielsen, 1978). To ensure that probabilities and expectations are property defined for the exponential family, it is assumed that $\mathbf{T}(\mathbf{Y})$ is a random vector whenever $\mathbf{Y}$ is a random vector with values in S. If $\mathbf{Y}$ has a distribution from the exponential family generated from $\mathbf{X}$ and $\mathbf{T}$, then $\mathbf{T}(\mathbf{Y})$ is a sufficient statistic in the sense that the conditional distribution of $\mathbf{Y}$ given $\mathbf{T}(\mathbf{Y})$ is the same no matter which distribution from the exponential family is associated with $\mathbf{Y}$.

In discrete exponential families, the set S is finite or countably infinite, and $\mathbf{T}$ can be any q-dimensional vector function on S. In continuous exponential families, the set S has a nonempty interior, and $\mathbf{X}$ is a continuous random vector.

Generation of an exponential family requires inner products of a q-dimensional parameter vector and the vector function $\mathbf{T}$. For q-dimensional vectors $\mathbf{v}$ and $\mathbf{w}$ with respective elements v_k and w_k, $1 \le k \le q$, let the vector product $\mathbf{v}'\mathbf{w}$ be $\sum_{k=1}^{q} v_k w_k$.

Associated with the random vector $\mathbf{X}$ and the function $\mathbf{T}$ is an exponential family. This family has natural parameters in the natural parameter space $\Omega(\mathbf{X},\mathbf{T})$. This natural parameter space provided is also the domain of the moment generating function $\mathrm{Mgf}(\bullet,\mathbf{X},\mathbf{T})$ of $\mathbf{T}(\mathbf{X})$. Thus, $\Omega(\mathbf{X},\mathbf{T})$ is the set of all q-dimensional vectors $\boldsymbol{\omega}$ such that $\exp[\boldsymbol{\omega}'\mathbf{T}(\mathbf{X})]$ has a finite expectation. The set $\Omega(\mathbf{X},\mathbf{T})$ is a nonempty convex set that includes the q-dimensional vector $\mathbf{0}_q$ with all elements 0. If $\mathbf{T}$ is bounded, as is surely true for S finite, then $\Omega(\mathbf{X},\mathbf{T})$ is the entire space R^q of q-dimensional vectors. In typical exponential families, $\Omega(\mathbf{X},\mathbf{T})$ is an open set. As a consequence, in this chapter it is assumed for convenience that the natural parameter space $\Omega(\mathbf{X},\mathbf{T})$ is open.

The definition of an exponential family requires use of the moment generating function of the random vector $\mathbf{T}(\mathbf{X})$. This moment generating function $\mathrm{Mgf}(\bullet,\mathbf{X},\mathbf{T})$ is the function on the natural parameter space $\Omega(\mathbf{X},\mathbf{T})$ with value

$$\mathrm{Mgf}(\boldsymbol{\omega},\mathbf{X},\mathbf{T}) = E(\exp[\boldsymbol{\omega}'\mathbf{T}(\mathbf{X})]) \tag{4.1}$$

for $\boldsymbol{\omega}$ in $\Omega(\mathbf{X},\mathbf{T})$.

A member of the exponential family associated with the random vector $\mathbf{X}$ on the space S and the vector-valued function $\mathbf{T}$ is associated with a natural parameter $\boldsymbol{\omega}$ in the natural parameter space $\Omega(\mathbf{X},\mathbf{T})$. This member $P(\boldsymbol{\omega},\mathbf{X},\mathbf{T})$ of the exponential family is associated with the density kernel $f(\bullet;\boldsymbol{\omega},\mathbf{X},\mathbf{T})$. This density kernel is a real function on S proportional to the exponential function $\exp[\boldsymbol{\omega}'\mathbf{T}(\mathbf{X})]$. It has value

$$f(\mathbf{s};\boldsymbol{\omega},\mathbf{X},\mathbf{T}) = \frac{\exp[\boldsymbol{\omega}'\mathbf{T}(\mathbf{s})]}{\mathrm{Mgf}(\boldsymbol{\omega},\mathbf{X},\mathbf{T})} \tag{4.2}$$

for $\mathbf{s}$ in S. Given the density kernel $f(\bullet; \boldsymbol{\omega}, \mathbf{X}, \mathbf{T})$, the probability distribution $P(\boldsymbol{\omega}, \mathbf{X}, \mathbf{T})$ is the distribution of a random vector $\mathbf{Y}(\boldsymbol{\omega}, \mathbf{X}, \mathbf{T})$ with values in S such that, for each bounded real continuous function g on S, the expectation of the function $g(\mathbf{Y}(\boldsymbol{\omega}, \mathbf{X}, \mathbf{T}))$ of the random vector $\mathbf{Y}(\boldsymbol{\omega}, \mathbf{X}, \mathbf{T})$ is the same as the expectation of the function $g(X)f(\mathbf{X}; \boldsymbol{\omega}, \mathbf{X}, \mathbf{T})$ of the random vector $\mathbf{X}$, so that

$$E(g(\mathbf{Y}(\boldsymbol{\omega}, \mathbf{X}, \mathbf{T})) = E(g(X)f(\mathbf{X}; \boldsymbol{\omega}, \mathbf{X}, \mathbf{T})). \tag{4.3}$$

The random vector $\mathbf{Y}(\boldsymbol{\omega}, \mathbf{X}, \mathbf{T})$ is not uniquely defined, but it can be useful in discussions of properties of the probability distribution $P(\boldsymbol{\omega}, \mathbf{X}, \mathbf{T})$. Because, $P(\mathbf{0}_q, \mathbf{X}, \mathbf{T})$ is the probability distribution of $\mathbf{X}$, one may let $\mathbf{Y}(\mathbf{0}_q, \mathbf{X}, \mathbf{T})$ be $\mathbf{X}$. In addition, one may let $Y_i(\boldsymbol{\omega}, \mathbf{X}, \mathbf{T})$ denote element i of $\mathbf{Y}(\boldsymbol{\omega}, \mathbf{X}, \mathbf{T})$ for $1 \le i \le n$. The natural exponential family $\mathcal{P}(\mathbf{X}, \mathbf{T})$ consists of the distributions $P(\boldsymbol{\omega}, \mathbf{X}, \mathbf{T})$, $\boldsymbol{\omega}$ in Ω. The natural parameter $\boldsymbol{\omega}$ is uniquely determined by $P(\boldsymbol{\omega}, \mathbf{X}, \mathbf{T})$ if, and only if, no q-dimensional vector $\mathbf{v}$ exists such that $\mathbf{v} \neq \mathbf{0}_q$ and $\mathbf{v}'\mathbf{T}(\mathbf{X})$ has variance 0. For convenience, it is assumed in this chapter that $\boldsymbol{\omega}$ is uniquely determined by $P(\boldsymbol{\omega}, \mathbf{X}, \mathbf{T})$. The function $\mathbf{T}$ has the sufficiency property that, for all $\boldsymbol{\omega}$ in $\Omega(\mathbf{X}, \mathbf{T})$, the conditional distribution of $\mathbf{Y}(\boldsymbol{\omega}, \mathbf{X}, \mathbf{T})$ given $\mathbf{T}(\mathbf{Y}(\boldsymbol{\omega}, \mathbf{X}, \mathbf{T}))$ is the same as the conditional distribution of $\mathbf{X}$ given $\mathbf{T}(\mathbf{X})$.

For moment computations and maximum likelihood estimation of $\boldsymbol{\omega}$, it is helpful to note that, for $\boldsymbol{\omega}$ in Ω, the natural parameter space $\Omega(\mathbf{Y}(\boldsymbol{\omega}, \mathbf{X}, \mathbf{T}), \mathbf{T})$ associated with $\mathbf{Y}(\boldsymbol{\omega}, \mathbf{X}, \mathbf{T})$ is the set of differences $\mathbf{v} - \boldsymbol{\omega}$ for $\mathbf{v}$ is in $\Omega(\mathbf{X}, \mathbf{T})$, and, for $\mathbf{v}$ in $\Omega(\mathbf{X}, \mathbf{T})$,

$$\text{Mgf}(\mathbf{v} - \boldsymbol{\omega}, \mathbf{Y}(\boldsymbol{\omega}, \mathbf{X}, \mathbf{T}), \mathbf{T}) = \frac{\text{Mgf}(\mathbf{v}, \mathbf{X}, \mathbf{T})}{\text{Mgf}(\boldsymbol{\omega}, \mathbf{X}, \mathbf{T})}. \tag{4.4}$$

The moment generating function $\text{Mgf}(\bullet, \mathbf{X}, \mathbf{T})$ is infinitely differentiable and, for a natural parameter $\boldsymbol{\omega}$ in $\Omega(\mathbf{X}, \mathbf{T})$, $\text{Mgf}(\bullet, \mathbf{X}, \mathbf{T})$ determines all moments of the function $\mathbf{T}(\mathbf{Y}(\boldsymbol{\omega}, \mathbf{X}, \mathbf{T}))$ of the random vector $\mathbf{Y}(\boldsymbol{\omega}, \mathbf{X}, \mathbf{T})$. Let $\text{Cgf}(\bullet, \mathbf{X}, \mathbf{T}) = \log[\text{Mgf}(\bullet, \mathbf{X}, \mathbf{T})]$ be the cumulant generating function of $\mathbf{T}(\mathbf{X})$. Then

$$\text{Cgf}(\mathbf{v} - \boldsymbol{\omega}, \mathbf{Y}(\boldsymbol{\omega}, \mathbf{X}, \mathbf{T}), \mathbf{T}) = \text{Cgf}(\mathbf{v}, \mathbf{X}, \mathbf{T}) - \text{Cgf}(\boldsymbol{\omega}, \mathbf{X}, \mathbf{T}) \tag{4.5}$$

if $\mathbf{v}$ is in $\Omega(\mathbf{X}, \mathbf{T})$. The expectation $\boldsymbol{\mu}(\boldsymbol{\omega}, \mathbf{X}, \mathbf{T})$ of $\mathbf{T}(\mathbf{Y}(\boldsymbol{\omega}, \mathbf{X}, \mathbf{T}))$ is the gradient $\nabla\text{Cgf}(\boldsymbol{\omega}, \mathbf{X}, \mathbf{T})$ at $\boldsymbol{\omega}$ of the cumulant generating function $\text{Cgf}(\bullet, \mathbf{X}, \mathbf{T})$. The covariance matrix $\boldsymbol{\Sigma}(\boldsymbol{\omega}, \mathbf{X}, \mathbf{T})$ of $\mathbf{Y}(\boldsymbol{\omega}, \mathbf{X}, \mathbf{T})$ is the Hessian matrix $\nabla^2\,\text{Cgf}(\boldsymbol{\omega}, \mathbf{X}, \mathbf{T})$ at $\boldsymbol{\omega}$ of $\text{Cgf}(\bullet, \mathbf{X}, \mathbf{T})$. Under the assumption in this chapter that $\boldsymbol{\omega}$ is uniquely determined by $P(\boldsymbol{\omega}, \mathbf{X}, \mathbf{T})$, $\boldsymbol{\Sigma}(\boldsymbol{\omega}, \mathbf{X}, \mathbf{T})$ is positive definite.

In the case of exponential families, there is a one-to-one relationship between expectations and natural parameters. There is a unique closed convex set $C(\mathbf{X}, \mathbf{T})$ called the convex support of $\mathbf{T}(\mathbf{X})$ such that $\mathbf{T}(\mathbf{X})$ is in C with probability 1 and such that any closed convex set C' such that $\mathbf{T}(\mathbf{X})$ is in C' with probability 1 satisfies the constraint that C is included in C'. Under the assumptions concerning $\mathbf{T}(\mathbf{X})$ in this chapter, C has a nonempty interior $C^0(\mathbf{X}, \mathbf{T})$. Any expectation $\boldsymbol{\mu}(\boldsymbol{\omega}, \mathbf{X}, \mathbf{T})$ is in $C^0(\mathbf{X}, \mathbf{T})$. On the other hand, if $\mathbf{t}$ is in $C^0(\mathbf{X}, \mathbf{T})$, then there is a unique natural parameter $\boldsymbol{\omega}$ in $\Omega(\mathbf{X}, \mathbf{T})$ such that $\boldsymbol{\mu}(\boldsymbol{\omega}, \mathbf{X}, \mathbf{T}) = \mathbf{t}$.

In many probability models based on exponential families, it is simply assumed that an n-dimensional random vector $\mathbf{Y}$ with values in S has a distribution $P(\boldsymbol{\omega}, \mathbf{X}, \mathbf{T})$ for some $\boldsymbol{\omega}$ in $\Omega(\mathbf{X}, \mathbf{T})$. In other cases, a restricted model is applied to the natural parameter $\boldsymbol{\omega}$. The model is defined in terms of a new parameter space and a transformation from the new parameter space to the natural parameter space. Thus, for some positive integer m, the new parameter space is a nonempty open set Γ in R^m and the transformation is an infinitely differentiable function $\mathbf{Q}$ from Γ to $\Omega(\mathbf{X},\mathbf{T})$. It is not really necessary to require that Γ is open and $\mathbf{Q}$ be infinitely differentiable, but the assumptions simplify exposition. For $1 \le k \le q$, Q_k is element k of $\mathbf{Q}$. The model has the added condition that $\boldsymbol{\omega} = \mathbf{Q}(\boldsymbol{\gamma})$ for some $\boldsymbol{\gamma}$ in Γ. The resulting family $\mathcal{Q}(\mathbf{X},\mathbf{T},\mathbf{Q})$ is the set of distributions $P(\mathbf{Q}(\boldsymbol{\gamma}),\mathbf{X},\mathbf{T})$, $\boldsymbol{\gamma}$ in Γ. If $\mathbf{Q}$ is the identity function on $\Gamma = \Omega(X,T)$, then $\mathcal{Q}(\mathbf{X},\mathbf{T},\mathbf{Q})$ is the original exponential family $\mathcal{P}(\mathbf{X},\mathbf{T})$. The Jacobian matrix of $\mathbf{Q}$ at $\boldsymbol{\omega}$ is denoted by $\nabla\mathbf{Q}(\boldsymbol{\omega})$, and $\nabla^2\mathbf{Q}(\boldsymbol{\omega})$ denotes the array of Hessian matrices $\nabla^2 Q_k(\boldsymbol{\omega})$ of Q_k at $\boldsymbol{\omega}$, $1 \le k \le q$.

An elementary example of use of a function $\mathbf{Q}$ involves linear transformations. In this case, the restricted model is also a natural exponential family. For some q by r matrix $\mathbf{Z}$ of rank $r \le q$, let Γ be the set of all m-dimensional vectors $\boldsymbol{\gamma}$ such that $\mathbf{Z}\boldsymbol{\gamma}$ is in $\Omega(\mathbf{X},\mathbf{T})$ and let $\mathbf{Q}(\boldsymbol{\gamma}) = \mathbf{Z}\boldsymbol{\gamma}$ for all $\boldsymbol{\gamma}$ in Γ. Let $\mathbf{Z}'$ be the transpose of $\mathbf{Z}$. Then $\Gamma = \Omega(\mathbf{X},\mathbf{Z}'\mathbf{T})$ and $P(\mathbf{Q}(\boldsymbol{\gamma}),\mathbf{X},\mathbf{Z})$ is $P(\boldsymbol{\gamma},\mathbf{X},\mathbf{Z}'\mathbf{T})$. Thus $\mathcal{Q}(\mathbf{Q},\mathbf{X},\mathbf{T})$ is $\mathcal{P}(\mathbf{X},\mathbf{Z}'\mathbf{T})$.

4.2.1 Polytomous Random Vectors

In the simplest case of exponential families, the set S is finite, and the exponential family $\mathcal{P}(\mathbf{X},\mathbf{T})$ contains distributions of polytomous random vectors with values in S. The resulting distributions satisfy log-linear models (Haberman, 1978, 1979). The function $\mathbf{T}$ may be any q-dimensional vector function on S, and the natural parameter space $\Omega(\mathbf{X},\mathbf{T})$ is R^q. For $\boldsymbol{\omega}$ in R^q, $P(\boldsymbol{\omega},\mathbf{X},\mathbf{T})$ is the distribution on S of the random vector $\mathbf{Y}(\boldsymbol{\omega},\mathbf{X},\mathbf{T})$ such that

$$P\{\mathbf{Y}(\boldsymbol{\omega},\mathbf{X},\mathbf{T}) = \mathbf{s}\} = P\{X = \mathbf{s}\} f(\mathbf{s};\boldsymbol{\omega},\mathbf{X},\mathbf{T}) \tag{4.6}$$

for $\mathbf{s}$ in S, where the density kernel

$$f(\mathbf{s};\boldsymbol{\omega},\mathbf{X},\mathbf{T}) = \frac{\exp[\boldsymbol{\omega}'\mathbf{T}(s)]}{\mathrm{Mgf}(\boldsymbol{\omega},\mathbf{X},\mathbf{T})} \tag{4.7}$$

and the moment generating function

$$\mathrm{Mgf}(\boldsymbol{\omega},\mathbf{X},\mathbf{T}) = \sum_{\mathbf{s}\in S} P\{X = \mathbf{s}\}\exp[\boldsymbol{\omega}'\mathbf{T}(\mathbf{s})]. \tag{4.8}$$

The distribution $P(\boldsymbol{\omega},\mathbf{X},\mathbf{T})$ determines the natural parameter $\boldsymbol{\omega}$ if no q-dimensional vector $\mathbf{v} \ne \mathbf{0}_q$ exists such that $\mathbf{v}'\mathbf{T}(\mathbf{s})$ is constant over $\mathbf{s}$ in S such that the probability $P\{\mathbf{X} = \mathbf{s}\}$ is positive. To any nonnegative weights $w(\mathbf{s})$, $\mathbf{s}$ in S, with sum 1 corresponds some random vector $\mathbf{X}$ with $P\{\mathbf{X} = \mathbf{s}\} = w(\mathbf{s})$ for $\mathbf{s}$ in S. In many log-linear models in common use, the random vector $\mathbf{X}$ is uniformly distributed on S, so that $P\{\mathbf{X} = \mathbf{s}\} = 1/N(S)$, where $N(S)$ is the number of elements of S.

In many common probability models relevant to item response theory, $n \ge 1$ independently distributed polytomous responses Y_i, $1 \le i \le n$, are considered. In the simplest cases, each response Y_i is 0 or 1, and the probability $P\{Y_i = 1\}$ is neither 0 nor 1. These cases include logistic regression models and probit regression models. Logistic regression models impose linear restrictions on the logits $\log[P\{Y_i = 1\}/P\{Y_i = 0\}]$. Probit regression models impose linear restrictions on the probits $\Phi^{-1}(P\{Y_i = 1\})$, where Φ^{-1} is the inverse of the cumulative distribution function of the standard normal distribution. Both these models can be described in terms of exponential families.

In a logistic regression model (Garwood, 1941; Cox and Snell, 1989), for some integer $n \ge 1$, S is the set of all n-dimensional vectors with each element 0 or 1. A positive integer q

is given. To each positive integer $i \leq n$ corresponds a q-dimensional vector $\mathbf{Z}_i$. To identify the natural parameter, it is assumed that the $\mathbf{Z}_i$, $1 \leq i \leq n$, span R^q. The random vector $\mathbf{X}$ has a uniform distribution on S. The q-dimensional vector function $\mathbf{T}$ on S satisfies

$$\mathbf{T}(\mathbf{s}) = \sum_{i=1}^{n} s_i \mathbf{Z}_i \tag{4.9}$$

for each vector $\mathbf{s}$ in S with elements s_i, $1 \leq i \leq n$. It follows that

$$f(\mathbf{s}; \boldsymbol{\omega}, \mathbf{X}, \mathbf{T}) = \prod_{i=1}^{n} \frac{2\exp(s_i \boldsymbol{\omega}'\mathbf{Z}_i)}{1 + \exp(\boldsymbol{\omega}'\mathbf{Z}_i)} \tag{4.10}$$

and

$$P\{\mathbf{Y}(\boldsymbol{\omega}, \mathbf{X}, \mathbf{T}) = \mathbf{s}\} = \prod_{i=1}^{n} \frac{\exp(s_i \boldsymbol{\omega}'\mathbf{Z}_i)}{1 + \exp(\boldsymbol{\omega}'\mathbf{Z}_i)} \tag{4.11}$$

for $\boldsymbol{\omega}$ in $\Omega(\mathbf{X}, \mathbf{T}) = R^q$. Thus the elements $Y_i(\boldsymbol{\omega}, \mathbf{X}, \mathbf{T})$ of $\mathbf{Y}(\boldsymbol{\omega}, \mathbf{X}, \mathbf{T})$, $1 \leq i \leq n$, are mutually independent Bernoulli random variables, and the probability that $Y_i(\boldsymbol{\omega}, \mathbf{X}, \mathbf{T})$ is 1 is

$$\frac{\exp(\boldsymbol{\omega}'\mathbf{Z}_i)}{1 + \exp(\boldsymbol{\omega}'\mathbf{Z}_i)}.$$

Note that the logit transformation of this probability is $\boldsymbol{\omega}'\mathbf{Z}_i$.

In a probit regression model (Bliss, 1935; Finney, 1971), a somewhat different approach is appropriate. Here S, $\mathbf{X}$, and $\mathbf{Z}_i$ are defined as in the logistic regression model, and $\mathbf{T}$ is the identity transformation on S. Thus, the density kernel

$$f(\mathbf{s}; \boldsymbol{\omega}, \mathbf{X}, \mathbf{T}) = \prod_{i=1}^{n} \frac{2\exp(s_i \omega_i)}{1 + \exp(\omega_i)} \tag{4.12}$$

for $\boldsymbol{\omega}$ in R^n. It is assumed that $\boldsymbol{\omega} = \mathbf{Q}(\boldsymbol{\gamma})$, where $\mathbf{Q}$ is the function on $\Gamma = R^q$ with element i of $\mathbf{Q}(\boldsymbol{\gamma})$ such that

$$Q_i(\boldsymbol{\gamma}) = \tau(\boldsymbol{\gamma}'\mathbf{Z}_i), \tag{4.13}$$

$\tau = \log[\Phi/(1 - \Phi)]$, and Φ is the cumulative distribution function of a random variable with a standard normal distribution. In this model, the elements $Y_i(\mathbf{Q}(\boldsymbol{\gamma}), \mathbf{X}, \mathbf{T})$ of $\mathbf{Y}(\mathbf{Q}(\boldsymbol{\gamma}), \mathbf{X}, \mathbf{T})$ are Bernoulli random variables that are mutually independent for $1 \leq i \leq n$. For each i, $Y_i(\mathbf{Q}(\boldsymbol{\gamma}), \mathbf{X}, \mathbf{T}) = 1$ with probability $\Phi(\boldsymbol{\gamma}'\mathbf{Z}_i)$, so that the probit transformation of this probability is $\boldsymbol{\gamma}'\mathbf{Z}_i$. The parameter $\boldsymbol{\gamma}$ is determined by $P(\mathbf{Q}(\boldsymbol{\gamma}), \mathbf{X}, \mathbf{T})$ (Haberman, 1974, pp. 308–309).

In IRT, multinomial response models also arise (Haberman, 1974, 1979). These models apply to independent polytomous responses Y_i, $1 \leq i \leq n$, with respective integer values 0 to $r_i - 1 > 0$. It is assumed that the relative logits $\log[P\{Y_i = k\}/P\{Y_i = k'\}]$ satisfy linear models for integers $0 \leq k < k' < r_i$. In the exponential family formulation, the set S is the set of all n-dimensional vectors $\mathbf{s}$ with each element s_i a nonnegative integer not greater than r_i. For a positive integer q, to each positive integer $i \leq n$ and each nonnegative integer $k \leq r_i$ corresponds a q-dimensional vector $\mathbf{Z}_{ik}$. The function $\mathbf{T}$ on S satisfies

$$\mathbf{T}(\mathbf{s}) = \sum_{i=1}^{n} \mathbf{Z}_{is_i} \tag{4.14}$$

for $\mathbf{s}$ in S. The elements of $\mathbf{X}$ are independent, and the probability $P\{X_i = k\} > 0$ for $0 \leq k \leq r_i$ and $1 \leq i \leq n$. Thus, the density kernel

$$f(\mathbf{s}; \boldsymbol{\omega}, \mathbf{X}, \mathbf{T}) = \frac{\exp[\boldsymbol{\omega}'\mathbf{T}(\mathbf{s})]}{\prod_{i=1}^{n} \sum_{k=0}^{r_i} P\{X_i = k\} \exp(\boldsymbol{\omega}'\mathbf{Z}_{ik})} \tag{4.15}$$

and

$$P\{\mathbf{Y}(\boldsymbol{\omega}, \mathbf{X}, \mathbf{T}) = \mathbf{s}\} = f(\mathbf{s}; \boldsymbol{\omega}, \mathbf{X}, \mathbf{T}) \prod_{i=1}^{n} P\{X_i = s_i\} \tag{4.16}$$

for $\mathbf{s}$ in S and $\boldsymbol{\omega}$ in $\Omega(\mathbf{X}, \mathbf{T}) = R^q$. The elements $Y_i(\boldsymbol{\omega}, \mathbf{X}, \mathbf{T})$ of $\mathbf{Y}(\boldsymbol{\omega}, \mathbf{X}, \mathbf{T})$, $1 \leq i \leq n$, are mutually independent, and the probability that $Y_i(\boldsymbol{\omega}, \mathbf{X}, \mathbf{T}) = k$ is

$$\frac{P\{X_i = k\} \exp(\boldsymbol{\omega}'\mathbf{Z}_{ik})}{\sum_{k'=0}^{r_i} P\{X_i = k'\} \exp(\boldsymbol{\omega}'\mathbf{Z}_{ik'})}$$

for $0 \leq k \leq r_i$ and $1 \leq i \leq n$.

If $\mathbf{0}_q$ is the only q-dimensional vector such that $\mathbf{v}'\mathbf{T}(\mathbf{s})$ is constant for $\mathbf{s}$ in S, then $P(\boldsymbol{\omega}; \mathbf{X}, \mathbf{T})$ uniquely determines $\boldsymbol{\omega}$. The multinomial response model generalizes the logit model in the case of each $r_i = 1$, $\mathbf{Z}_{i0} = \mathbf{0}_q$, and $\mathbf{Z}_{i1} = \mathbf{Z}_i$.

Exponential families can also be used to obtain general parameterizations for all distributions on a finite nonempty set S that assign positive probability to each member of S. Let $q = N(S) - 1$ be one less than the number of elements of S, let $\mathbf{X}$ be a random vector with values on S such that $P\{\mathbf{X} = \mathbf{s}\} > 0$ for all $\mathbf{s}$ in S, and let $\mathbf{T}$ be any function from S to R^q such that $\mathbf{0}_q$ is the only vector $\mathbf{v}$ in R^q such that $\mathbf{v}'\mathbf{T}(\mathbf{s})$ is constant over $\mathbf{s}$ in S. Then the exponential family $\mathcal{P}(\mathbf{X}, \mathbf{T})$ includes all distributions on S that assign a positive probability to each member of S. There is a one-to-one correspondence between the distribution $P(\boldsymbol{\omega}, \mathbf{X}, \mathbf{T})$ and the natural parameter $\boldsymbol{\omega}$.

4.2.2 Continuous Exponential Families

Continuous exponential families are families of continuous distributions. They include beta distributions, Dirichlet distributions, exponential distributions, gamma distributions, normal distributions, and Weibull distributions. In continuous exponential families, the set S has a nonempty interior and $\mathbf{X}$ is a continuous random vector on S with probability density function w, so that w is nonnegative and has integral $\int_S w(\mathbf{s})d\mathbf{s} = 1$ over S. The natural parameter space $\Omega(\mathbf{X}, \mathbf{T})$ consists of q-dimensional vectors $\boldsymbol{\omega}$ such that $w \exp(\boldsymbol{\omega}'\mathbf{T})$ is integrable. If $\boldsymbol{\omega}$ is in $\Omega(\mathbf{X}, \mathbf{T})$, then $P(\boldsymbol{\omega}, \mathbf{X}, \mathbf{T})$ is the distribution on S of the continuous random vector $\mathbf{Y}(\boldsymbol{\omega}, \mathbf{X}, \mathbf{T})$ with probability density function $wf(\bullet; \boldsymbol{\omega}, \mathbf{X}, \mathbf{X})$, where

$$f(\mathbf{s}; \boldsymbol{\omega}, \mathbf{X}, \mathbf{T}) = \frac{\exp[\boldsymbol{\omega}'\mathbf{T}(\mathbf{s})]}{\mathrm{Mgf}(\boldsymbol{\omega}, \mathbf{X}, \mathbf{T})} \tag{4.17}$$

for $\mathbf{s}$ in S. Here the moment generating function $\text{Mgf}(\boldsymbol{\omega}, \mathbf{X}, \mathbf{T})$ at $\boldsymbol{\omega}$ in $\Omega(\mathbf{X}, \mathbf{T})$ is the integral over S of $w \exp(\boldsymbol{\omega}'\mathbf{T})$.

In IRT, an important example of a continuous exponential family is a normal probability model. Let S be the set R^n of n-dimensional vectors, let $q = (n+3)n/2$, and let $\mathbf{X}$ be a random vector with elements X_i, $1 \leq i \leq n$, that are mutually independent standard normal random variables. Thus, $\mathbf{X}$ has probability density function $w(\mathbf{s}) = \prod_{i=1}^{n} \phi(s_i)$, where ϕ is the density function of the standard univariate normal distribution. Let $\mathbf{T}(\mathbf{s})$ have elements $T_k(\mathbf{s})$, $1 \leq k \leq q$, such that $T_k(\mathbf{s}) = s_k$ for $1 \leq k \leq n$ and $T_k(\mathbf{s}) = s_i s_j$, for $1 \leq j \leq i \leq n$ and $k = n + j + i(i-1)/2$. For $\boldsymbol{\omega}$ in R^q with elements ω_i, $1 \leq i \leq q$, let $\boldsymbol{\omega}_n$ be the n-dimensional vector with elements ω_i for $1 \leq i \leq n$, and let $D_{ij}(\boldsymbol{\omega})$ and $D_{ji}(\boldsymbol{\omega})$ be defined for $1 \leq i \leq j \leq n$ and $k = n + j + i(i-1)/2$ so that $D_{ij}(\boldsymbol{\omega}) = 1 - 2\omega_k$ if $i = j$ and $D_{ij}(\boldsymbol{\omega}) = D_{ji}(\boldsymbol{\omega}) = -\omega_k$ if $j < i$. Let $\mathbf{D}(\boldsymbol{\omega})$ be the n by n symmetric matrix with row i and column j equal to $D_{ij}(\boldsymbol{\omega})$ for $1 \leq i \leq n$ and $1 \leq j \leq n$. Then, $\Omega(\mathbf{X}, \mathbf{T})$ consists of vectors $\boldsymbol{\omega}$ in R^q such that $\mathbf{D}(\boldsymbol{\omega})$ is positive definite. The distribution $P(\boldsymbol{\omega}, \mathbf{X}, \mathbf{T})$ is a multivariate normal distribution with mean $[\mathbf{D}(\boldsymbol{\omega})]^{-1}\boldsymbol{\omega}_n$ and covariance matrix $\boldsymbol{\Sigma}(\boldsymbol{\omega}) = [\mathbf{D}(\boldsymbol{\omega})]^{-1}$. If $n = 1$, then $q = 2$, the natural parameter space $\Omega(\mathbf{X}, \mathbf{T})$ consists of vectors $\boldsymbol{\omega}$ in R^2 with $\omega_2 < 1/2$, the mean of the one element $Y(\boldsymbol{\omega}, \mathbf{X}, \mathbf{T})$ of $\mathbf{Y}(\boldsymbol{\omega}, \mathbf{X}, \mathbf{T})$ is $(1 - 2\omega_2)^{-1}\omega_1$, and the variance of $Y(\boldsymbol{\omega}, \mathbf{X}, \mathbf{T})$ is $(1 - 2\omega_2)^{-1}$.

A restricted model generated from a normal probability model is a normal regression model. In a simple normal regression model, for some integer $m > 1$, a q by m matrix $\mathbf{Z}$ is used with elements Z_{kl}, $1 \leq k \leq q$, $1 \leq l \leq m$, such that Z_{kl} is 0 for $k \leq n$ and $l = m$ and for $k = n + j + i(i-1)/2$ for $1 \leq j < i \leq n$ and $1 \leq l \leq m$ and $Z_{ki} = 1$ for $k = n + i(i+1)/2$ and $1 \leq i \leq n$. The model requires that $\boldsymbol{\omega} = \mathbf{Z}\boldsymbol{\gamma}$. Under these conditions, Γ contains all m-dimensional vectors $\boldsymbol{\gamma}$ with $\gamma_m < 1/2$, and the the matrix $\boldsymbol{\Sigma}$ is $\sigma^2\mathbf{I}_n$, where $\mathbf{I}_n$ is the n by n identity matrix and $\sigma^2 = (1 - 2\gamma_m)^{-1}$. Thus the elements $Y_i(\mathbf{Z}\boldsymbol{\gamma}, \mathbf{X}, \mathbf{T})$ of $\mathbf{Y}(\mathbf{Z}\boldsymbol{\gamma}, \mathbf{X}, \mathbf{T})$ are mutually independent normal random variables for $1 \leq i \leq n$. For $1 \leq i \leq n$, the expectation of $Y_i(\mathbf{Z}\boldsymbol{\gamma}, \mathbf{X}, \mathbf{T})$ is $\sigma^2 \sum_{l=1}^{m-1} Z_{il}\gamma_l$, and its variance is σ^2. Thus, the $\sigma^2\gamma_l$, $1 \leq l \leq m-1$, are regression coefficients for the linear regression of $Y_i(\mathbf{Z}\boldsymbol{\gamma}, \mathbf{X}, \mathbf{T})$ on Z_{il}, $1 \leq l \leq m-1$. If $\mathbf{Q}$ is the function on Γ such that $\mathbf{Q}(\boldsymbol{\gamma}) = \mathbf{Z}\boldsymbol{\gamma}$ for all $\boldsymbol{\gamma}$ in Γ, then $\mathcal{Q}(\mathbf{X}, \mathbf{T}, \mathbf{Q})$ is $\mathcal{P}(\mathbf{X}, \mathbf{Z}'\mathbf{T})$, and $P(\mathbf{Q}(\boldsymbol{\gamma}), \mathbf{X}, \mathbf{T})$ is $P(\boldsymbol{\gamma}, \mathbf{X}, \mathbf{Z}'\mathbf{T})$.

4.2.3 Functions of Random Vectors

In many statistical applications, a function of an underlying variable is observed rather than the variable itself. For example, one may only observe the sign of a continuous real random variables. Such applications also apply to exponential families. For an exponential family $\mathcal{P}(\mathbf{X}, \mathbf{T})$ generated by the random vector $\mathbf{X}$ with values in S and the function $\mathbf{T}$ on S, no random vector $\mathbf{Y}$ is actually observed with a distribution in the exponential family. Instead a random variable $\mathbf{h}(\mathbf{Y})$ is observed for a known function $\mathbf{h}$ from S onto a subset $\mathbf{h}(S)$ of vectors of dimension v. It is assumed that $\mathbf{h}(\mathbf{Y})$ is a random vector whenever $\mathbf{Y}$ is a random vector with values in S. For each $\boldsymbol{\omega}$ in $\Omega(\mathbf{X}, \mathbf{T})$, $P_M(\boldsymbol{\omega}, \mathbf{X}, \mathbf{T}, \mathbf{h})$ denotes the distribution of $\mathbf{h}(\mathbf{Y}(\boldsymbol{\omega}, \mathbf{X}, \mathbf{T}))$. If $\mathbf{h}$ is the identity function on S, then $P_M(\boldsymbol{\omega}, \mathbf{X}, \mathbf{T}, \mathbf{h})$ is $P(\boldsymbol{\omega}, \mathbf{X}, \mathbf{T})$.

The distribution of $\mathbf{h}(\mathbf{Y}(\boldsymbol{\omega}, \mathbf{X}, \mathbf{T}))$ may be described in terms of exponential families by use of conditional distributions (Sundberg, 1974). For $\mathbf{u}$ in $\mathbf{h}(S)$, let $\mathbf{X}[\mathbf{h} = \mathbf{u}]$ be a random vector with a distribution equal to the conditional distribution of $\mathbf{X}$ given $\mathbf{h}(\mathbf{X}) = \mathbf{u}$. For simplicity, assume that $\Omega(\mathbf{X}[\mathbf{h} = \mathbf{u}], \mathbf{T})$ includes $\Omega(\mathbf{X}, \mathbf{T})$ for each $\mathbf{u}$ in $\mathbf{h}(S)$. Given the assumption that $\Omega(\mathbf{X}, \mathbf{T})$ is open, it is always possible to define conditional distributions so that this condition holds (Lehmann, 1959). The probability density kernel $f_M(\bullet; \boldsymbol{\omega}, \mathbf{X}, \mathbf{T}, \mathbf{h})$

is defined for $\mathbf{u}$ in $\mathbf{h}(S)$ to be

$$f_M(\mathbf{u};\omega,\mathbf{X},\mathbf{T},\mathbf{h}) = \frac{\mathrm{Mgf}(\omega,\mathbf{X}[\mathbf{h}=\mathbf{u}],\mathbf{T})}{\mathrm{Mgf}(\omega,\mathbf{X},\mathbf{T})}. \tag{4.18}$$

For any continuous and bounded real function g on $\mathbf{h}(S)$, the conditional expectation of $g(\mathbf{Y}(\omega,\mathbf{X},\mathbf{T}))$ given $\mathbf{h}(\mathbf{Y}(\omega,\mathbf{X},\mathbf{T}))=\mathbf{u}$ in $\mathbf{h}(S)$ is

$$\begin{aligned} &E(g(\mathbf{Y}(\omega,\mathbf{X},\mathbf{T}))|\mathbf{h}(\mathbf{Y}(\omega,\mathbf{X},\mathbf{T}))=\mathbf{u}) \\ &\quad = E(g(\mathbf{h}(\mathbf{X}))f_M(\mathbf{h}(\mathbf{X});\omega,\mathbf{X},\mathbf{T},\mathbf{h})|\mathbf{h}(\mathbf{X})=\mathbf{u}). \end{aligned} \tag{4.19}$$

The moment generating function $\mathrm{Mgf}_M(\bullet,\omega,\mathbf{X},\mathbf{T},\mathbf{h})$ and cumulant generating function

$$\mathrm{Cgf}_M(\bullet,\omega,\mathbf{X},\mathbf{T},\mathbf{h}) = \log[\mathrm{Mgf}_M(\bullet,\omega,\mathbf{X},\mathbf{T},\mathbf{h})]$$

of $\mathbf{h}(\mathbf{Y}(\omega,\mathbf{X},\mathbf{T}))$ are defined on the convex set $\Omega_M(\omega,\mathbf{X},\mathbf{T},\mathbf{h})$.

To describe these functions, it is helpful to use concatenation of vectors. Define the concatenation $\mathbf{u}\#\mathbf{y}$ of a k-dimensional vector $\mathbf{u}$ with elements u_i, $1\le i\le k$, and an l-dimensional vector $\mathbf{z}$ with elements z_i, $1\le i\le l$, be the vector of dimension $k+l$ with element i equal to u_i for $1\le i\le k$ and element $k+i$ equal to z_i for $1\le i\le l$. If A is a set of k-dimensional vectors and B is a set of l-dimensional vectors, let $A\#B$ be the set of all vectors $\mathbf{u}\#\mathbf{v}$ such that $\mathbf{u}$ is in A and $\mathbf{v}$ is in B.

Then $\boldsymbol{\delta}$ is in $\Omega_M(\omega,\mathbf{X},\mathbf{T},\mathbf{h})$ if, and only if, $\boldsymbol{\delta}\#\omega$ is in $\Omega(\mathbf{X},\mathbf{h}\#\mathbf{T})$. If $\boldsymbol{\delta}$ is in $\Omega_M(\omega,\mathbf{X},\mathbf{T},\mathbf{h})$, then $\mathrm{Mgf}_M(\bullet,\omega,\mathbf{X},\mathbf{T},\mathbf{h})$ has value

$$\mathrm{Mgf}_M(\boldsymbol{\delta},\omega,\mathbf{X},\mathbf{T},\mathbf{h}) = \frac{\mathrm{Mgf}(\boldsymbol{\delta}\#\omega,\mathbf{X},\mathbf{h}\#\mathbf{T})}{\mathrm{Mgf}(\omega,\mathbf{X},\mathbf{T})} \tag{4.20}$$

and $\mathrm{Cgf}_M(\bullet,\omega,\mathbf{X},\mathbf{T},\mathbf{h})$ has value

$$\mathrm{Cgf}_M(\boldsymbol{\delta},\omega,\mathbf{X},\mathbf{T},\mathbf{h}) = \mathrm{Cgf}(\boldsymbol{\delta}\#\omega,\mathbf{X},\mathbf{h}\#\mathbf{T}) - \mathrm{Cgf}(\omega,\mathbf{X},\mathbf{T}) \tag{4.21}$$

at $\boldsymbol{\delta}$. If $\mathbf{0}_v\#\omega$ is in the interior of $\Omega(\mathbf{X},\mathbf{h}\#\mathbf{T})$, then the expectation $\boldsymbol{\mu}_M(\omega,\mathbf{X},\mathbf{T},\mathbf{h})$ of $\mathbf{h}(\mathbf{Y}(\omega,\mathbf{X},\mathbf{T}))$ is the vector consisting of the first m elements of the gradient at $\mathbf{0}_v\#\omega$ of $\mathrm{Cgf}(\bullet,\mathbf{X},\mathbf{h}\#\mathbf{T})$, while the covariance matrix of $\mathbf{h}(\mathbf{Y}(\omega,\mathbf{X},\mathbf{T}))$ is the v by v matrix $\boldsymbol{\Sigma}_M(\omega,\mathbf{X},\mathbf{T},\mathbf{h})$ formed from the first v rows and first v columns of the Hessian matrix at $\mathbf{0}_v\#\omega$ of $\mathrm{Cgf}(\bullet,\mathbf{X},\mathbf{h}\#\mathbf{T})$.

A special case of this result arises in which S is divided into $v\ge 2$ disjoint subsets S_k, $1\le k\le v$. Let χ_k, $1\le k\le v$, be the function on S such that $\chi_k(\mathbf{s})$ is 1 for $\mathbf{s}$ in S_k and 0 for $\mathbf{s}$ in S but not S_k. Assume that $\chi_k(\mathbf{Y})$ is a random variable for all random vectors $\mathbf{Y}$ with values in S. Let $\mathbf{h}$ be the v-dimensional vector function on S such that element k of $\mathbf{h}(\mathbf{s})$ is $\chi_k(\mathbf{s})$ for $1\le k\le v$ and $\mathbf{s}$ in S. Then, element k of the gradient at $\mathbf{0}_v\#\omega$ of $\mathrm{Cgf}(\bullet,\mathbf{X},\mathbf{h}\#\mathbf{T})$ is the probability that $\mathbf{Y}(\omega,\mathbf{X},\mathbf{T})$ is in S_k.

The family $\mathcal{P}_M(\mathbf{X},\mathbf{T},\mathbf{h})$ of distributions $P_M(\omega,\mathbf{X},\mathbf{T},\mathbf{h})$, ω in $\Omega(\mathbf{X},\mathbf{T})$, in some cases is an exponential family. In other cases, for some positive integers m, m^*, and q^*, some nonempty open subset Γ of R^m, some open subset Γ^* in R^{m^*}, some q-dimensional vector function $\mathbf{Q}$ on Γ, some q^*-dimensional vector function $\mathbf{Q}^*$ on Γ^*, and some function $\mathbf{T}^*$ from S^* to R^{q^*}, the family $\mathcal{Q}_M(\mathbf{X},\mathbf{T},\mathbf{Q},\mathbf{h})$ of distributions $P_M(\mathbf{Q}(\boldsymbol{\gamma}),\mathbf{X},\mathbf{T},\mathbf{h})$ is equal to $\mathcal{Q}(\mathbf{X},\mathbf{T}^*,\mathbf{Q}^*)$. This situation always applies if S^* is finite.

A good example of use of functions of random vectors with distributions in an exponential family is provided by the relationship of the normal regression model to the probit regression model (Haberman, 1974, Chapter 8). The probit model can be obtained from the normal regression model by recording whether the elements of the random vector are positive or negative. To generate the desired function of the random vector associated with the normal regression model, let S^* be the set of n-dimensional vectors with all elements 0 or 1. Let $\mathbf{h}$ be the n-dimensional vector function on R^n such that, if $\mathbf{z}$ is an n-dimensional vector with elements z_i, $1 \le i \le n$, then $\mathbf{h}(\mathbf{z})$ is the n-dimensional vector with elements $h_i(\mathbf{z})$ equal to 0 for $z_i > 0$ and 1 for $z_i \le 0$. Define S, $\mathbf{X}$, $\mathbf{T}$, m, Γ, and $\mathbf{Q}$ as in the normal regression model, so that for γ in Γ, $\mathbf{Y}(\mathbf{Q}(\gamma), \mathbf{X}, \mathbf{T})$ is a multivariate normal random vector with independent elements with common variance $\sigma^2 = (1 - \gamma_m)^{-1}$ and $Y_i(\mathbf{Q}(\gamma), \mathbf{X}, \mathbf{T})$, $1 \le i \le n$, has expectation $\sigma^2 \sum_{l=1}^{m-1} Z_{il}\gamma_l$. To obtain a corresponding probit model, let $\mathbf{X}^*$ have a uniform distribution on S^*, let $\mathbf{T}^*$ be the identity function on S^*, and let $\mathbf{Q}^*$ be the function on R^{m-1} with element i of $\mathbf{Q}^*(\gamma)$ such that

$$Q_i^*(\gamma^*) = \tau([\gamma^*]'\mathbf{Z}_i^*), \tag{4.22}$$

where γ^* is in R^{m-1} and $\mathbf{Z}_i^*$ is the vector of dimension $m - 1$ with elements Z_{il}, $1 \le l \le m - 1$. Let γ^* be the vector with dimension $m - 1$ with element γ_k^* equal to $\sigma\gamma_k$ for $1 \le k \le m - 1$. Then, $\mathbf{h}(\mathbf{Y}(\mathbf{Q}(\gamma), \mathbf{X}, \mathbf{T}))$ satisfies the probit model $P(\mathbf{Q}^*(\gamma^*), \mathbf{X}^*, \mathbf{T}^*)$. On the other hand, let γ^* be an arbitrary vector of dimension $m - 1$, and let σ be an arbitrary positive number. Let γ be the vector of dimension m such that γ_k is equal to γ_k^*/σ for $1 \le k \le m - 1$ and $\gamma_m = \frac{1}{2} - \frac{1}{\sigma^2}$. Then, the probit model $P(\mathbf{Q}^*(\gamma^*), \mathbf{X}^*, \mathbf{T}^*)$ is satisfied by $\mathbf{h}(\mathbf{Y}(\mathbf{Q}(\gamma), \mathbf{X}, \mathbf{T}))$. Note that the same probit regression model parameter γ^* can be produced by multiple normal regression parameters γ. This situation, which is not uncommon, is often encountered in IRT.

4.2.4 Conditional Distributions

A conditional exponential family may be generated from an exponential family. For example, random vectors with multinomial distributions may be generated from independent Poisson random variables by conditioning on the sum of the Poisson variables. In general, conditioning on a linear function of the sufficient statistic is involved. Consider a natural exponential family $\mathcal{P}(\mathbf{X}, \mathbf{T})$ such that the conditions of Section 4.2 hold for the n-dimensional random vector $\mathbf{X}$ with values in S and the q-dimensional vector function $\mathbf{T}$ on S. Let the dimension q of the vector-valued function $\mathbf{T}$ be at least 2, and let $q = q_1 + q_2$, where both q_1 and q_2 are positive integers. Let $\mathbf{T} = \mathbf{T}_1\#\mathbf{T}_2$, where $\mathbf{T}_1$ is a vector function on S with dimension q_1 and $\mathbf{T}_2$ is a vector function on S with dimension q_2. If $\mathbf{Y}$ has a distribution in the exponential family $\mathcal{P}(\mathbf{X}, \mathbf{T})$, then the distribution of $\mathbf{Y}$ given $\mathbf{T}_2(\mathbf{Y})$ is considered. This distribution also belongs to an exponential family. To describe this family in straightforward cases, let $\Omega(\mathbf{X}, \mathbf{T}_1|\mathbf{T}_2)$ be the convex set of q_1-dimensional vectors ω_1 such that for all ω_2 in $\Omega(\mathbf{X}, \mathbf{T}_2)$, $\omega_1\#\omega_2$ is in $\Omega(\mathbf{X}, \mathbf{T})$. Observe that $\Omega(\mathbf{X}, \mathbf{T}_1|\mathbf{T}_2)$ is nonempty and convex. Assume for convenience that $\Omega(\mathbf{X}, \mathbf{T}_1|\mathbf{T}_2)$ is open.

The conditional distribution of $\mathbf{Y}$ given $\mathbf{T}_2(\mathbf{Y})$ only depends on part of the natural parameter ω in $\Omega(\mathbf{X}, \mathbf{T})$ if $\mathbf{Y}$ has distribution $P(\omega, \mathbf{X}, \mathbf{T})$. Let ω be $\omega_1\#\omega_2$, where ω_1 is in $\Omega(\mathbf{X}, \mathbf{T}_1|\mathbf{T}_2)$ and ω_2 is in $\Omega(\mathbf{X}, \mathbf{T}_2)$. Then, only ω_1 affects the conditional distribution of $\mathbf{Y}$ given $\mathbf{T}_2(\mathbf{Y})$. Let $\mathbf{t}$ be in $\mathbf{T}_2(S)$. Then, the conditional distribution of $\mathbf{Y}(\omega, \mathbf{X}, \mathbf{T})$ given $\mathbf{T}_2(\mathbf{Y}(\omega, \mathbf{X}, \mathbf{T}) = \mathbf{t}$ is $P(\omega_1, \mathbf{X}[\mathbf{T}_2 = \mathbf{t}], \mathbf{T}_1)$. Thus, the conditional distribution depends only on ω_1 and has a form associated with an exponential family. The density kernel $f_C(\mathbf{s}; \omega_1, \mathbf{X}, \mathbf{T}_1|\mathbf{T}_2)$ at $\mathbf{s}$ in S is then

$$f_C(\mathbf{s}; \omega_1, \mathbf{X}, \mathbf{T}_1|\mathbf{T}_2) = \frac{\exp[\omega_1'\mathbf{T}_1(\mathbf{s})]}{\mathrm{Mgf}(\omega_1, \mathbf{X}[\mathbf{T}_2 = \mathbf{T}_2(\mathbf{s})], \mathbf{T}_1)}. \tag{4.23}$$

In the polytomous case of S finite, if the probability that $\mathbf{T}_2(\mathbf{X}) = \mathbf{t}$ is positive, $\mathbf{T}_2^{-1}(\mathbf{t})$ is the set of $\mathbf{s}$ in S such that $\mathbf{T}_2(\mathbf{s}) = \mathbf{t}$, and $\mathbf{T}_2(\mathbf{s}) = \mathbf{t}$ for some $\mathbf{s}$ in S, then the conditional probability that $\mathbf{Y}(\omega, \mathbf{X}, \mathbf{T}) = \mathbf{s}$ given $\mathbf{T}_2(\mathbf{Y}(\omega, \mathbf{X}, \mathbf{T})) = \mathbf{t}$ is

$$P\{\mathbf{Y}(\omega, \mathbf{X}, \mathbf{T}) = \mathbf{s}|\mathbf{T}_2(\mathbf{Y}(\omega, \mathbf{X}, \mathbf{T})) = \mathbf{t}\}$$
$$= \frac{P\{\mathbf{X} = \mathbf{s}\}\exp[\omega_1'\mathbf{T}_1(\mathbf{s})]}{\sum_{\mathbf{s}' \in \mathbf{T}_2^{-1}(\mathbf{t})} P\{\mathbf{X} = \mathbf{s}'\}\exp[\omega_1'\mathbf{T}_1(\mathbf{s}')]}. \tag{4.24}$$

In this case,

$$f_C(\mathbf{s} : \mathbf{X}, \mathbf{T}_1|\mathbf{T}_2) = \frac{P\{\mathbf{T}_2(\mathbf{X}) = \mathbf{T}_2(\mathbf{s})\}\exp[\omega_1'\mathbf{T}_1(\mathbf{s})]}{\sum_{\mathbf{s}' \in \mathbf{T}_2^{-1}(\mathbf{T}_2(\mathbf{s}))} P\{\mathbf{X} = \mathbf{s}'\}\exp[\omega_1'\mathbf{T}_1(\mathbf{s}')]}. \tag{4.25}$$

The use of conditional distributions is quite common in conditional inference, an approach often encountered in IRT, especially in connection with the Rasch (Rasch, 1960) and partial credit (Masters, 1982) models.

4.2.5 Mixtures of Exponential Families

Exponential families are often used to construct families of distributions which involve mixtures. Such families are important in description of latent structure models used in IRT. In a typical case, for one exponential family, a random vector $\mathbf{Y}_1$ has a conditional distribution given a random vector $\mathbf{Y}_2$ that belongs to the exponential family for all values of $\mathbf{Y}_2$. For a second exponential family, $\mathbf{Y}_2$ has a distribution within that family. In addition, $\mathbf{Y}_1$ or a function of $\mathbf{Y}_1$ is observed but neither $\mathbf{Y}_2$ nor a function of $\mathbf{Y}_2$ is available. In IRT applications, $\mathbf{Y}_2$ is a latent vector and $\mathbf{Y}_1$ or a function of $\mathbf{Y}_1$ is a manifest vector. The distribution of the observed vector need not belong to an exponential family and need not belong to a family of distributions generated as in Section 4.2.3.

In general, the exponential family models are $\mathcal{P}(\mathbf{X}_1, \mathbf{T}_1)$ and $\mathcal{P}(\mathbf{X}_2, \mathbf{T}_2)$. The first family defines the conditional distribution of $\mathbf{Y}_1$ given $\mathbf{Y}_2$. The second family defines the conditional distribution of $\mathbf{Y}_2$. These exponential families satisfy standard assumptions. Thus, for $1 \le k \le 2$, $\mathbf{X}_k$ is a random vector with values in S_k, where S_k is a set of n_k-dimensional vectors and n_k is a positive integer. For some positive integer q_k, $\mathbf{T}_k$ is a q_k-dimensional vector function on S_k. In addition, for integers $v_k > 0$, $\mathbf{h}_k$ is a v_k-dimensional function on S_k. It is assumed that $\mathbf{X}_1$ and $\mathbf{X}_2$ are independent. For any random vector $\mathbf{Y}_k$ with values in S_k, $\mathbf{T}_k(\mathbf{Y}_k)$ and $\mathbf{h}_k(\mathbf{Y}_k)$ are both random vectors. In the cases under study, the underlying random vectors are $\mathbf{Y}_1$ and $\mathbf{Y}_2$. The observed vector is $\mathbf{h}_1(\mathbf{Y}_1)$. The conditional distribution of $\mathbf{Y}_1$ given $\mathbf{Y}_2$ depends only on $\mathbf{h}_2(\mathbf{Y}_2)$ and belongs to for all values of $\mathbf{Y}_2$. The distribution of $\mathbf{Y}_2$ belongs to $\mathcal{P}(\mathbf{X}_2, \mathbf{T}_2)$.

Parametric models are applied to both conditional and unconditional distributions. An open parameter space Γ of dimension m is considered. For some open set Δ of $m + v_2$-dimensional vectors such that $\Gamma\#\mathbf{h}_2(S_2)$ is included in Δ, there is an infinitely differentiable function $\mathbf{Q}_1$ from Δ to the natural parameter space $\Omega(\mathbf{X}_1, \mathbf{T}_2)$ and an infinitely differentiable function $\mathbf{Q}_2$ from Γ to the natural parameter space $\Omega(\mathbf{X}_2, \mathbf{T}_2)$. The model assumptions are that the conditional distribution of $\mathbf{Y}_1$ given $\mathbf{h}_2(\mathbf{Y}_2)$ is $P(\mathbf{Q}_1(\gamma\#\mathbf{h}_2(\mathbf{Y}_2)), \mathbf{X}_2, \mathbf{T}_2)$ and the unconditional distribution of $\mathbf{Y}_2$ is $P(\mathbf{Q}_2(\gamma), \mathbf{X}_2, \mathbf{T}_2)$.

To permit effective use of the model, some technical requirements are made to ensure that standard arguments apply to the relationships between conditional and unconditional distributions. The initial technical requirements are that the natural parameter space $\Omega(\mathbf{X}_k[\mathbf{h}_k = \mathbf{u}_k], \mathbf{T}_k)$ for the conditional distribution of $\mathbf{X}_k$ given $\mathbf{h}_k(\mathbf{X}_k) = \mathbf{u}_k$ includes the natural parameter space $\Omega(\mathbf{X}_k, \mathbf{T}_k)$ for each $\mathbf{u}_k$ in $\mathbf{h}_k(S_k)$ and some polynomial function H_2 on the space of q_2-dimensional vectors and some vector norm $|\bullet|$ on the space of v_2-dimensional vectors exist such that $|\mathbf{h}_2(\mathbf{s}_2)| \le H_2(\mathbf{T}_2(\mathbf{s}_2))$ for all $\mathbf{s}_2$ in S_2. These requirements have no practical impact in IRT.

To ensure that standard results apply concerning expectations, the technical assumption is made that a polynomial π on Δ exists such that for $\boldsymbol{\delta}$ in Δ, each element of the vector $\mathbf{Q}_1(\boldsymbol{\delta})$, the Jacobian matrix $\nabla\mathbf{Q}_1(\boldsymbol{\delta})$, or the array $\nabla^2\mathbf{Q}_2(\boldsymbol{\delta})$ of Hessian matrix has absolute value no greater than $\pi(\boldsymbol{\delta})$. Again this technical requirement does not have much practical impact in IRT.

To simplify notation for the mixture, $\mathbf{X}$ will denote $\mathbf{X}_1\#\mathbf{X}_2$, $\mathbf{T}$ will denote the function on $S_1\#S_2$ with value $\mathbf{T}_1(\mathbf{s}_1)\#\mathbf{T}_2(\mathbf{s}_2)$ at $\mathbf{s}_1\#\mathbf{s}_2$ for $\mathbf{s}_1$ in S_1 and $\mathbf{s}_2$ in S_2, $\mathbf{h}$ will denote the function on $S_1\#S_2$ with value $\mathbf{h}_1(\mathbf{s}_1)\#\mathbf{h}(\mathbf{s}_2)$ at $\mathbf{s}_1\#\mathbf{s}_2$ for $\mathbf{s}_1$ in S_1 and $\mathbf{s}_2$ in S_2, and $\mathbf{Q}$ will denote the function on Δ with value $\mathbf{Q}_1(\boldsymbol{\delta})\#\mathbf{Q}_2(\gamma)$ at $\boldsymbol{\delta}$ in Δ, where γ is the m-dimensional vector that consists of the first m elements of $\boldsymbol{\delta}$. In the mixture case, random vectors $\mathbf{Y}_{MIk}(\gamma, \mathbf{X}, \mathbf{T}, \mathbf{Q}, \mathbf{h})$, $1 \le k \le 2$, with values in S_k satisfy the conditions that the conditional distribution of $\mathbf{Y}_{MI1}(\gamma, \mathbf{X}, \mathbf{T}, \mathbf{Q}, \mathbf{h})$ given $\mathbf{Y}_{MI2}(\gamma, \mathbf{X}, \mathbf{T}, \mathbf{Q}, \mathbf{h}) = \mathbf{u}_2$ in $\mathbf{h}_2(S_2)$ is $P_M(\mathbf{Q}_1(\gamma\#\mathbf{u}_2), \mathbf{X}_1, \mathbf{T}_1, \mathbf{h}_1)$ and the distribution of $\mathbf{Y}_{MI2}(\gamma, \mathbf{X}, \mathbf{T}, \mathbf{Q}, \mathbf{h})$ is $P_M(\mathbf{Q}_2(\gamma), \mathbf{X}_2, \mathbf{T}_2, \mathbf{h}_2)$. The notation $\mathbf{Y}_{MI}(\gamma, \mathbf{X}, \mathbf{T}, \mathbf{Q}, \mathbf{h})$ is used for $\mathbf{Y}_{MI1}(\gamma, \mathbf{X}, \mathbf{T}, \mathbf{Q}, \mathbf{h})\#\mathbf{Y}_{MI2}(\gamma, \mathbf{X}, \mathbf{T}, \mathbf{Q}, \mathbf{h})$. The observed random vector $\mathbf{Y}_{MI1}(\gamma, \mathbf{X}, \mathbf{T}, \mathbf{Q}, \mathbf{h})$ has distribution $P_{MI1}(\gamma, \mathbf{X}, \mathbf{T}, \mathbf{Q}, \mathbf{h})$.

To obtain the density kernel for the mixture case involves two steps. For $\mathbf{u}_1$ in $\mathbf{h}_1(S_1)$ and $\mathbf{u}_2$ in $\mathbf{h}_2(S_2)$, let the density kernel $f_{MI}(\bullet; \gamma, \mathbf{X}, \mathbf{T}, \mathbf{Q}, \mathbf{h})$ of $\mathbf{Y}_{MI}(\gamma, \mathbf{X}, \mathbf{T}, \mathbf{Q}, \mathbf{h})$ be the function on $\mathbf{h}_1(S_1)\#\mathbf{h}_2(S_2)$ with value

$$\begin{aligned} &f_{MI}(\mathbf{u}_1\#\mathbf{u}_2; \gamma, \mathbf{X}, \mathbf{T}, \mathbf{Q}, \mathbf{h}) \\ &\quad = f_M(\mathbf{u}_1; \mathbf{Q}_1(\gamma\#\mathbf{u}_2), \mathbf{X}_1, \mathbf{T}_1, \mathbf{h}_1) f_M(\mathbf{u}_2; \mathbf{Q}_2(\gamma), \mathbf{X}_2, \mathbf{T}_2, \mathbf{h}_2). \end{aligned} \tag{4.26}$$

At $\mathbf{u}_1$, the density kernel $f_{MI1}(\bullet; \gamma, \mathbf{X}, \mathbf{T}, \mathbf{Q}, \mathbf{h})$ of $\mathbf{Y}_{MI1}(\gamma, \mathbf{X}, \mathbf{T}, \mathbf{Q}, \mathbf{h}))$ is then

$$f_{MI1}(\mathbf{u}_1; \gamma, \mathbf{X}, \mathbf{T}, \mathbf{Q}, \mathbf{h}) = E(f_{MI}(\mathbf{u}_1\#\mathbf{h}_2(\mathbf{X}_2); \gamma, \mathbf{X}, \mathbf{T}, \mathbf{Q}, \mathbf{h})). \tag{4.27}$$

In practice, the distribution of the random vector $\mathbf{Y}_{MI2}(\gamma, \mathbf{X}, \mathbf{T}, \mathbf{Q}, \mathbf{h})$ given the random vector $\mathbf{Y}_{MI1}(\gamma, \mathbf{X}, \mathbf{T}, \mathbf{Q}, \mathbf{h})$ is often needed. By Bayes's theorem, the conditional density kernel $f_{MB}(\bullet|\mathbf{u}_1, \gamma, \mathbf{X}, \mathbf{T}, \mathbf{Q}, \mathbf{h})$ for the conditional distribution of $\mathbf{Y}_{MI2}(\gamma, \mathbf{X}, \mathbf{T}, \mathbf{Q}, \mathbf{h}))$ given $\mathbf{Y}_{MI1}(\gamma, \mathbf{X}, \mathbf{T}, \mathbf{Q}, \mathbf{h})) = \mathbf{u}_1$ has value

$$f_{MB}(\mathbf{u}_2|\mathbf{u}_1; \gamma, \mathbf{X}, \mathbf{T}, \mathbf{Q}, \mathbf{h}) = \frac{f_{MI}(\mathbf{u}_1\#\mathbf{u}_2; \gamma, \mathbf{X}, \mathbf{T}, \mathbf{Q}, \mathbf{h})}{f_{MI1}(\mathbf{u}_1; \gamma, \mathbf{X}, \mathbf{T}, \mathbf{Q}, \mathbf{h})} \tag{4.28}$$

at $\mathbf{u}_2$.

In a latent class model, S_2 is a finite set (Lazarsfeld and Henry, 1968). In many common latent structure models, S_2 and $\mathbf{h}_2(S_2)$ are nonempty open sets and $\mathbf{X}_2$ is a continuous random vector. In mixture distribution IRT models (Volume One, Chapter 23), $\mathbf{h}_2(S_2)$ is finite but S_2 is the space of all n_2-dimensional vectors and $\mathbf{X}_2$ is a continuous random vector.

4.3 Application to IRT

To discuss applications of exponential families to IRT, a simple case of $n \geq 1$ items may be considered. Let item i, $1 \leq i \leq n$, have possible integer scores from 0 to $r_i \geq 1$. For a random selected person, let $U_i, 1 \leq i \leq n$, be the response score for item i, and let $\mathbf{U}$ be the n-dimensional response vector for person p, so that element i of $\mathbf{U}$ is U_i. For some integer $D \geq 1$, let a D-dimensional unobserved (latent) random vector $\boldsymbol{\theta}$ of person parameters be associated with the person. The vector $\boldsymbol{\theta}$ is assumed to be in a set Θ. The set Θ may consist of all vectors in R^D or it may be more restricted. As in a latent structure model, it is assumed that, conditional on $\boldsymbol{\theta}$, the response scores $U_i, 1 \leq i \leq n$, are mutually independent random variables. In latent trait models, $\boldsymbol{\theta}$ is a continuous random vector. In the case of latent class models, $\boldsymbol{\theta}$ is a polytomous random vector and Θ is finite.

In a very general IRT model expressed in terms of exponential families, an open and nonempty parameter space Γ of finite and positive dimension m is given. There is an associated open parameter space Δ of dimension $m + D$ that includes $\Gamma\#\Theta$. To each item i corresponds an infinitely differentiable function $\mathbf{Q}_{i1}$ on Δ, where $\mathbf{Q}_{i1}$ and its Jacobian matrix and array of Hessian matrices satisfy the polynomial bounding condition for $\mathbf{Q}_1$ in Section 4.2.5. The function $\mathbf{Q}_{i1}$ has dimension $q_i \geq 1$. In addition, associated with item i is a positive integer d_i, a set S_{i1} of d_i-dimensional vectors, a random vector $\mathbf{X}_{i1}$ with values in S_{i1}, a q_i-dimensional function $\mathbf{T}_{i1}$ on S_{i1}, and a function h_{i1} from S_{i1} to the set of integers from 0 to r_i. It is assumed that, if $\mathbf{Y}$ is a random vector with values in S_{i1}, then $\mathbf{T}_i(\mathbf{Y})$ is a random vector and $h_{i1}(\mathbf{Y})$ is a random variable. It is assumed that the $\mathbf{X}_{i1}$, $1 \leq i \leq n$, are mutually independent.

For each item i, the conditional distribution of the item score U_i given the person parameter $\boldsymbol{\theta}$ is specified by a parameter $\boldsymbol{\gamma}$ in Γ. The IRT model assumes that, for each $\boldsymbol{\theta}'$ in Θ, the conditional probability distribution of U_i given $\boldsymbol{\theta} = \boldsymbol{\theta}'$ is $P_M(\mathbf{Q}_{i1}(\boldsymbol{\gamma}\#\boldsymbol{\theta}'), \mathbf{X}_{i1}, \mathbf{T}_{i1}, \mathbf{h}_{i1})$, where $\mathbf{h}_{i1}$ is the one-dimensional vector wiih element h_{i1}.

Concatenation is employed to produce a model based on exponential families for the complete vector $\mathbf{U}$ of item scores. If k_i, $1 \leq i \leq n$, are positive integers and $\mathbf{u}_i$ is a k_i-dimensional vector, then let $\#_{i=1}^n \mathbf{u}_i$ be the vector defined by the recurrence relationship $\#_{i=1}^1 \mathbf{u}_i = \mathbf{u}_1$ and $\#_{i=1}^j \mathbf{u}_i = (\#_{i=1}^{j-1} \mathbf{u}_i)\#\mathbf{u}_j$ for $1 < j \leq n$. If A_i is a nonempty k_i-dimensional set, then $\#_{i=1}^n A_i$ is the set of $\#_{i=1}^n \mathbf{u}_i$ such that $\mathbf{u}_i$ is in A_i for $1 \leq i \leq n$. For $\mathbf{s}_{i1}$ in S_{i1}, $1 \leq i \leq n$ and $\mathbf{s}_1 = \#_{i=1}^n \mathbf{s}_{i1}$, let $\mathbf{X}_1$ be the function on $S_1 = \#_{i=1}^n S_{i1}$ such that $\mathbf{X}(\mathbf{s}) = \#_{i=1}^n \mathbf{X}_{i1}(\mathbf{s}_{i1})$, let $\mathbf{T}_1$ be the function on S_1 such that $\mathbf{T}_1(\mathbf{s}) = \#_{i=1}^n \mathbf{T}_{i1}(\mathbf{s}_{i1})$, let $\mathbf{h}_1$ be the function on S_1 such that $\mathbf{h}_1(\mathbf{s}_1) = \#_{i=1}^n \mathbf{h}_{i1}(\mathbf{s}_{i1})$, and let $\mathbf{Q}_1$ be the function on Δ such that $\mathbf{Q}_1(\boldsymbol{\delta}) = \#_{i=1}^n \mathbf{Q}_{i1}(\boldsymbol{\delta})$ for $\boldsymbol{\delta}$ in Δ. For $\boldsymbol{\theta}'$ in Θ, the conditional distribution of $\mathbf{U}$ given $\boldsymbol{\theta} = \boldsymbol{\theta}'$ is $P_M(\mathbf{Q}_1(\boldsymbol{\gamma}\#\boldsymbol{\theta}'), \mathbf{X}_1, \mathbf{T}_1, \mathbf{h}_1)$.

In IRT models associated with marginal estimation (Bock and Lieberman, 1970; Bock and Aitkin, 1981; see Chapter 11), an exponential family is associated with the distribution of $\boldsymbol{\theta}$. Let n_2, q_2, and v_2 be positive integers. Let S_2 be a nonempty subset of n_2-dimensional vectors. Let $\mathbf{X}_2$ be a random vector with values in S_2 that is independent of $\mathbf{X}_1$. Let $\mathbf{T}_2$ be a q_2-dimensional function on S_2, and let $\mathbf{h}_2$ be a function from S_2 to Θ. Assume that

$\mathbf{T}_2(\mathbf{Y}_2)$ and $\mathbf{h}_2(\mathbf{Y}_2)$ are random vectors for all random vectors $\mathbf{Y}_2$ with values in Θ. Assume that $\Omega(\mathbf{X}_2, \mathbf{T}_2)$ is open. Let $\mathbf{h}_2$ sastisfy the polynomial bounding condition of Section 4.2.5. Let $\mathbf{Q}_2$ be an infinitely differentiable function from Γ to $\Omega(\mathbf{X}_2, \mathbf{T}_2)$. It is assumed that θ has the distribution $P_M(\mathbf{Q}_2(\gamma), \mathbf{X}_2, \mathbf{T}_2, \mathbf{h}_2)$. It is common to encounter the log-linear case of $\Theta = S_2$ finite, $\mathbf{h}_2$ the identity transformation, and $\mathbf{Q}_2$ a linear function (Heinen, 1996; Haberman et al., 2008) or the normal case with $\Theta = S_2 = R^D$, $\mathbf{h}_2$ the identify function, and θ with some multivariate normal distribution. Much more elaborate possibilities can be brought into this same framework.

For example, in the case of a two-parameter logistic (2PL) model for a one-dimensional person parameter with a standard normal distribution (Lord and Novick, 1968, p. 400), $D = 1$, each $d_i = 1$, $m = 2n$, Γ is the set of γ in R^m with elements $\gamma_{2i} > 0$ for $1 \le i \le n$, each $q_i = 1$, $\mathbf{T}_{i1}$ is the identity function on the set S_{i1} with elements $\mathbf{0}_1$ and $\mathbf{1}_1$, $\mathbf{X}_{i1}$ is equal to each element of S_{i1} with probability 1/2, $\mathbf{Q}_{i1}(\gamma\#\theta')$ is the vector with the single element $\gamma_{2i-1} + \gamma_{2i}\theta'_1$ for γ in Γ and θ' in $\Theta = R^1$, and h_{i1} is the one-to-one function that maps $\mathbf{k}_1$ to k for k equal 0 or 1. Thus

$$P\{U_i = k|\theta_1 = \theta'_1\} = \frac{\exp[k(\gamma_{2i-1} + \gamma_{2i}\theta_1)]}{1 + \exp(\gamma_{2i-1} + \gamma_{2i}\theta_1)}. \tag{4.29}$$

The parameter γ_{2i} is the item discrimination a_i for item i, and the item difficulty b_i of item i is $-\gamma_{2i-1}/\gamma_{2i}$. The marginal probability that $\mathbf{U} = \mathbf{s}$ in S is then

$$\int \phi(\theta) \left[\prod_{i=1}^{n} \frac{\exp[s_i(\gamma_{2i-1} + \gamma_{2i}\theta)]}{1 + \exp(\gamma_{2i-1} + \gamma_{2i}\theta)} \right] d\theta.$$

The distribution of θ_1 is a standard normal distribution.

The use of functions h_{i1} that are not one-to-one is illustrated by the three-parameter logistic (3PL) model with a constant guessing parameter for a one-dimensional person parameter with a normal distribution (Lord and Novick, 1968, p. 405; Volume One, Chapter 2). Here, $D = 1$, each $d_i = 2$, $m = 2n + 1$, Γ is the subset γ in R^m with elements $\gamma_{2i} > 0$ for $1 \le i \le n$, $q_i = 1$, S_{i1} is the set of two-dimensional vectors with each element 0 or 1, $\mathbf{T}_{i1}$ is the identity function on the set S_{i1}, $\mathbf{X}_{i1}$ is uniformly distributed on S_{i1}, $\mathbf{Q}_{i1}(\gamma\#\theta')$ is the vector with elements $Q_{i1}(\gamma\#\theta') = \gamma_{2i-1} + \gamma_{2i}\theta'_1$ and $Q_{i2}(\gamma\#\theta') = \gamma_m$ for γ in Γ and θ' in R^1, and h_{i1} is the function on S_i such that $h_{i1}(\mathbf{s})$ is 1 if $\mathbf{s}$ in S_{i1} is not $\mathbf{0}_2$ and $h_{i1}(\mathbf{0}_2)$ is 0. If the common guessing probability for the items is

$$c = \frac{\exp(\gamma_m)}{1 + \exp(\gamma_m)}, \tag{4.30}$$

then

$$P\{U_i = 0|\theta_1 = \theta'_1\} = \begin{cases} (1-c)\dfrac{1}{1 + \exp(\gamma_{2i-1} + \gamma_{2i}\theta'_1)}, & k = 0, \\[2ex] c + \dfrac{1-c}{[1 + \exp(\gamma_{2i-1} + \gamma_{2i}\theta'_1)}, & k = 1. \end{cases} \tag{4.31}$$

The parameter γ_{2i} is the item discrimination a_i for item i, and the item difficulty b_i of item i is $-\gamma_{2i-1}/\gamma_{2i}$. The distribution of θ_1 is a standard normal distribution.

Numerous other common IRT models can be described within this framework. Cases include the traditional normal ogive model (Lord and Novick, 1968) with the equation

$$P\{U_i = k|\theta_1 = \theta_1'\} = \begin{cases} \Phi(\gamma_{2i-1} + \gamma_{2i}\theta_1), & k = 1, \\ 1 - \Phi(\gamma_{2i-1} + \gamma_{2i}\theta_1), & k = 0 \end{cases} \tag{4.32}$$

and the normal ogive model with a constant guessing parameter (Lord and Novick, 1968, p. 404) with

$$P\{U_i = 0|\theta_1 = \theta_1'\} = \begin{cases} (1-c)[1 - \Phi(\gamma_{2i-1} + \gamma_{2i}\theta_1)], & k = 0, \\ c + (1-c)\Phi(\gamma_{2i-1} + \gamma_{2i}\theta_1), & k = 1. \end{cases} \tag{4.33}$$

In the one-parameter logistic (1PL) model (Rasch, 1960), all γ_{2i} are equal in Equation 4.29. The nominal model (Bock, 1972), the partial credit model (Masters, 1982), the one-parameter logistic (OPLM) model (Verhelst and Glas, 1995), the generalized partial credit models (Muraki, 1992), and the multidimensional random coefficients multinomial logit model (Adams et al., 1997) are a few of the models readily incorporated into the framework considered for exponential families.

In discussions of statistical estimation, independent and identically distributed random vectors $\mathbf{U}_p, 1 \leq p \leq P$, are considered for $P \geq 1$ persons. Each $\mathbf{U}_p$ has the same distribution as $\mathbf{U}$. The observed item score for person p on item i is then U_{pi}, and the proficiency vector $\boldsymbol{\theta}_p$ for person p has elements θ_{pd} for $1 \leq d \leq D$. Treatment of estimation will be considered in terms of general results related to maximum likelihood in the context of exponential families.

4.4 Inference with Exponential Families

In practice, exponential families must be studied by use of data. For this purpose, exponential families have advantages in terms of simple sufficient statistics and simplified maximum likelihood estimation. In addition, in some cases, conditional inference may be applied effectively. In other cases, properties of exponential families may be applied even though the observed vectors are only functions of vectors with distributions in an exponential family or a distributed as in a mixture derived from exponential families. In this section, the applicability of common results for exponential families is explored within the context of IRT. The discussion here is somewhat simplified to the extent that it is confined to a fixed model and a large number of observations rather than to a model which becomes increasingly complex as the number of observations increases (Haberman, 1977a). In addition, problems of incorrect models are avoided even though useful large-sample approximations may still be available (Berk, 1972; Haberman, 1989).

4.4.1 Natural Parameters

In the most elementary case of inferences for an exponential family (Berk, 1972), S, $\mathbf{X}$, and $\mathbf{T}$ are defined as in Section 4.2, and the covariance matrix of $\mathbf{T}(\mathbf{X})$ is positive definite. For some positive integer P and some $\boldsymbol{\omega}$ in $\Omega(\mathbf{X}, \mathbf{T})$, observations $\mathbf{Y}_p$ with values in S, $1 \le p \le P$, are independent and identically distributed with common distribution $P(\boldsymbol{\omega}, \mathbf{X}, \mathbf{T})$. The statistic

$$\bar{\mathbf{T}}_P = P^{-1} \sum_{p=1}^{P} \mathbf{T}(\mathbf{Y}_p) \tag{4.34}$$

is sufficient for all inferences concerning $\boldsymbol{\omega}$. At $\boldsymbol{\omega}$, the log likelihood kernel ℓ_P per observation satisfies

$$\ell_P(\boldsymbol{\omega}) = \boldsymbol{\omega}'\bar{\mathbf{T}}_P - \mathrm{Cgf}(\boldsymbol{\omega}, \mathbf{X}, \mathbf{T}). \tag{4.35}$$

The gradient $\nabla \ell_P$ of ℓ_P satisfies

$$\nabla \ell_P(\boldsymbol{\omega}) = \bar{\mathbf{T}}_P - \boldsymbol{\mu}(\hat{\boldsymbol{\omega}}_P, \mathbf{X}, \mathbf{T}). \tag{4.36}$$

The Hessian matrix $\nabla^2 \ell_P$ of ℓ_P satisfies

$$\nabla^2 \ell_P(\boldsymbol{\omega}) = -\boldsymbol{\Sigma}(\hat{\boldsymbol{\omega}}_P, \mathbf{X}, \mathbf{T}). \tag{4.37}$$

It follows that ℓ_P is a strictly concave function. For any $\boldsymbol{\upsilon}$ in $\Omega(\mathbf{X}, \mathbf{T})$, the expected value of $\ell_P(\boldsymbol{\upsilon})$ is the Kullback–Leibler discriminant information

$$K(\boldsymbol{\omega}, \boldsymbol{\upsilon}) = \boldsymbol{\upsilon}'\boldsymbol{\mu}(\boldsymbol{\omega}, \mathbf{X}, \mathbf{T}) - \mathrm{Cgf}(\boldsymbol{\upsilon}, \mathbf{X}, \mathbf{T}) \tag{4.38}$$

for discrimination between the probability distributions $P(\boldsymbol{\omega}, \mathbf{X}, \mathbf{T})$ and $P(\boldsymbol{\upsilon}, \mathbf{X}, \mathbf{T})$ (Kullback and Leibler, 1951; see Chapter 7). One has $K(\boldsymbol{\omega}, \boldsymbol{\upsilon}) \ge 0$, with equality if, and only if, $\boldsymbol{\omega} = \boldsymbol{\upsilon}$.

The maximum likelihood estimate (MLE) $\hat{\boldsymbol{\omega}}_P$ of $\boldsymbol{\omega}$ exists if, and only if, $\bar{\mathbf{T}}_P$ is in $C^0(\mathbf{X}, \mathbf{T})$. If $\hat{\boldsymbol{\omega}}_P$ exists, then it is uniquely determined by the equation

$$\boldsymbol{\mu}(\hat{\boldsymbol{\omega}}_P, \mathbf{X}, \mathbf{T}) = \bar{\mathbf{T}}_P. \tag{4.39}$$

With probability 1, as P goes to ∞, $\hat{\boldsymbol{\omega}}_P$ is defined for all but a finite number of P and $\hat{\boldsymbol{\omega}}_P$ converges to $\boldsymbol{\omega}$. Thus $\hat{\boldsymbol{\omega}}_P$ is a strongly consistent estimator of $\boldsymbol{\omega}$. In addition, the normalized difference $P^{1/2}(\hat{\boldsymbol{\omega}}_P - \boldsymbol{\omega})$ converges in distribution to a random vector with a multivariate normal distribution $N(\mathbf{0}_q, [\boldsymbol{\Sigma}(\boldsymbol{\omega}, \mathbf{X}, \mathbf{T})]^{-1})$. The asymptotic covariance matrix $[\boldsymbol{\Sigma}(\boldsymbol{\omega}, \mathbf{X}, \mathbf{T})]^{-1}$ may then be estimated consistently by substitution of $\hat{\boldsymbol{\omega}}_P$ for $\boldsymbol{\omega}$.

Use of the Newton–Raphson algorithm is often quite straightforward. Let $\boldsymbol{\omega}_0$ be an initial approximation to $\hat{\boldsymbol{\omega}}_P$. Then the Newton–Raphson algorithm has the form

$$\boldsymbol{\omega}_{t+1} = \boldsymbol{\omega}_t + [\boldsymbol{\Sigma}(\boldsymbol{\omega}_t, \mathbf{X}, \mathbf{T})]^{-1}[\bar{\mathbf{T}}_p - \boldsymbol{\mu}(\boldsymbol{\omega}_t, \mathbf{X}, \mathbf{T})] \tag{4.40}$$

for all nonnegative integers t. Normally $\boldsymbol{\omega}_t$ converges to $\hat{\boldsymbol{\omega}}$. This result certainly holds for $\boldsymbol{\omega}_0$ sufficiently close to $\hat{\boldsymbol{\omega}}_P$. For any vector norm $|\bullet|$, constants $\tau_P \ge 0$ and $\epsilon_P > 0$ exist such that the quadratic convergence property

$$|\boldsymbol{\omega}_{t+1} - \hat{\boldsymbol{\omega}}_P| \le \tau_P |\boldsymbol{\omega}_t - \hat{\boldsymbol{\omega}}_P|^2, \tag{4.41}$$

holds whenever $|\boldsymbol{\omega}_t - \hat{\boldsymbol{\omega}}_P| < \epsilon_P$. The Newton–Raphson algorithm can be slightly modified to ensure convergence and to preserve quadratic convergence (Haberman, 1974, Chapter 3); however, such modifications are not generally required for the case of estimation of natural parameters.

4.4.2 Parametric Models for Natural Parameters

Unfortunately the very simple results associated with inferences on the natural parameter are not readily applied to IRT. Nonetheless, if $\mathbf{Q}$ is defined as in Section 4.2, and $\boldsymbol{\omega} = \mathbf{Q}(\boldsymbol{\gamma})$ for a unique $\boldsymbol{\gamma}$ in Γ, then some useful results are worth note that can be applied to an IRT model in which the distribution of $\boldsymbol{\theta}$ is assumed to be known or to be generated from a member of an exponential family. Assume that $\mathbf{Q}(\boldsymbol{\gamma}')$ only converges to $\mathbf{Q}(\boldsymbol{\gamma})$ for $\boldsymbol{\gamma}'$ in Γ that converges to $\boldsymbol{\gamma}$. The statistic $\bar{\mathbf{T}}_p$ remains sufficient for all inferences concerning $\boldsymbol{\omega}$ and $\boldsymbol{\gamma}$. Assume that $\nabla\mathbf{Q}(\boldsymbol{\gamma})$ has rank m. By the chain rule, the gradient $\nabla\ell_{\mathbf{Q}P}$ of $\ell_{\mathbf{Q}P} = \ell_P(\mathbf{Q})$ at $\boldsymbol{\gamma}$ is

$$\nabla\ell_{\mathbf{Q}P}(\boldsymbol{\gamma}) = \nabla\mathbf{Q}(\boldsymbol{\gamma})\nabla\ell_P(\mathbf{Q}(\boldsymbol{\gamma})). \tag{4.42}$$

The formula for the Hessian matrix $\nabla^2\ell_{\mathbf{Q}P}$ of $\ell_{\mathbf{Q}P}$ is somewhat more complicated than the formula for the Hessian matrix of ℓ_P. For an array $\mathbf{A}$ of q matrices $\mathbf{A}_k$ with m rows and m columns, $1 \le k \le q$, and a vector $\mathbf{b}$ of dimension q with elements b_k for $1 \le k \le q$, let $\mathbf{Ab}$ be $\sum_{k=1}^{q} b_k\mathbf{A}_k$. Let the information matrix at $\boldsymbol{\gamma}$ for $\ell_{\mathbf{Q}P}$ be

$$\mathbf{I}(\boldsymbol{\gamma},\mathbf{X},\mathbf{T},\mathbf{Q}) = \nabla\mathbf{Q}(\boldsymbol{\gamma})\boldsymbol{\Sigma}(\mathbf{Q}(\boldsymbol{\omega}),\mathbf{X},\mathbf{T})[\nabla\mathbf{Q}(\boldsymbol{\gamma})]'. \tag{4.43}$$

Then the Hessian matrix $\nabla^2\ell_{\mathbf{Q}P}$ of $\ell_P(\mathbf{Q})$ has value at $\boldsymbol{\gamma}$ of

$$\nabla^2\ell_{\mathbf{Q}P}(\boldsymbol{\gamma}) = -\mathbf{I}(\boldsymbol{\gamma},\mathbf{X},\mathbf{T},\mathbf{Q}) + \nabla^2\mathbf{Q}(\boldsymbol{\gamma})[\bar{\mathbf{T}}_P - \boldsymbol{\mu}(\mathbf{Q}(\boldsymbol{\gamma}),\mathbf{X},\mathbf{T})]. \tag{4.44}$$

Any MLE $\hat{\boldsymbol{\gamma}}_P$ of $\boldsymbol{\gamma}$ and any corresponding MLE $\hat{\boldsymbol{\omega}}_P = \mathbf{Q}(\hat{\boldsymbol{\gamma}}_P)$ of $\boldsymbol{\omega}$ must satisfy the equation

$$\nabla\mathbf{Q}(\hat{\boldsymbol{\gamma}}_P)\boldsymbol{\mu}(\hat{\boldsymbol{\omega}}_P,\mathbf{X},\mathbf{T}) = \nabla\mathbf{Q}(\hat{\boldsymbol{\gamma}}_P)\bar{\mathbf{T}}_P. \tag{4.45}$$

With probability 1, $\hat{\boldsymbol{\gamma}}_P$ is uniquely defined for all but a finite number of P, $\hat{\boldsymbol{\gamma}}_P$ converges to $\boldsymbol{\gamma}$, and $\hat{\boldsymbol{\omega}}_P$ converges to $\boldsymbol{\omega}$. In addition, $P^{1/2}(\hat{\boldsymbol{\omega}}_P - \boldsymbol{\omega})$ converges in distribution to a random vector with a multivariate normal distribution with mean $\mathbf{0}_m$ and covariance matrix $[\mathbf{I}(\boldsymbol{\gamma},\mathbf{X},\mathbf{T},\mathbf{Q})]^{-1}$. The estimated information matrix $\mathbf{I}(\hat{\boldsymbol{\gamma}}_P,\mathbf{X},\mathbf{T},\mathbf{Q})$ converges with probability 1 to the information matrix $\mathbf{I}(\boldsymbol{\gamma},\mathbf{X},\mathbf{T},\mathbf{Q})$; however, computation of $\mathbf{I}(\boldsymbol{\gamma},\mathbf{X},\mathbf{T},\mathbf{Q})$ is often inconvenient in the case of IRT. It is often worth noting that $-\nabla^2\ell_{\mathbf{Q}P}(\hat{\boldsymbol{\gamma}}_P)$ also converges to the information matrix $\mathbf{I}(\boldsymbol{\gamma},\mathbf{X},\mathbf{T},\mathbf{Q})$ with probability 1. The Louis (Louis, 1982) approach provides an added alternative that is often attractive in IRT applications. For each p from 1 to P, let

$$\nabla\ell_{\mathbf{Q}[p]}(\boldsymbol{\gamma}) = \nabla\mathbf{Q}(\boldsymbol{\gamma})[\mathbf{T}(\mathbf{Y}_p - \boldsymbol{\mu}(\mathbf{Q}(\boldsymbol{\gamma}),\mathbf{X},\mathbf{T})]. \tag{4.46}$$

Let

$$\mathbf{I}_{LP}(\boldsymbol{\gamma},\mathbf{X},\mathbf{T},\mathbf{Q}) = P^{-1}\sum_{p=1}^{P}\nabla\ell_{\mathbf{Q}[p]}(\boldsymbol{\gamma})[\nabla\ell_{\mathbf{Q}[p]}(\boldsymbol{\gamma})]'. \tag{4.47}$$

Then $\mathbf{I}_{LP}(\hat{\boldsymbol{\gamma}}_P,\mathbf{X},\mathbf{T},\mathbf{Q})$ converges to $\mathbf{I}(\boldsymbol{\gamma},\mathbf{X},\mathbf{T},\mathbf{Q})$ with probability 1.

If $\nabla^2\ell_{\mathbf{Q}P}$ is always negative definite, then application of the Newton–Raphson algorithm is generally straightforward. A starting approximation $\boldsymbol{\gamma}_0$ for $\hat{\boldsymbol{\gamma}}_P$ generates the sequence of approximations

$$\boldsymbol{\gamma}_{t+1} = \boldsymbol{\gamma}_t + [-\nabla^2\ell_{\mathbf{Q}P}(\boldsymbol{\omega}_t)]^{-1}\nabla_{\mathbf{Q}P}(\boldsymbol{\gamma}_t) \tag{4.48}$$

for integers $t \geq 0$. Convergence properties are quite simiilar to those found in the estimation of a natural parameter without restrictions. Two alternatives based on the Newton–Raphson algorithm are worth note. In the scoring algorithm (Fisher, 1925), $\mathbf{I}(\boldsymbol{\gamma}_t, \mathbf{X}, \mathbf{T}, \mathbf{Q})$ replaces $-\nabla^2 \ell_{\mathbf{Q}P}(\boldsymbol{\gamma}_t)$. This substitution only has impact if $\mathbf{Q}$ is not a linear function. Scoring is normally employed to simplify computations; however, this situation often does not apply in IRT. When the scoring algorithm and Newton–Raphson algorithm do not coincide, then constants $\tau_P \geq 0$, and $\epsilon_P > 0$ exist such that the linear convergence property

$$|\boldsymbol{\gamma}_{t+1} - \hat{\boldsymbol{\gamma}}_P| \leq \tau_P |\boldsymbol{\gamma}_t - \hat{\boldsymbol{\gamma}}_P|, \tag{4.49}$$

holds whenever $|\boldsymbol{\omega}_t - \hat{\boldsymbol{\omega}}_P| < \epsilon_P$. If $\boldsymbol{\gamma}_0$ is a random vector such that $P^{1/2}(\boldsymbol{\gamma}_0 - \boldsymbol{\gamma})$ converges in distribution to a multivariate normal random vector with mean $\mathbf{0}_r$, then τ_P and ϵ_P can be selected so that τ_P is a random variable such that $P^{1/2}\tau_P$ converges in distribution to some random variable. Thus, convergence is quite rapid in large samples. As in the Newton–Raphson algorithm, the scoring algorithm can be slighted modified to ensure convergence and to preserve rapid convergence in large sample (Haberman, 1974, Chapter 3); however, such modifications are not generally required for the case of estimation of natural parameters. A variation of the scoring algorithm that is valuable in IRT involves the Louis approach in which $\mathbf{I}_{LP}(\boldsymbol{\gamma}_t, \mathbf{X}, \mathbf{Q})$ replaces $-\nabla^2 \ell_{\mathbf{Q}P}(\boldsymbol{\gamma}_t)$. Results are similar to those for the scoring algorithm.

If $\nabla^2 \ell_{\mathbf{Q}P}$ need not be negative definite, then analysis is somewhat more complicated due to problems of parameter identification. Given that $\mathbf{I}(\boldsymbol{\gamma}, \mathbf{X}, \mathbf{T}, \mathbf{Q})$ is positive definite, large-sample approximations and convergence properties are typically rather similar to those for the case of negative-definite Hessian matrices; however, much more care is commonly required for computations (Haberman, 1988).

Because of the relationship of random vectors with values in a finite set to log-linear models that was noted in Section 4.2.1, it is normally possible to describe IRT models in which the distribution of $\boldsymbol{\theta}$ is known or generated from a known exponential family in terms of some random vector $\mathbf{X}$ and some functions $\mathbf{T}$ and $\mathbf{Q}$. The independent observations $\mathbf{U}_p$, $1 \leq p \leq P$, then have common distribution $P(\mathbf{Q}(\boldsymbol{\gamma}), \mathbf{X}, \mathbf{T})$ for some $\boldsymbol{\gamma}$ in Γ. The nontrivial complication involves the identification requirements for $\mathbf{Q}$. In typical cases, rigorous verification that these conditions are met appears to be quite difficult. The conditions definitely cannot hold if m is not less than $N(S)$, the number of members of S (Haberman, 2005). For the 2PL example in Section 4.3 with θ_1 with a standard normal distribution, $2n$ must be less than 2^n, so that $n > 2$. In the 3PL case with constant guessing parameter, $2n + 1$ must be less than 2^n, so that $n > 2$.

4.4.3 Functions of Random Vectors

In the case of functions of random vectors, maximum likelihood results involve differences between conditional and unconditional moments. In many typical cases, log likelihood functions are not strictly concave, so that both large-sample theory and numerical work is more difficult than in the case of direct observation of an exponential family. Under the conditions of Section 4.2.3, consider the case of independent vectors $\mathbf{Y}_p$, $1 \leq p \leq P$, with common distribution $P_M(\mathbf{Q}(\boldsymbol{\gamma}, \mathbf{X}, \mathbf{T}, \mathbf{h})$. Then, the log likelihood kernel at $\boldsymbol{\gamma}$ per observation is

$$\ell_{\mathbf{Qh}P}(\boldsymbol{\gamma}) = P^{-1} \sum_{p=1}^{P} \ell_{Qh[p]}(\boldsymbol{\gamma}), \tag{4.50}$$

where

$$\ell_{\mathbf{Qh}[p]}(\boldsymbol{\gamma}) = \mathrm{Cgf}(\mathbf{Q}(\boldsymbol{\gamma}), \mathbf{X}[\mathbf{h} = \mathbf{Y}_p], \mathbf{T}) - \mathrm{Cgf}(\mathbf{Q}(\boldsymbol{\gamma}), \mathbf{X}, \mathbf{T}). \tag{4.51}$$

The gradient of $\ell_{\mathbf{Qh}P}$ at $\boldsymbol{\gamma}$ is then

$$\nabla \ell_{\mathbf{Qh}P}(\boldsymbol{\gamma}) = P^{-1} \sum_{p=1}^{P} \nabla \ell_{\mathbf{Qh}[p]}(\boldsymbol{\gamma}), \tag{4.52}$$

where

$$\nabla \ell_{\mathbf{Qh}[p]}(\boldsymbol{\gamma}) = \nabla \mathbf{Q}(\boldsymbol{\gamma})[\boldsymbol{\mu}(\mathbf{Q}(\boldsymbol{\gamma}), \mathbf{X}[\mathbf{h} = \mathbf{Y}_p], \mathbf{T}) - \boldsymbol{\mu}(\mathbf{Q}(\boldsymbol{\gamma}), \mathbf{X}, \mathbf{T})] \tag{4.53}$$

and the Hessian matrix of $\ell_{\mathbf{Qh}P}$ at $\boldsymbol{\gamma}$ is then

$$\nabla^2 \ell_{\mathbf{Qh}P}(\boldsymbol{\gamma}) = P^{-1} \sum_{p=1}^{P} \nabla^2 \ell_{\mathbf{Qh}[p]}(\boldsymbol{\gamma}), \tag{4.54}$$

where

$$\begin{aligned} \nabla^2 \ell_{\mathbf{Qh}[p]}(\boldsymbol{\gamma}) = {} & \nabla \mathbf{Q}(\boldsymbol{\gamma})[\boldsymbol{\Sigma}(\mathbf{Q}(\boldsymbol{\gamma}), \mathbf{X}[\mathbf{h} = \mathbf{Y}_p], \mathbf{T}) - \boldsymbol{\Sigma}(\mathbf{Q}(\boldsymbol{\gamma}), \mathbf{X}, \mathbf{T})][\nabla \mathbf{Q}(\boldsymbol{\gamma})]' \\ & + \nabla^2 \mathbf{Q}(\boldsymbol{\gamma})[\boldsymbol{\mu}(\mathbf{Q}(\boldsymbol{\gamma}), \mathbf{X}[\mathbf{h} = \mathbf{Y}_p], \mathbf{T}) - \boldsymbol{\mu}(\mathbf{Q}(\boldsymbol{\gamma}), \mathbf{X}, \mathbf{T})] \end{aligned} \tag{4.55}$$

(Sundberg, 1974). The information matrix at $\boldsymbol{\gamma}$ is the expectation $\mathbf{I}_M(\boldsymbol{\gamma}, \mathbf{X}, \mathbf{T}, \mathbf{Q}, \mathbf{h})$ of

$$\nabla \mathbf{Q}(\boldsymbol{\gamma})[\boldsymbol{\Sigma}(\mathbf{Q}(\boldsymbol{\gamma}), \mathbf{X}, \mathbf{T}) - \boldsymbol{\Sigma}(\mathbf{Q}(\boldsymbol{\gamma}), \mathbf{X}[\mathbf{h} = \mathbf{h}(\mathbf{Y}(\mathbf{Q}(\boldsymbol{\gamma}), \mathbf{X}, \mathbf{T})], \mathbf{T})][\nabla \mathbf{Q}(\boldsymbol{\gamma})].$$

Equivalently $\mathbf{I}_M(\boldsymbol{\gamma}, \mathbf{X}, \mathbf{T}, \mathbf{Q}, \mathbf{h})$ is the covariance matrix of

$$\nabla \mathbf{Q}(\boldsymbol{\gamma})[\boldsymbol{\mu}(\mathbf{Q}(\boldsymbol{\gamma}), \mathbf{X}[\mathbf{h} = \mathbf{h}(\mathbf{Y}(\mathbf{Q}(\boldsymbol{\gamma}), \mathbf{X}, \mathbf{T}))], \mathbf{T}) - \boldsymbol{\mu}(\mathbf{Q}(\boldsymbol{\gamma}), \mathbf{X}, \mathbf{T})].$$

For $\boldsymbol{\upsilon}$ in Γ, the Kullback–Leibler information for discrimination between the distributions $P_M(\mathbf{Q}(\boldsymbol{\omega}), \mathbf{X}, \mathbf{T}, \mathbf{h})$ and $P_M(\boldsymbol{\upsilon}, \mathbf{X}, \mathbf{T}, \mathbf{h})$ is

$$\begin{aligned} K_{\mathbf{Qh}} M(\boldsymbol{\gamma}, \boldsymbol{\omega}) = {} & E(\mathrm{Cgf}(\mathbf{Q}(\boldsymbol{\upsilon}), \mathbf{X}[\mathbf{h} = \mathbf{h}(\mathbf{Y}(\mathbf{Q}(\boldsymbol{\gamma}, \mathbf{X}, \mathbf{T}))], \mathbf{T}) \\ & - \mathrm{Cgf}(\mathbf{Q}(\boldsymbol{\upsilon}), \mathbf{X}, \mathbf{T}). \end{aligned} \tag{4.56}$$

This information is nonnegative, and it is 0 if $\boldsymbol{\gamma} = \boldsymbol{\upsilon}$. A basic requirement for satisfactory large-sample results is that $K_{\mathbf{Qh}}(\boldsymbol{\gamma}, \boldsymbol{\upsilon})$ only approaches 0 if $\boldsymbol{\upsilon}$ approaches $\boldsymbol{\gamma}$.

Any MLE $\hat{\boldsymbol{\gamma}}_P$ of $\boldsymbol{\gamma}$ must satisfy the equation

$$\nabla \mathbf{Q}(\hat{\boldsymbol{\gamma}}_P) P^{-1} \sum_{p=1}^{P} \boldsymbol{\mu}(\mathbf{Q}(\hat{\boldsymbol{\gamma}}_P), \mathbf{X}[\mathbf{h} = \mathbf{Y}_p], \mathbf{T}) = \nabla \mathbf{Q}(\hat{\boldsymbol{\gamma}}_P) \boldsymbol{\mu}(\mathbf{Q}(\hat{\boldsymbol{\gamma}}_P), \mathbf{X}, \mathbf{T}). \tag{4.57}$$

Note that the equation involves a comparison of conditional and unconditional expectations of the statistic $\mathbf{T}(\mathbf{Y}_p)$.

Large-sample approximations are a bit more complex in this case (Sundberg, 1974). In addition to the condition on the Kullback–Leibler information, the requirement is also imposed that the Fisher information $\mathbf{I}_M(\boldsymbol{\gamma}, \mathbf{X}, \mathbf{T}, \mathbf{Q}, \mathbf{h})$ be positive definite. Unless $\ell_{\mathbf{Qh}}$ is always concave, some further conditions are required to ensure that some closed and bounded subset D of Γ exists such that the probability is one that $\hat{\boldsymbol{\gamma}}_P$ is in D for all but a finite number of P (Haberman, 1989). The latter condition on $\hat{\boldsymbol{\gamma}}_P$ can always be ensured by restriction of Γ to a sufficiently small open set that contains $\boldsymbol{\gamma}$; however, the restriction may be unattractive. The standard result is then that the probability approaches 1 that $\hat{\boldsymbol{\gamma}}_P$ is uniquely defined and converges to $\boldsymbol{\gamma}$ and $P^{1/2}(\hat{\boldsymbol{\gamma}}_P - \boldsymbol{\gamma})$ converges in distribution to a random variable with a multivariate normal distribution with mean $\mathbf{0}_r$ and covariance matrix $[\mathbf{I}_M(\boldsymbol{\gamma}, \mathbf{X}, \mathbf{T}, \mathbf{Q}, \mathbf{h})]^{-1}$. Estimation of the Fisher information matrix $\mathbf{I}_M(\boldsymbol{\gamma}, \mathbf{X}, \mathbf{T}, \mathbf{Q}, \mathbf{h})$ may then be accomplished as in Section 4.4.2.

As in Section 4.4.2, versions of the Newton–Raphson or scoring algorithm may be employed for computations, and the use of the Louis approximation can also provide a computational alternative to the scoring algorithm (see Chapter 11). One somewhat different option, often called the EM algorithm, is also available (Sundberg, 1974; Dempster et al., 1977; see Chapter 12). This option begins with an initial approximation $\boldsymbol{\gamma}_0$ to $\hat{\boldsymbol{\gamma}}_P$. For step $t \geq 0$ of the algorithm, there is an expectation (E) step and a maximization (M) step. Hence, the name EM comes from the combination of the E and M steps. In the expectation step,

$$\mathbf{T}_{tP} = P^{-1} \sum_{p=1}^{P} \boldsymbol{\mu}(\boldsymbol{\gamma}_t, \mathbf{X}[\mathbf{h} = \mathbf{Y}_p], \mathbf{T}) \tag{4.58}$$

is computed. In the maximization step, $\boldsymbol{\gamma}_t$ is obtained by substitution of $\mathbf{T}_{tP}$ for $\bar{\mathbf{T}}_P$ in a procedure to maximize $\ell_{\mathbf{Q}P}$. Thus the function $\ell_{\mathbf{Q}tP}$ maximized at the expectation step has value

$$\ell_{\mathbf{Q}tP}(\boldsymbol{\upsilon}) = [\mathbf{Q}(\boldsymbol{\upsilon})]'\mathbf{T}_{tP} - \mathrm{Cgf}(\mathbf{Q}(\boldsymbol{\upsilon}), \mathbf{X}, \mathbf{T}) \tag{4.59}$$

for $\boldsymbol{\upsilon}$ in Γ. The EM algorithm is quite stable; however, it does not provide the rapid convergence results associated with the Newton–Raphson, scoring, and Louis algorithms. Let F be the set of $\boldsymbol{\upsilon}$ in Γ such that $\ell_{\mathbf{Qh}P}(\boldsymbol{\upsilon})$ is at least $\ell_{\mathbf{Qh}P}(\boldsymbol{\gamma}_0)$. Let F be a closed and bounded subset of Γ, and let $\nabla\ell_{\mathbf{Qh}P}(\boldsymbol{\upsilon})$ be $\mathbf{0}_m$ for $\boldsymbol{\upsilon}$ in F only if $\boldsymbol{\upsilon} = \hat{\boldsymbol{\gamma}}_P$. Then, $\boldsymbol{\gamma}_t$ converges to $\hat{\boldsymbol{\gamma}}_P$ (Haberman, 1977b). In typical cases, all that can be shown concerning the rate of convergence is that for some vector norm $|\bullet|$ and some $\tau > 0$, $|\boldsymbol{\gamma}_{t+1} - \hat{\boldsymbol{\gamma}}_P|$ is less than $\tau|\boldsymbol{\gamma}_t - \hat{\boldsymbol{\gamma}}_P|$ for t sufficiently large. A version of the EM algorithm is very commonly encountered in IRT (Bock and Aitkin, 1981); however, the details are not quite the same as those presented here due to the use of mixtures. For more discussion of this issue, see Section 4.4.5.

4.4.4 Conditional Distributions

Conditional maximum likelihood is quite commonly encountered in IRT in cases based on the conditions of Section 4.2.4. Conditional maximum likelihood is relatively straightforward in terms of conditional likelihood equations and large-sample results. Numerical work is relatively straightforward.

Let $\mathbf{Y}_p$, $1 \leq p \leq P$, be independent and identically distributed random variables with values in S. Assume that the conditional distribution of $\mathbf{T}_1(\mathbf{Y}_p)$ given $\mathbf{T}_2(\mathbf{Y}_p)$ is the same as the conditional distribution of $\mathbf{T}_1(\mathbf{Y}(\boldsymbol{\omega}, \mathbf{X}, \mathbf{T}))$ given $\mathbf{T}_2(\mathbf{Y}(\boldsymbol{\omega}, \mathbf{X}, \mathbf{T}))$ for some $\boldsymbol{\omega}$ in $\Omega(\mathbf{X}, \mathbf{T})$.

Let $\boldsymbol{\omega} = \boldsymbol{\omega}_1 \# \boldsymbol{\omega}_2$, where $\boldsymbol{\omega}_k$ has dimension fq_k for k equal 1 or 2. The conditional log likelihood kernel per observation then has value at $\boldsymbol{\omega}_1$ of

$$\ell_{CP}(\boldsymbol{\omega}_1) = \boldsymbol{\omega}'\bar{T}_{1P} - P^{-1}\sum_{p=1}^{P} \mathrm{Cgf}(\boldsymbol{\omega}_1, \mathbf{X}[\mathbf{T}_2 = \mathbf{T}_2(\mathbf{Y}_p)], \mathbf{T}_1). \tag{4.60}$$

A conditional maximum-likelihood estimate (CMLE) $\hat{\boldsymbol{\omega}}_{1P}$ of $\boldsymbol{\omega}_1$ is a member of $\Omega(\mathbf{X}, \mathbf{T}_1|\mathbf{T}_2)$ that maximizes ℓ_{CP}. The gradient of ℓ_{CP} at $\boldsymbol{\omega}_1$ is

$$\nabla\ell_{CP}(\boldsymbol{\omega}_1) = \bar{T}_{1P} - P^{-1}\sum_{p=1}^{P} \boldsymbol{\mu}(\boldsymbol{\omega}_1, \mathbf{X}[\mathbf{T}_2 = \mathbf{T}_2(\mathbf{Y}_p)], \mathbf{T}_1) \tag{4.61}$$

and the corresponding Hessian matrix is

$$\nabla^2\ell_{CP}(\boldsymbol{\omega}_1) = P^{-1}\sum_{p=1}^{P} \boldsymbol{\Sigma}(\boldsymbol{\omega}_1, \mathbf{X}[\mathbf{T}_2 = \mathbf{T}_2(\mathbf{Y}_p)], \mathbf{T}_1). \tag{4.62}$$

The function ℓ_{CP} is concave, and any CMLE $\hat{\boldsymbol{\omega}}_{1P}$ satisfies

$$P^{-1}\sum_{p=1}^{P} \boldsymbol{\mu}(\hat{\boldsymbol{\omega}}_{1P}, \mathbf{X}[\mathbf{T}_2 = \mathbf{T}_2(\mathbf{Y}_p)], \mathbf{T}_1) = \bar{T}_{1P}. \tag{4.63}$$

To simplify large-sample results, assume that $\boldsymbol{\Sigma}(\boldsymbol{\omega}', \mathbf{X}[\mathbf{T}_2 = \mathbf{T}_2(\mathbf{Y}_p)], \mathbf{T}_1)$ has a finite expectation $\mathbf{J}(\boldsymbol{\omega}')$ for $\boldsymbol{\omega}'$ in $\Omega(\mathbf{X}, \mathbf{T}_1|\mathbf{T}_2)$. This assumption is certainly true if the common distribution of the $\mathbf{Y}_p$ is $\mathbf{Y}(\boldsymbol{\omega}_1\#\boldsymbol{\omega}_2, \mathbf{X}, \mathbf{T})$ for some $\boldsymbol{\omega}_2$ in $\Omega(\mathbf{X}, \mathbf{T}_1|\mathbf{T}_2)$. Let $\mathbf{J}(\boldsymbol{\omega}_1)$ be positive definite. Then, the probability is 1 that $\hat{\boldsymbol{\omega}}_{1P}$ is uniquely defined for all but a finite number of P and $\hat{\boldsymbol{\omega}}_{1P}$ converges to $\boldsymbol{\omega}_1$. In addition, $P^{1/}(\hat{\boldsymbol{\omega}}_{1P} - \boldsymbol{\omega}_1)$ converges in distribution to a multivariate normal random variable with mean $\mathbf{0}_{q_1}$ and covariance matrix $[\mathbf{J}(\boldsymbol{\omega}_1)]^{-1}$ (Andersen, 1970; Haberman, 1989). In typical cases, computations may be performed with the Newton–Raphson algorithm, and estimation of the asymptotic covariance is a straightforward matter. These results readily apply to such IRT models as the Rasch model and the partial credit model.

4.4.5 Mixtures

Inferences for the mixture models of Section 4.2.5 are similar to inferences in Section 4.4.3 for functions of random vectors that have distributions in exponential families; however, the details are a bit more complicated. This case is basic in IRT because marginal estimation models involve mixtures. Let $\mathbf{Y}_p$, $1 \le p \le P$, be independent random vectors with values in $\mathbf{h}_2(S_2)$ and common distribution $P_{MI1}(\boldsymbol{\gamma}, \mathbf{X}, \mathbf{T}, \mathbf{Q}, \mathbf{h})$. The log likelihood kernel per observation has value at $\boldsymbol{\gamma}$ of

$$\ell_{MIP}(\boldsymbol{\gamma}) = P^{-1}\sum_{p=1}^{P} \ell_{MI[p]}(\boldsymbol{\gamma}), \tag{4.64}$$

where

$$\ell_{MI[p]}(\boldsymbol{\gamma}) = \log f_{MI1}(\mathbf{Y}_p; \boldsymbol{\gamma}, \mathbf{X}, \mathbf{T}, \mathbf{Q}, \mathbf{h}). \tag{4.65}$$

The gradient of ℓ_{MIP} at $\boldsymbol{\gamma}$ is then

$$\nabla \ell_{MIP}(\boldsymbol{\gamma}) = P^{-1} \sum_{p=1}^{P} \nabla \ell_{MI[p]}(\boldsymbol{\gamma}), \tag{4.66}$$

where the gradient $\nabla \ell_{MI[p]}(\boldsymbol{\gamma})$ of ℓ_{MIP} at $\boldsymbol{\gamma}$ is $\nabla \ell_{MI[p]}(\boldsymbol{\gamma})$. To determine $\nabla \ell_{MI[p]}(\boldsymbol{\gamma})$, for $\mathbf{u}_1$ in $\mathbf{h}_1(S_1)$, $\mathbf{u}_2$ in $\mathbf{h}_2(S_2)$, and $\mathbf{u} = \mathbf{u}_1\#\mathbf{u}_2$, let

$$\begin{aligned} \mathbf{z}(\mathbf{u}; \boldsymbol{\gamma}) = {} & \nabla_1 \mathbf{Q}(\boldsymbol{\gamma}\#\mathbf{u}_2)[\boldsymbol{\mu}(\mathbf{Q}(\boldsymbol{\gamma}\#\mathbf{u}_2), \mathbf{X}[\mathbf{h} = \mathbf{u}], \mathbf{T}) \\ & - \boldsymbol{\mu}(\mathbf{Q}(\boldsymbol{\gamma}\#\mathbf{u}_2), \mathbf{X}, \mathbf{T})]. \end{aligned} \tag{4.67}$$

Let $\mathbf{U}_k$ in $\mathbf{h}_k(S_k)$, $1 \le k \le 2$, be random vectors such that $\mathbf{U} = \mathbf{U}_1\#\mathbf{U}_2$ has the same distribution as $\mathbf{Y}_{MI}(\boldsymbol{\gamma}, \mathbf{X}, \mathbf{T}, \mathbf{Q}, \mathbf{h})$. Let $\mathbf{z}_1(\mathbf{u}_1; \boldsymbol{\gamma})$ be the conditional expectation of $\mathbf{z}(\mathbf{U}; \boldsymbol{\gamma})$ given $\mathbf{U}_1 = \mathbf{u}_1$. Then the gradient $\nabla \ell_{MI[p]}(\boldsymbol{\gamma})$ is $\mathbf{z}_1(\mathbf{Y}_p; \boldsymbol{\gamma})$. The information matrix $\mathbf{I}_{MI1}(\boldsymbol{\gamma}, \mathbf{X}, \mathbf{T}, \mathbf{Q}, \mathbf{h})$ is then the covariance matrix of $\mathbf{z}_1(\mathbf{U}_1; \boldsymbol{\gamma})$. The Hessian matrix of ℓ_{MIP} at $\boldsymbol{\gamma}$ is then

$$\nabla^2 \ell_{MIP}(\boldsymbol{\gamma}) = P^{-1} \sum_{p=1}^{P} \nabla^2 \ell_{MI[p]}(\boldsymbol{\gamma}), \tag{4.68}$$

where the Hessian $\nabla^2 \ell_{MI[p]}$ of ℓ_{MI} at $\boldsymbol{\gamma}$ is $\nabla^2 \ell_{MI[p]}(\boldsymbol{\gamma})$. To evaluate this Hessian matrix, let

$$\begin{aligned} \mathbf{Z}(\mathbf{u}; \boldsymbol{\gamma}) = {} & \nabla_1^2 \mathbf{Q}(\boldsymbol{\gamma}\#\mathbf{u}_2)[\boldsymbol{\mu}(\mathbf{Q}(\boldsymbol{\gamma}\#\mathbf{u}_2), \mathbf{X}[\mathbf{h} = \mathbf{u}], \mathbf{T}) - \boldsymbol{\mu}(\mathbf{Q}(\boldsymbol{\gamma}\#\mathbf{u}_2), \mathbf{X}, \mathbf{T})] \\ & + \mathbf{Q}(\boldsymbol{\gamma}\#\mathbf{u}_2)[\boldsymbol{\Sigma}(\mathbf{Q}(\boldsymbol{\gamma}\#\mathbf{u}_2), \mathbf{X}[\mathbf{h} = \mathbf{u}], \mathbf{T}) \\ & - \boldsymbol{\Sigma}(\mathbf{Q}(\boldsymbol{\gamma}\#\mathbf{u}_2), \mathbf{X}, \mathbf{T})][\mathbf{Q}(\boldsymbol{\gamma}\#\mathbf{u}_2)]' \\ & + [\mathbf{z}(\mathbf{u}; \boldsymbol{\gamma}) - \mathbf{z}_1(\mathbf{u}_1; \boldsymbol{\gamma})][\mathbf{z}(\mathbf{u}; \boldsymbol{\gamma}) - \mathbf{z}_1(\mathbf{u}_1; \boldsymbol{\gamma})]' \end{aligned} \tag{4.69}$$

and let $\mathbf{Z}_1(\mathbf{u}_1; \boldsymbol{\gamma})$ be the conditional expectation of $\mathbf{Z}(\mathbf{U}; \boldsymbol{\gamma})$ given $\mathbf{U}_1 = \mathbf{u}_1$. Then $\nabla^2 \ell_{MI[p]}(\boldsymbol{\gamma})$ is $\mathbf{Z}_1(\mathbf{Y}_p; \boldsymbol{\gamma})$.

Assume that the information matrix $\mathbf{I}_{MI1}(\boldsymbol{\gamma}, \mathbf{X}, \mathbf{T}, \mathbf{Q}, \mathbf{h})$ is positive definite. An MLE $\hat{\boldsymbol{\gamma}}_P$ in Γ for $\boldsymbol{\gamma}$ is defined to maximize ℓ_{MI}. Under conditions somewhat similar to those required in Section 4.4.3, the probability is 1 that $\hat{\boldsymbol{\gamma}}_P$ is uniquely defined for all but a finite number of P and converges to $\boldsymbol{\gamma}$ and $P^{1/2}(\hat{\boldsymbol{\gamma}}_P - \boldsymbol{\gamma})$ converges in distribution to a normal random vector with mean $\mathbf{0}_m$ and covariance matrix $[\mathbf{I}_{MI1}(\boldsymbol{\gamma}, \mathbf{X}, \mathbf{T}, \mathbf{Q}, \mathbf{h})]^{-1}$. Development of the Newton–Raphson, scoring, and Louis algorithms is straightforward, but the EM algorithm requires attention. The commonly used version (Bock and Aitkin, 1981) involves a sequence of approximations $\boldsymbol{\gamma}_t$, $t \ge 0$, for $\hat{\boldsymbol{\gamma}}_P$. At stage t, let $w_t(\boldsymbol{\gamma}; \mathbf{u}_1)$ be the conditional expectation of $\log f_{MI}(\mathbf{u}_1\#\mathbf{U}_{2t}; \boldsymbol{\gamma}_t, \mathbf{X}, \mathbf{T}, \mathbf{Q}, \mathbf{h})$ given $\mathbf{U}_{1t} = \mathbf{u}_1$, where the random vectors $\mathbf{U}_{kt}$ are in $\mathbf{h}_k(S_k)$ for $1 \le k \le 2$ and $\mathbf{U}_t = \mathbf{U}_{1t}\#\mathbf{U}_{2t}$ has the same distribution as $\mathbf{Y}_{MI}(\boldsymbol{\gamma}_t, \mathbf{X}, \mathbf{T}, \mathbf{Q}, \mathbf{h})$. Then $\boldsymbol{\gamma}_{t+1}$ in Γ is found by maximization of $P^{-1} \sum_{p=1}^{P} w_t(\boldsymbol{\gamma}_{t+1}; \mathbf{Y}_p)$ for $\boldsymbol{\gamma}_{t+1}$ in Γ. Properties of $\boldsymbol{\gamma}_t$ are similar to those for the traditional EM algorithm described in Section 4.4.3.

4.5 Conclusion

Exponential families provide a basis for IRT. These families can be employed to describe conditional distributions of item responses given person parameters and the distributions of person parameters. In addition to the role of exponential families in description of distributions, these families also play a role in simplification of inferences, especially in the case of conditional inferences for Rasch, partial credit, and OPLM models.

The coverage here of the role of exponential families has been quite selective. It is worth considering some roads not taken. Item responses can be considered that have Poisson distributions, normal distributions, or Weibull distributions. It is not necessary to confine attention to polytomous responses. The examples have involved one-dimensional person parameters; however, much more complex cases are well known. Testing of hypotheses concerning models is a significant topic that can be examined in terms of exponential families. Measurement of model error may also be examined in the context of exponential families.

References

Adams, R. J., Wilson, M. R., and Wang, W. C. 1997. The multidimensional random coefficients multinomial logit model. *Applied Psychological Measurement, 21*, 1–23.

Andersen, E. B. 1970. Asymptotic properties of conditional maximum-likelihood estimators. *Journal of the Royal Statistical Society, Series B, 32*, 283–301.

Barndorff-Nielsen, O. 1978. *Information and Exponential Families in Statistical Theory*. Chichester: John Wiley.

Berk, R. H. 1972. Consistency and asymptotic normality of MLE's for exponential models. *Annals of Mathematical Statistics, 43*, 193–204.

Bliss, C. I. 1935. The calculation of the dosage-mortality curve. *Annals of Applied Biology, 22*, 134–167. (Appendix by R. A. Fisher.)

Bock, R. D. 1972. Estimating item parameters and latent ability when responses are scored in two or more nominal categories. *Psychometrika, 37*, 29–51.

Bock, R. D. and Aitkin, M. 1981. Marginal maximum likelihood estimation of item parameters: Application of an EM algorithm. *Psychometrika*, 46, 443–459.

Bock, R. D. and Lieberman, M. 1970. Fitting a response curve model for dichotomously scored items. *Psychometrika, 35*, 179–198.

Cox, D. R. and Snell, E. J. 1989. *Analysis of Binary Data* (2nd ed.). London: Chapman and Hall.

Dempster, A. P., Laird, N. M., and Rubin, D. B. 1977. Maximum likelihood from incomplete data via the EM algorithm (with discussion). *Journal of the Royal Statistical Society, Series B, 39*, 1–38.

Finney, D. J. 1971. *Probit Analysis* (3rd ed.). Cambridge: Cambridge University Press.

Fisher, R. A. 1925. Theory of statistical estimation. *Proceedings of the Cambridge Philosophical Society, 22*, 700–725.

Garwood, F. 1941. The application of maximum likelihood to dosage-mortality curves. *Biometrika, 32*, 46–58.

Haberman, S. J. 1974. *The Analysis of Frequency Data*. Chicago: University of Chicago Press.

Haberman, S. J. 1977a. Maximum likelihood estimates in exponential response models. *The Annals of Statistics, 5*, 815–841.

Haberman, S. J. 1977b. Product models for frequency tables involving indirect observation. *The Annals of Statistics, 5*, 1124–1147.

Haberman, S. J. 1978. *Analysis of Qualitative Data, Volume 1: Introductory Topics*. New York: Academic Press.

Haberman, S. J. 1979. *Analysis of Qualitative Data, Volume 2: New Developments*. New York: Academic Press.

Haberman, S. J. 1988. A stabilized Newton–Raphson algorithm for log-linear models for frequency tables derived by indirect observation. *Sociological Methodology, 18*, 193–211.

Haberman, S. J. 1989. Concavity and estimation. *The Annals of Statistics, 17*, 1631–1661.

Haberman, S. J. 2005. *Identifiability of Parameters in Item Response Models with Unconstrained Ability Distributions* (Research Rep. No. RR-05-24). Princeton, NJ: ETS.

Haberman, S. J., von Davier, M., and Lee, Y. 2008. *Comparison of Multi-Dimensional Item Response Models: Multivariate Normal Ability Distributions versus Multivariate Polytomous Distributions* (Research Rep. No. RR-08-45). Princeton, NJ: ETS.

Heinen, T. 1996. *Latent Class and Discrete Latent Trait Models*. Thousand Oaks, CA: Sage.

Kotz, S., Johnson, N. L., and Balakrishnan, N. 2000. *Continuous Multivariate Distributions: Models and Applications*. New York: John Wiley.

Kullback, S. and Leibler, R. A. 1951. On information and sufficiency. *Annals of Mathematical Statistics, 22*, 79–86.

Lazarsfeld, P. F. and Henry, N. W. 1968. *Latent Structure Analysis*. Boston: Houghton Mifflin.

Lehmann, E. L. 1959. *Testing Statistical Hypotheses*. New York: John Wiley.

Lord, F. M. and Novick, M. R. 1968. *Statistical Theories of Mental Test Scores*. Reading, MA: Addison-Wesley. (With contributions by A. Birnbaum.)

Louis, T. 1982. Finding the observed information matrix when using the EM algorithm. *Journal of the Royal Statistical Society, Series B, 44*, 226–233.

Masters, G. N. 1982. A Rasch model for partial credit scoring. *Psychometrika, 47*, 149–174.

McCullagh, P. and Nelder, J. A. 1989. *Generalized Linear Models* (2nd ed.). Boca Raton, FL: Chapman and Hall.

Morris, C. M. 1982. Natural exponential families with quadratic variance functions. *Annals of Statistics, 10*, 65–80.

Muraki, E. 1992. A generalized partial credit model: Application of an EM algorithm. *Applied Psychological Measurement, 16*, 159–176.

Rasch, G. 1960. *Probabilistic Models for Some Intelligence and Attainment Tests*. Copenhagen: Danish Institute for Educational Research.

Sundberg, R. 1974. Maximum likelihood theory for incomplete data from an exponential family. *Scandinavian Journal of Statistics, 1*, 49–58.

Verhelst, N. D. and Glas, C. A. W. 1995. The one parameter logistic model. In G. H. Fischer and I. W. Molenaar (Eds.), *Rasch Models: Foundations, Recent Developments, and Applications* (pp. 215–238). New York: Springer-Verlag.

5

Loglinear Models for Observed-Score Distributions

Tim Moses

CONTENTS

5.1 Introduction

Loglinear models have useful applications with the discrete distributions typically encountered in psychometrics. These applications are primarily focused on the estimation of observed-score distributions (Hanson, 1990; Holland and Thayer, 1987, 2000; Rosenbaum and Thayer, 1987). The test score distributions obtained from fitting loglinear models to sample distributions can be used in place of the sample distribution to improve the stability, interpretability, and smoothness of estimated univariate distributions, conditional distributions, and equipercentile equating functions (von Davier et al., 2004; Kolen and Brennan, 2004; Liou and Cheng, 1995; Livingston, 1993). This chapter reviews fundamental statistical concepts and applications of loglinear models for estimating observed-score distributions.

5.2 Assumptions and Estimation Properties of Loglinear Models

Descriptions of loglinear modeling for observed-score distributions usually begin with assumptions about the sample frequencies corresponding to the scores of a single test, X, with possible scores X_j, where $j = 1, \ldots, J$ (Holland and Thayer, 1987, 2000). The sample frequencies of X can be written as a transposed row vector, $\mathbf{n} = (n_1, \ldots, n_J)^t$, with J entries that sum to the total examinee sample size, N. When $\mathbf{n}$ is assumed to reflect N independent observations of discrete random variable X, then $\mathbf{n}$ is said to have a multinomial distribution with probability function

$$\mathrm{f}(\mathbf{n}) = \frac{N!}{\prod_j n_j!} \prod_j \pi_j^{n_j} \tag{5.1}$$

where the J population probabilities, π_j, form the J by 1 column vector $\boldsymbol{\pi}$. Each of the J π_js is assumed to be greater than 0, and all of the π_js are assumed to sum to 1. Under these assumptions, $\mathbf{n}$ has a mean vector $N\boldsymbol{\pi} = \mathbf{m}$ and a covariance matrix given by $Cov(\mathbf{n}) = N(\mathbf{D}_\pi - \boldsymbol{\pi}\boldsymbol{\pi}^t)$, where $\mathbf{D}_\pi$ is the J by J diagonal matrix of $\boldsymbol{\pi}$ and $\boldsymbol{\pi}^t$ denotes the 1 by J transposition of $\boldsymbol{\pi}$.

Loglinear models enter into discussions about multinomial assumptions and the observed-score distributions of tests when the natural log of the *jth* entry of $\boldsymbol{\pi}$ can be expressed as a polynomial function of the test scores

$$\ln(\pi_j) = \sum_{r=0}^{R} X_j^r \beta_r = \mathbf{X}_j \boldsymbol{\beta} \tag{5.2}$$

In Equation 5.2, the X_j^rs are functions of the test scores (i.e., usually power functions such as $X_j^1, X_j^2, X_j^3, \ldots, X_j^R$ but they can also be orthogonal polynomials), β_0 is a normalizing constant that forces the sum of the π_j probabilities to equal 1, and the remaining β_rs are parameters. The matrix version of Equation 5.2, $\mathbf{X}_j\boldsymbol{\beta}$, includes a 1 by R row vector of the X_j^rs, $\mathbf{X}_j$, and an R by 1 column vector of the βs, $\boldsymbol{\beta}$. When the log of Equation 5.1 is taken and the loglinear model in Equation 5.2 is used for $\boldsymbol{\pi}$, the resulting log likelihood (ignoring constant terms) can be shown to contain the R sample moments of the observed-score distribution of X, $\sum_j n_j X_j^r$, as sufficient statistics

$$L = \sum_j n_j \ln(\pi_j) = \sum_j n_j \left(\sum_{r=0}^{R} \beta_r X_j^r \right) \tag{5.3}$$

Loglinear models are usually estimated in the observed-score distributions of the samples using maximum likelihood methods. The β_rs in loglinear models are estimated so that L is maximized, and the first derivative of L with respect to the β_rs, $\partial L/\partial(\beta_0, \ldots, \beta_R)$, is set to zero. The maximum likelihood estimates for a loglinear model are considered to have the following desirable statistical properties (Agresti, 2002; Bishop et al., 1975; Cox, 1984; Holland and Thayer, 1987, 2000):

- *Integrity:* Loglinear model estimates that are obtained with maximum likelihood methods have *integrity* (von Davier et al., 2004, p. 48) in the sense that the first R moments of the estimated observed-score distribution for a loglinear model will equal those of the sample distribution. Integrity is also described as *moment matching* (Holland and Thayer, 2000, p. 137), a property of the maximum likelihood estimation result where $\partial L/\partial(\beta_0, \ldots, \beta_R) = 0$, which means that $\sum_j (n_j/N) X_j^r = \sum_j \widehat{\pi}_j X_j^r$ for $r = 1$ to R.
- *Consistency:* As sample sizes increase, loglinear model estimates converge to the population values.
- *Efficiency:* Given the sample sizes, the deviations of loglinear model estimates from the population values are as small as possible.

- *Asymptotic normality:* Loglinear model estimates are asymptotically normally distributed with covariances that can be estimated.
- *Positivity:* Like the population values in π, the corresponding $\widehat{\pi}_j$ values of the estimated model are greater than zero.

The developments and applications reviewed in this chapter exploit the integrity, consistency, efficiency, asymptotic normality, and positivity properties of loglinear model estimates.

5.3 Types of Loglinear Models

Equation 5.2 describes a set of several *univariate* loglinear models that contain parameters ranging from $R = 0$ to $J - 1$, (i.e., R must be less than the available degrees of freedom in π, $J - 1$). The moment-matching property of loglinear models indicates that univariate models based on different R values fit different numbers of moments for a sampled observed-score distribution. If $R = 4$, then the modeled distribution preserves the sample size of the sample distribution, the means of the scores, the squared scores, the cubed scores, the quartic scores, and the central moments (i.e., mean, variance, skewness, and kurtosis). Because psychometric tests are usually long so that the js in X_j may reference 20, 50, or more possible scores, loglinear models of the form in Equation 5.2 could potentially use large R values to fit very high moments in the distribution.

Additional complexities in observed-score distributions can be modeled with extensions to the loglinear model in Equation 5.2, such as by incorporating a second categorical variable, U_i,

$$\ln(\pi_{j,U_i}) = \beta_0 + \sum_{r=1}^{R} \beta_{X,r} X_j^r + \sum_{s=1}^{S} \beta_{U,s} U_i^s + \sum_{t}^{T} \sum_{v}^{V} \beta_{XU,tv} X_j^t U_i^v \tag{5.4}$$

Models having the form shown in Equation 5.4 describe the joint distribution of X and U. Loglinear models such as Equation 5.4 have been used to address interests such as the modeling of observed-score distributions of examinee subgroups (where U_i denotes membership in an examinee subgroup; Hanson, 1996; Moses et al., 2010), of polytomous items (where multiple dichotomous variables such as U_i indicate the polytomous response categories of questionnaire items; Agresti, 1993), of score-specific structures in the distribution of X (where U_i denotes a set of X scores with a distribution that departs from the overall distribution of X; Holland and Thayer, 2000), and the modeling of the joint bivariate distributions of two tests (where U_i denotes the scores on another test; von Davier et al., 2004; Holland and Thayer, 1987, 2000; Rosenbaum and Thayer, 1987). An application of Equation 5.4 that is considered in the example in this chapter is one in which U_i represents responses to a dichotomously scored item from test X.

5.3.1 Assessing the Fit of Loglinear Models

The large number of available loglinear models suggested in Equation 5.2 means that many plausible models can usually be considered for fitting a sampled observed-score distribution and that the process of choosing a model can be intensive. Several statistical strategies for selecting model parameterizations have been developed to inform the model selection process (Agresti, 2002; Bishop et al., 1975; Holland and Thayer, 2000). These selection strategies are usually based on evaluations of how well the estimated frequencies of the models, $\widehat{m}_j = N\widehat{\pi}_j$, fit the sample frequencies, n_j.

Assuming that the sample frequencies follow a multinomial distribution, two well-known asymptotically equivalent and approximately χ-squared distributed statistics are available, the $J - R - 1$ degree-of-freedom likelihood ratio χ-square,

$$G^2 = 2\sum_j n_j \ln\left(n_j/\widehat{m}_j\right) \tag{5.5}$$

and the $J - R - 1$ degree-of-freedom Pearson χ-square,

$$\sum_j \widehat{m}_j^{-1}\left(n_j - \widehat{m}_j\right)^2 \tag{5.6}$$

Consider a set of nested loglinear models such as Equation 5.2 in which each model in the set is fit using a different R value so that each model has χ-square fit statistics computed as in Equations 5.5 and 5.6. Model selection can proceed with a nested sequence of χ-square significance tests in which, beginning with the most complex model in the set (i.e., the model with the largest considered R), the difference in the χ-square statistics of the complex model and the model with one fewer parameters is evaluated as a 1 degree-of-freedom χ-square test, indicating whether the more complex model has a significantly better fit than the simpler model (Haberman, 1974). In addition to χ-square significance tests, several information criteria have been developed that encourage a model selection process based on evaluating model fit (G^2) with respect to the number of parameters needed to achieve that fit ($R + 1$). A well-known information criterion is the Akaike information criterion (AIC) (Akaike, 1981), which suggests a model from a set of loglinear models such as Equation 5.2 with the smallest value of

$$AIC = G^2 + 2(R + 1) \tag{5.7}$$

Simulation studies compared model selection results based on several χ-square significance tests and information criteria and found that *AIC* minimization and nested hypothesis tests using the G^2 statistics of the models are particularly accurate (Moses and Holland, 2010).

Score-level residuals can be used to evaluate the fit of a loglinear model in more detail than the global measures shown in Equations 5.5 through 5.7. The most well-known score-level residual for loglinear models is probably the standardized residual, the square root of the *jth* part of the Pearson χ-square statistic

$$e_j = \widehat{m}_j^{-1/2}\left(n_j - \widehat{m}_j\right) \tag{5.8}$$

For bivariate models such as Equation 5.4, model evaluations can proceed using bivariate n_{j,U_i} and $\widehat{m}_{j,U_i}$ estimates in Equations 5.5 through 5.8, but these evaluations can also consider aspects of the conditional distributions and moments of the bivariate models (Holland and Thayer, 2000; von Davier et al., 2004). For situations of interest in this chapter in which U_i represents correct and incorrect responses to a dichotomous item ($U_i = 1$

and 0), a useful evaluation of the conditional moments of Equation 5.4 is obtained as the estimated conditional means of U_i at each X score

$$\widehat{\Pr}\{U_i = 1|X_j\} = \left(\sum_{U_i} \widehat{m}_{j,U_i}\right)^{-1} \sum_{U_i} U_i \widehat{m}_{j,U_i} \tag{5.9}$$

Equation 5.9 can be used to describe the functional relationship of an item with X, so that the equation serves as a method of item analysis and item model evaluation (Livingston and Dorans, 2004; Sinharay, 2006; Turnbull, 1946; Wainer, 1988). When the estimated frequencies of Equation 5.9 are obtained from loglinear models, the resulting conditional mean estimates can be directly interpreted with respect to the fit of the moments of the sampled observed-score distributions of the loglinear model. The example in this chapter shows how Equation 5.9 can be useful for interpreting the conditional means of items that are estimated from loglinear and item response theory (IRT) models.

5.4 Covariance Estimation for Loglinear Models

Established results show that when using maximum likelihood estimation, the βs for a loglinear model converge to an asymptotic normal distribution with means equal to their population values and a covariance matrix approximated by

$$Cov(\widehat{\boldsymbol{\beta}}) = \left[-\frac{\partial \partial L}{\partial \boldsymbol{\beta} \partial \boldsymbol{\beta}}\right]^{-1} = \left[\mathbf{X}^t N(\mathbf{D}_\pi - \pi\pi^t)\mathbf{X}\right]^{-1} \tag{5.10}$$

where $\mathbf{X}$ is a J by R matrix containing all J of the $\mathbf{X}_j$s in Equation 5.2 (Agresti, 2002; Bishop et al., 1975; Cox, 1984). Several covariance estimates of the results from loglinear models are of interest in psychometric applications and in general loglinear-modeling work. As summarized in this section and in Agresti (2002), Bishop et al. (1975), and Cox (1984), the additional covariance estimates can be obtained with applications of the delta method using $Cov(\widehat{\boldsymbol{\beta}})$ in Equation 5.10 and exploiting continuously differentiable functions of the βs for loglinear models.

Covariance matrices of the estimated frequencies of loglinear models are of interest in psychometric applications in which estimated distributions of loglinear models are used to improve the stability of equipercentile equating functions (Kolen and Brennan, 2004; Liou and Cheng, 1995; Livingston, 1993; von Davier et al., 2004). These covariance matrices can be estimated as follows:

$$\begin{aligned} Cov(\widehat{\mathbf{m}}) &= \left[\frac{\partial \mathbf{m}}{\partial \boldsymbol{\beta}}\right]\left[Cov(\widehat{\boldsymbol{\beta}})\right]\left[\frac{\partial \mathbf{m}}{\partial \boldsymbol{\beta}}\right]^t \\ &= [Cov(\mathbf{n})\mathbf{X}]\left[Cov(\widehat{\boldsymbol{\beta}})\right][\mathbf{X}^t Cov(\mathbf{n})] \\ &= \left[N(\mathbf{D}_\pi - \pi\pi^t)\mathbf{X}\right]\left[\mathbf{X}^t N(\mathbf{D}_\pi - \pi\pi^t)\mathbf{X}\right]^{-1}\left[\mathbf{X}^t N(\mathbf{D}_\pi - \pi\pi^t)\right] \end{aligned} \tag{5.11}$$

Equation 5.11 can be used to estimate the standard errors of equipercentile equating functions calculated from the estimated frequencies of loglinear models and also to assess the extent to which these standard errors are smaller than those of equipercentile functions obtained directly from sample frequencies of test scores (Liou and Cheng, 1995; Moses and Holland, 2007, 2008).

Interpretations of the standardized residuals of a loglinear model (Equation 5.8) are another interest that can be informed by covariance estimates. Theoretical results indicate that standardized residuals have a somewhat misleading name, as their asymptotic variances are actually less than one, and can be considerably less than one if the loglinear model has a large number of parameters and/or if the number of test score categories being modeled is large (Agresti, 2002; Haberman, 1973). Adjusted residuals "adjust" the standardized residuals in Equation 5.8 so that the residuals are actually standardized and have asymptotic variances of one. This adjustment is made by dividing the standardized residuals by the square root of their asymptotic variance, which is estimated using the delta method. The $e_j = \widehat{m}_j^{-1/2}(n_j - \widehat{m}_j)$s are functions of **n** and **m** and, assuming the loglinear model is correct, **n** and **m** are functions of $\boldsymbol{\pi}$. These assumptions ultimately suggest that the covariance matrix of the e_js can be estimated using the following application of the delta method:

$$Cov(\widehat{\mathbf{e}}) = \left[\frac{\partial \mathbf{e}}{\partial \mathbf{n}}, \frac{\partial \mathbf{e}}{\partial \mathbf{m}}\right] \begin{bmatrix} Cov(\mathbf{n}) & Cov(\mathbf{n}, \widehat{\mathbf{m}}) \\ Cov(\mathbf{n}, \widehat{\mathbf{m}}) & Cov(\widehat{\mathbf{m}}) \end{bmatrix} \left[\frac{\partial \mathbf{e}}{\partial \mathbf{n}}, \frac{\partial \mathbf{e}}{\partial \mathbf{m}}\right]^{t} \tag{5.12}$$

where $\partial\mathbf{e}/\partial\mathbf{n} = \mathbf{D}_{\pi}^{-1/2}$, $\partial\mathbf{e}/\partial\mathbf{m} = -\mathbf{D}_{\pi}^{-1/2}$, $Cov(\mathbf{n}, \widehat{\mathbf{m}}) = N^{-1}\mathbf{D}_{\pi}^{1/2}\mathbf{H}Cov(\widehat{\boldsymbol{\beta}})\mathbf{H}^{t}\mathbf{D}_{\pi}^{-1/2}Cov(\mathbf{n})$, and $\mathbf{H} = \mathbf{D}_{\pi}^{-1/2}Cov(\mathbf{n})\mathbf{X}$. Substituting the estimated $N^{-1}\widehat{\mathbf{m}}$ values of a loglinear model for $\boldsymbol{\pi}$ in Equation 5.12 and $\mathbf{H}$ and performing the matrix calculations results in the formula for the covariance matrix given in Agresti (2002) and Haberman (1973), $\mathbf{I} - N^{-1}\sqrt{\widehat{\mathbf{m}}}\sqrt{\widehat{\mathbf{m}}}^{t} - \mathbf{H}(\mathbf{H}^{t}\mathbf{H})^{-1}\mathbf{H}^{t}$, where **I** is a J by J identity matrix.

Standard error estimates can be useful for interpreting the estimated conditional means of an item (Equation 5.9). Moses et al. (2010) used estimated frequencies in Equation 5.9 obtained from fitting the loglinear model in Equation 5.4 with $S = V = 1$ and $T = R$. Doing so essentially expresses Equation 5.4 as two loglinear models of the form in Equation 5.2, which fit the sample sizes and R moments of the observed-score distributions of X for examinee groups that had correct and incorrect responses to item i. Because this approach models the examinee groups responding correctly and incorrectly to item i as independent, standard errors for the conditional means in Equation 5.9 can be estimated by applying the delta method using the covariance matrices of the two loglinear models

$$\begin{aligned} Var\left(\widehat{\Pr}\{U_i = 1|X_j\}\right) &= \sum_{U_i}\left[\frac{\partial \Pr\{U_i = 1|X_j\}}{\partial m_{j,U_i}}\right] Var(\widehat{m}_{j,U_i}) \left[\frac{\partial \Pr\{U_i = 1|X_j\}}{\partial m_{j,U_i}}\right] \\ &= \left[\left(\sum_{U_i} m_{j,U_i}\right)^{-2} (m_{j,U_i=0})\right]^2 Var(\widehat{m}_{j,U_i=1}) \\ &\quad + \left[\left(\sum_{U_i} m_{j,U_i}\right)^{-2} (-m_{j,U_i=1})\right]^2 Var(\widehat{m}_{j,U_i=0}) \end{aligned} \tag{5.13}$$

where the $Var(\widehat{m}_{j,U_i})$s are obtained as diagonal entries from the $Cov(\widehat{\mathbf{m}})$ matrices of the $\widehat{m}_{j,U_i=1}$ and $\widehat{m}_{j,U_i=0}$ models (Equation 5.11). Simulations suggest that the estimates of Equation 5.13 are generally accurate when the n_js are not too small (Moses and Holland, 2007; Moses et al., 2010).

5.5 Example

To illustrate loglinear modeling for observed-score distributions, a reanalysis is performed for the results that Kolen and Brennan (2004) obtained from a dataset of the responses of 1655 examinees to a test containing 36 dichotomously scored items. The original analyses featured a comparison of the fits of one- and three-parameter logistic (1PL and 3PL) IRT models to the observed-score distribution of the total test score. For the reanalysis, in this section, the loglinear models and results presented in the previous sections are used to demonstrate loglinear modeling and also to interpret the results of the 1PL and 3PL IRT models.

In the first step of this reanalysis, an appropriately parameterized loglinear model is found for the sample distribution of the test (shown in Figure 5.1a). The sample distribution in Figure 5.1a was modeled with nine loglinear models of the form in Equation 5.2 using R values to fit 0 through 8 moments. The fit statistics of the loglinear models and the moments of their estimated distributions are presented in Table 5.1. Table 5.1 shows that the differences in the G^2 and Pearson χ-square statistics for the $R = 8$ versus $R = 7$, $R = 7$ versus $R = 6$, $R = 6$ versus $R = 5$, and $R = 5$ versus $R = 4$ models are less than 1 and, by most criteria for evaluating 1 degree-of-freedom χ-square statistics, statistically insignificant. However, the differences in the χ-square statistics for the $R = 3$ and $R = 4$ models are greater than 60 and, by most criteria, statistically significant. This suggests that the fit of the $R = 4$ model is not significantly worse than the fits of the more complex models ($R > 4$) but is significantly better than those of simpler models ($R < 4$). This suggestion is further supported by the $R = 4$ model having the smallest AIC statistic.

Figure 5.1b graphically depicts the distributions of the $R = 0$ through $R = 4$ loglinear models, for which the series' numbers denote results from models with specific R values. Figure 5.1b, which is consistent with the results in Table 5.1, shows that the fit of the sample frequencies for the test is poorest for the uniform distribution of the $R = 0$ model and is increasingly improved for models with R values of 1, 2, 3, and 4. The distributions for the $R = 5$ through $R = 8$ models are not shown in Figure 5.1b, but as implied in the χ-square values for Table 5.1, these distributions are not distinguishable from the distribution of the $R = 4$ model when graphed. Altogether the results in Table 5.1 and Figure 5.1 suggest that the $R = 4$ model produces a fitted distribution that is smooth, stable, and preserves the most important moments of the sample distribution.

The results of the $R = 4$ model can be useful for reconsidering one of the original interests of Kolen and Brennan (2004) by comparing the fit of 1PL and 3PL IRT models to the sample observed-score distribution (Figure 5.1a). The IRT results from Kolen and Brennan (2004) can be replicated by using their implementation choices with Bilog to obtain the item parameter estimates for 1PL and 3PL models (Zimonowski et al., 2003) and then using these estimates with the recursion algorithm of Lord and Wingersky (1984) to produce the estimated observed-score distributions of the IRT models. To comparatively evaluate the fitted distributions of the 1PL and 3PL models with respect to the distributions of the $R = 4$

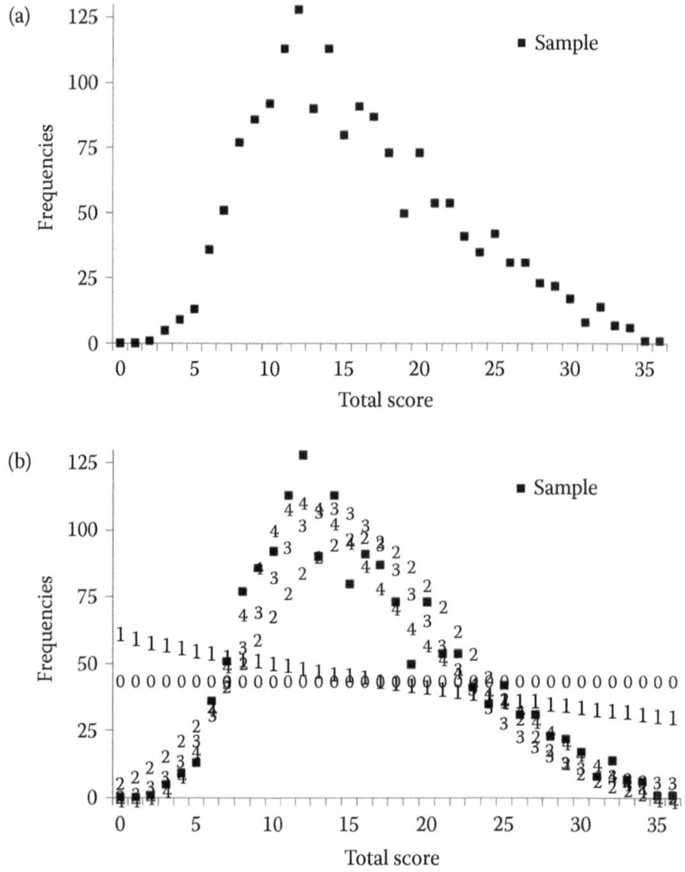

FIGURE 5.1
Display of (a) the sample frequencies of test X and (b) sample frequencies of X with the distributions obtained from loglinear models fitting $R = 0$–4 moments of the distribution of X.

loglinear model, Table 5.2 presents the fitted frequencies of the adjusted residuals of the loglinear model. Table 5.2 also presents the results and the generalized residuals of the IRT models (Haberman, 2009). Generalized residuals (described in the appendix) allow for the evaluation of the fitted frequencies and moments of the IRT models in terms of statistics that, such as the adjusted residuals of loglinear models, are asymptotically normal and have variances of one.

The primary result in Table 5.2 is that the estimated distribution for the loglinear model fits the sample distribution more closely than the distributions of the 1PL and 3PL models. In terms of score-level frequencies, the estimated distribution of the loglinear model has two adjusted residuals that might be considered statistically "large" (i.e., that exceed two in absolute value), whereas the 1PL and 3PL models have seven and six generalized residuals exceeding two in absolute value. The nonzero residuals in the fit of the mean, variance, skewness, and kurtosis of the observed-score distribution of the 1PL and 3PL models are additional indications that the IRT models fit the sample distribution less closely than the loglinear model, though the only generalized residual of the moment exceeding two in absolute value was due to the underestimated skewness of the lPL model (0.37 vs. 0.58).

TABLE 5.1

Summary of the Fits of Nine Loglinear Models to the Example Observed-Score Distribution from $N = 1655$ Examinees

		Results from Loglinear Models Based on *R* Values								
		0	1	2	3	4	5	6	7	8
G^2		1287.70	1218.45	215.71	109.31	31.55	30.96	30.77	30.33	30.01
Pearson χ-square		1149.96	1023.16	196.80	97.32	30.50	30.08	29.93	29.26	28.96
AIC		1289.70	1222.45	221.71	117.31	41.55	42.96	44.77	46.33	48.01
Degrees of freedom		36	35	34	33	32	31	30	29	28
Sample distribution										
Mean	15.82	18.00	15.82	15.82	15.82	15.82	15.82	15.82	15.82	15.82
Variance	42.61	114.00	111.15	42.61	42.61	42.61	42.61	42.61	42.61	42.61
Skew	0.58	0.00	0.25	0.09	0.58	0.58	0.58	0.58	0.58	0.58
Kurtosis	2.72	1.80	1.88	2.73	3.27	2.72	2.72	2.72	2.72	2.72

The findings that the 1PL and 3PL models fit the sample observed-score distribution less closely than the loglinear model and that the fit of the 3PL model is more accurate than that of the 1PL model may seem unsurprising. Kolen and Brennan (2004) suggested that the worse fit of the 1PL model might be due to its inability to account for item-level issues such as guessing effects and unequal item discriminations (pp. 198–199). The consideration of the IRT models' estimated item-level results is the basis for follow-up analyses of the individual items' functional relationships with the test scores. For each of the 36 items on the test, loglinear models in the form in Equation 5.4 were used to fit the $R = T = 4$ moments of the X distributions, in which U_i indicates subgroups of examinees who obtained correct and incorrect responses to each item. The conditional means and ± 2 standard error estimates of the items were then obtained using Equations 5.9 and 5.13. The estimates of the conditional means of the items for the 1PL and 3PL IRT models were obtained by applying the Lord and Wingersky (1984) recursion algorithm to estimate observed-score distributions for examinees that provided correct and incorrect responses to each item and using these estimated frequency distributions in Equation 5.9. In this way, the functional relationships of the 36 items to the total test score could be considered in terms of estimates from the loglinear, 1PL, and 3PL models, and the results of each model could be directly related to the fit of the observed-score distributions of the models for the U_i subgroups. Graphical presentations were produced for each item, with the conditional means of the loglinear model and the ± 2 standard error bands of these estimates plotted in solid and dashed lines and the estimates of the 1PL and 3PL models denoted as a series of 1s and 3s.

A survey of the results of all 36 items revealed a subgroup of items for which the estimated conditional means of the 3PL model were usually within the ± 2 standard error bands for the estimates of the loglinear model, and the estimated conditional means of the 1PL model were usually outside of the ± 2 standard error bands. Three representative items from this group are shown in Figure 5.2 (Items 21, 31, and 34). Figure 5.2 indicates that these items are relatively difficult (i.e., their conditional means are low for several X scores). The conditional means of the 1PL model deviate from those of the 3PL and loglinear models in ways that are consistent with the suggestion by Kolen and Brennan (2004, pp. 198–199) that the single discrimination value of the 1PL model may not accurately reflect the functional relationships of these items with the total test scores. Assessments of the estimated

TABLE 5.2

Sample and Estimated Distributions and Residuals for the Loglinear, 1PL, and 3PL Models

		Loglinear Model		1PL Model		3PL Model	
Score	Sample Counts	Estimated Counts	Adjusted Residuals	Estimated Counts	Generalized Residuals	Estimated Counts	Generalized Residuals
36	1	1.05	−0.05	0.27	0.74	0.51	0.52
35	1	2.19	−0.89	1.12	−0.12	2.00	−1.08
34	6	4.00	1.11	2.67	1.38	4.51	0.65
33	7	6.57	0.18	4.93	0.80	8.03	−0.43
32	14	9.83	1.44	7.85	1.71	11.84	0.62
31	8	13.60	−1.63	11.33	−1.20	14.34	−2.46*
30	17	17.64	−0.16	15.07	0.49	15.24	0.47
29	22	21.73	0.06	18.93	0.68	16.87	1.30
28	23	25.70	−0.57	23.23	−0.05	21.83	0.29
27	31	29.49	0.30	28.35	0.50	30.09	0.19
26	31	33.11	−0.39	33.97	−0.55	38.54	−1.50
25	42	36.65	0.94	39.20	0.45	43.38	−0.23
24	35	40.25	−0.88	43.60	−1.50	43.25	−1.48
23	41	44.05	−0.49	47.99	−1.14	40.51	0.08
22	54	48.25	0.87	53.79	0.03	39.75	2.13*
21	54	53.00	0.15	61.52	−1.07	44.75	1.39
20	73	58.45	2.01*	69.86	0.39	55.94	2.23*
19	50	64.72	−1.95	76.60	−3.86*	70.09	−3.01*
18	73	71.85	0.15	80.66	−0.93	82.36	−1.18
17	87	79.75	0.87	83.32	0.41	89.46	−0.28
16	91	88.16	0.32	87.21	0.42	91.77	−0.09
15	80	96.55	−1.80	93.66	−1.60	92.72	−1.51
14	113	104.09	0.93	100.86	1.20	95.84	1.77
13	90	109.62	−2.01*	104.78	−1.62	101.78	−1.35
12	128	111.74	1.66	102.30	2.43*	107.69	2.00
11	113	109.14	0.40	93.81	1.95	109.29	0.38
10	92	100.96	−0.98	82.66	1.05	103.66	−1.33
9	86	87.32	−0.16	72.03	1.63	90.65	−0.56
8	77	69.64	0.96	62.37	1.80	72.40	0.60
7	51	50.42	0.09	52.01	−0.15	52.17	−0.19
6	36	32.60	0.66	39.97	−0.70	33.25	0.53
5	13	18.49	−1.42	27.47	−4.13*	18.25	−1.63
4	9	9.02	−0.01	16.76	−2.67*	8.31	0.27
3	5	3.71	0.73	9.03	−1.84	3.00	1.04
2	1	1.26	−0.24	4.13	−3.14*	0.80	0.21
1	0	0.34	−0.60	1.43	−40.68*	0.14	−120.92*
0	0	0.07	−0.28	0.28	−40.68*	0.01	−120.92*
Mean	15.82	15.82	0.00	15.82	−0.01	15.82	0.07
Variance	42.61	42.61	0.00	42.43	0.15	42.57	0.04
Skewness	0.58	0.58	0.00	0.37	3.93*	0.58	−0.07
Kurtosis	2.72	2.72	0.00	2.60	0.75	2.72	−0.02

*Denotes an adjusted or generalized residual greater than 2 in absolute value.

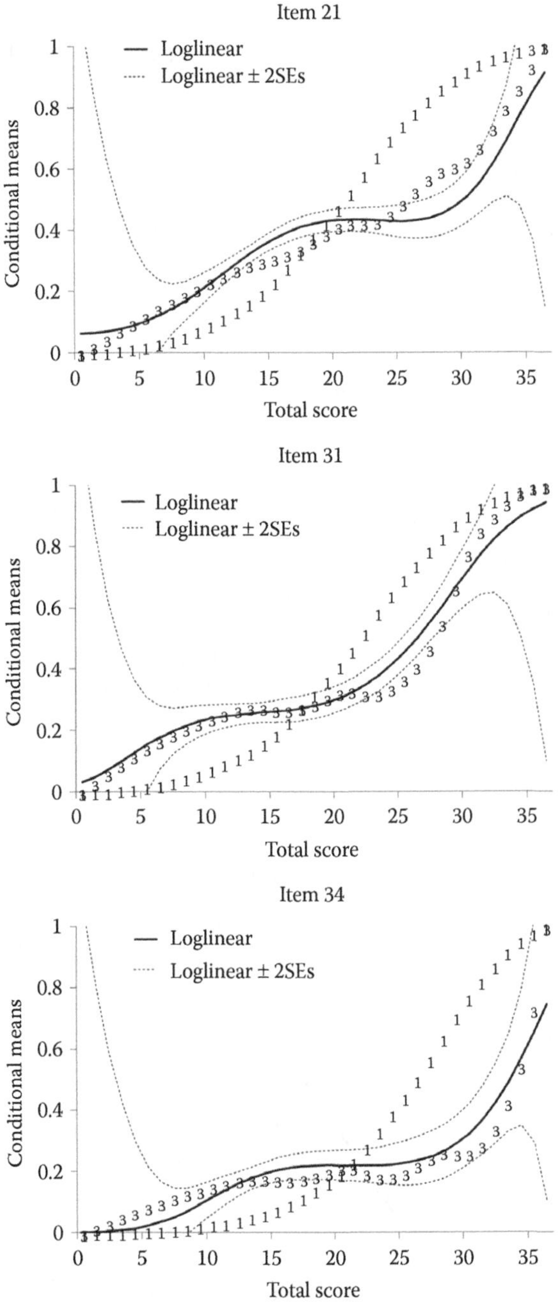

FIGURE 5.2
Conditional means at each total test score for items fit better by the 3PL model than the 1PL model, items 21, 31, and 34.

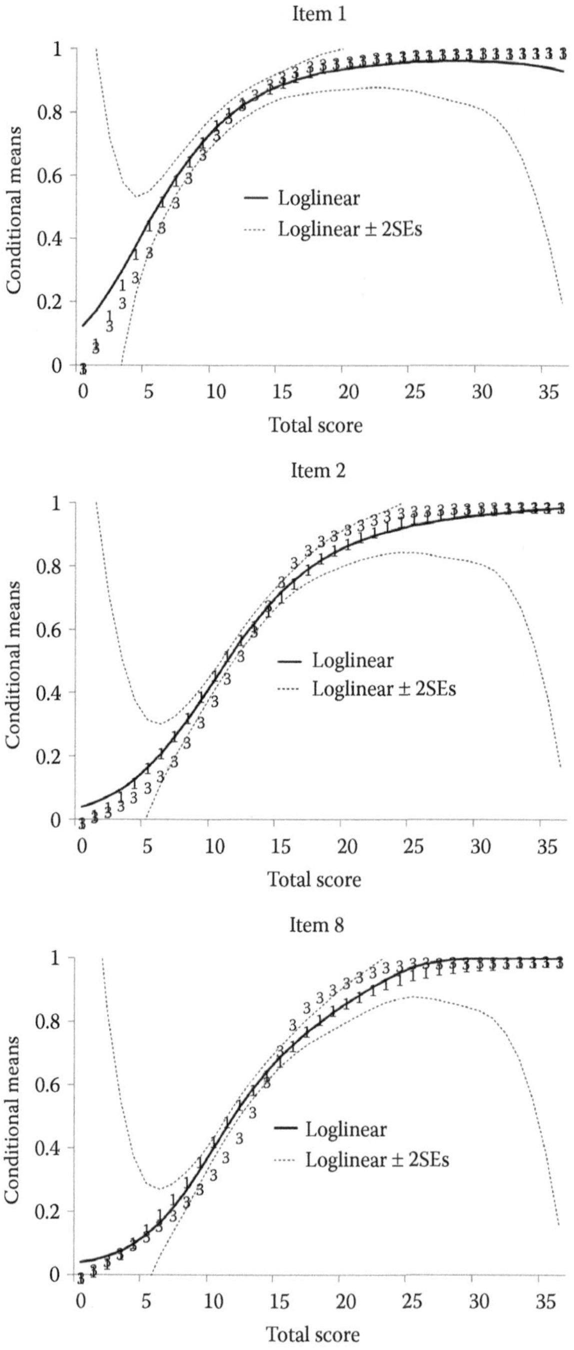

FIGURE 5.3
Conditional means at each total test score for items fit slightly better by the 1PL model than the 3PL model, items 1, 2, and 8.

observed-score distributions of the loglinear, 1PL, and 3PL models for examinees obtaining correct and incorrect scores on the items in Figure 5.2 are consistent with Table 5.2, indicating that the results of the 1PL model could be attributed to less accurately fitting the variances and skewness of the sample distributions.

For another subgroup of the 36 items, the estimated conditional means of the 1PL model were within the ± 2 standard error bands of the estimates of the loglinear model, and the estimated conditional means of the 3PL model were slightly outside of the ± 2 standard error bands. Three representative items from this second group are the relatively easy items shown in Figure 5.3 (Items 1, 2, and 8). An examination of estimated observed-score distributions of the models for examinees obtaining correct and incorrect scores to the items in Figure 5.3 indicated that the results of the 3PL model could be attributed to less accurately fitting the means of the sample distributions (as suggested in the generalized residuals in Table 5.2).

5.6 Summary

The application of loglinear models to observed-score distributions encountered in psychometrics is statistical and flexible. For most psychometric test data, a large number of models are usually available that may be considered and evaluated using a large number of fit statistics. Covariance estimates are available for gauging the accuracy of the estimates of a loglinear model (i.e., model frequencies, residuals, and conditional distributions). From this chapter, research efforts are continuing to clarify the accuracy implications of various statistical practices associated with loglinear modeling.

As shown in the example in this chapter, the results of loglinear models can be useful not only for providing accurate and stable estimates of observed-score distributions, but also for interpreting the estimated distributions and conditional item scores from IRT models (Haberman, 2009; Kolen and Brennan, 2004; Sinharay, 2006). The example results in Table 5.2 and Figures 5.2 and 5.3 indicate that IRT models vary in terms of how accurately they fit the functional relationships of different items with total test scores. The IRT model fit results can be interpreted with respect to the accuracy of the models in fitting the means, variances, and skewness of the observed-score distributions used to compute the conditional means of the items. Future investigations exploring these findings over a wider range of test data would be informative, and the results of the loglinear models can be useful for interpreting the results of these investigations.

Appendix 5A: Definition of Generalized Residuals

To obtain Haberman's (2009) generalized residuals, let the unstandardized residual be denoted as $O-E$, the simple difference between the statistic of an observed-score distribution calculated as a function of the response profiles in the sample data, $O = N^{-1}\sum_p f(\mathbf{u}_p)$, and that statistic estimated from an IRT model over all possible item response profiles ($\varphi = 1$ to Γ) using a theta distribution approximated with $q = 1$ to Q

quadrature points, $E \approx \sum_{\varphi}^{\Gamma} f(\mathbf{u}_{\varphi})(\sum_q \Pr\{\mathbf{u}_{\varphi};\theta_q\}\Pr\{\theta_q\})$. The estimated variance of O–E is

$$\tau^2 = N^{-2}\sum_{p}^{N}\left[f(\mathbf{u}_p) - E - (\mathbf{W}^{-1}\mathbf{y})^t \partial L(\mathbf{u}_p)\right]^2 \tag{5A.1}$$

where $\mathbf{W} = \sum_p^N \partial L(\mathbf{u}_p)\left[\partial L(\mathbf{u}_p)\right]^t, \mathbf{y} = \sum_p^N \left[f(\mathbf{u}_p) - E\right]\partial L(\mathbf{u}_p)$, and $\partial L(\mathbf{u}_p)$ is a vector of the relevant derivatives for the IRT model and examinee p. Assuming the IRT model holds, Equation 5A.1 can be used to produce the generalized residual τ^{-1} $(O–E)$. Like the adjusted residuals of loglinear models, the generalized residuals of an IRT model are approximately distributed as standard normal variables (Haberman, 2009). Generalized indicates that residuals are available for evaluating many aspects of the fit of an estimated distribution, five of which are considered in this chapter:

- For evaluating the IRT model fit of the observed-score distribution of a sample, define $f(\mathbf{u}_p)$ as 1 when $\sum_i U_{i,p} = X_j$ and as 0 when $\sum_i U_{i,p} \neq X_j$, so that $O = N^{-1}\sum_p f(\mathbf{u}_p) = N^{-1}n_j$ are the sample probabilities.
- For evaluating the IRT model fit of the mean of a distribution, define $f(\mathbf{u}_p)$ as $\sum_i U_{i,p} = X_{j,p}$ so that $O = N^{-1}\sum_p f(\mathbf{u}_p) = \bar{X}$ is the sample mean.
- For evaluating the IRT model fit of the variance of a distribution, define $f(\mathbf{u}_p)$ as $\left(\sum_i U_{i,p} - \bar{X}\right)^2$ so that $O = N^{-1}\sum_p f(\mathbf{u}_p) = s_X^2$ is the sample variance.
- For evaluating the IRT model fit of the skewness of a distribution, define $f(\mathbf{u}_p)$ as $(s_X^2)^{-3/2}\left(\sum_i U_{i,p} - \bar{X}\right)^3$ so that $O = N^{-1}\sum_p f(\mathbf{u}_p)$ is the sample skewness.
- For evaluating the IRT model fit of the kurtosis of a distribution, define $f(\mathbf{u}_p)$ as $(s_X^2)^{-2}\left(\sum_i U_{i,p} - \bar{X}\right)^4$ so that $O = N^{-1}\sum_p f(\mathbf{u}_p)$ is the sample kurtosis.

References

Agresti, A. 1993. Computing conditional maximum likelihood estimates for generalized Rasch models using simple loglinear models with diagonals parameters. *Scandinavian Journal of Statistics*, 20, 63–71.

Agresti, A. 2002. *Categorical Data Analysis* (2nd ed.). New York, NY: Wiley.

Akaike, H. 1981. Likelihood of a model and information criteria. *Journal of Econometrics*, 16, 3–14.

Bishop, Y. M. M., Fienberg, S. E., and Holland, P. W. 1975. *Discrete Multivariate Analysis*. Cambridge, MA: MIT Press.

Cox, C. 1984. An elementary introduction to maximum likelihood estimation for multinomial models: Birch's theorem and the delta method. *The American Statistician*, 38, 283–287.

von Davier, A. A., Holland, P. W., and Thayer, D. T. 2004. *The Kernel Method of Test Equating*. New York, NY: Springer-Verlag.

Haberman, S. J. 1973. The analysis of residuals in cross-classification tables. *Biometrics*, 29, 205–220.

Haberman, S. J. 1974. Log-linear models for frequency tables with ordered classifications. *Biometrics*, 30, 589–600.

Haberman, S. J. 2009. *Use of Generalized Residuals to Examine Goodness of Fit of Item Response Models* (Research Report 09-15). Princeton, NJ: Educational Testing Service.

Hanson, B. A. 1990. *An Investigation of Methods for Improving Estimation of Test Score Distributions* (Research Report 90-4). Iowa City, IA: American College Testing Program.

Hanson, B. A. 1996. Testing for differences in test score distributions using log-linear models. *Applied Measurement in Education*, 9, 305–321.

Holland, P. W. and Thayer, D. T. 1987. *Notes on the Use of Loglinear Models for Fitting Discrete Probability Distributions* (Research Report 87-31). Princeton, NJ: Educational Testing Service.

Holland, P. W. and Thayer, D. T. 2000. Univariate and bivariate loglinear models for discrete test score distributions. *Journal of Educational and Behavioral Statistics*, 25, 133–183.

Kolen, M. J. and Brennan, R. L. 2004. *Test Equating, Scaling, and Linking: Methods and Practices* (2nd ed.). New York, NY: Springer-Verlag.

Liou, M. and Cheng, P. E. 1995. Asymptotic standard error of equipercentile equating. *Journal of Educational and Behavioral Statistics*, 20, 259–286.

Livingston, S. A. 1993. Small-sample equating with log-linear smoothing. *Journal of Educational Measurement*, 30, 23–39.

Livingston, S. A. and Dorans, N. J. 2004. *A Graphical Approach to Item Analysis* (Research Report 04-10). Princeton, NJ: Educational Testing Service.

Lord, F. M. and Wingersky, M. S. 1984. Comparison of IRT true-score and equipercentile observed-score "equatings." *Applied Psychological Measurement*, 8, 453–461.

Moses, T. and Holland, P. 2007. *Kernel and Traditional Equipercentile Equating with Degrees of Presmoothing* (Research Report 07-15). Princeton, NJ: Educational Testing Service.

Moses, T. and Holland, P. 2008. *Notes on a General Framework for Observed Score Equating* (Research Report 08-59). Princeton, NJ: Educational Testing Service.

Moses, T. and Holland, P. W. 2010. A comparison of statistical selection strategies for univariate and bivariate log-linear models. *British Journal of Mathematical and Statistical Psychology*, 63, 557–574.

Moses, T., Miao, J., and Dorans, N. J. 2010. A comparison of strategies for estimating conditional DIF. *Journal of Educational and Behavioral Statistics*, 35, 726–743.

Rosenbaum, P. R. and Thayer, D. 1987. Smoothing the joint and marginal distributions of scored two-way contingency tables in test equating. *British Journal of Mathematical and Statistical Psychology*, 40, 43–49.

Sinharay, S. 2006. Bayesian item fit analysis for unidimensional item response theory models. *British Journal of Mathematical and Statistical Psychology*, 59, 429–449.

Turnbull, W. W. 1946. A normalized graphic method of item analysis. *The Journal of Educational Psychology*, 37, 129–141.

Wainer, H. 1988. *The Future of Item Analysis* (Research Report 88-50). Princeton, NJ: Educational Testing Service.

Zimonowski, M. F., Muraki, E., Mislevy, R. J., and Bock, R. D. 2003. *BILOG-MG [Computer Software]*. Lincolnwood, IL: Scientific Software International.

6

Distributions of Sums of Nonidentical Random Variables

Wim J. van der Linden

CONTENTS

6.1 Introduction

An obvious example of the subject of this chapter is the probability distribution of the number-correct score of a test taker on a test with dichotomous or polytomous items. However, similar types of distributions are met in statistical tests of cheating on tests, for instance, as null distributions of the number of matching responses between two test takers suspected of answer copying or the number of answer changes by a single test taker reviewing his answers in the detection of fraudulent erasures on answer sheets. In each of these cases, the desired distribution is found as the convolution of more elementary distributions. Although they may seem to lead to complicated combinatorics, a simple recursive algorithm exists that allows us to evaluate the convolutions efficiently and accurately. The same algorithm can be used to evaluate the posterior odds of certain events given sum scores or the combinatorial expressions that arise in conditional maximum-likelihood (CML) estimation of the parameters in the Rasch model. In addition to these discrete cases, the chapter addresses the distribution of the total time spent by a test taker on a test, which is found as the convolution of the distributions of his response times (RTs) on the items.

6.2 Distributions of Number-Correct Scores

Let U_i be the random variable that records a test taker's response on the dichotomous items $i = 1, \ldots, n$ in a test. Each of these variables has a Bernoulli distribution with probability mass function (pmf)

$$f(u;p_i) \equiv \Pr\{U_i = u; p_i\} \equiv \begin{cases} p_i^u(1-p_i)^u, & u = 0,1; \\ 0, & \text{otherwise,} \end{cases} \tag{6.1}$$

where p_i the success parameter for the test taker on item i (see Chapter 2). As is well known, a Bernoulli distribution has mean and variance are equal to

$$\mu_{U_i} = p_i \tag{6.2}$$

and

$$\sigma^2_{U_i} = p_i(1-p_i), \tag{6.3}$$

respectively. We will also need the probability generating function for the distribution, which is equal to

$$\varphi_{U_i}(t) \equiv \mathcal{E}(t^{U_i}) = q_i + p_i t \tag{6.4}$$

with $q_i \equiv 1 - p_i$. Observe how $\varphi_{U_i}(t)$ returns the two probabilities q_i and p_i as the coefficients of t^u for $u = 0$ and 1, respectively.

Item response theory (IRT) models for dichotomous items explain success parameter p_i in Equation 6.1 as a function of test taker and item parameters. In the current context, the choice of model is inconsequential. For the sake of illustration, we choose the three-parameter logistic (3PL) model, which explains p_i as

$$p_i \equiv p_i(\theta) \equiv c_i + (1-c_i)\Psi(a_i(\theta - b_i)), \tag{6.5}$$

where $\theta \in \mathbb{R}$ is the ability parameter for the test taker, $a_i \in \mathbb{R}^+$, $b_i \in \mathbb{R}$, and $c_i \in [0,1]$ are the discrimination, difficulty, and guessing parameters for item i, respectively, and $\Psi(\cdot)$ denotes the logistic distribution function (see Volume One, Chapter 2).

Upon the usual assumption of conditional (or "local") independence, the probability of a response pattern $\mathbf{U} \equiv (U_1, \ldots, U_n)$ is equal to

$$\Pr\{\mathbf{U} = \mathbf{u}\} = \prod_{i=1}^{n} p_i^{u_i} q_i^{1-u_i}. \tag{6.6}$$

The pmf of the test taker's total score $X \equiv U_1 + \cdots + U_n$ follows immediately from Equation 6.6 as

$$\begin{aligned} f(x;n,\mathbf{p}) &\equiv \Pr\{X = x; n, \mathbf{p}\} \\ &= \begin{cases} \sum\limits_{\Sigma u_i = x} \prod\limits_{i=1}^{n} p_i^{u_i} q_i^{1-u_i}, & x = 0,1,\ldots,n; \\ 0, & \text{otherwise} \end{cases} \end{aligned} \tag{6.7}$$

with $\mathbf{p} \equiv (p_1, \ldots, p_n)$.

For the special case of identically distributed response variables, that is, $p_i = p$, $i = 1, \ldots, n$, the function specialized to the pmf of the binomial distribution

$$f(x; n, p\} = \begin{cases} \binom{n}{x} p^x q^{n-x}, & x = 0, 1, \ldots, n; \\ 0, & \text{otherwise} \end{cases} \tag{6.8}$$

reviewed by Casabianca and Junker (see Chapter 2). Note how the more complicated combinatorial sum in Equation 6.7 reduces to the simple binomial coefficient in Equation 6.8.

Interest in the family of distributions in Equation 6.7 is not restricted to test theory but is present in the statistical literature on cases more generally known as sums of independent Bernoulli variables, multiple Bernoulli trials, random binary indicators, etc. In some texts, Equation 6.7 is known as the family of *compound binomial* distributions—a name that appears to be motivated by the replacement of the replications of the simple Bernoulli trial that generate Equation 6.8 by the compound experiment of observing the joint outcome of multiple independent trials with different success probabilities (e.g., Feller, 1968, Section VI.9). In their landmark text, Lord and Novick (1968, Section 16.2) adopted the distribution under the same name. But it has also been named the *Poisson binomial distribution* (e.g., Hodges and Le Cam, 1960). Both names are less fortunate, however; either wrongly suggests the operation of compounding (or marginalizing) the binomial distribution with respect to a given probability distribution of its success parameter p—the Poisson distribution in the case of the latter. With this cautionary note in mind, we will refer to Equation 6.7 as the compound binomial distribution.

Because of the independence assumption, the mean and variance of the distribution of total score X in Equation 6.7 follow directly from Equations 6.2 and 6.3 as

$$\mu_X = \sum_{i=1}^{n} p_i \tag{6.9}$$

and

$$\sigma_X^2 = \sum_{i=1}^{n} p_i q_i. \tag{6.10}$$

From the same assumption, its probability-generating function can be obtained as a product of the functions for the Bernoulli distributions in Equation 6.4. That is,

$$\begin{aligned} \varphi_X(t) &= \prod_{i=1}^{n} \varphi_{U_i}(t) \\ &= \prod_{i=1}^{n} (q_i + p_i t), \end{aligned} \tag{6.11}$$

where, upon expansion of the product, $\Pr\{X = x\}$ is the coefficient of t^x for $x = 0, 1, \ldots, n$. Unfortunately, use of the function does not enable us to escape the combinatorial complications involved in the calculation of the sum in Equation 6.7; they now return in the form of the product in Equation 6.11.

A possible way out of the complications might seem to use a binomial approximation with the average of the Bernoulli parameters

$$\xi = n^{-1}\sum_{i=1}^{n} p_i \tag{6.12}$$

as success parameter. Figure 6.1 shows an example of the approximation. The number-correct score distributions are for three test takers with $\theta = -2, 0$, and 2 on a 30-item tests with means and standard deviations of the a_i, b_i, and c_i parameters equal to (0.66, 0.21), (0.50, 1.01), and (0.16, 0.09), respectively. The distributions were generated sampling 10,000 responses at each of these θ values on each of the items and counting the numbers of correct responses. The binomial approximations are for the average response probabilities equal to $\xi = 0.35$, 0.55, and 0.79. Although the fit may seem reasonable at first side, the approximations misplace the modes and have variances generally greater than that of the compound binomial. Specifically, it holds for the two that

$$\sum_{i=1}^{n} p_i q_i \leq n\xi(1-\xi) \tag{6.13}$$

with equality when $p_i = \xi$, a fact easily established substituting Equation 6.12 into Equation 6.13.

A more complete image of the accurateness of the binomial approximation follows from a finite series expansion of the compound bionomial by Walsh (1953, 1963). The expansion, which had a prominent place in the earlier test-theory literature (e.g., Lord and Novick, 1968, Section 23.10) but seems to have lost its attractiveness later on, is in the powers of $p_1 - \xi, p_2 - \xi, \ldots, p_n - \xi$. It shows that

$$\begin{aligned}\Pr\{X = x; n, \xi\} &= \pi_n(x) + \frac{n}{2}V_2C_2(x) + \frac{n}{3}V_3C_3(x) \\ &\quad + \left(\frac{n}{4}V_4 - \frac{n^2}{8}V_2^2\right)C_4(x) \\ &\quad + \left(\frac{n}{5}V_5 - \frac{5n^2}{6}V_2V_3\right)C_5(x) \\ &\quad + \cdots \qquad x = 0, 1, \ldots, n,\end{aligned} \tag{6.14}$$

where

$$\pi_n(x) = \binom{n}{x}\xi^x(1-\xi)^{n-x}; \tag{6.15}$$

$$C_r(x) = \sum_{v=0}^{r}(-1)^{v+1}\binom{r}{v}\pi_{n-3}(x-v), \quad r = 2, \ldots, n; \tag{6.16}$$

$$V_r = n^{-1}\sum_{i=1}^{n}(p_i - \xi)^r, \quad r = 2, \ldots, n. \tag{6.17}$$

With all n terms included, Equation 6.14 is an exact identity. Also, note that its first term is just the binomial pmf. Consequently, the sum of all higher-order terms represents the error

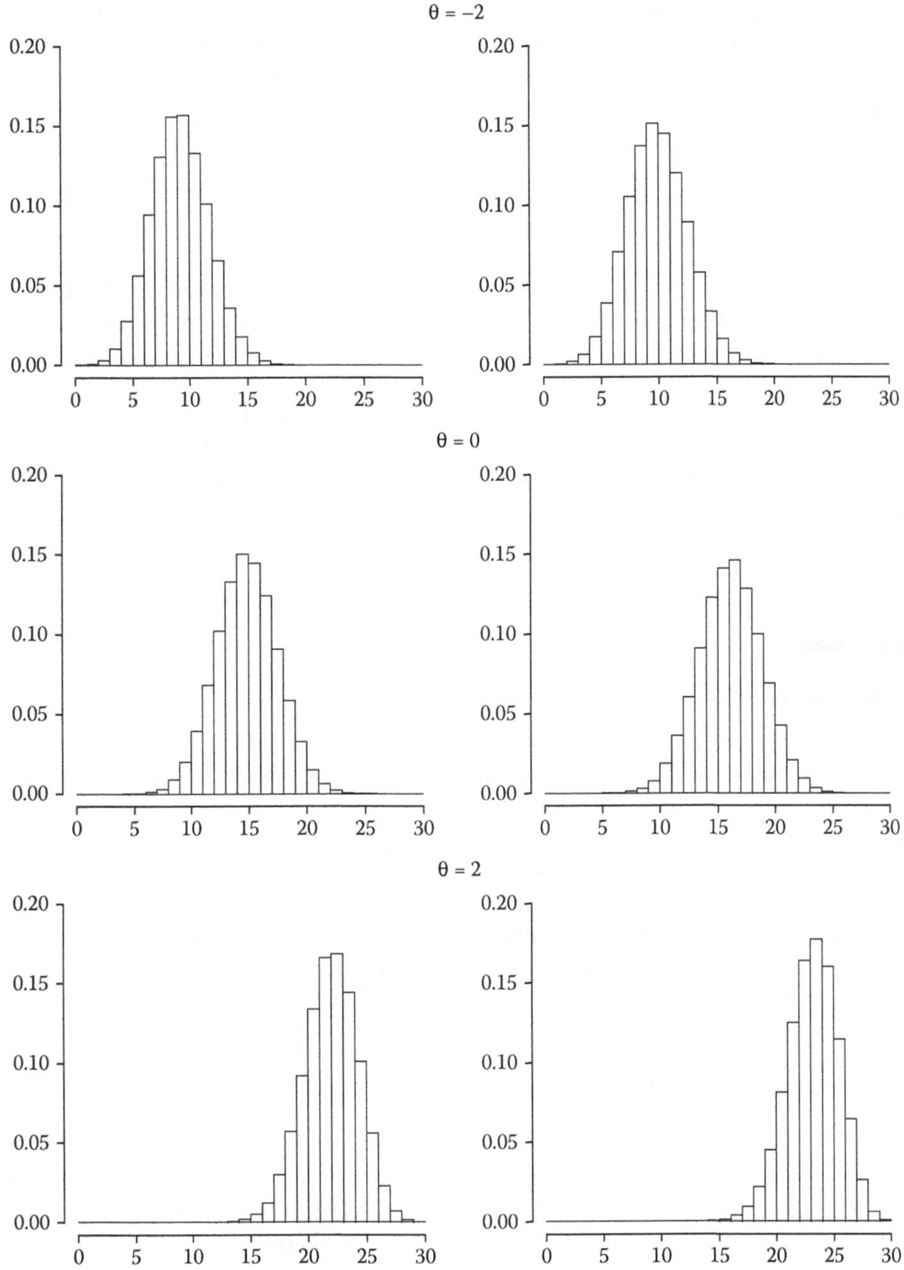

FIGURE 6.1
Histograms with number-correct score distributions on a 30-item test at $\theta = -2, 0$, and 2 (left panes) along with their binomial approximations (right panes).

involved in the binomial approximation. However, as these higher-order terms quickly become negligible, the first one or two of them already capture nearly all of the error.

The Walsh series has been used to establish an extremely practical result in the form of a simple set of necessary and sufficient conditions for two number-correct score distributions to be identical. Let $p_i(\theta)$ and $p_j(\theta), i, j = 1, \ldots, n$, be the success probabilities provided by the response model for a test taker with ability θ on two different test forms each existing of n dichotomous item. The choice of model does not matter. From Equations 6.14 through 6.17, it then follows that, for any θ, the number-correct score distributions on the two forms are identical if and only if

$$\sum_{i=1}^{n} p_i^r(\theta) = \sum_{j=1}^{n} p_j^r(\theta),\ r = 1, \ldots, n; \tag{6.18}$$

that is, when the sums of the first n powers of the success probabilities on the items in the two forms are equal (van der Linden and Luecht, 1998, proposition 1). As Equation 6.18 holds for any θ, it equally holds for the marginal distribution of X given any distribution of θ.

The identity can be used to constrain the sums for a new test form to be equal to those of a reference form, preventing the necessity of empirical studies to find the equating transformation between the observed scores on the two. Also, as Equations 6.14 and 6.18 are termwise equivalent, use of the sums of the powers of the first two or three orders already suffices (see Volume Three, Chapter 8).

6.3 Discrete Convolutions

One of the reasons that the early interest in approximations to the compound binomial distribution has waned is a simple recursive algorithm for the computation of its probabilities introduced in the test-theory literature by Lord and Wingersky (1984). In order to appreciate the algorithm, we need to represent the compound binomial pmf as the result of a series of convolution operations on its underlying Bernoulli functions.

Convolution operations are operations on two functions that produce a third function. They are used widely in several branches of mathematics as well as in such applied fields as engineering. We first review the more general case of any two functions f_1 and f_2 defined on the nonnegative integers $x = 0, 1, \ldots$. Their convolution is defined as the function

$$(f_2 * f_1)(x) \equiv \sum_{m=0}^{\infty} f_1(x - m) f_2(m), \tag{6.19}$$

where $0 \leq m \leq x$, with m integer as well. Thus, the new function is on the same domain and has as its argument the sum of the arguments of the two given functions. The use of these (discrete) convolutions in statistics is for the special choice of f_1 and f_2 as pmfs. For this choice, they provide us with an operation of "addition" on probability distributions that parallels the one on their random variables.

To illustrate the point, consider the Bernoulli functions in Equation 6.1 for two items. For these functions, the operation gives us

$$(f_2 * f_1)(x) = \begin{cases} \Pr\{U_1 = 0\}\Pr\{U_2 = 0\}; \\ \Pr\{U_1 = 1\}\Pr\{U_2 = 0\} + \Pr\{U_1 = 0\}\Pr\{U_2 = 1\}; \\ \Pr\{U_1 = 1\}\Pr\{U_2 = 1\}; \end{cases}$$

$$= \begin{cases} q_1 q_2, & \text{for } x = 0; \\ p_1 q_2 + q_1 p_2, & \text{for } x = 1; \\ p_1 p_2, & \text{for } x = 2, \end{cases} \tag{6.20}$$

which is the pmf of the sum score $X = U_1 + U_2$.

For n functions, Equation 6.19 generalizes to

$$f(x) = f_n * (f_{n-1} * \cdots * (f_2 * f_1)(x) \tag{6.21}$$

with each successive step defined according to Equation 6.19. For the choice of Bernoulli functions, the generalization thus produces the pmf of the distribution of the number-correct score for a test taker on n dichotomous test items. As convolution operations are both commutative and distributive, the order in which the steps are taken does not matter. For simplicity, we will always assume the order in which the items appear in the test.

Direct use of Equation 6.21 would not take us much further than Equation 6.7 though; both involve the same numbers of additions and multiplications. However, its repeated application of the same operation suggests the use of a rather simple recursive algorithm: Obviously, for the first item, the only probabilities of its "sum score" are q_1 for $X = 0$ and p_1 for $X = 1$. Addition of the second item leads to the first convolution, which produces the result in Equation 6.20. Adding the third item means a new operation, but now on the function for the first two items in Equation 6.20 and the Bernoulli function for the third item. The result can be written as

$$f_3 * (f_2 * f_1)(x) = \begin{cases} q_3(f_2 * f_1)(x) & \text{for } x = 0; \\ q_3(f_2 * f_1)(x) + p_3(f_2 * f_1)(x-1), & \text{for } x = 1,\ 2; \\ p_3(f_2 * f_1)(x) & \text{for } x = 3. \end{cases} \tag{6.22}$$

The first and third case in Equation 6.22 are obvious; the second tell us that the event of x successes after $n = 3$ items can only be result of (i) x successes after two items and no success on the third or (ii) $x - 1$ successes after two items but now with an additional success on the third. Table 6.1 shows the products of all q_i and p_i parameters in the convolution of the response variables of three test items. For reasons that will become clear below, the third and fourth column show the partial products for the two kinds of parameters separately. The elements in the last column are just the products of those in the third and fourth column.

The step to a recursive algorithm is now straightforward. Beginning with the Bernoulli function for the first item ($s = 1$), the algorithm runs through iterative steps $s = 2, \ldots, n$ each time performing a convolution on the result from the previous step and the function for the next item. Let $\pi_s(z)$, $z = 0, \ldots, s$, denote the values of the new function obtained at step s (that is, the pmf of the distribution of the sum-score on the first s items). More formally, the algorithm can then be written as follows:

1. For $s = 1$, set

$$\pi_1(z) \equiv \begin{cases} q_i, & z = 0; \\ p_i, & z = 1. \end{cases} \tag{6.23}$$

TABLE 6.1

Products of q_i and p_i Parameters Involved in the Convolution of the Response Variables for Three Items

	Items	Products of q_i and p_i		
$x=0$	$\varnothing$	$q_1q_2q_3$	1	$q_1q_2q_3$
	{1}	q_2q_3	p_1	$p_1q_2q_3$
$x=1$	{2}	q_1q_3	p_2	$q_1p_2q_3$
	{3}	q_1q_2	p_3	$q_1q_2p_3$
	{1,2}	q_3	p_1p_2	$p_1p_2q_3$
$x=2$	{1,3}	q_2	p_1p_3	$p_1q_2p_3$
	{2,3}	q_1	p_2p_3	$q_1p_2p_3$
$x=3$	{1,2,3}	1	$p_1p_2p_3$	$p_1p_2p_3$

2. For $s=2,\ldots,n$, calculate

$$\pi_s(z) \equiv \begin{cases} q_s\pi_{s-1}(z), & z=0; \\ q_s\pi_{s-1}(z)+p_s\pi_{s-1}(z-1), & z=1,\ldots,s-1; \\ p_s\pi_{s-1}(z), & z=s; \\ 0, & \text{otherwise.} \end{cases} \tag{6.24}$$

The algorithm is extremely efficient as it does not suffer from the explosion of the numbers of additions and multiplications involved in the use of Equation 6.7 or Equation 6.11. At each step s, the total of number of these operations is only equal to $3(s-1)$.

As already indicated, the algorithm was introduced in the test-theory literature by Lord and Wingersky (1984), almost as a casual note. However, it has been reinvented at multiple places. For instance, the same recursive scheme was already proposed by Fischer (1974, Section 14.3) to solve a similar problem in the calculation of the elementary symmetric functions involved in CML estimation of the parameters in the Rasch model.

Both Hanson (1994) and Thissen et al. (1995) generalized the algorithm for use with polytomous items. Their generalization can be motivated analogously to the dichotomous case. Let U_{ia} be the binary variable used to indicate whether ($U_{ia}=1$) or not ($U_{ia}=0$) a given test taker produces a response on item i in scoring category $a=0,\ldots,A$, with $\sum_{a=1}^{A} U_{ia}=1$ for each i. The probability of a response vector $\mathbf{U}_i=\mathbf{u}$ for item i is

$$f(\mathbf{u};\mathbf{p}_i) \equiv \Pr\{\mathbf{U}_i=\mathbf{u};\mathbf{p}_i\} \equiv \begin{cases} \prod_{a=1}^{A} p_{ia}^{u_{ia}}, & u_{ia}=0,1; \quad a=0,\ldots,A; \\ 0, & \text{otherwise,} \end{cases} \tag{6.25}$$

where $\mathbf{p}_i \equiv (p_{i1},\ldots,p_{iA})$ with $\sum_{a=1}^{A} p_{ia}=1$. The distribution of the total score $X \equiv \sum_{i=1}^{n}\sum_{a=1}^{A} U_{ia}$ is now a compound multinomial with a pmf that can be given as

$$\begin{aligned} f(x;n,\mathbf{p}) &\equiv \Pr\{X=x;n,\mathbf{p}\} \\ &= \begin{cases} \sum_{\Sigma\Sigma u_{ia}=x} \prod_{i=1}^{n}\prod_{a=1}^{A} p_{ia}^{u_{ia}}, & x=0,1,\ldots,n; \\ 0, & \text{otherwise,} \end{cases} \end{aligned} \tag{6.26}$$

where $\mathbf{p} \equiv (\mathbf{p}_1,\ldots,\mathbf{p}_n)$.

Conceiving of Equation 6.26 as the result of $n-1$ successive convolutions of the functions in Equation 6.25, its values can be calculated using the following generalization of Equations 6.23 and 6.24:

1. For $s = 1$, set

$$\pi_1(z) \equiv p_{1a}, \text{ for } z = 0, \ldots, A. \tag{6.27}$$

2. For $s = 2, \ldots, n$, calculate

$$\pi_s(z) \equiv \begin{cases} q_s \pi_{s-1}(z), & z = 0; \\ \sum_{v=1}^{z} p_{sv} \pi_{s-1}(z-v), & z = 1, \ldots, sA-1; \\ p_{sA} \pi_{s-1}(z), & z = sA; \\ 0, & \text{otherwise.} \end{cases} \tag{6.28}$$

Further generalization to the case of items with unequal numbers of score categories requires some careful accounting but is otherwise straightforward.

6.4 Continuous Convolutions

For real-valued functions $f_1, \ldots, f_n$, the convolution sums in Equation 6.21 are replaced by integrals

$$(f_2 * f_1)(x) \equiv \int_{-\infty}^{\infty} f_1(x-m) f_2(m) dm \tag{6.29}$$

with $x \in \mathbb{R}$. An obvious application in educational and psychological testing is the derivation of the distribution of the total time spent on a test from the (nonidentical) distributions of the RTs on the items.

As RT model, we use the lognormal model, which assumes the following pdfs for the items:

$$f(t_i; \tau, \alpha_i, \beta_i) = \frac{\alpha_i}{t_i\sqrt{2\pi}} \exp\left\{-\frac{1}{2}[\alpha_i(\ln t_i - (\beta_i - \tau))]^2\right\}, \tag{6.30}$$

where $\tau \in \mathbb{R}$ is the speed of the test taker, parameter $\beta_i \in [-\infty, \infty]$ represents the amount of labor required by item i, and α_i can be interpreted as its discrimination parameter (see Volume One, Chapter 16).

Just as in the preceding case of item scores, RTs can be assumed to be conditionally independent between items given a test taker working at a fixed speed. Consequently, the distribution of the total time on a test of n items

$$T \equiv \sum_{i=1}^{n} T_i$$

can be viewed as the result of series of $n-1$ convolutions beginning with

$$(f_{T_2} * f_{T_1})(t) = \int f_{T_2}(z) f_{T_1}(t-z)dz. \tag{6.31}$$

The problem of the distribution of sums of lognormals has a long tradition, dating back at least to Naus (1969), with a wide range of applications including signal-processing in engineering, distributions of claims in actuarial science, and income distributions in economics. In spite of its practical relevance, exact solutions are unknown to exist. Particularly, a sum of lognormals is not lognormal itself; unlike a few well-known families of distributions as the normal, γ, and χ-square, the lognormal family is not closed under convolution. But even worse, accurate numerical evaluation of the pdf of sums of lognormals is far beyond our current computational capabilities, even for cases involving only a handful of items.

As a first alternative one might consider making a call to a central limit theorem for nonidentical variables (e.g., Liapuonov's theorem; Lehman, 1999, Section 2.7) and use a normal approximation. But convergence to normality is generally slow. Each of the histograms with the total-time distributions in Figure 6.2 is for the sum of the RTs of $N = 10,000$ test takers sampled from the distributions on $n = 20$, 40, 60, and 80 items randomly taken from a large pool. The pool was for a real-world testing program calibrated under the lognormal model and the test takers were taken to operate at a speed equal to the average for its population, one standard deviation below the average or one standard deviation above it. Even after 80 items the shape is still not that of a normal distribution, no matter the speed of the test taker. However, the longer upper tails of all these distributions suggest something close to a lognormal. Indeed, an old idea, originated in engineering (Fenton, 1960), is fitting a lognormal to sums of lognormals matching a pair of their lower-order cumulants. The author of this chapter has used the approximation with remarkable success, both as for its fit and the applications it has made possible.

It is important to make a careful distinction between the lognormal pdfs for the items with the special parameterization in Equation 6.30 and the member of the standard lognormal family used to approximate their sum. The latter has the pdf

$$f(x;\mu,\sigma^2) = \frac{1}{x\sigma\sqrt{2\pi}} \exp\left\{-\frac{1}{2}\left(\frac{\ln x - \mu}{\sigma}\right)^2\right\} \tag{6.32}$$

with parameters μ and $\sigma^2 > 0$ (see Chapter 3).

The moments about zero of the family of standard lognormals are equal to

$$m_k \equiv \mathcal{E}(X^k) = \exp(k\mu + k^2\sigma^2/2) \tag{6.33}$$

(Kotz and Johnson, 1985; note that the square for the last k in the equation is erroneously omitted in this reference). We want to find the member with the first two cumulants (mean and variance) that match those of the sum of the RTs (for an alternative based on the second and third cumulant, see van der Linden, 2011, Appendix A). From Equation 6.33, the mean and variance of the standard lognormal are seen to be equal to

$$\mathcal{E}(X) = \exp(\mu + \sigma^2/2); \tag{6.34}$$

$$\mathcal{E}[X - \mathcal{E}(X)]^2 = \exp(2\mu + \sigma^2)[\exp(\sigma^2) - 1]. \tag{6.35}$$

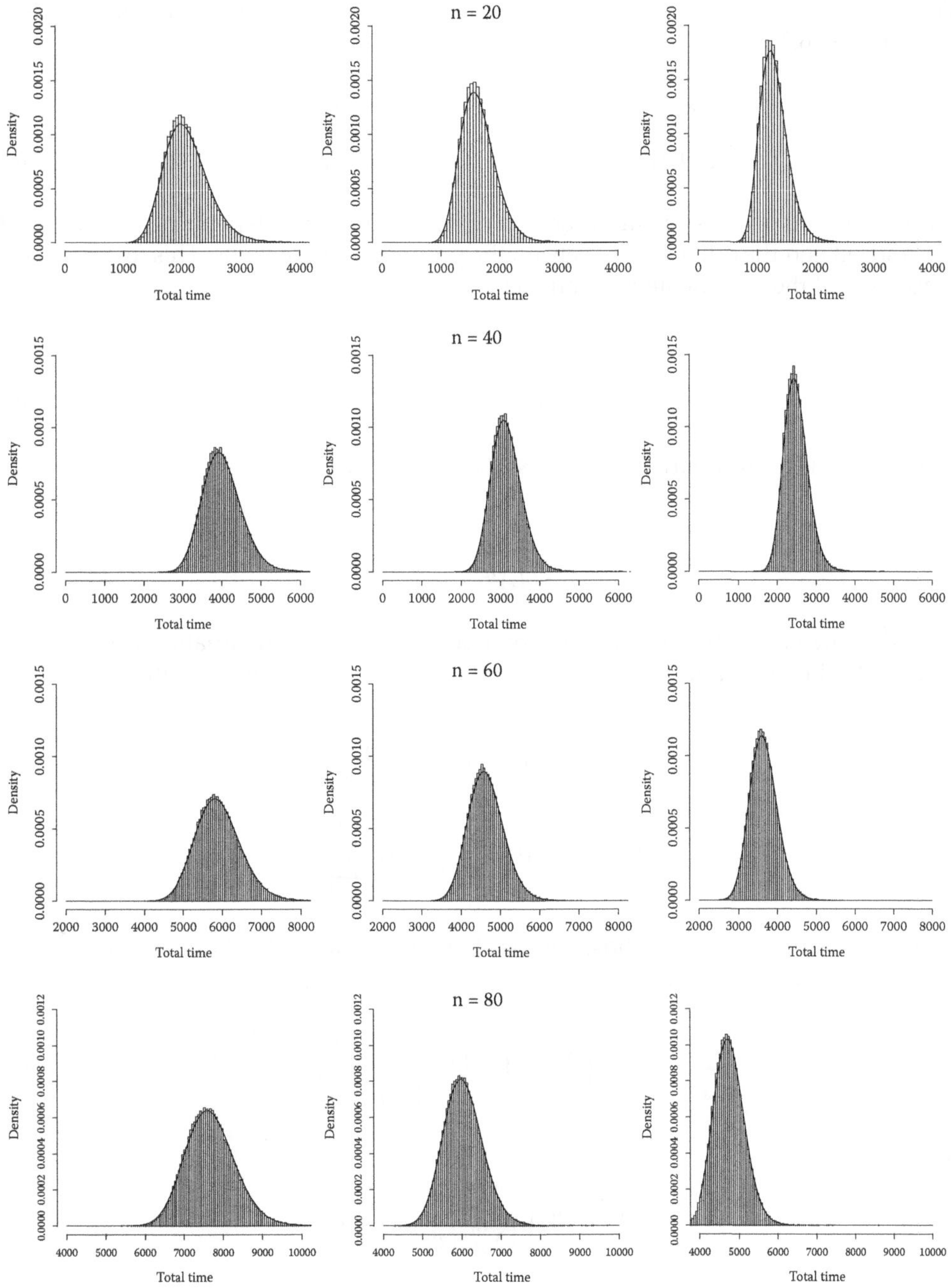

FIGURE 6.2
Histograms with total-time distributions of samples of $N = 10{,}000$ test takers operating at average (middle panes), one standard deviation below average (left panes), and one standard deviation above average speed (right panes) on tests of $n = 20, 40, 60$, and 80 items, along with the standard lognormal curves with the parameters in Equations 6.44 and 6.45 fitted to them.

For the specific parameterization of the lognormal model in Equation 6.30, it follows from Equation 6.33 that

$$\mathcal{E}(T_i) = \exp(-\tau)\exp(\beta_i + \alpha_i^{-2}/2); \tag{6.36}$$

$$\mathcal{E}[T_i - \mathcal{E}(T_i)]^2 = \exp(-2\tau)\exp(2\beta_i + \alpha_i^{-2})[\exp(\alpha_i^{-2}) - 1]. \tag{6.37}$$

Observe that both expression are products of two factors, one depending only on the test taker's speed and the other exclusively on the properties of the items. This feature suggests replacement of the two the item parameters by

$$q_i \equiv \exp(\beta_i + \alpha_i^{-2}/2); \tag{6.38}$$

$$r_i \equiv \exp(2\beta_i + \alpha_i^{-2})[\exp(\alpha_i^{-2}) - 1], \tag{6.39}$$

which allows us to summarize Equations 6.36 and 6.37 as

$$\mathcal{E}(T_i) = \exp(-\tau)q_i; \tag{6.40}$$

$$\mathcal{E}[T_i - \mathcal{E}(T_i)]^2 = \exp(-2\tau)r_i. \tag{6.41}$$

A well-known property of cumulants of independent random variables is their additivity. Thus, for the mean and variance of the total time T on a test it holds that

$$\mathcal{E}(T) = \exp(-\tau)\sum_{i=1}^{n} q_i; \tag{6.42}$$

$$\mathcal{E}[T - \mathcal{E}(T)]^2 = \exp(-2\tau)\sum_{i=1}^{n} r_i. \tag{6.43}$$

Equating these expressions to those for the standard lognormal in Equations 6.34 and 6.35 and solving for μ and σ^2 gives

$$\mu = -\tau + \ln\left(\sum_{i=1}^{n} q_i\right) - \ln\left(\frac{\sum_{i=1}^{n} r_i}{[\sum_{i=1}^{n} q_i]^2} + 1\right)/2 \tag{6.44}$$

and

$$\sigma^2 = \ln\left(\frac{\sum_{i=1}^{n} r_i}{[\sum_{i=1}^{n} q_i]^2} + 1\right). \tag{6.45}$$

The curves fitted to the histograms with the total-time distributions in Figure 6.2 are for standard lognormals with these parameters. Except for minor systematic underestimation of the densities near the mode, the fit is generally good. Observe especially the excellent fit to both tails, which is where the exclusive interest is in applications of RT modeling to the detection of aberrant test behavior, test speededness, or the selection of time limits on a test. The reason for this fit lies in the features of the lognormal distribution captured by its two parameters. Remember that μ and σ^2 are not the mean and variance of a lognormal variate X but of its log. However, μ can be shown to be equal to the median of the distribution.

From Equation 6.33, for $k = 1$, it follows that the mean of a lognormal is always larger than its median by an amount that depends directly on σ^2. Thus, together these two parameters not only capture the location and spread but also the typical right skew of total-time distributions.

Analogous to the case of the number-correct score distributions in Equation 6.18, the expressions in Equations 6.42 and 6.43 allow us to derive a simple set of conditions for the identity of two total-time distributions. Let $i = 1, \ldots, n$ and $j = 1, \ldots, m$ denote the items in two different test forms. It then follows that, for any value of speed parameter τ, the two forms have identical distributions of total time T if and only if

$$\sum_{i=1}^{n} q_i = \sum_{j=1}^{m} q_j; \tag{6.46}$$

$$\sum_{i=1}^{n} r_i = \sum_{j=1}^{m} r_j. \tag{6.47}$$

As the identity holds for any τ, it automatically holds for the marginal distribution of T given any distribution of τ.

Note that the conditions follow directly from the lognormal RT model in Equation 6.30; they do not involve any approximation to the total-time distributions. Also, the conditions are actually less stringent than those for the number-correct score distributions in Equation 6.18; it is no longer necessary for the two forms to have equal numbers of items. The conditions can easily be imposed on the assembly of a new test form to guarantee the same degree of speededness for each test taker as a reference form; for an application of the methodology that allows us to do so, see van der Linden (see Volume Three, Chapter 12).

6.5 Other Applications

The earlier results for the compound binomial distribution are indifferent as to the nature of the underlying Bernoulli trials. Hence, they can be applied to any instance in educational and psychological testing where we need to count the results from such trials. We demonstrate the possibilities for a few applications in forensic analysis of test behavior to detect cheating. As the subject is addressed more completely elsewhere in this handbook (see Volume Three, Chapter 13), we focus on its distributional aspects only.

The first example is the detection of answer copying between test takers. Let c be a test taker suspected of copying the answers on multiple-choice items during testing using s as his source. Obviously, the number of matching responses between c and s is a relevant statistic. But matches also arise at random when two test takers work completely independently of each other. So, the problem of detecting answer copying can be conceived of as a right-sided statistical hypothesis test with the number of matching responses as statistic.

We use

$$I_{csi} \equiv \begin{cases} 1, & U_c = U_s; \\ 0, & U_c \neq U_s \end{cases} \tag{6.48}$$

as the (random) indicator of matching answer choices by c and s on item i, and let $a = 1, \ldots, A$ denote the response alternatives for the items. Assume a response model that gives us the probabilities $p_{cia}(\theta_c)$ and $p_{sia}(\theta_s)$ of c and s choosing alternative a, respectively. If they work independently, the probability of a matching choice on item i is equal to

$$p_{csi} \equiv \Pr\{I_{csi} = 1\} = \sum_{a=1}^{A} p_{cia}(\theta_c) p_{sia}(\theta_s) \tag{6.49}$$

(van der Linden and Sotaridona, 2006). It follows that, under the null hypothesis of no answer copying, the total number of matching responses by c and s,

$$M_{cs} \equiv \sum_{i=1}^{n} I_{csi} \tag{6.50}$$

has a compound binomial distribution with the success parameters p_{csi} in Equation 6.49 for the items. The distribution can be evaluated using the algorithm in Equations 6.23 and 6.24 with the substitution of $(p_{csi}, 1 - p_{csi})$ for (p_s, q_s).

If c actually copies on an item, the probability of a match on it is equal to one. If the subset of items copied, Γ, is claimed to be known, the alternative distribution is a compound binomial with parameters p_{cis} for $i \notin \Gamma$ and 1 for $i \in \Gamma$. Otherwise, the test is against a general alternative.

The same framework can be applied to detect other types of cheating. van der Linden and Jeon (2012) applied it to detect fraudulent erasures on answer sheets detected by optical scanners; for instance, erasures by teachers or school administrators trying to increase the test scores of their students. As response model they used a two-stage IRT model for the probabilities of a wrong-to-right change of an initial answer on an item by a test taker who reviews his answer sheet. The probabilities imply a compound binomial as null distribution for the total number of wrong-to-right changes. Another example with a similar result is the detection of whether or not a test taker had any preknowledge of items stolen from the testing agency prior to the administration of the test.

Instead of classical statistical hypothesis testing, detection of cheating on tests can be based on Bayesian analysis. We then need to find the posterior odds of cheating given the test taker's response vector. This alternative has the attractive option of allowing us to introduce existing prior evidence in the analysis, for instance, in the form of a report by a test proctor on suspicious behavior observed while the test taker worked on a specific section of the test. We only highlight the combinatorial nature of the posterior odds, which appears to be similar to that of all earlier cases.

Again, consider a test taker c suspected of copying answers. Let M denote the observed set of matching responses and Γ the unknown subset of items that were actually copied (of course, it may hold that $\Gamma = \varnothing$). The posterior odds of c having copied any of the answers in M can be shown to be equal to

$$\frac{1 - \Pr\{\Gamma = \varnothing \mid \theta_c, \mathbf{u}_c, \mathbf{u}_s\}}{\Pr\{\Gamma = \varnothing \mid \theta_c, \mathbf{u}_c, \mathbf{u}_s\}} = \frac{\sum_{\varnothing \neq \Gamma \subset M} \prod_{i \in \Gamma} \gamma_i \prod_{i \in M \setminus \Gamma} \xi_{ci}}{\prod_{i \in M} \xi_{ci}}, \tag{6.51}$$

where γ_i is the prior probability of the test taker copying on item i and $\xi_{ci} \equiv (1 - \gamma_i) p_{ci}$ is the product of the prior probability of no copying on item i times the regular response probability for the chosen alternative given the ability level of s (van der Linden and Lewis,

2015). Because of norming constants that cancel, the right-hand side of Equation 6.51 can be interpreted as the ratio of (i) the sum of the posterior probabilities associated with all possible combinations of c having copied some of the items in M but not on any of its other items and (ii) the posterior probability of c not having copied on any of the item in M at all.

Table 6.2 illustrates the combinatorics in Equation 6.51 for the observation of $M = 3$ matching responses. The first column lists all possible combinations of the three items in M, the second and third, all possible products of their γs and ξs, respectively, and the last column, the complete product of all these parameters. The posterior odds in Equation 6.51 are just the ratio of the first entry in the last column (denominator) and the sum of all its other entries (numerator). The structure of this table is formally identical that of the compound binomial probabilities in Table 6.1. In fact, it can be shown to be the result of successive convolutions of functions with the two possible values of ξ_{ci} and γ_i. Consequently, the posterior odds in Equation 6.51 can be calculated using the recursive algorithm in Equations 6.23 and 6.24 with the substitution of (γ_i, ξ_{ci}) for (p_i, q_i).

Observe, however, that, unlike all earlier cases, ξ_{ci} and γ_i are *no* probabilities. As noted earlier, convolutions of pmfs are only special cases; discrete convolutions are more generally defined as operations on any two pairs of functions defined on nonnegative integers (or just two sequences of numbers). Another application of this more general case arises in CML estimation of the Rasch model. For the version of the model with the original parameterization introduced in Rasch (1960)—that is,

$$\Pr\{U_{pi} = 1; \vartheta_p, \varepsilon_i\} = \frac{\vartheta_p}{\vartheta_p + \varepsilon_i} \tag{6.52}$$

with ability parameter ϑ_p and item parameter ε_i (see Volume One, Chapter 3)—the likelihood equations for the item parameters given their marginal totals $U_{.i} = \sum_p u_{pi}$ can be written as follows:

$$u_{.i} = \sum_{r=1}^{n-1} N_r \frac{\varepsilon_i \sum_{\{r-1;v\neq i\}} \prod_v \varepsilon_v}{\sum_{\{r\}} \prod_v \varepsilon_v}, \quad i = 1, \ldots, n, \tag{6.53}$$

where N_r is the number of test takers with r items correct. The ratio in the right-hand side of Equation 6.53 has as denominator the sum of all possible products of length r of the n item parameters and as numerator the same sum but now with the restriction that ε_i be included in each of them.

TABLE 6.2

Example of All Products of γ_i and ξ_{ci} in the Posterior Odds for $M = 3$ Matching Responses

Items	Products of γ_i and ξ_{ci}		
∅	1	$\xi_{c1}\xi_{c2}\xi_{c3}$	$\xi_{c1}\xi_{c2}\xi_{c3}$
{1}	γ_1	$\xi_{c2}\xi_{c3}$	$\gamma_1\xi_{c2}\xi_{c3}$
{2}	γ_2	$\xi_{c1}\xi_{c3}$	$\xi_{c1}\gamma_2\xi_{c3}$
{3}	γ_3	$\xi_{c1}\xi_{c2}$	$\xi_{c1}\xi_{c2}\gamma_3$
{1,2}	$\gamma_1\gamma_2$	ξ_{c3}	$\gamma_1\gamma_2\xi_{c3}$
{1,3}	$\gamma_1\gamma_3$	ξ_{c2}	$\gamma_1\xi_{c2}\gamma_3$
{2,3}	$\gamma_2\gamma_3$	ξ_{c1}	$\xi_{c1}\gamma_2\gamma_3$
{1,2,3}	$\gamma_1\gamma_2\gamma_3$	1	$\gamma_1\gamma_2\gamma_3$

For a fixed value of r, combinatorial expressions as $\sum_{\{r\}} \prod_v \varepsilon_v$ are known as rth-order elementary symmetric functions in $(\varepsilon_1, \ldots, \varepsilon_n)$. However, observe how their structure corresponds with the elements of the fourth column of Table 6.1. For example, for $r = 2$ the elementary symmetric function in $(\varepsilon_1, \varepsilon_2, \varepsilon_3)$ is defined as $\varepsilon_1\varepsilon_2 + \varepsilon_1\varepsilon_3 + \varepsilon_2\varepsilon_3$, which matches the sum of the entries in this column for $x = 2$. It follows that we can calculate all the expressions in Equation 6.53 for $i = 1, \ldots, n$ by a single run of the recursive algorithm in Equations 6.23 and 6.24, this time with the substitution of $(\varepsilon_i, 1)$ for (p_i, q_i). In fact, this version of the algorithm was exactly the one proposed as "summation algorithm" by Fischer (1974, Section 14.3). For more on its history, see Formann (1986), Gustaffson (1980), Jiang et al. (2013), Liou (1994), and Verhelst et al. (1984).

6.6 Conclusion

A fundamental feature of test items is different response distributions for each test taker on different items. Consequently, the number-correct score of a given test taker on a test with dichotomous items has a distribution that is the result of an $(n - 1)$-fold convolution of the different Bernoulli distributions for the items. A similar results holds for the case of polytomous items. The combinatorial structure underlying such convolutions is found not just for convolutions of probability functions but of any set of discrete functions. A prominent example of the more general case are the elementary symmetric functions emerging in CML estimation of the Rasch model. For a given order, these functions have a structure entirely identical to that of the probability of a number-correct score of the same size on a test of dichotomous items.

References

Feller, W. 1968. *An Introduction to Probability Theory and Its Applications* (3rd ed.; vol. 1). New York: Wiley.

Fenton, L. 1960. The sum of log-normal probability distributions in scatter transmission systems. *IRE Transactions on Communication Systems, 8*, 57–67.

Gustafsson, J.-E. 1980. A solution of the conditional estimation problem for long tests in the Rasch model for dichotomous items. *Educational and Psychological Measurement, 40*, 377–385.

Fischer, G. H. 1974. *Einführung in Die Theorie Psychologischer Tests*. Bern, Switzerland: Huber.

Formann, A. K. 1986. A note on the computation of the second-order derivatives of the elementary symmetric functions in the Rasch model. *Psychometrika, 51*, 335–339.

Hanson, B. A. 1994. *Extension of Lord–Wingersky Algorithm to Computing Test Score Distributions for Polytomous Items* (Research Note). Retrieved December 18, 2014, from http://www.b-a-h.com/papers/note9401.html.

Hodges, Jr. J. L. and Le Cam, L. 1960. The Poisson approximation to the Poisson-binomial distribution. *The Annals of Mathematical Statistics, 31*, 737–740.

Jiang, H., Graillat, S., and Barrio, R. 2013. Accurate and fast evaluation of elementary symmetric functions. *Proceedings of the 21st IEEE Symposium on Computer Arithmetic* (pp. 183–190). Austin, TX.

Kotz, S. and Johnson, N. L. 1985. *Encyclopedia of Statistical Science* (vol. 5). New York: Wiley.

Lehman, E. L. 1999. *Elements of Large Sample Theory*. New York: Springer.

Liou, M. 1994. More on the computation of higher-order derivatives of the elementary symmetric functions in the Rasch model. *Applied Psychological Measurement, 18*, 53–62.

Lord, F. M. and Novick, M. R. 1968. *Statistical Theories of Mental Test Scores*. Reading, MA: Addison-Wesley.

Lord, F. M. and Wingersky, M. S. 1984. Comparison of IRT true-score and equipercentile observed-score "equatings." *Applied Psychological Measurement, 8*, 453–461.

Naus, J. I. 1969. The distribution of the logarithm of the sum of two log-normal variates. *Journal of the American Statistical Society, 64*, 655–659.

Rasch, G. 1960. *Probabilistic Models for Some Intelligence and Attainment Tests*. Chicago, IL: University of Chicago Press.

Thissen, D., Pommerich, M., Billeaud, K., and Williams, V. 1995. Item response theory for scores on tests including polytomous items with ordered responses. *Applied Psychological Measurement, 19*, 39–49.

van der Linden, W. J. 2011. Setting time limits on tests. *Applied Psychological Measurement, 35*, 183–199.

van der Linden, W. J. and Jeon, M. 2012. Modeling answer changes on test items. *Journal of Educational and Behavioral Statistics, 37*, 180–199.

van der Linden, W. J. and Lewis, C. 2015. Bayesian checks on cheating on tests. *Psychometrika, 80*, 689–706.

van der Linden, W. J. and Luecht, F. M. 1998. Observed-score equating as a test assembly problem. *Psychometrika, 63*, 401–418.

van der Linden, W. J. and Sotaridona, L. 2006. Detecting answer copying when the regular response process follows a known response model. *Journal of Educational and Behavioral Statistics, 31*, 283–304.

Verhelst, N. D., Glas, C. A. W., and van der Sluis, A. 1984. Estimation problems in the Rasch model: The basic symmetric functions. *Computational Statistics Quarterly, 1*, 245–262.

Walsh, J. E. 1953. Approximate probability values for observed numbers of successes. *Sankhya, Series A, 15*, 281–290.

Walsh, J. E. 1963. Corrections to two papers concerned with binomial events. *Sankhya, Series A, 25*, 427.

7

Information Theory and Its Application to Testing

Hua-Hua Chang, Chun Wang, and Zhiliang Ying

CONTENTS

7.1 Introduction

Information was first introduced by R.A. Fisher (1925) in a mathematically precise formulation for his statistical estimation theory (Kullback, 1959). For an observation Y assumed to follow density $f(\cdot;\theta)$ parametrized by θ, the Fisher information is defined by

$$I(\theta) = E_\theta\left[\frac{\partial \log f(Y;\theta)}{\partial\theta}\right]^2,$$

where E_θ denotes the expectation under density $f(\cdot;\theta)$. Fisher showed that the asymptotic variance of the maximum-likelihood estimator (MLE) is the inverse of $I(\theta)$, which also serves as the lower bound to the variance of any other unbiased estimator. In other words, the MLE is optimal in terms of minimizing variance. An alternative expression for the Fisher information is

$$I(\theta) = -E_\theta\left[\frac{\partial^2 \log f(Y;\theta)}{\partial\theta^2}\right]$$

giving its geometric interpretation as the expected curvature of the log-likelihood function. The Fisher information is a nonnegative number when θ is a scalar and it is a nonnegative definite matrix when θ is a vector. It provides a lower bound, also known as the Cramer–Rao inequality, to the variance (or covariance matrix in the multidimensional case) of any unbiased estimator of θ.

A second important concept in the information theory that is of statistical nature is the Kullback–Leibler (KL) distance (divergence or information), introduced in 1951 by Solomon Kullback and Richard Leibler. Their motivation for this concept partly arises from "the statistical problem of discrimination," as they put it (Kullback and Leibler, 1951, p.79). They believed that statistically, "two populations differ more or less according as to how difficult it is to discriminate between them with the best test." Indeed, the KL distance gives precisely the asymptotic power for the likelihood ratio test, which is the best (most powerful) according to the Neyman–Pearson theory.

The KL distance between two probability density (mass) functions, f_1 and f_2, is defined as

$$KL(f_1, f_2) = \int f_1(y) \log\left[\frac{f_1(y)}{f_2(x)}\right] d\mu(y),$$

where μ is typically the Lebesgue measure for the continuous case and the counting measure for the discrete case. The KL distance is always nonnegative and attains zero if and only if the two density functions are identical. But it is not, strictly speaking, a distance in the usual sense since, among other things, it is not a symmetric function and does not satisfy the triangle inequality.

Suppose our statistical problem is to test null hypothesis H_0: $\theta = \theta_0$ against alternative H_1: $\theta = \theta_1$ for two specified values θ_0 and θ_1. From the Neyman–Pearson theory, it is well known that the optimal test is the likelihood ratio based test: rejecting H_0 in favor of H_1 if and only if the likelihood ratio statistic

$$LR = \frac{f(Y;\theta_1)}{f(Y;\theta_0)}$$

exceeds certain threshold. The KL distance, $KL(H_1, H_0)$ may be thought of, as least for the case of a discrete random variable, as the expected log odds of H_1 against H_0, with expectation taken under H_1.

The KL distance also has a natural Bayesian interpretation. Following Kullback (1959), we place prior probabilities, π_0 and π_1, on H_0 and H_1, respectively. Then the KL distance is simply the expected logarithm of the posterior odds minus the logarithm of the prior odds.

$$E_{f(y|\theta_1)}\left[\log\frac{f(y|\theta_1)}{f(y|\theta_0)}\right] = E_{f(y|\theta_1)}\left[\log\frac{f(y|\theta_1)\pi_1}{f(y|\theta_0)\pi_0} - \log\frac{\pi_0}{\pi_1}\right].$$

In other words, the farther away the posterior odds is from the prior odds, the larger the KL is. When $f(y|\theta_1) = f(y|\theta_0)$ for all y, the KL distance is zero and the posterior odds is equal to the prior odds. In this case, the observation does not provide any additional information to discriminate between the two hypotheses. An alternative way of utilizing the KL information is to explicitly compute the KL distance between two posterior distributions of θ as follows:

$$E_{p(\theta|\boldsymbol{u},u)}\left[\log\frac{p(\theta|\boldsymbol{u},u_j)}{p(\theta|\boldsymbol{u})}\right],$$

which quantifies the amount of useful information gathered about θ by adding a new response u_j to the test (Mulder and van der Linden, 2010; Wang and Chang, 2011).

Another important development in the information theory is that of entropy, introduced by Claude Shannon in his 1948 paper. The Shannon entropy quantifies the uncertainty in a probability density function or the corresponding random variable. For a probability mass function $p(y)$ on $\{y_1, \ldots, y_m\}$, the Shannon entropy is defined as

$$S_p = -\sum_{i=1}^{m} p(y_i) \log p(y_i).$$

Let u_m be the uniform density over $\{y_1, \ldots, y_m\}$. Then, S_p can be rewritten as follows:

$$S_p = \log m - \sum_{i=1}^{m} p(y_i) \log \frac{p(y_i)}{u_m(y_i)},$$

which is equal to $\log m - KL(p, u_m)$, a constant $\log m$ subtracting off the KL distance between p and u_m. It attains its maximum value, $\log m$, when $KL(p, u_m) = 0$ or $p(y_i) = m^{-1}, i = 1, \ldots, m$, and its minimum value, 0, when the random variable Y is degenerate, that is, Y is a nonrandom constant.

The Fisher information, KL distance, and Shannon entropy share a number of properties, an important one being additive when independent observations are added to Y, the observed random variable. Specifically, suppose that Y is an m-vector and that its m components are mutually independent random variables. The Fisher information of Y is the sum of the information of its components, so are the KL distance and Shannon entropy. In addition, the KL distance and Shannon entropy are always nonnegative and only become 0 in a degenerate case. The Fisher information is also nonnegative and becomes a nonnegative definite matrix when θ is multidimensional.

7.2 One-Dimensional Continuous Binary IRT Models

When item response theory (IRT) was initially developed, Birnbaum (1968) introduced the concept of information to quantify power of ability-level classification and precision of ability-level estimation. Birnbaum's approach borrowed heavily from the development of asymptotic efficiency by E. J. Pitman (1949) and other statisticians; see Serfling (1980). It contains formal definitions of scoring system-specific information, test information, and item information, as well as their connections to the Fisher information. A succinct account of Birnbaum's work was given by Lord (1980), who advocated use of the MLE and incorporated the Fisher information into the conceptualization of computerized adaptive testing (CAT).

7.2.1 Item Information, Test Information, and Fisher Information

Suppose that the latent ability of interest can be assessed using a test composed of items that can be accurately modeled by a unidimensional IRT model and that the true latent ability for a person is denoted by θ. Suppose further that a scoring system for the test is specified with the observed score for the person denoted by y. If the same test was given to the person repeatedly and independently, the resulting observed score y would form a distribution with mean $\mu_{y|\theta}$ and standard deviation $\sigma_{y|\theta}$ which is also known in the classical test theory as the conditional standard error of measurement. Following Lord (1980), for two individuals with different levels θ_1 and θ_2, an intuitive way to measure how well the test can differentiate the two levels of θ is through the following "signal-to-noise" ratio

$$\frac{(\mu_{y;\theta_2} - \mu_{y;\theta_1})/\sigma_{y;\theta_1,\theta_2}}{\theta_2 - \theta_1},$$

where $\sigma_{y;\theta_1,\theta_2}$ is the pooled standard error of measurement at the two θ values. Use of $\theta_2 - \theta_1$ in the denominator is based on an assumption that the distance between individuals on the expected observed score scale increases *linearly* with an increase in distance on the θ-scale, and this assumption will be satisfied only when the two θ values are very close to each other, that is, when both θ_2 and θ_1 approaching a same value, say θ, resulting in the following expression:

$$\lim_{\theta_2 \to \theta \leftarrow \theta_1} \frac{(\mu_{y;\theta_2} - \mu_{y;\theta_1})/\sigma_{y;\theta_1,\theta_2}}{\theta_2 - \theta_1} = \frac{\frac{\partial \mu_{y;\theta}}{\partial \theta}}{\sigma_{y;\theta}}. \tag{7.1}$$

Geometrically, Equation 7.1 may be viewed as the slope of the expected test score curve at a single point θ (Lord, 1980). It is closely related to the well-known Pitman's asymptotic efficiency (Serfling, 1980), introduced initially by Pitman (1949) for statistical hypothesis testing. It indicates how well differences in θ can be detected using the test with observed score y (Reckase, 2009). Birnbaum (1968) named the squared value of Equation 7.1,

$$I(\theta, y) = \frac{\left[\frac{\partial E(y;\theta)}{\partial \theta}\right]^2}{\sigma^2_{y;\theta}}, \tag{7.2}$$

the information function of score (scoring system) y. He showed that Equation 7.2 is always bounded above by the Fisher information $I(\theta)$, a fact also known as the Cramer–Rao inequality in mathematical statistics. When y is a single binary response U_i to the ith item, Equation 7.2 simplifies to

$$I(\theta, u_i) = \frac{\left[\frac{\partial E(u_i|\theta)}{\partial \theta}\right]^2}{P_i(\theta)Q_i(\theta)} = \frac{\left[\frac{\partial P_i(\theta)}{\partial \theta}\right]^2}{P_i(\theta)Q_i(\theta)}, \tag{7.3}$$

where $P_i(\theta) = 1 - Q_i(\theta)$ is the probability of correct response to the ith item. Birnbaum (1968) showed that Equation 7.3 attains the Fisher information given by observation $U_i = u_i$, and named Equation 7.3 the item information function. Because of local independence and additivity property of the Fisher information, it follows that the Fisher information from a test consisting of $i = 1, \ldots, n$ items is the sum of individual item information, that is,

$$I(\theta) = \sum_{i=1}^{n} \frac{\left[\frac{\partial P_i(\theta)}{\partial \theta}\right]^2}{P_i(\theta)Q_i(\theta)}. \tag{7.4}$$

Birnbaum further showed that the Fisher information can be achieved with a weighted scoring system with weights proportional to $P_i'(\theta)/P_i(\theta)Q_i(\theta)$ and he called Equation 7.4 the test information function. Therefore, an optimal way to combine the u_i for an overall score would be to use

$$y = \sum_{i=1}^{n} \frac{P_i'(\theta)}{P_i(\theta)Q_i(\theta)} u_i. \tag{7.5}$$

However, this approach has an implementation problem in that the weights contain the latent ability θ which is unobserved. On the other hand, the MLE of θ, $\hat{\theta}$, can be shown through the use of the Taylor series expansion to be a linear function of Equation 7.5, and thus asymptotically equivalent to the latter. In other words, the MLE $\hat{\theta}$ in effect achieves the optimal scoring system given by Equation 7.5. Lord consistently advocated the use of MLE in the estimation of θ.

The Fisher information function can be readily used to assess the accuracy/power of an IRT based test, as long as the MLE is used for scoring. Its dependence on θ entails that testing accuracy varies from individual to individual. A test can be informative for one examinee but much less so for another. Hence, it is desirable to match tests to individual ability levels so that θ can be efficiently measured. Such a consideration led F. Lord, in a series of papers published around 1970, to conceptualize what he called tailored testing and what now known as the CAT (Volume Three, Chapter 10). Lord envisioned that items could be selected sequentially and tailored to an examinee's ability level θ. He further argued that such a testing format could be realized on computer-based tests and on a large scale when computing technology became ready.

Use of the Fisher information function is a centerpiece in Lord's tailored testing. Ideally, at each step, it is the item information that dictates which item is to be selected. The amount of information may be sensitive to the specification of the ability level θ, especially for highly discriminating items, as pointed out by Chang and Ying (1999). Figure 7.1 shows two item-information functions of the two-parameter logistic (2PL) model with the same difficulty but different discrimination parameters. When the difficulty parameter matches the ability level, the item information of the more discriminating item is larger than that of the less discrimination item. However, the opposite is true when the difficulty parameter is away from the ability level. Thus, if the knowledge of θ is substantial, then a

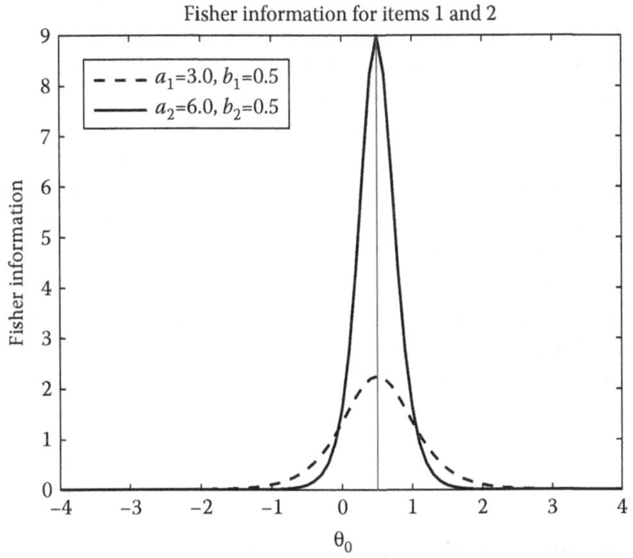

FIGURE 7.1
Fisher information curves for two items with different a-parameters.

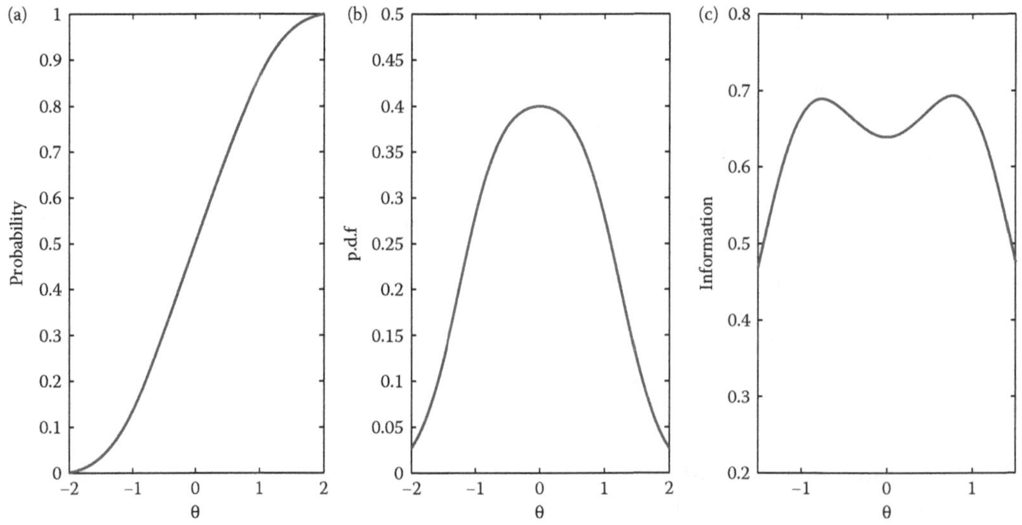

FIGURE 7.2
(a) Item response function ($P(\theta)$), (b) density function ($P'(\theta)$), and (c) item information function ($I(\theta)$).

highly discriminating item should be used; but if that knowledge is limited, then a lowly discriminating item may be more suitable.

An intuitive explanation for maximizing the Fisher (item) information in CAT item selection is that in doing so items will be selected to match the examinee's ability level. Such a connection is appealing because it endows a theoretically backed selection criterion with a common sense name. In one of his early papers on tailored testing, Lord (1971) introduced the measurement community to the technique of stochastic approximation, a sequential method developed by Robbins and Monro (1951). In the context of IRT-based adaptive testing, the Robbins–Monro method means that if a correct response is observed, then a more difficult item may be selected while if the response is incorrect, then the selection will be towards a less difficult item. In other words, the Robbins–Monro approach results in matching difficulty with ability.

Simple derivations show that for the 2PL and normal ogive models (three-parameter models require a shift due to nonsymmetry), item information reaches the maximum value when difficulty parameter b is matched to ability θ. Bickel et al. (2001) investigated whether or not such a result would continue to hold for link functions other than the logistic and normal curves. Rather surprisingly, they found that the answer is not always true, even for distribution functions from a family of unimodal distributions. They gave an example of normal-uniform mixture whose Fisher information does not reach the maximum value when b and θ are matched; see Figure 7.2. Another example provided by them has a flat Fisher information, meaning that for a range of b around θ, the Fisher information remains the same.

7.2.2 KL Information and Its Application in Testing

While the Fisher information quantifies the power to differentiate two nearby ability levels, it is the KL information that provides the discrimination power for any two ability levels.

The KL information is closely related to the likelihood-ratio tests in statistical hypothesis testing. For testing simple null hypothesis $\theta = \theta_0$ versus simple alternative hypothesis $\theta = \theta_1$, it is known that the likelihood-ratio test is most powerful. The KL information function is the average (expectation) of the log-likelihood ratio.

The application of the KL information to testing was first introduced by Chang and Ying (1996), who called it global information in reference to the fact that it does not require that two ability levels to be close to each other and provides a global profile of the discrimination power for any two ability levels. Because the KL information is also additive under the local independence assumption, the test KL information is the sum of item KL information. For binary item i, its KL information takes form

$$KL_i(\theta_0, \theta_1) = P_i(\theta_0) \log\left(\frac{P_i(\theta_0)}{P_i(\theta_1)}\right) + Q_i(\theta_0) \log\left(\frac{Q_i(\theta_0)}{Q_i(\theta_1)}\right), \tag{7.6}$$

which is a bivariate function (of θ_0 and θ_1). In contrast, the Fisher information is a function of θ_0 only and represents discrimination power around θ_0 (Hambleton and Swaminathan, 1985, p.102). For a test of n items, the corresponding KL information can therefore be written as

$$KL^n(\theta_0, \theta_1) = \sum_{i=1}^{n} KL_i(\theta_0, \theta_1). \tag{7.7}$$

From Equation 7.1, it is clear that the Fisher information is a rate (per unit change in θ scale) function. In contrast, the KL information represents an absolute (not relative) measure of power. As θ_1 approaches θ_0, the KL information approaches zero, as it becomes more and more difficult to distinguish between the two ability levels. However, the two concepts are related and in fact the Fisher information function can be recovered from the KL information function through differentiation as shown by the following equation:

$$\frac{\partial^2}{\partial \theta_1^2} KL(\theta_0, \theta_1)_{\theta_1 = \theta_0} = I(\theta_0).$$

Therefore, for any given θ_0, the Fisher information $I(\theta_0)$ is simply the curvature of the KL information function (viewed as a function of θ_1) at θ_0.

As a bivariate function, the KL information is more complicated to handle, but also more informative. Figure 7.3a displays the KL information surface for a single item. As it shows, for any fixed pair (θ_0, θ_1), $KL(\theta_0, \theta_1)$ (the z-coordinate) represents the discrimination power at this point. Cutting the surface by different vertical planes, such as a vertical plane at $\theta_0 = 0$ shown in Figure 7.3b, results in a KL information curve that presents the discrimination power of an item in distinguishing θ_0 from all possible θ_1 values.

The Fisher information at $\theta_0 = 0$ is represented by the curvature of the KL information curve at $\theta_0 = 0$. From Figure 7.3a and b, one observes that KL information function changes its shape as θ_0 changes its values. But no matter how it changes, the KL information is always zero along the entire 45° line ($\theta_0 = \theta_1$). Figure 7.1c and d plot the KL information functions for two items with $\theta_0 = 0$. For each function, the curvature at 0 is equal to the value of the Fisher information at 0. Note that in terms of the Fisher information, item 1 provides more information than item 2; however, this is not the case for the KL information.

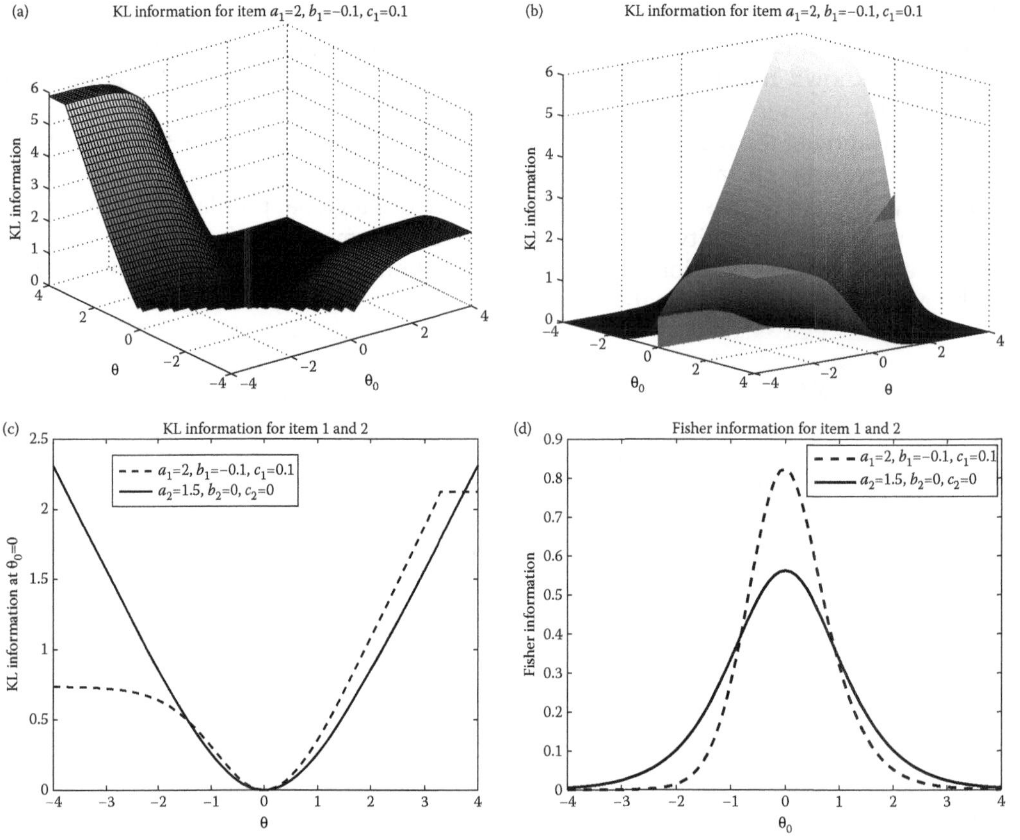

FIGURE 7.3
Fisher and KL information for unidimensional items. (a) KL information surface for one item, (b) KL information surface cut by a plane $\theta_0 = 0$, (c) Fisher information curve for two items, and (d) Fisher information curve for two items.

7.2.3 Applications

The applications of information theory in testing mainly focus on two aspects: (1) evaluating the quality of items/tests and facilitating parallel test assembly; (2) selecting items sequentially in computer adaptive testing. With the IRT gaining popularity, the definition of parallel tests have gradually shifted from equal reliability in classical test theory to the same test information curve. Lord (1977) following Birnbaum (1968), proposed to use the test information function as the quality control criterion of test assembly. His procedure includes first deciding the target shape of the test information curve and then sequentially select items to fill up the area underneath the target curve. Figure 7.4a illustrates the "target" test information curve for a 25-item test from ASVAB. By controlling the test-information curve, the level of measurement error is controlled. In the same year, Samejima (1977) published her paper with similar ideas. She proposed the concept of "weakly parallel tests," which is defined as "tests measuring the same ability or latent ability whose test information functions are identical" (p. 194). What is so significant about both Equations 7.4 and 7.7 is that test information is additive. This feature enables test developers to

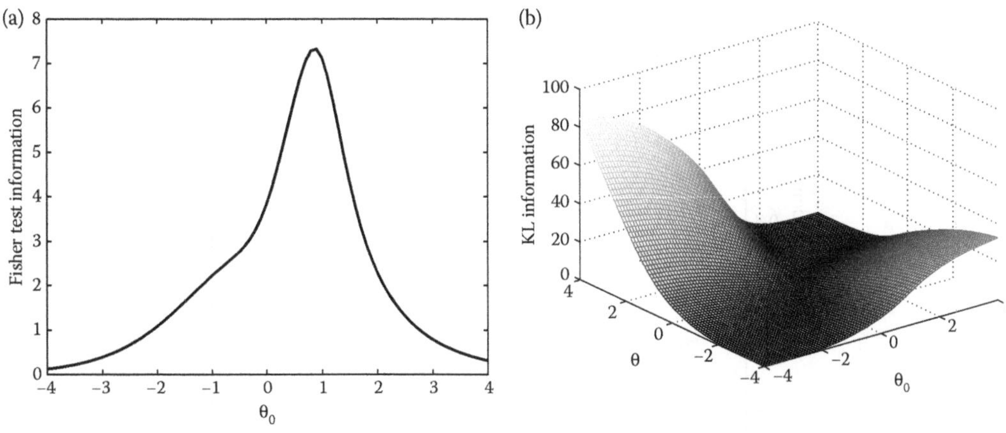

FIGURE 7.4
Target Fisher information curve and KL information surface for 25-item ASVAB test. (a) Target test Fisher information curve and (b) Target test KL information surface.

separately calculate the information for each item and combine them to form updated test information at each stage.

Mimicking Lord's procedure (1977) for test construction based on the Fisher information, Chang et al. (1997) proposed another test construction procedure, which can briefly describe as follows: (1) decide on the shape desired for the test KL information function; that is, a target KL information surface as shown in Figure 7.4b; (2) select items with the item KL information surfaces that will fill the hard-to-fill volumes under the target information surface; (3) cumulatively add the item information surfaces; and (4) continue until the volume under the target information surface is filled to a satisfactory approximation.

If we view this KL information-based procedure as test construction in a three-dimensional (3D) space, test construction using Fisher information can be considered as construction in a two-dimensional (2D) space. 3D test construction differs from 2D test construction in that the latter is solely based on knowing the examinee's true ability θ_0 and emphasizing the test is constructed for examinees with θ_0, whereas, at every fixed θ_0, test construction based on the KL information uses global information to evaluate the discrimination power of the item when the estimator is both near and away from θ_0. The curvature along the 45° line represents the local information on the θ_0 scale. Thus, the 2D test construction can also be carried out in the 3D space; that is, matching Fisher information is equivalent to matching curvatures. On the other hand, the 3D method puts emphasis not only on the curvature but on the entire surface. In that sense, the 3D information surface produces a more comprehensive information profile than the 2D Fisher information curve does.

Item selection in adaptive testing is another realm that could benefit tremendously from information theory. The maximum-information approach, proposed by Lord (1977), is one of the earliest and the most popular item selection method in adaptive testing. The idea is to select each next item as the one with maximum Fisher information. Let $\hat{\theta}$ denote the interim ability estimate after n items. The next candidate item is the one that maximizes $I_i(\hat{\theta})$. Under the 3PL model, maximizing Fisher information means intuitively matching item difficulty parameter with the latent ability level of an examinee. Because the latent ability is unknown, this optimal item selection rule cannot be implemented but may be

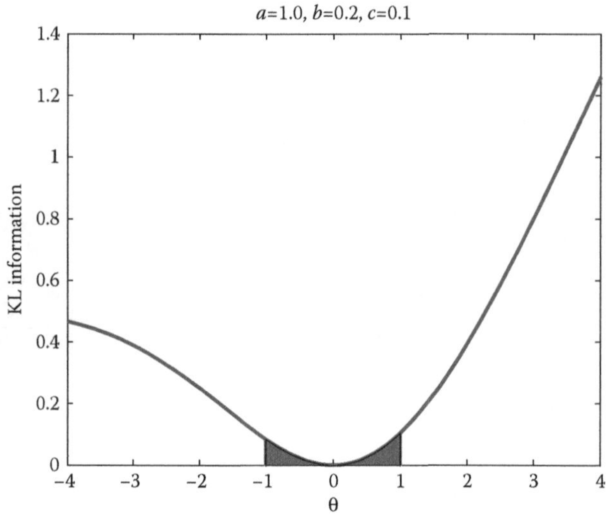

FIGURE 7.5
Illustration of the KL index.

approximated using the interim estimate $\hat{\theta}$ each time a new item is selected. Under this maximum-information approach, items with highly discrimination parameters will preferentially be selected. The statistical rationale behind maximizing the Fisher information is to make $\hat{\theta}$ most efficient.

Almost a decade after the maximum-information approach was first proposed, and much later implemented in the first real-world large-scale testing porgrams, Chang and Ying (1996) identified a statistical imperfection behind this approach. They argued that the Fisher information should not be used at the beginning of the test when $\hat{\theta}$ is likely not close to its destination θ_0 and suggested that the KL information be used instead. As the test moves along, when there are enough items in the test to ensure that $\hat{\theta}$ is close enough to θ_0, the Fisher information should then be used (Chang and Ying, 1999, 2008). They proposed a KL information index (called KL index or KI hereafter), defined as

$$KI_i(\hat{\theta}) = \int_{\hat{\theta}-\frac{r}{n}}^{\hat{\theta}+\frac{r}{n}} KL_i(\hat{\theta}, \theta)d\theta. \tag{7.8}$$

Geometrically, the KL index is simply the area below the KL information curve, bounded by the integration interval, as shown in Figure 7.5. Items with the maximum KL index value are selected, where r is chosen to be a constant, usually equal to 3. This KL index summarizes the discrimination power of an item in differentiating $\hat{\theta}$ from all its neighboring levels. In real-world adaptive testing, the examinee's true θ_0 is unknown and $\hat{\theta}$ is used as a point estimate; the integration interval is large at the beginning of the test when n is small so as to cover θ_0 that might be far away from $\hat{\theta}$; toward the end of the test when n is large, the interval shrinks and when n approximates infinity, maximizing the KL index is the same as maximizing the item Fisher information at $\hat{\theta}$. Simulation studies have shown that the KL index method provides more robust item selection especially when test length is short.

7.3 Multidimensional Continuous Binary IRT Models

Multidimensional item response theory (MIRT) deals with situations in which multiple skills or attributes are being tested simultaneously. It has gained much recent attention in both educational and psychological assessment. For instance, in educational testing, many certification and admission boards are interested in combining regular tests with diagnostic services to allow candidates to obtain more informative diagnostic profiles of their abilities (Mulder and van der Linden, 2009). In this section, we discuss extensions of both Fisher and KL information to cover MIRT and present some interesting applications.

7.3.1 Fisher Information in MIRT

The concept of item information for the unidimensional binary IRT models is readily extended to the multidimensional case. Let $\theta = (\theta_1, \ldots, \theta_d)^T$ be the vector of d latent abilities. For dichotomous item i, the response probability is denoted by $P_i(\theta) = 1 - Q_i(\theta)$. Analogous to Equation 7.3, the Fisher information $\boldsymbol{I}_i(\theta)$ is a $d \times d$ matrix whose (j,k)th entry is given by

$$I_i^{(j,k)}(\theta) = -E\left[\frac{\partial^2 \log L(\theta)}{\partial\theta_j \partial\theta_k}\right] = \frac{\frac{\partial P_i(\theta)}{\partial\theta_j}\frac{\partial P_i(\theta)}{\partial\theta_k}}{P_i(\theta)Q_i(\theta)}, \tag{7.9}$$

where $L(\theta)$ is the likelihood function. Moreover, it follows from the assumption of local independence and additivity property that the Fisher information matrix for a test consisting of n items is $\sum_{i=1}^{n} \boldsymbol{I}_i(\theta)$. For the 3PL multidimensional compensatory model with item response function $P_i(\theta) = c_i + (1 - c_i)\frac{1}{1+\exp(a_i'\theta - b_i)}$, the (j,k)th entry of its item information matrix is

$$I_i^{(j,k)}(\theta) = \frac{Q_i(\theta)}{P_i(\theta)}\left(\frac{P_i(\theta) - c_i}{1 - c_i}\right)^2 a_{ij}a_{ik}. \tag{7.10}$$

In addition, as a generalization of the earlier geometric argument on the Fisher information, in the multidimensional context, the discrimination power of an item in differentiating two adjacent θ points ($\theta_2 \rightarrow \theta \leftarrow \theta_1$) along the direction β can be expressed as a quadratic form

$$\cos(\beta)^T \boldsymbol{I}_i(\theta)\cos(\beta),$$

where β is a vector of angles from the coordinate axes defining the direction taken from the θ-point. Again with the 3PL multidimensional compensatory model, the item information becomes

$$\frac{Q_i(\theta)}{P_i(\theta)}\left(\frac{P_i(\theta) - c_i}{1 - c_i}\right)^2\left(\sum_{j=1}^{d} a_{ij}\cos\beta_j\right)^2. \tag{7.11}$$

To graphically illustrate the item information, suppose an item measures two dimensions. Then the item information in the direction specified by the β-vector can be represented by an information surface. Figure 7.6 shows the item response surface for a particular item on the left panel, and the item information with three different angles in the right three panel—angles of 45°, 90, and 0°. In the latter, the information surfaces have

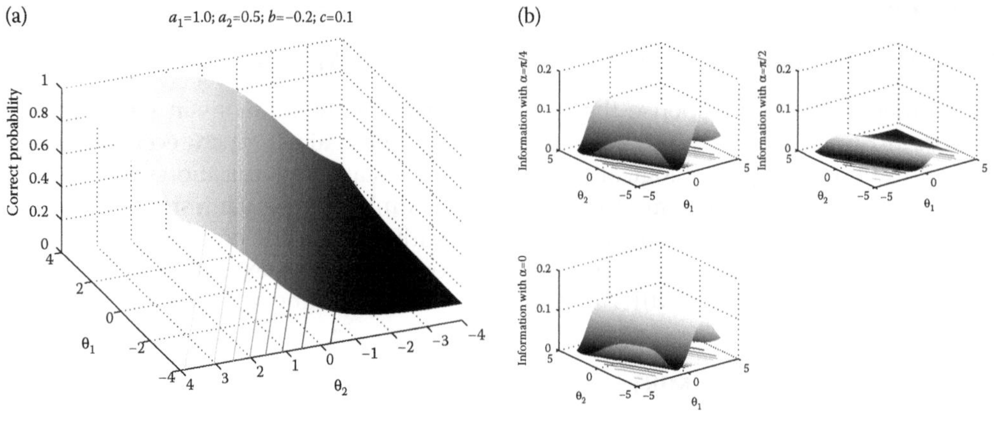

FIGURE 7.6
Item response surface for a single item in (a) and item Fisher information surface (see Equation 7.11) at three different angles in (b).

the shape of a ridge that has its highest region over the 0.5 probability contour line for the item—this is also the line in the θ-space where the slope of the item response surface is the greatest as well. This figure indicates that the amount of information an item contains depend on the item difficulty as well as the angle from which we compute the information. When the angle is 90°, the height of the information surface is much lower because a_2 is smaller than a_1 for this particular item.

7.3.2 KL Information in MIRT

The extension of the KL information to multidimensional IRT models is more straightforward, as the KL information in Equation 7.6 involves the likelihood function only at two parameter values and is itself dimensionless. In other words, when dichotomous items are considered, Equation 7.6 remains the same for the MIRT when the two-parameter values are understood as vector valued. The item KL information can again be interpreted as representing the discrimination power of an item in differentiating two latent ability levels. In addition, the Fisher information (matrix) can be obtained from the KL information through differentiation (Wang et al., 2011)

$$I^{(j,k)}(\theta_0) = \frac{\partial^2}{\partial\theta_j\theta_k} KL(\theta_0, \theta) \mid_{\theta=\theta_0} . \tag{7.12}$$

In the 2D case, the diagonal elements in the Fisher information matrix relates to the curvature of the KL information curve resulting from cutting the 2D KL information surface by a vertical plane that is parallel to either side coordinate plane. For instance in Figure 7.7, the left panel shows the KL information surface for an item assuming $\theta_0 = (-1, 1)$. If we cut the surface by a plane $\theta_{02} = 1$, the resulting curve is the KL information curve in unidimensional case, and the curvature of the curve at $\theta_{01} = -1$ equals to the second diagonal element in the Fisher information matrix. The off-diagonal elements, however, have no explicit geometric interpretations.

The KL information is also related to the item information defined in Equation 7.11. Wang et al. (2011) showed analytically that if the vertical cutting plane is not parallel to either

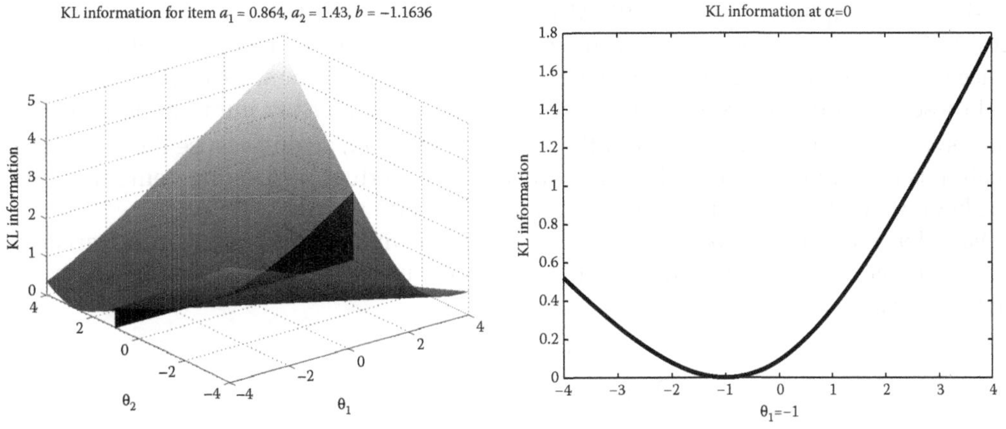

FIGURE 7.7
KL information surface for an item and its relationship with the item Fisher information matrix.

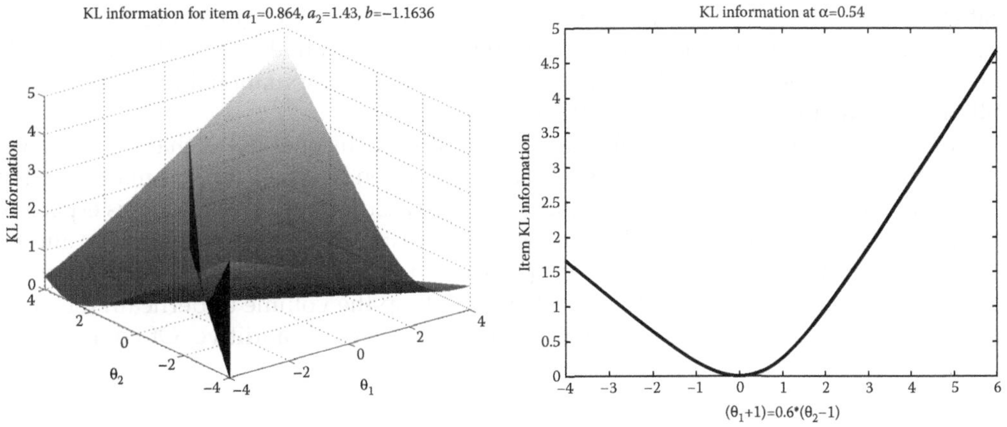

FIGURE 7.8
KL information surface for an item and its relationship with the item information defined from the "discrimination power" perspective.

side coordinate plane, such as in Figure 7.8, the resulting KL information curve has the curvature at θ_0 equal to $P_i(\theta_0)Q_i(\theta_0)(a_{i1}\cos\beta + a_{i2}\sin\beta)$, which is exactly the same as the item information in Equation 7.11 when $c_i = 0$.

7.3.3 Applications

Information theory is also crucial to multidimensional test assembly and adaptive testing. Multidimensional adaptive testing (MAT) offers at least two advantages over unidimensional adaptive testing (UAT): (a) MAT includes more information than UAT because the multiple subscales being measured are often correlated; (b) MAT can balance content coverage automatically without fully resorting to content balancing techniques (Segall, 1996). For MAT, various Fisher information matrix based indices have been proposed for item selection. They include (1) its determinant, which is also known as D-optimality in optimal design (see Chapter 16; Segall, 1996; Mulder and van der Linden, 2009; Wang and Chang, 2011); (2) its trace, known as A-optimality (Mulder and van der Linden, 2009); and (3) the trace of inverse of the Fisher information matrix (van der Linden, 1996; Veldkamp, 2002). All these built upon the statistical connection between the Fisher information matrix and the asymptotic covariance matrix of the MLE, $\hat{\theta}$. However, again because the Fisher information is only a local information measure, it might produce misleading item selections at the beginning of the test, the KL information comes into rescue.

Van der Linden and Veldkamp (2002) were the first to propose a multivariate KL index that is simply the multiple integral of the KL information over a multidimensional hypercube as

$$KI_i(\hat{\theta}, \theta) = \int_{\hat{\theta}_1-\frac{3}{n}}^{\hat{\theta}_1+\frac{3}{n}} \cdots \int_{\hat{\theta}_d-\frac{3}{n}}^{\hat{\theta}_d+\frac{3}{n}} KL_i(\hat{\theta}, \theta)\partial\theta.$$

An illustration of the KL index in 2D case, where it is the volume under the KL information surface enclosed by a square, is presented in Figure 7.9. Wang et al. (2011) proved that maximizing the KL index in a limiting case (i.e., when test length n goes to infinity) is equivalent to maximizing the trace of the Fisher information matrix, and they proposed a simplified version of the KL index that largely reduces the computational complexity while still maintaining the efficiency of the algorithm.

Another advantage of the KL information in multidimensional IRT is that for any given two latent ability levels, θ_0 and θ_1, the test KL information is simply the summation of item KL information, as a result of the local independence assumption. This feature is especially convenient in parallel test assembly. Suppose that a test measures two dimensions and the target KL information surface at a fixed θ_0 is chosen in advance. Then, one simply needs to select appropriate items from the item bank to fill up the volume underneath the target KL information surface, and the contribution of every item is additive. Of course this is

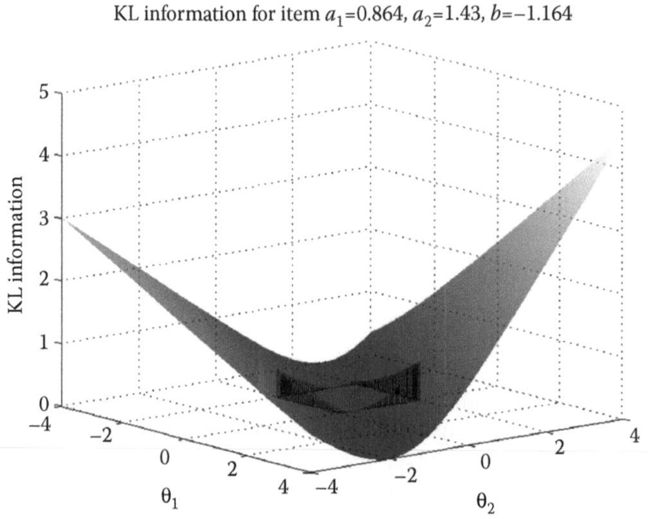

FIGURE 7.9
Illustration of the KL index in 2D case.

much more complicated than in the unidimensional case because, for each different θ_0, the target information surface will take on a different shape, so that the entire test-assembly problem becomes a multiobjective maximization problem. When the Fisher information matrix is considered, the contribution of every item can no longer be isolated, which adds complexity to the test assembly (van der Linden, 1996). For instance, both van der Linden (1996) and Veldkamp (2002) used weighted sum of

$$\text{var}(\hat{\theta}_1|\theta) = \frac{\sum_{i=1}^{n} a_{i2}^2 P_i(\theta) Q_i(\theta)}{\left[\sum_{i=1}^{n} a_{i1}^2 P_i(\theta) Q_i(\theta)\right]\left[\sum_{i=1}^{n} a_{i2}^2 P_i(\theta) Q_i(\theta)\right] - \left[\sum_{i=1}^{n} a_{i1} a_{i2} P_i(\theta) Q_i(\theta)\right]^2}$$

and

$$\text{var}(\hat{\theta}_2|\theta) = \frac{\sum_{i=1}^{n} a_{i1}^2 P_i(\theta) Q_i(\theta)}{\left[\sum_{i=1}^{n} a_{i1}^2 P_i(\theta) Q_i(\theta)\right]\left[\sum_{i=1}^{n} a_{i2}^2 P_i(\theta) Q_i(\theta)\right] - \left[\sum_{i=1}^{n} a_{i1} a_{i2} P_i(\theta) Q_i(\theta)\right]^2}$$

as their target objective functions to be maximized in the test assembly. Each variance component is obtained from the inverse-Fisher information matrix. The sums are taken over the items in the test, and, apparently, the contribution of each item to the variance terms is dependent on the remaining items in the test.

7.4 Diagnostic Classification Models

Diagnostic classification models (DCMs) are important psychometric tools in cognitive assessment that have recently received a great deal of attention (Rupp et al., 2010). In a typical setting of cognitive diagnosis, there are d attributes or skills and possession or lack of it for an individual is specified through a d-vector $\boldsymbol{\alpha}$ of 1's and 0's. Clearly, there are 2^d possible configurations (values) for $\boldsymbol{\alpha}$. The discrete nature of the DCMs makes use of the Fisher information typically inappropriate. Thus, the focus here will be on the KL information and the Shannon entropy. Throughout this section, $P_i(\boldsymbol{\alpha})$ ($Q_i(\boldsymbol{\alpha})$) will denote the probability of a correct (incorrect) response to the ith item for an individual with attributes configuration $\boldsymbol{\alpha}$.

Under the formulation of a diagnostic classification model, there are a finite number of possible configurations an examinee's latent abilities belong to. The goal of classification is to identify the examinee's latent configuration of mastery levels. Thus, the classification error rates are of primary concern. The application of information theory in this context mainly focuses on designing appropriate item selection algorithms for cognitive diagnostic computerized adaptive testing (CD-CAT). CD-CAT incorporates cognitive diagnosis into a sequential design of adaptive testing. It has the potential to determine students' mastery levels more efficiently, along with other advantages offered in traditional adaptive testing. Sequentially selecting items based on an examinee's current classification is particularly appropriate for developing assessment to address today's challenges in learning. The goal of CD-CAT is to tailor a test to each individual examinee via an item selection algorithm that allows the test to hone in on the examinees' true status of $\boldsymbol{\alpha}$ in an interactive manner (Chang, 2012). Information theory can also play a pivotal role in designing such tests. We describe below two approaches, one based on the KL information and the other on the Shannon entropy.

Let α_0 denote the true cognitive profile for an examinee. A good item should be sensitive to differentiate α_0 from all other possible cognitive profiles. The average KL information between α_0 and the remaining $2^d - 1$ attribute configurations can be defined as

$$AKL_i(\alpha_0) = \sum \left[P_i(\alpha_0) \log \left(\frac{P_i(\alpha_0)}{P_i(\alpha)}\right) + Q_i(\alpha_0) \log \left(\frac{Q_i(\alpha_0)}{Q_i(\alpha)}\right) \right], \tag{7.13}$$

where the summation is over all $\alpha \neq \alpha_0$. Since α_0 is unknown, we can replace it with its current estimator $\hat{\alpha}$ and use $AKL_i(\hat{\alpha})$ as the selection criterion, that is, select the item that maximizes the average KL information; see Xu et al. (2003) for details.

A second approach is to use the Shannon entropy, which measures the uncertainty associated with a distribution. Taking a Bayesian viewpoint, an item is more informative if adding the item results in a more concentrated posterior distribution; or, in terms of the Shannon entropy, a smaller entropy for the posterior distribution. This idea was first introduced in Tatsuoka (2002) and Tatsuoka and Ferguson (2003), and it is briefly presented below. Suppose that $n-1$ items have been administered with responses denoted by an $n-1$ dimensional vector $\boldsymbol{u}_{n-1}$. Let u_i denote the response to the next item that is selected. Furthermore, let $\pi(\alpha|\boldsymbol{u}_{n-1}, u_i)$ denote the posterior probability density for the latent state α given n responses $\boldsymbol{u}_{n-1}$ and u_i and $\pi(\alpha|\boldsymbol{u}_{n-1}, k) = \pi(\alpha|\boldsymbol{u}_{n-1}, u_i = k)$, $k = 0, 1$. We define expected posterior Shannon entropy

$$\begin{aligned} EPSE_i = \sum_{\alpha} \Bigg[& \pi(\alpha|\boldsymbol{u}_{n-1}, 1) \log \left(\frac{1}{\pi_n(\alpha|\boldsymbol{u}_{n-1}, 1)} \right) P_i(\boldsymbol{u}_{n-1}) \\ & + \pi(\alpha|\boldsymbol{u}_{n-1}, 0) \log \left(\frac{1}{\pi_n(\alpha|\boldsymbol{u}_{n-1}, 0)} \right) Q_i(\boldsymbol{u}_{n-1}) \Bigg], \end{aligned}$$

where $P_i(\boldsymbol{u}_{n-1}) = 1 - Q_i(\boldsymbol{u}_{n-1})$ is the posterior probability of $u_i = 1$ given $\boldsymbol{u}_{n-1}$. The resulting selection rule is to choose item i so that $EPSE_i$ is minimized. As mentioned earlier, the Shannon entropy can be viewed as a constant minus the KL distance between the uniform distribution and the posterior distribution of α. Thus, minimizing the Shannon entropy is tantamount to maximizing the divergence between the posterior distribution of α and the uniform distribution.

While traditional CAT and CD-CAT are tailored to the examinee's latent ability evel θ and cognitive mastery profile α, respectively, it may be desirable to incorporate both goasl into a single item-selection criterion so that both θ and α are accurately estimated (Chang, 2012). Being applicable to both cases, the KL information becomes a natural candidate to construct a summary criterion function which can serve the dual purpose. Indeed, Cheng and Chang (2007) developed a dual-information index, which is a weighted average between $\mathrm{KI}(\hat{\theta})$ and $\mathrm{KI}(\hat{\alpha})$, that is,

$$\mathrm{KI}_i(\hat{\theta}, \hat{\alpha}) = w\mathrm{KI}_i(\hat{\theta}) + (1-w)\mathrm{KI}_i(\hat{\alpha}), \tag{7.14}$$

where $0 \leq w \leq 1$ is a weight assigned to the KL index of $\hat{\theta}$. Therefore, the next item can be selected to maximize the weighted dual information defined in Equation 7.14. Wang et al. (2011) extended the dual information algorithm to a method called Aggregate Ranked Information (ARI), after observing that the two KL indices may not be on the same scale. Their idea is to transform the two information indices to ordinal scales in such a way that each item will have two percentile ranks for $KI(\hat{\theta})$ and $KI(\hat{\alpha})$ separately. ARI is therefore calculated as

$$\mathrm{ARI}_i = w\,R(\mathrm{KI}_i(\hat{\theta})) + (1 - w)\,R(\mathrm{KI}_i(\hat{\alpha})), \tag{7.15}$$

where $R(\cdot)$ is the rank function. As such, the scale difference between $KI(\hat{\theta})$ and $KI(\hat{\theta})$ is removed. Simulation studies indicate that this ARI method is capable of getting more precise latent ability estimates, along with accurate attribute mastery classification (Wang et al., 2014).

7.5 Conclusion

The concept of information for testing, initiated by A. Birnbaum and F. Lord, is at the core of IRT. It has given rise to various applications central to the field of measurement (Hambleton and Swaminathan, 1985, p. 101). The Fisher information, due to its connection to the maximum-likelihood estimation and the associated asymptotic theory, has been widely used to define item and test information. Alternative approaches to the use of information, especially the work by Chang and Ying (1996), Tatsuoka and Ferguson (2003), Xu et al. (2005), Wang et al. (2011), and Wang and Chang (2011) that rely on the KL distance and the Shannon entropy, have recently been developed. These approaches either complement the weakness of or fill the gap left by the Fisher information method. Some recent large-scale applications of CAT and cognitive diagnostic assessment (e.g., Chang, 2012; Liu et al., 2013) have shown that both KL information and Shannon entropy can work very effectively on innovative tasks such as classifying students' mastery levels on a set of skills pertinent to learning. Knowledge which skills students have or have not mastered, teachers can target their instructional interventions to the areas where students need most improvement. As assessment objectives and tasks are getting more complex and challenging, adding new members to the current information family might still be welcome. For example, the concept of mutual information proposed by Wang and Chang (2011) certainly needs further study and development.

Despite the wide range of applications of the Fisher information in test development, quality control, linking designs, automated test assembly, and item selection designs in CAT, issues of fundamental nature remain to be resolved. In particular, the bimodal phenomenon given by Bickel et al. (2001) and discussed in this chapter shows that theory may not always give what we expect from the intuition. It would be interesting to investigate its practical implications, including robustness of the Fisher information-based methods when the underlying IRT model is misspecified. For the KL distance and the Shannon entropy, more research is needed for their application to diagnostic assessment, particularly on how such measures can guide us in designing/delivering more reliable and efficient tests. Finally, the information theory has yet to be incorporated into the setting when response times are also included in latent trait analysis (Gulliksen, 1950; van der Linden, 2007; Volume One, Chapter 29; Wang et al., 2012), that is, when there are simultaneous considerations of power and speed.

References

Baker, F. M. and Kim, S. H. 2004. *Item Response Theory: Parameter Estimation Techniques* (2nd ed.). New York: Marcel Dekker.

Bickel, P., Buyske, S., Chang, H., and Ying, Z. 2001. On maximizing item information and matching difficulty with ability. *Psychometrika, 66*, 69–77.

Birnbaum, A. 1968. Some latent trait models and their use in inferring an examinee's ability. In F. M. Lord and M. R. Novick (eds.), *Statistical Theories of Mental Test Scores*. Reading, MA: Addison-Wesley.

Chang, H. 1996. The asymptotic posterior normality of the latent trait for polytomous IRT models. *Psychometrika, 61*, 445–463.

Chang, H. H. (2012). Making computerized adaptive testing diagnostic tools for schools. In R. W. Lissitz and H. Jiao (eds.), *Computers and Their Impact on State Assessment: Recent History and Predictions for the Future* (pp. 195–226). Charlotte, NC: Information Age Publisher.

Chang, H. Wang, S., and Ying, Z. 1997. *Three Dimensional Visulization of Item/Test Information*. Paper Presented at the Annual Meeting of American Educational Research Association, Chicago, IL.

Chang, H. and Ying, Z. 1996. A global information approach to computerized adoptive testing. *Applied Psychological Measurement, 20*, 213–229.

Chang, H. and Ying, Z. 1999. A-stratified multistage computerized adaptive testing. *Applied Psychological Measurement, 23*, 211–222.

Chang, H. and Ying, Z. 2008. To weight or not to weight? Balancing influence of initial items in adaptive testing. *Psychometrika, 73*, 441–450.

Chang, Y. and Chang, H. 2007. *Dual Information Method in Cognitive Diagnostic Computerized Adaptive Testing*. Paper Presented at the the Annual Meeting of National Council on Measurement in Education, Chicago, IL.

Fisher, R. A. 1925. Theory of statistical estimation. *Proceedings of the Cambridge Philpsophical Society, 22*, 700–725.

Gulliksen, H. 1950. *Theory of Mental Tests*. New York: Wiley.

Hambleton, R. K. and Swaminathan, H. 1985. *Item Response Theory: Principles and Applications*. Boston: Kluwer Nijhoff Publishing.

Kullback, S. 1959. *Information Theory and Statistics*. New York: Wiley.

Kullback, S. and Leibler, R. A. 1951. On information and sufficiency. *Annals of Mathematical Statististics, 22*, 79–86.

van der Linden, W. J. 1996. Assembling tests for the measurement of multiple traits. *Applied Psychological Measurement, 20*, 373–388.

van der Linden, W. J. 2007. A hierarchical framework for modeling speed and accuracy on test items. *Psychometrika, 72*, 287–308.

Liu, H., You, X., Wang, W., Ding, S., and Chang, H. 2013. The development of computerized adaptive testing with cognitive diagnosis for an English achievement test in China. *Journal of Classification, 30*, 152–172.

Lord, F. M. 1971. Robbins–Monro procedures for testing. *Educational and Psychological Measurement, 31*, 3–31.

Lord, F. M. 1977. Practical applications of item characteristic curve theory. *Journal of Educational Measurement, 14*, 117–138.

Lord, F. M. 1980. *Applications of Item Response Theory to Practical Testing Problems*. Hillsdale, NJ: Erlbaum.

Mulder, J. and van der Linden, W. J. 2009. Multidimensional adaptive testing with optimal design criteria for item selection. *Psychometrika, 74*, 273–296.

Mulder, J. and van der Linden, W. J. 2010. Multidimensional adaptive testing with Kullback–Leibler information item selection. In W. J. van der Linden and C. A. W. Glas (eds.), *Elements of Adaptive Testing* (pp. 77–101). New York: Springer.

Pitman, E. J. G. 1949. *Letcture Notes on Nonparametric Statistical Inference.* New York: Columbia University.
Reckase, M. D. 2009. *Multidimensional Item Response Theory*. New York: Springer.
Robbins, H. and Monro, S. 1951. A stochastic approximation method. *Annals of Mathematical Statistics, 22*, 400–407.
Rupp, A. A., Templin, J., and Henson, R. A. 2010. *Diagnostic Measurement: Theory, Methods, and Applications*. New York, NY: Guilford Press.
Samejima, F. 1977. Weakly parallel tests in latent triat theory with some criticisms of classical test theory. *Psychometrika, 42*, 193–198.
Segall, D. O. 1996. Multidimensional adaptive testing. *Psychometrika, 61*, 331–354.
Serfling, R. J. 1980. *Approximation Theorems of Mathematical Statistics*. New York: Wiley.
Shannon, C. E. 1948. A mathematical theory of communication. *Bell System Technology Journal, 27*, 379–423; 623–656.
Tatsuoka, C. 2002. Data analytic methods for latent partially ordered classification models. *Journal of the Royal Statistical Society, Series C, 51*, 337–350.
Tatsuoka, C. and Ferguson, T. 2003. Sequential classification on partially ordered sets. *Journal of Royal Statistics, Series B, 65*, 143–157.
Veldkamp, B. P. 2002. Multidimensional constrained test assembly. *Applied Psychological Measurement, 26*, 133–146.
Veldkamp, B. P. and van der Linden, W. J. 2002. Multidimensional adaptive testing with constraints on test content. *Psychometrika, 67*, 575–588.
Wang, C. and Chang, H. 2011. Item selection in multidimensional computerized adaptive tests—Gaining information from different angles. *Psychometrika, 76*, 363–384.
Wang, C., Chang, H., and Boughton, K. 2011. Kullback–Leibler information and its application in multidimensional adaptive tests. *Psychometrika, 76*, 13–39.
Wang, C., Chang, H., and Douglas, J. 2013. The linear transformation model with frailties for the analysis of item response times. *British Journal of Mathematical and Statistical Psychology, 66*, 144–168.
Wang, C., Chang, H., and Wang, X. 2011. *An Enhanced Approach to Combine Item Response Theory with Cognitive Diagnosis in Adaptive Tests.* Paper Presented in National Council of Educational Measurement annual meeting, New Orleans, Louisiana.
Wang, C., Zheng, C., and Chang, H. 2014. An enhanced approach to combine item response theory with cognitive diagnosis in adaptive testing. *Journal of Educational Measurement, 51*, 358–380.
Xu, X. Chang, H., and Douglas, J. 2005. *Computerized Adaptive Testing Strategies for Cognitive Diagnosis.* Paper Presented at the Annual Meeting of National Council on Measurement in Education, Montreal, Canada.

Section II

Modeling Issues

8

Identification of Item Response Theory Models

Ernesto San Martín

CONTENTS

8.1 Introduction

Why identification is relevant for model construction? In 1922, R. A. Fisher pointed out that, in spite of the large amount of fruitful applications of statistics, its basic principles were still in a "state of obscurity" and that "during the recent rapid development of practical methods, fundamental problems have been ignored" (Fisher, 1922, p. 310). This evaluation led Fisher to define the scope of statistical methods as the reduction of data. "No human mind is capable of grasping in its entirety the meaning of any considerable quantity of numerical data" (Fisher, 1973, p. 6). Consequently, data should be replaced by a relatively few quantity of values that should contain the *relevant information* provided by the original data.

Conceptually speaking, the reduction in data is accomplished by constructing a hypothetical infinite population, of which the actual data are considered a random sample. This hypothetical population is fully characterized by a probability distribution, which in turn is specified by relatively few parameters. The parameters "are sufficient to describe it exhaustively in respect of all qualities under discussion." Thus, any information provided by a sample, which is used to estimate those parameters, is relevant information. This conceptual scheme leads to distinguish three problems of statistics (Fisher, 1922, 1973, Chapter 1): *Problems of specification, problems of estimation,* and *problems of distribution*. Problems of specification arise in the choice of the mathematical form of the probability distribution which characterizes the observed population. Its choice "is not arbitrary, but requires an understanding of the way in which the data are supposed to, or did in fact, originate" (Fisher, 1973, p. 8).

A quarter of century later, Koopmans and Reiersøl (1950) suggested a reformulation of the specification problem as defined by Fisher, which is "appropriate to many applications of statistical methods." Such a reformulation is based on a distinction between population and structure:

> In many fields the objective of the investigator's inquisitiveness is not just a "population" in the sense of a distribution of observable variables, *but a physical structure projected behind this distribution, by which the latter is thought to be generated.* The word "physical" is used merely to convey that the structure concept is based on the investigator's ideas as to the "explanation" or "formation" of the phenomena studied, briefly, *on his theory of these phenomena,* whether they are classified as physical in the literal sense, biological, psychological, sociological, economic or otherwise (Koopmans and Reiersøl, 1950, p. 165; the italics are ours).

A structure is determined on the basis of a substantive theory, combined with systematically collected data for the relevant variables. The collected data are represented by observed variables, whereas (some) aspects of a substantive theory are represented by unobservable or latent variables—or at least unobserved ones. Consequently, a structure is composed of a particular probability distribution of the latent variables, combined with a relationship between the observed and latent variables. Koopmans and Reiersøl's (1950) reformulation of the specification problem is accordingly defined as follows:

> The specification is therefore concerned with the mathematical forms of both the distribution of the latent variables and the relationships connecting observed and latent variables (p. 168).

The distribution H of the observed variables is regarded as produced by an operation performed on the distribution of the latent variables. The specification problem is, therefore, concerned with specifying a model which, by hypothesis, contains the structure S generating the distribution H of the observed variables. The result of this reformulation is the emergence of the identification problem: Can the distribution of H, generated by a given structure S, be generated by only one structure in that model? When the answer is affirmative, it is said that the structure S is identified by the observations. In such a case, the structure S has an empirical meaning since it is supported by the observations (Koopmans, 1949; Hurwicz, 1950; Clifford, 1982; Manski, 1995).

The identification problem should, therefore, be regarded "as a necessary part of the specification problem" (Koopmans and Reiersøl, 1950, p. 169), particularly in fields where a substantive theory plays a role in the construction of a statistical model. In other words, when structural modeling is being considered, that is, when the model formalizes a certain phenomenon, the identification condition is more than a simple technical assumption but covers a more fundamental aspect, namely, the adequacy of a theoretical statistical model for an observed process.

This chapter intends to discuss the identification problems of binary item response theory (IRT) models. A basic substantive aspect underlying this class of models is that the probability that a person correctly answers an item not only depends on a characteristic of the person, but also on characteristics of the item. Consequently, we investigate under which conditions those characteristics have an empirical meaning. Before entering into the details, we first review the concepts of statistical model and identification. The specification of the class of binary IRT models is then presented and their identification is discussed.

8.2 Fundamental Concepts

Both Fisher's (1922) perspective on the specification problem and its reformulation due to Koopmans and Reiersøl (1950) share sampling distribution as a common element. In both cases, the sampling distribution describes the data-generating process*. The difference is based on the way in which the parameters of the sampling distribution are specified. In Fisher's (1922) perspective, those parameters exhaustively describe the observed qualities of a population, whereas in Koopmans and Reiersøl (1950) view, the parameters of the sampling distribution correspond to the parameters of both the distribution of the latent variables and those of the structural relationships between observed and latent variables. These common elements are formalized through the concept of *statistical model,* which we discuss next.

8.2.1 Statistical Model

Let us start by the following example. Suppose we are analyzing the response pattern that a person obtains in a multiple-choice test composed of I items. Each item has only one correct answer. Accordingly, we define a binary random variable U_{pi} as $U_{pi} = 1$ if person p answers correctly an item i, and $U_{pi} = 0$ otherwise. The random variable U_{pi} is distributed according to a Bernoulli distribution; its parameter may be specified in three different ways:

$$\text{(i) } U_{pi} \sim \text{Bern}(\pi); \quad \text{(ii) } U_{pi} \sim \text{Bern}(\pi_p); \quad \text{(iii) } U_{pi} \sim \text{Bern}(\pi_{pi}). \tag{8.1}$$

These specifications are completed by assuming that $\{U_{pi} : p = 1, \ldots, P;\ i = 1, \ldots, I\}$ are mutually independent.

The first specification means that all persons answer correctly each item with the same probability. The second specification means that the probability to answer correctly an item is person-specific, but once a person is considered, such a probability is the same for all the items of the test. The third possibility means that the probability to answer correctly an item is person/item specific. Each of these models can be considered as a plausible specification for analyzing the responses patterns. Using these models, the probability of events of interest can be characterized in terms of the corresponding parameters. Thus, for instance, if we want to compute the probability that the score X_p of person p is equal to k (for $0 \leq k \leq I$), we first characterize the event $\{X_p = k\}$ as the union of the $\binom{I}{k}$ responses

* In order to relate the concept of *sampling distribution* with the standard use of *sampling distribution as the distribution of sample statistics,* see Remark 8.1 below.

patterns having exactly k ones. Based on them, the probability that $\{X_p = k\}$ is given by

$$\binom{I}{k}\pi^k(1-\pi)^{I-k} \quad \text{if the specification in Equation 8.1.i is used;}$$

$$\binom{I}{k}\pi_p^k(1-\pi_p)^{I-k} \quad \text{if the specification in Equation 8.1.ii is used;}$$

$$\sum_{\{\mathcal{J}\subset\{1,\ldots,I\}:|\mathcal{J}|=k\}} \prod_{i\in\mathcal{J}} \pi_{pi} \prod_{i\in\mathcal{J}^c} (1-\pi_{pi}) \quad \text{if the specification in Equation 8.1.iii is used,}$$

where $|\mathcal{J}|$ denotes the cardinality of subset $\mathcal{J}$, and $\mathcal{J}^c$ its complement.

This example allows us to recognize the elements necessary to describe the data-generating process:

1. The *sample space* $(S, \mathcal{S})$, where the set S contains the observations and the class $\mathcal{S}$ is a σ-field of subsets of S. The elements of $\mathcal{S}$ are the events of interest. In our example (considering only one person), $S = \{0,1\}^I$ and $\mathcal{S}$ corresponds to the subsets of $\{0,1\}^I$.
2. A family of *sampling distributions* $P(\cdot;\boldsymbol{\alpha})$ defined on the sample space $(S, \mathcal{S})$; the index $\boldsymbol{\alpha}$ corresponds to a *parameter*, which can be a scalar, a vector or a more complex entity. In our example (considering only one person), for the specification in Equation 8.1.i, the sampling distribution is the product of I's Bern (q); for the specification in Equation 8.1.ii, it is the product of I's Bern (q_p); for the specification in Equation 8.1.iii, it is the product of Bern $(q_{p1}), \ldots,$ Bern (q_{pI}).
3. The *parameter space* A, which corresponds to the set of possible values of the parameter $\boldsymbol{\alpha}$. In our example (considering only one person), for the specifications in Equation 8.1.i and Equation 8.1.ii, $A = [0,1]$; for the specification in Equation 8.1.iii, $A = [0,1]^I$.

A statistical model is, therefore, a family of sampling distributions indexed by a parameter (Cox and Hinkley, 1974; Raoult, 1975; Basu, 1975; Barra, 1981; Bamber and Santen, 2000; McCullagh, 2002), and can compactly be written as

$$\mathcal{E} = \{(S, \mathcal{S}),\ P\{\cdot;\boldsymbol{\alpha}\} : \boldsymbol{\alpha} \in A\}. \tag{8.2}$$

When the sampling probabilities can be represented through density functions $f(\cdot;\boldsymbol{\alpha})$, (8.2) can be rewritten by replacing $P\{\cdot;\boldsymbol{\alpha}\}$ by the density $f(\cdot;\boldsymbol{\alpha})$. It may be noted that the parameter space A might be an Euclidean as well as a functional space, as it is typically specified for nonparametric statistical models.

Remark 8.1

In the statistical literature, *sampling distribution* typically refers to the distribution of sample statistics. This usage agrees with the formal definition given above. As a matter of fact, let X_i be an observed real random variable distributed according to $P\{\cdot;\boldsymbol{\alpha}\}$, where $\alpha \in A$. Let $\mathbf{X} = (X_1, \ldots, X_n)'$ be a sample of size n. Assuming, for simplicity, that the X_i's are *iid*, the statistical model of $\mathbf{X}$ is given by

$$\{(S^n, \mathcal{S}^n),\ P^n\{\cdot;\boldsymbol{\alpha}\} : \boldsymbol{\alpha} \in A\}. \tag{8.3}$$

The observed sample $\boldsymbol{X}$ is, therefore, described by the sampling probabilities $P^n\{\cdot;\boldsymbol{\alpha}\}$, with $\boldsymbol{\alpha} \in A$. Now, if the interest focuses on a specific sample statistics $T(\boldsymbol{X})$, where $T : (S^n, \mathcal{S}^n) \longrightarrow (\mathbb{R}, \mathcal{B})$ is a measurable function, the statistical model of the sample statistic $T(\boldsymbol{X})$ is given by

$$\left\{\left(S^n, T^{-1}(\mathcal{B})\right),\ P^{(T)}\{\cdot;\boldsymbol{\alpha}\} : \boldsymbol{\alpha} \in A\right\}, \tag{8.4}$$

where $P^{(T)}\{C;\boldsymbol{\alpha}\} = P^n\{T^{-1}(C);\boldsymbol{\alpha}\}$ for $C \in \mathcal{B}$; these sampling probabilities describe the sample statistics $T(\boldsymbol{X})$. As an example, consider $X_i \overset{\text{iid}}{\sim} \mathcal{N}(\mu, \sigma^2)$ for $i = 1, \ldots, n$. The sampling probabilities describing the generation of the sample $\boldsymbol{X} = (X_1, \ldots, X_n)'$ correspond to

$$\left\{\mathcal{N}_n(\mu\imath_n, \sigma^2 I_n)\ :\ (\mu, \sigma^2) \in \mathbb{R} \times \mathbb{R}_+\right\},$$

where $\imath_n = (1, \ldots, 1)' \in \mathbb{R}^n$. The sampling probabilities describing the generation of the sample statistics $T(\boldsymbol{X}) = \overline{X}_n$ are consequently given by

$$\{\mathcal{N}(\mu, \sigma^2/\sqrt{n}) : (\mu, \sigma^2) \in \mathbb{R} \times \mathbb{R}_+\}. \qquad \blacksquare$$

Often a model user is not interested in the whole parameter $\boldsymbol{\alpha} \in A$, but in a noninjective function of it, namely, $\psi \doteq h(\boldsymbol{\alpha})$, where $h : A \longrightarrow \Psi$. This function is called *parameter of interest* (Engle et al., 1983). If the function f is bijective, ψ corresponds to a *reparametrization* of $\boldsymbol{\alpha}$.

Example 8.1

Consider a 2×2 contingency table $\boldsymbol{X}$ which, conditionally on the grand total X_{++}, is generated by a multinomial distribution $(\boldsymbol{X} \mid X_{++} = n) \sim \text{Mult}\,(n;\boldsymbol{\theta})$, where $\boldsymbol{X} = (X_{11}, X_{12}, X_{21}, X_{22})$ and $\boldsymbol{\theta} = (\theta_{11}, \theta_{12}, \theta_{21}, \theta_{22})$. The statistical model corresponds to a family of multinomial distributions of parameter $\boldsymbol{\theta} \in [0,1]^4$. A typical parameter of interest is the cross product ratio

$$\psi = \frac{\theta_{11}\theta_{22}}{\theta_{12}\theta_{21}},$$

the value 1 of which characterizes the row–column independence.

8.2.2 Parameter Identification

Parameters and observations are related through sampling probabilities. Thus, when a parameter $\boldsymbol{\alpha} \in A$ is given, the sampling probability $P(\cdot;\boldsymbol{\alpha})$ is fully determined and, therefore, the observations generated by $P(\cdot;\boldsymbol{\alpha})$ provide information about it, not about other sampling probabilities. However, as discussed in Section 8.2, the question is to know whether a parameter can be determined from a sufficient number of observations (possibly, an infinite number of them) in the sense that the knowledge of a sampling probability leads to one and only one parameter. As pointed out by Koopmans (1949), instead of reasoning from a sufficiently large number of observations, the discussion should be based on hypothetical knowledge of the sampling probabilities. It is clear that exact knowledge of this family of distributions cannot be derived from any finite number of observations.

Such knowledge is the limit approachable but not attainable by extended observation. By hypothesizing nevertheless the full availability of such knowledge, we obtain a clear separation between problems of statistical inference arising from the variability of finite samples, and problems of identification in which we explore the limits to which inference, even from an infinite number of observations is suspected (McHugh, 1956).

Thus, in the context of a statistical model of the form (8.2), a parameter $\boldsymbol{\alpha} \in A$ is said to be *identified by the observations* if the mapping $\boldsymbol{\alpha} \longmapsto P\{\cdot;\boldsymbol{\alpha}\}$ is injective; that is, for $\boldsymbol{\alpha}, \boldsymbol{\alpha}' \in A$, the following implication is true:

$$P\{E;\boldsymbol{\alpha}\} = P\{E;\boldsymbol{\alpha}'\} \quad \text{for all events } E \text{ in } \mathcal{S} \Longrightarrow \boldsymbol{\alpha} = \boldsymbol{\alpha}'$$

or, equivalently,

$$\boldsymbol{\alpha} \neq \boldsymbol{\alpha}' \Longrightarrow P\{E;\boldsymbol{\alpha}\} \neq P\{E;\boldsymbol{\alpha}'\} \quad \text{for at least one event } E \text{ in } \mathcal{S}.$$

Similarly, a parameter of interest $h(\boldsymbol{\alpha})$, with h a noninjective function, is identified by the observations if the mapping $h(\boldsymbol{\alpha}) \longmapsto P\{\cdot\,;\boldsymbol{\alpha}\}$ is injective.

In order to grasp the impact of the lack of identifiability on the relationship between parameters and observations, it is convenient to define on the parameter space A the following binary relation. Let $\boldsymbol{\alpha}_1, \boldsymbol{\alpha}_2 \in A$; it is said that $\boldsymbol{\alpha}_1$ and $\boldsymbol{\alpha}_2$ are *observationally equivalent* if and only if $P\{E;\boldsymbol{\alpha}_1\} = P\{E;\boldsymbol{\alpha}_2\}$ for all $E \in \mathcal{S}$ (Rothenberg, 1971; Hsiao, 1983). In such a case, we write $\boldsymbol{\alpha}_1 \sim \boldsymbol{\alpha}_2$. The relation $\sim$ is an equivalence relation (i.e., $\sim$ is reflexive, symmetric, and transitive) and, consequently induces a partition on the parameter space:

$$A = \bigcup_{\boldsymbol{\alpha} \in A} [\boldsymbol{\alpha}], \quad [\boldsymbol{\alpha}_1] \cap [\boldsymbol{\alpha}_2] = \emptyset \quad \text{if } \boldsymbol{\alpha}_1 \neq \boldsymbol{\alpha}_2,$$

where $[\boldsymbol{\alpha}] = \{\boldsymbol{\alpha}' \in A : \boldsymbol{\alpha} \sim \boldsymbol{\alpha}'\}$ corresponds to the equivalence class under $\sim$.

This partition of A is used as follows. Take a parameter $\boldsymbol{\alpha}_0 \in A$. By the definition of a statistical model, $\boldsymbol{\alpha}_0$ is related to the sampling distribution $P\{\cdot;\boldsymbol{\alpha}_0\}$ and, therefore, the observed population characterized by $P\{\cdot;\boldsymbol{\alpha}_0\}$ provides information about $\boldsymbol{\alpha}_0$. However, it should be questioned that if observed population provides information about a parameter other than $\boldsymbol{\alpha}_0$. To answer it, the equivalence class $[\boldsymbol{\alpha}_0]$ of $\boldsymbol{\alpha}_0$ should be considered; if it contains a parameter $\boldsymbol{\alpha}_1 \neq \boldsymbol{\alpha}_0$, then the observed population also provides information about $\boldsymbol{\alpha}_1$ and, consequently, the parameters that belong to $[\boldsymbol{\alpha}_0]$ cannot be distinguished through the sampling probabilities. This observational ambiguity can be avoided by imposing that $[\boldsymbol{\alpha}] = \{\boldsymbol{\alpha}\}$ for all $\boldsymbol{\alpha} \in A$. This condition is equivalent to the identification of $\boldsymbol{\alpha} \in A$ by the observations.

Example 8.2

As an example, consider $Y \sim \mathcal{N}(\alpha + \beta, 1)$, where $(\alpha, \beta) \in \mathbb{R}^2$ is the parameter of interest. In this case, the statistical model should be written as follows:

$$\{(\mathbb{R}, \mathcal{B}),\ \mathcal{N}(\alpha + \beta, 1) : (\alpha, \beta) \in \mathbb{R}^2\},$$

where $\mathcal{B}$ denotes the Borel σ-field. The parameter of interest (α, β) is not identified because $(\alpha, \beta) \sim (\alpha + c, \beta - c)$ for all $c \in \mathbb{R}$. For a fixed (α_0, β_0), its equivalence class is accordingly given by $\{(\alpha_0 + c, \beta_0 - c) : c \in \mathbb{R}\}$. Thus, if an infinite quantity of data is generated by a normal distribution $\mathcal{N}(\alpha_0 + \beta_0, 1)$, these data are not enough to distinguish between the original parameter (α_0, β_0) and any other parameter of the form $(\alpha_0 + c, \beta_0 - c)$ with $c \neq 0$.

8.3 Specification of Binary IRT Models

The basic assumption underlying binary IRT models is that the probability that a person correctly answers an item depends on a characteristic of the person as well as on characteristics of the item. The characteristic of the person, often called ability, is denoted by θ_p. The characteristics of the item will be denoted by ω_i. Depending on the role the person-specific ability θ_p, it is possible to specify this class of models into three different ways, that we introduce next.

8.3.1 Fixed-Effects Specification

Let U_{pi} be a binary random variable such that $U_{pi} = 1$ when person p correctly answers item i, and $U_{pi} = 0$ otherwise. Fixed-effects IRT models consider the person-specific ability θ_p as a fixed, but unknown parameter. The probability that $U_{pi} = 1$ is accordingly specified as $F(\theta_p; \omega_i)$, where $\theta_p \in \mathbb{R}$, $\omega_i \in \mathbb{R}^K$ and F is a strictly increasing cumulative distribution function (cdf). The model is completed assuming that $\{U_{pi} : p = 1, \ldots, P;\ i = 1, \ldots, I\}$ are mutually independent. We focus our attention on the following four specifications, where $\theta_{1:P} = (\theta_1, \ldots, \theta_P)$, and similarly for $\boldsymbol{a}_{1:I}, \boldsymbol{b}_{1:I}$, and $\boldsymbol{c}_{1:I}$:

1PL-type models:

$$P\{U_{pi} = 1; \theta_p, \omega_i\} = F(\theta_p - b_i); \tag{8.5}$$

$$\text{Parameters of interest}: (\theta_{1:P}, \boldsymbol{b}_{1:I}) \in \mathbb{R}^P \times \mathbb{R}^I.$$

2PL-type models:

$$P\{U_{pi} = 1; \theta_p, \omega_i\} = F(a_i(\theta_p - b_i)); \tag{8.6}$$

$$\text{Parameters of interest}: (\theta_{1:P}, \boldsymbol{a}_{1:I}, \boldsymbol{b}_{1:I}) \in \mathbb{R}^P \times \mathbb{R}_+^I \times \mathbb{R}^I.$$

1PL-G-type models:

$$P\{U_{pi} = 1; \theta_p, \omega_i\} = c_i + (1 - c_i)\, F(\theta_p - b_i); \tag{8.7}$$

$$\text{Parameters of interest}: (\theta_{1:P}, \boldsymbol{b}_{1:I}, \boldsymbol{c}_{1:I}) \in \mathbb{R}^P \times \mathbb{R}^I \times [0,1]^I.$$

3PL-type models:

$$P\{U_{pi} = 1; \theta_p, \omega_i\} = c_i + (1 - c_i)\, F(a_i(\theta_p - b_i)); \tag{8.8}$$

$$\text{Parameters of interest}: (\theta_{1:P}, \boldsymbol{a}_{1:I}, \boldsymbol{b}_{1:I}, \boldsymbol{c}_{1:I}) \in \mathbb{R}^P \times \mathbb{R}_+^I \times \mathbb{R}^I \times [0,1]^I.$$

For the 1PL-type model, $\omega_i = a_i$; for the 2PL-type model, $\omega_i = (a_i, b_i)$; for the 1PL-G-type model, $\omega_i = (b_i, c_i)$; and for the 3PL-type model, $\omega_i = (a_i, b_i, c_i)$.

When F is equal to the logistic distribution, the standard 1PL, 2PL, and 3PL models are recovered. It should also be mentioned that Rasch (1960) introduced the 1PL model by considering the person-specific characteristics θ_p as an unknown parameter. Concerning to the 1PL-G model, see San Martin et al. (2006).

The parameters of interest indexing the fixed-effects IRT models can be estimated using the joint maximum-likelihood (JML) estimator. However, it is known that this method produces biased estimations for the 1PL or Rasch model due to the incidental parameter problem (see, e.g., Andersen, 1980; Lancaster, 2000; Del Pino et al., 2008, and finally, see Chapter 7). A solution to this problem is to estimate IRT models using a marginal maximum-likelihood (MML) estimator (Molenaar, 1995; Thissen, 2009), where the person-specific abilities are assumed to be iid distributed random variables. This leads to specify IRT models under a random-effects perspective.

8.3.2 Random-Effects Specification

In modern item response theory, θ_p is usually considered as a latent or unobserved variable and, therefore, its probability distribution is an essential part of the model. It is in fact assumed to be a location-scale distribution $G(\cdot; \mu, \sigma)$ defined as

$$P\{\theta_i \leq x; \mu, \sigma\} = G((-\infty, x]; \mu, \sigma) \doteq G\left(\left(-\infty, \frac{x-\mu}{\sigma}\right]\right), \tag{8.9}$$

where $\mu \in \mathbb{R}$ is the location parameter and $\sigma \in \mathbb{R}_+$ is the scale parameter. In applications, G is typically chosen as a standard normal distribution.

It is also assumed that for each person p, his/her responses pattern $\boldsymbol{U}_p = (U_{p1}, \ldots, U_{pI})'$ satisfy the axiom of local independence, namely, that $U_{p1}, \ldots, U_{pI}$ are mutually independent conditionally on θ_p. The conditional probability that $U_{pi} = 1$ given θ_p is accordingly specified as $F(\theta_p; \omega_i)$, where $\theta_p \in \mathbb{R}$, $\omega_i \in \mathbb{R}^K$ and F is a strictly increasing cumulative distribution function (cdf) such that $F(\theta; \cdot)$ is continuous in ω_i for all $\theta_p \in \mathbb{R}$.

The statistical model, that is, the sampling distribution generating the data, is obtained integrating out the random effects θ_p's. The above-mentioned hypotheses imply that the responses patterns $\boldsymbol{U}_1, \ldots, \boldsymbol{U}_P$ are mutually independent, with a common probability distribution defined as

$$\begin{aligned} P\{\boldsymbol{U}_p = \boldsymbol{u}_p; \boldsymbol{\omega}_{1:I}, \mu, \sigma\} \\ = \int_{\mathbb{R}} \prod_{i=1}^{I} \{F(\theta; \omega_i)\}^{u_{pi}} \{1 - F(\theta; \omega_i)\}^{1-u_{pi}} \, dG(\theta; \mu, \sigma), \end{aligned} \tag{8.10}$$

where $\boldsymbol{u}_p = (u_{p1}, \ldots, u_{pI})' \in \{0,1\}^I$. The random-effects specifications corresponding to the fixed-effects specifications (8.5)–(8.8) are obtained after considering the function $F(\theta_p; \omega_i)$ as a conditional probability of $\{U_{pi} = 1\}$ given θ_p, where θ_p is distributed according to $G(\cdot\,;\ \mu, \sigma)$. The parameters of interest are the following:

1PL-type models:

$$\text{Parameters of interest: } (\boldsymbol{b}_{1:I}, \mu, \sigma) \in \mathbb{R}^I \times \mathbb{R} \times \mathbb{R}_+. \tag{8.11}$$

2PL-type models:

$$\text{Parameters of interest: } (\boldsymbol{a}_{1:I}, \boldsymbol{b}_{1:I}, \mu, \sigma) \in \mathbb{R}_+^I \times \mathbb{R}^I \times \mathbb{R} \times \mathbb{R}_+. \tag{8.12}$$

1PL-G-type models:

$$\text{Parameters of interest: } (\boldsymbol{b}_{1:I}, \boldsymbol{c}_{1:I}, \mu, \sigma) \in \mathbb{R}^I \times [0,1]^I \times \mathbb{R} \times \mathbb{R}_+. \tag{8.13}$$

3PL-type models:

$$\text{Parameters of interest: } (\boldsymbol{a}_{1:I}, \boldsymbol{b}_{1:I}, \boldsymbol{c}_{1:I}, \mu, \sigma) \in \mathbb{R}_+^I \times \mathbb{R}^I \times [0,1]^I \times \mathbb{R} \times \mathbb{R}_+. \tag{8.14}$$

Random-effects specification of IRT models is the standard way to introduce these models; see, among many others, Lord and Novick (1968); Fischer and Molenaar (1995); Van der Linden and Hambleton (1997); Embretson and Reise (2000); Boomsma et al. (2001); De Boeck and Wilson (2004); and Millsap and Maydeu-Olivares (2009).

8.3.3 Semiparametric Specification

Let us finish this section by introducing semi parametric IRT models. In practice, many IRT models are fitted under the assumption that θ_p is normally distributed with an unknown variance. The normal distribution is convenient to work with, specially because it is available in statistical package such as SAS (Proc NLMIXED) or R (lme4, ltm). However, as pointed out by Woods and Thissen (2006) and Woods (2006), there exist specific fields, such as personality and psychopathology, in which the normality assumption is not realistic (for references, see Woods, 2006). In these fields, it could be argued that psychopathology and personality variables are likely to be positively skewed, because most persons in general population have low pathology, and fewer persons have severe pathology. However, the distribution G of θ_p is unobservable and, consequently, though a researcher may hypothesize about it, it is not known in advance of an analysis. Therefore, any *a priori* parametric restriction on the shape of the distribution G could be considered as a misspecification.

These considerations lead to extend parametric IRT models by considering the distribution G as a parameter of interest and, therefore, to estimating it by using nonparametric techniques. Besides the contributions of Woods and Thissen (2006) and Woods (2006, 2008), Bayesian nonparametric methods applied to IRT models should also be mentioned; see, among others, Roberts and Rosenthal (1998); Karabatsos and Walker (2009); and Miyazaki and Hoshino (2009). In spite of these developments, it is relevant to investigate whether the item parameters as well as the distribution G of an IRT model are identified by the observations or not. If such parameters were identified, then semiparametric extensions of IRT models would provide greater flexibility than does the assumption of nonnormal parametric form for G.

Semiparametric IRT models are specified as random-effects IRT model, but hypothesis (8.9) should be replaced by the following hypothesis:

$$\theta_p \overset{\text{iid}}{\sim} G, \tag{8.15}$$

where $G \in \mathcal{P}(\mathbb{R}, \mathcal{B})$ is the set of probability distributions defined on the Borel sets $(\mathbb{R}, \mathcal{B})$. The statistical model is characterized by the mutual independence of the responses patterns $\boldsymbol{U}_1, \ldots, \boldsymbol{U}_P$, with a common distribution defined as

$$\begin{aligned} &P\{\boldsymbol{U}_p = \boldsymbol{u}_p; \boldsymbol{\omega}_{1:I}, G\} \\ &\quad = \int_{\mathbb{R}} \prod_{i=1}^{I} \{F(\theta; \omega_i)\}^{u_{pi}} \{1 - F(\theta; \omega_i)\}^{1-u_{pi}} \, G(d\theta), \end{aligned} \tag{8.16}$$

where $\boldsymbol{u}_p = (u_{p1}, \ldots, u_{pI})' \in \{0,1\}^I$. The corresponding parameters of interest are the following:

1PL-type models:

$$\text{Parameters of interest: } (\boldsymbol{b}_{1:I}, G) \in \mathbb{R}^I \times \mathcal{P}(\mathbb{R}, \mathcal{B}). \tag{8.17}$$

2PL-type models:

$$\text{Parameters of interest :} (\boldsymbol{a}_{1:I}, \boldsymbol{b}_{1:I}, G) \in \mathbb{R}^I_+ \times \mathbb{R}^I \times \mathcal{P}(\mathbb{R}, \mathcal{B}). \tag{8.18}$$

1PL-G-type models:

$$\text{Parameters of interest :} (\boldsymbol{b}_{1:I}, \boldsymbol{c}_{1:I}, G) \in \mathbb{R}^I \times [0,1]^I \times \mathcal{P}(\mathbb{R}, \mathcal{B}). \tag{8.19}$$

3PL-type models:

$$\text{Parameters of interest: } (\boldsymbol{a}_{1:I}, \boldsymbol{b}_{1:I}, \boldsymbol{c}_{1:I}, G) \in \mathbb{R}^I_+ \times \mathbb{R}^I \times [0,1]^I \times \mathcal{P}(\mathbb{R}, \mathcal{B}). \tag{8.20}$$

8.4 Identification under a Fixed-Effects Specification

Under a fixed-effects specification, the identification of binary IRT models reduces to establishing (under restrictions, if necessary) an injective relationship between $\{F(\theta_p; \omega_i) : p = 1, \ldots, P;\ i = 1, \ldots, I\}$ and the parameters of interest $(\theta_{1:P}, \boldsymbol{\omega}_{1:I})$. As a matter of fact, fixed-effects IRT models consist of mutually independent Bernoulli random variable U_{pi}, each of them parametrized by $F(\theta_p; \omega_i)$, where F is a known strictly increasing cdf. Since the parameter of a Bernoulli distribution is identified, it follows that, for each pair (p, i), $F(\theta_p; \omega_i)$ is identified by the observation U_{pi}. Thanks to the mutual independence of the U_{pi}'s, it follows that $\{F(\theta_p; \omega_i) : p = 1, \ldots, P;\ i = 1, \ldots, I\}$ is identified by $\boldsymbol{U} = \{\boldsymbol{U}_p : p = 1, \ldots, P;\ i = 1, \ldots, I\}$; for the validity of this implication, see Mouchart and San Martín (2003). Thus, the parameters of interest $(\theta_{1:P}, \boldsymbol{\omega}_{1:I})$ are identified by $\boldsymbol{U}$ if a bijective relationship is established with $\{F(\theta_p; \omega_i) : p = 1, \ldots, P; i = 1, \ldots, I\}$. This strategy of identification deserves two comments:

1. The parameters $\{F(\theta_p; \omega_i) : p = 1, \ldots, P; i = 1, \ldots, I\}$ are called identified parameters. All statistical models have an identified parametrization, which is the parametrization indexing in a one-to-one way the sampling probabilities; for a proof, see Florens et al. (1990) and Chapter 4. However, such an identified parametrization does not necessarily need to coincide with the parametrization of interest, as it is the case for IRT models. The identification problem is precisely due to this distinction.
2. The identified parameter $F(\theta_p; \omega_i)$ has a precise statistical meaning, which is based on the sampling process. It corresponds to the probability that a person p correctly answer an item i. Consequently, the statistical meaning of the parameters of interest relies on the statistical meaning of the identified parameter once a bijective relation between them is established. This explains why an unidentified parameter is empirically meaningless.

8.4.1 Identification of the Fixed-Effects 1PL Model

Let us illustrate the preceding identification strategy following the arguments developed by Rasch (1966) himself. Consider the 1PL model; the probability that a person correctly answer an item is given by

$$\pi_{pi} \doteq P\{U_{pi} = 1; \theta_p, b_i\} = \frac{\exp(\theta_p - b_i)}{1 + \exp(\theta_p - b_i)} = \frac{\epsilon_p/\eta_i}{1 + \epsilon_p/\eta_i},$$

where $(\epsilon_p, \eta_i) = (\exp(\theta_p), \exp(b_i))$; that is, (ϵ_p, η_i) is a reparametrization of (θ_p, b_i). It follows that

$$\frac{\epsilon_p}{\eta_i} = \frac{\pi_{pi}}{1 - \pi_{pi}} \quad \text{for all } p \text{ and all } i. \tag{8.21}$$

If ϵ'_p and η'_i is another set of solutions for Equation 8.21, then relation $\epsilon_p/\eta_i = \epsilon'_p/\eta'_i$ must holds for any combination of p and i. Thus

$$\frac{\epsilon_p}{\epsilon'_p} = \frac{\eta_i}{\eta'_i}$$

must be constant, and accordingly the general solution is

$$\epsilon'_p = \frac{1}{\alpha}\epsilon_p, \qquad \eta'_i = \frac{1}{\alpha}\eta_i \quad \text{for } \alpha > 0 \text{ arbitrary.}$$

The indeterminacy can be removed by the choice of one of the items, say $i = 1$, as the standard item having "a unit of difficulty," in multiples of which the degrees of difficulty of the other items are expressed. Therefore, if $\eta_1 = 1$, then

$$\epsilon_p = \frac{\pi_{p1}}{1 - \pi_{p1}}.$$

Consequently, ϵ_p is identified since it is a function of the identified parameter π_{p1}. Moreover, what it is typically called "ability of person p," represents the betting odds of a correct answer to the standard item 1. Thus, if $\epsilon_p > 1$ (resp. $\epsilon_p < 1$), then for person p, his/her probability to correctly answer the standard item 1 is greater (resp. lesser) than his/her probability to answer it incorrectly.

Pursuing Rasch's (1966) argument, the statistical meaning of η_i can be recovered. As a matter of fact, from Equation 8.21, it follows that

$$\eta_i = \epsilon_p \cdot \frac{1 - \pi_{pi}}{\pi_{pi}} = \frac{\pi_{p1}}{1 - \pi_{p1}} \cdot \frac{1 - \pi_{pi}}{\pi_{pi}}.$$

Thus, η_i is identified since it is written as a function of the identified parameters π_{p1} and π_{pi}. Furthermore, what it is called "difficulty of item i" corresponds to an odd ratio between item 1 and item i. Hence, the difficulty parameter η_i of item i actually measures an association between item i and the standard item 1.

From the preceding identification analysis, the following conclusion should be stated: Whatever we may think of the meaning of ϵ_p and η_i, their only valid statistical meaning becomes clear after identifying the statistical model in terms of the parameters of interest.

Let us finish this subsection pointing out an alternative but equivalent identification analysis. From Equation 8.21 it follows that, for each person p, $(\theta_p - b_1, \ldots, \theta_p - b_I)$ is identified. Therefore, $(\theta_p, b_1, \ldots, b_I)$ becomes identified after establishing a bijective relationship with $(\theta_p - b_1, \ldots, \theta_p - b_I)$. Such a relationship can be established under a linear restriction of the type

$$f' b_{1:I} = 0 \quad \text{for a known } f \in \mathbb{R}^I \text{ such that } f' \mathbb{1}_I \neq 0, \tag{8.22}$$

where $\mathbb{1}_I = (1, \ldots, 1)' \in \mathbb{R}^I$ and $x'y$ denotes the scalar product between the I-dimensional vectors x and y. Typical choices for f are $f = (1, 0, \ldots, 0)' \in \mathbb{R}^I$ or $f = \mathbb{1}_I$. Condition 8.22 is, therefore, a necessary and a sufficient identification restriction. It implies that $b_{1:I} \in \langle f \rangle^{\perp}$ and $\mathbb{1}_I \notin \langle f \rangle^{\perp}$, where $\langle f \rangle$ denotes the linear space generated by the vector f. That is, $b_{1:I}$ belongs to a linear space which *excludes* equal difficulties of all the items. Thus, in a fixed-effects specification of the 1PL model, the identification restriction not only fixes the scale of the parameters of interest, but also imposes a specific design on the multiple-choice test, namely, it should contain at least two items with different difficulties.

8.4.2 Identification of the Fixed-Effects 1PL-G Model

Maris and Bechger (2009) considered the identification problem of a particular case of the 3PL, namely, when the discrimination parameters a_i, with $i = 1, \ldots, I$, are equal to a. The parameter indeterminacies inherited from the Rasch model and the 2PL model that should be removed, are the location and scale ones. Maris and Bechger (2009) removed them by fixing $a = 1$ and constraining b_1 and c_1 in such a way that $b_1 = -\ln(1 - c_1)$. In this specific case, they show that unidentifiability of the parameters of interest persists, and conclude that "in contrast to the location and scale indeterminacy, this new form of indeterminacy involves not only the ability and the item difficulty parameters, but also the guessing parameter" (Maris and Bechger, 2009, p. 6).

The question is under which additional restrictions, the parameters of interest of the 1PL-G model (8.7) are identified by the observations. Using an identification strategy similar to that used in Section 8.4.1, it is possible to establish the following theorem:

Theorem 8.1

For the fixed-effects 1PL-G model, the parameters $(\theta_{1:P}, b_{1:I}, c_{1:I})$ are identified by the observations provided the following conditions hold:

1. There exists at least two persons such that their probabilities to correctly answer all the items are different.
2. c_1 and b_1 are fixed at 0.

For a proof, see San Martín et al. (2009).

The identification restriction $c_1 = 0$ means that the multiple-choice test must contain an item that persons answer *without guessing*. Moreover, under the constraints $b_1 = 0$ and $c_1 = 0$, the person-specific parameter $\epsilon_p = \exp(\theta_p)$ corresponds to the betting odds of a correct answer of person p to the standard item 1. Therefore, under the assumption that there exists an item (the standard item whose difficulty is fixed at 0) that persons cannot answer it correctly by guessing, abilities can be compared in the same way as it is done under a fixed-effects 1PL model. However, the meaning of the item parameter b_i is different to that established for the 1PL model. In fact, let $\eta_i = \exp(\beta_i)$; it can be verified that $\eta_i > \eta_1 = 1$ if and only if

$$\begin{cases} \text{(i) } (1-\pi_{1i})(1-\pi_{21}) > (1-\pi_{11})(1-\pi_{2i}), & \text{if } \pi_{21} < \pi_{11}; \\ \text{(ii) } (1-\pi_{1i})(1-\pi_{21}) < (1-\pi_{11})(1-\pi_{2i}), & \text{if } \pi_{21} > \pi_{11} \end{cases} \tag{8.23}$$

(the relation is given here for persons 1 and 2, but it holds for any pair of persons (p, p') with $p \neq p'$). Thus, in the case that the probability that person 2 incorrectly answers the standard item 1 is greater than the probability that person 1 incorrectly answers it, then item i is more difficult than the standard item 1 if the following cross-effect between the two items and the two persons holds: The probability that both person 1 incorrectly answers item i and person 2 incorrectly answer item 1, is greater than the probability that both person 2 incorrectly answers the item i and person 1 incorrectly answers item 1; see the inequality in Equation 8.23.i. Similar comments can be done for the inequality in Equation 8.23.ii.

We finish this section by four remarks. First, a similar parameter interpretation can be made for the standard 2PL model. Second, similar identification analysis for both the 1PL and 1PL-G can be developed for other known cumulative distribution functions F. Third, it is still an open problem the identification of the parameters of interest of a fixed-effects 3PL model. Fourth, the distribution F is sometimes considered as a parameter. Recently, Peress, (2012) has addressed this problem, showing that F is identified if an infinite number of persons and items is available, and that the parameter space is dense in $\mathbb{R}$.

8.5 Identification under a Random-Effects Specification

Under a random-effects specification of IRT models, the identification of the parameters of interest should be obtained from the statistical model in Equation 8.10. Since the responses patterns $\boldsymbol{U}_1, \ldots, \boldsymbol{U}_P$ define an *iid* process, the identification of the parameters of interest by one observation $\boldsymbol{U}_p$ is entirely equivalent to their identification by an infinite sequence of observations $\boldsymbol{U}_1, \boldsymbol{U}_2, \ldots$ (for a proof, see Florens et al., 1990, Theorem 7.6.6.). The identification problem we are dealing with corresponds, therefore, to the injectivity of the mapping $(\omega_{1:I}, \mu, \sigma) \longmapsto P\{\cdot; \omega_{1:I}, \mu, \sigma\}$, where $P\{\cdot; \omega_{1:I}, \mu, \sigma\}$ is given by Equation 8.10.

Similarly to the identification strategy described in Section 8.4, a way to solve this identification problem consists in distinguishing between parameters of interest and identified parameters. In the case of the statistical model 8.10, the parameters of interest are $(\omega_{1:I}, \mu, \sigma)$. Now, the probabilities of the 2^I different possible responses patterns are given by

$$\begin{aligned} \delta_{12\cdots I} &= P\{U_{p1} = 1, \ldots, U_{p,I-1} = 1, U_{pI} = 1; \omega_{1:I}, \mu, \sigma\} \\ \delta_{12\cdots \bar{I}} &= P\{U_{p1} = 1, \ldots, U_{p,I-1} = 1, U_{pI} = 0; \omega_{1:I}, \mu, \sigma\} \\ &\;\;\vdots \\ \delta_{\bar{1}\bar{2}\cdots \bar{I}} &= P\{U_{p1} = 0, \ldots, U_{p,I-1} = 0, U_{pI} = 0; \omega_{1:I}, \mu, \sigma\}; \end{aligned} \tag{8.24}$$

note that we are using a bar on the label of an item when its response is incorrect; otherwise, its label does not include a bar. The statistical model in Equation 8.10 corresponds, therefore, to a multinomial distribution $\boldsymbol{U}_p \sim \text{Mult}\,(2^I, \boldsymbol{\delta})$, where

$\delta = (\delta_{12\cdots I}, \delta_{12\cdots I-1,\bar{I}}, \ldots, \delta_{\bar{1}\bar{2},\ldots,\bar{I}})$. It is known that the parameter δ of a multinomial distribution is identified since the mapping $\delta \longmapsto \text{Mult}(2^I, \delta)$ is injective. It corresponds, therefore, to the identified parameter indexing the statistical model in Equation 8.10; the δ's with fewer than I subscripts are linear combinations of them and, consequently, are also identified. Thus, the identifiability of the parameters of interest $(\omega_{1:I}, \mu, \sigma)$ is obtained after establishing a bijective relation between $(\omega_{1:I}, \mu, \sigma)$ and functions of δ. By this way, the restrictions under which the identifiability of $(\omega_{1:I}, \mu, \sigma)$ is obtained are not only sufficient conditions, but also necessary conditions.

8.5.1 Identification of the Random-Effects 1PL-Type and 2PL-Type Models

Let us focus our attention on a 1PL-type model such that the person-specific abilities are distributed according to a distribution G known up to the scale parameter σ. In this case, the parameters of interest are $(\boldsymbol{b}_{1:I}, \sigma)$; see Equation 8.11. The identification of the parameters of interest by $\boldsymbol{U}_p$ follows from the following two steps:

1. Let the function $p(\sigma, b)$ be defined as

$$p(\sigma, b) = \int_{\mathbb{R}} F(\theta - b)dG(\theta; \sigma) = \int_{\mathbb{R}} F(\sigma x - b)G(dx).$$

For each item i, define the identified parameter δ_i as

$$\delta_i = P\{U_{pi} = 1; b_i, \sigma\} = p(\sigma, b_i).$$

The parameter δ_i is identified since it is a linear combination of $\boldsymbol{\delta}$; see the equalities in Equation 8.24. Assuming that F is a strictly continuous increasing cdf, it follows that $p(\sigma, b)$ is a continuous function in $(\sigma, b) \in \mathbb{R}_+ \times \mathbb{R}$ that is strictly decreasing in b. Therefore, if we define

$$\overline{p}(\sigma, \delta) = \inf\{b : p(\sigma, b) < \delta\},$$

it follows that $\overline{p}[\,\sigma, p(\sigma, b)\,] = b$ and, consequently, $b_i = \overline{p}(\sigma, \delta_i)$ for each item $i = 1, \ldots, I$. Thus, the item parameter b_i becomes identified once the scale parameter σ is identified.

2. In order to identified σ, suppose that $I \geq 2$ and define the identified parameter δ_{12} as

$$\delta_{12} = P\{U_{p1} = 1, U_{p2} = 1; b_1, b_2, \sigma\} = \int_{\mathbb{R}} F(\sigma x - b_1)F(\sigma x - b_2)G(dx).$$

Using the previous step, it follows that $\delta_{12} = r(\sigma, \delta_1, \delta_2)$, where

$$r(\sigma, \delta_1, \delta_2) = \int_{\mathbb{R}} F[\,\sigma x - \overline{p}(\sigma, \delta_1)\,]\,F[\,\sigma x - \overline{p}(\sigma, \delta_2)\,]G(dx).$$

If F is a strictly increasing cdf with a continuous density function strictly positive on $\mathbb{R}$, then it can be shown that the function $\delta_{12} = r(\sigma, \delta_1, \delta_2)$ is a strictly increasing continuous function of σ. It follows that $\sigma = \overline{r}(\delta_{12}, \delta_1, \delta_2)$, where

$$\overline{r}(\delta, \delta_1, \delta_2) = \inf\{\sigma : r(\sigma, \delta_1, \delta_2) > \delta\}.$$

Therefore, σ is identified. For a proof, see San Martín and Rolin (2013) and Appendix A.

Summarizing, we obtain the following theorem:

Theorem 8.2

Consider the random-effects 1PL-type models, where F is a continuous strictly increasing cdf, with a continuous density function strictly positive on $\mathbb{R}$. Suppose that the person-specific abilities θ_p are iid with a common distribution $G(\cdot; \sigma)$ known up to the scale parameter σ. If at least two items are available, then the item parameters $(\boldsymbol{b}_{1:I}, \sigma)$ are identified by one observation $\boldsymbol{U}_p$.

It seems relevant to compare the identification of the random-effects 1PL-type model with the fixed-effects 1PL model. Two differences deserve to be pointed out:

1. In a fixed-effects 1PL model, the difficulty parameters b_i corresponds to an odd ratio between the standard item and item i; see Section 8.4.1. In a random-effects 1PL mode, the statistical meaning of b_i is rather different. As a matter of fact, b_i is a function of both the marginal probability δ_i and the scale parameter σ. Now, σ represents the dependency between items 1 and 2 as captured by δ_{12}, along with δ_1 and δ_2. Consequently, in a random-effects 1PL model, the item parameter b_i not only captures information on the item itself, but also on items 1 and 2 through their dependency.
2. In a fixed-effects 1PL model, the identification restriction imposes a design on the multiple-choice test: it should contain at least two items with different difficulties. In a random-effects 1PL model, this is not the case since Theorem 8.2 is still valid if all the items parameters b_i are equal to a common unknown value b.

If the person-specific abilities are distributed according to a distribution $G(\cdot; \mu, \sigma)$ known up to the location parameter μ and the scale parameter σ, the identification of the parameters of interest $(\boldsymbol{b}_{1:I}, \mu, \sigma)$ is obtained under a constraint on the item parameters. More precisely, we obtain the following corollary to Theorem 8.2:

Corollary 8.1

Consider the random-effects 1PL-type models, where F is a continuous strictly increasing cdf, with a continuous density function strictly positive on $\mathbb{R}$. Suppose that the person-specific abilities θ_p are iid with a common distribution $G(\cdot\,;\, \mu, \sigma)$ known up to the location parameter μ and the scale parameter σ. If at least two items are available, then the item parameters $(\boldsymbol{b}_{1:I}, \mu, \sigma)$ are identified by one observation $\boldsymbol{U}_p$ provided that $\boldsymbol{f}'\boldsymbol{b}_{1:I} = 0$ for a known I-dimensional vector $\boldsymbol{f}$ such that $\boldsymbol{f}'\mathbb{1}_I \neq 0$.

The constraint imposed on the item parameters has a twofold role: On the one hand, it leads to distinguish between the location parameter and the item parameters; on the other hand, it excludes equal item parameters for all the items.

We finish this subsection by pointing out that the way in which the scale parameter is identified, provides the necessary insight to establish the identification of a random-effects 2PL-type model; see Equation 8.12. The results are summarized in the following theorem:

Theorem 8.3

Consider the random-effects 2PL-type models, where F is a continuous strictly increasing cdf, with a continuous density function strictly positive on $\mathbb{R}$.

1. Suppose that the person-specific abilities θ_p are iid with a common distribution $G(\cdot;\sigma)$ known up to the scale parameter σ. If at least three items are available, then the item parameters $(\boldsymbol{a}_{1:I}, \boldsymbol{b}_{1:I}, \sigma)$ are identified by one observation $\boldsymbol{U}_p$ provided that $\alpha_1 = 1$.
2. Suppose that the person-specific abilities θ_p are iid with a common distribution $G(\cdot;\mu,\sigma)$ known up to the location parameter μ and the scale parameter σ. If at least three items are available, then the item parameters $(\boldsymbol{a}_{1:I}, \boldsymbol{b}_{1:I}, \mu, \sigma)$ are identified by one observation $\boldsymbol{U}_p$ provided that $\alpha_1 = 1$ and $\beta_1 = 0$.

For a proof, see San Martín et al. (2013) and Appendix B.

Let us finish this subsection mentioning that the identification of the random-effects 3PL model is still an open problem.

8.5.2 Identification of the Random-Effects 1PL-G-Type Model

Let us consider the identification of the 1PL-G model, that is, when the cdf F corresponds to the logistic distribution. Assume that the person-specific abilities are distributed according to a distribution G known up to the scale parameter σ. The identification analysis is similar to that developed for identifying the 1PL-type model. By so doing, the parameters of interest $(\boldsymbol{b}_{1:I}, \boldsymbol{c}_{2:I}, \sigma)$ can be written as a function of c_1 and the identified parameter δ (see the equality in Equation 8.24) and, therefore, after fixing c_1, the parameters of interest become identified. More precisely,

Theorem 8.4

Consider the random-effects 1PL-G model. Suppose that the person-specific abilities θ_p are iid with a common distribution $G(\cdot;\sigma)$ known up to the scale parameter σ. The parameters of interest $(\boldsymbol{b}_{1:I}, \boldsymbol{c}_{2:I}, \sigma)$ are identified by one observation $\boldsymbol{U}_p$ provided that

1. At least three items are available.
2. $c_1 = 0$.

Moreover, the specification of the model entails that

$$0 \leq c_i \leq P\{U_{pi} = 1; \boldsymbol{b}_{1:I}, \boldsymbol{c}_{1:I}, \sigma\} \quad \text{for every } i = 2, \ldots, I \tag{8.25}$$

with

$$c_i = 1 - \frac{\delta_0\delta_{1k} - \delta_1\delta_k}{\delta_{ik}(\delta_0 - \delta_1) + \delta_i(\delta_{1k} - \delta_k)}, \qquad \text{for } k \neq 1, i, \tag{8.26}$$

where, for $i \neq 1$ and $k \neq i$ (with $i, k \leq I$),

$$\delta_0 = P\left[\bigcap_{1\leq i\leq I} \{U_{pi} = 0\} \mid \boldsymbol{b}_{1:I}, \boldsymbol{c}_{1:I}, \sigma\right];$$

$$\delta_i = P\left[\bigcap_{1\leq k\leq I,\, k\neq i} \{U_{pk} = 0\} \mid \boldsymbol{b}_{1:I}, \boldsymbol{c}_{1:I}, \sigma\right];$$

$$\delta_{ik} = P\left[\bigcap_{1\leq r\leq I,\, r\neq j,\, r\neq k} \{U_{pr} = 0\} \mid \boldsymbol{b}_{1:I}, \boldsymbol{c}_{1:I}, \sigma\right].$$

The probabilities δ_0, δ_i, and δ_{ik} are computed according to the statistical model in Equation 8.10, when $F(\theta; \omega_i) = c_i + (1 - c_i)\Psi(\theta - b_i)$, where $\Psi(x) = \exp(x)/(\exp(x) + 1)$. For a proof, see San Martín et al. (2013) and Section 8.3.

This theorem deserves the following comments:

1. The identification restriction $c_1 = 0$ means that the multiple-choice test must contain an item (labeled with the index 1) such that each person answers it without guessing. A possible way to ensure this restriction design, is including an open question, for which it is not possible to answer correctly by guessing. However, it should be said that guessing is possible for some types of open question; for example, what colors do we need to mix to get orange? Therefore, it is a matter of debate how to design a multiple-choice test such that $c_1 = 0$.
2. The inequality in Equation 8.25 provides an explicit meaning to the guessing parameter: At most, it is equal to the marginal probability to correctly answer an item. Furthermore, it can be used to evaluate if the practical rules of fixing an overall guessing parameter at A^{-1}, A being the number of response categories, is empirically adequate.
3. According to the equality in Equation 8.26, the guessing parameter c_i is a function of marginal probabilities, which in turn can be estimated from the data through relative frequencies. This suggests to use it as an estimator of c_i, the advantage being that it does not depend on the distribution $G(\cdot; \sigma)$ generating the person-specific abilities.

Similarly to Corollary 8.1, the following corollary to Theorem 8.4 can be established:

Corollary 8.2

Consider the random-effects 1PL-G model. Suppose that the person-specific abilities θ_p are iid with a common distribution $G(\cdot; \mu, \sigma)$ known up to the location parameter μ and the scale parameter σ. The parameters of interest $(\boldsymbol{b}_{1:I}, \boldsymbol{c}_{2:I}, \mu, \sigma)$ are identified by one observation U_p provided that

1. At least three items are available.
2. The guessing parameter $c_1 = 0$.
3. $f'b_{1:I} = 0$ for a known I-dimensional vector f such that $f'\mathbb{1}_I \neq 0$.

8.6 Identification under a Semiparametric Specification

The motivations leading to extend IRT models are far from being dubious. These extensions intend to estimate the distribution $G \in \mathcal{P}(\mathbb{R}, \mathcal{B})$ generating the person-specific abilities from the observations. However, it is relevant to known if the observations support the adequacy of semiparametric IRT models. Recently, San Martín et al. (2011) and San Martín et al. (2013) have addressed the identification problem of the semiparametric 1PL model and the semiparametric 1PL-G model, respectively. For the semiparametric 1PL model, the results are the following:

1. The item parameters $b_{2:I}$ are identified by one observation U_p if $b_1 = 0$. Moreover, the item parameters can be written as a function of marginal probabilities, namely

$$b_i = \ln\left(\frac{P\{U_p = e_1; b_{2:I}, G\}}{P\{U_p = e_i; b_{2:I}, G\}}\right),$$

   where e_i denotes an I-dimensional vector such that its ith coordinate is equal to 1 and the remaining coordinates are equal to 0; see San Martín et al. (2011) and Theorem 8.4. This equality allows us to grasp the meaning of the item parameter b_i with respect to the data-generating process. In fact, $b_i > b_j$ if and only if

$$P\left\{U_p = e_j;\ b_{2:I}, G\right\} > P\left\{U_p = e_i;\ b_{2:I}, G\right\}.$$

   Empirically, this inequality suggests that an item i is more difficult than an item j when the relative frequency of the response pattern e_j is greater than the relative frequency of the response pattern e_j.
2. The distribution G is not identified by the observations. The only information on G that can be identified are the following $I + 1$ functionals:

$$m_G(t) = \int_{\mathbb{R}} \frac{e^{t\theta}}{\prod_{1\leq i\leq I}(1 + e^{\theta - b_i})} G(d\theta), \quad \text{for } t = 0, 1, \ldots, I,$$

   which are jointly denoted as $\boldsymbol{m}_G$, that is, $\boldsymbol{m}_G = (m_G(0), m_G(1), \ldots, m_G(I))$. For details, see San Martín et al. (2011) and Theorem 8.5.

Regarding the semiparametric 1PL-G model, the following results have been obtained:

1. The item parameters $(b_{2:I}, c_{2:I})$ are identified by one observation U_p if at least three items are available, and b_1 and c_1 are fixed at 0. Moreover, the item parameters can

be written as a function of marginal probabilities, namely

$$b_i = \ln\left[\frac{\delta_i\delta_{1k} - \delta_1\delta_{ik}}{\delta_0\delta_{ik} - \delta_j\delta_k} + 1\right], \qquad \text{with } k \neq 1, i;$$

$$c_i = 1 - \frac{\delta_0\delta_{1k} - \delta_1\delta_k}{\delta_{ik}(\delta_0 - \delta_1) + \delta_i(\delta_{1k} - \delta_k)}, \qquad \text{with } k \neq 1, i,$$

where δ_0, δ_i, and δ_{ik} are defined as in Theorem 8.4 but these probabilities should be computed under the statistical model induced by the semiparametric 1PL-G model; for a proof, see San Martín et al. (2013) and Theorem 8.3.

2. The distribution G is not identified by the observations. The only information on G that can be identified are the following 2^I functionals: For $\mathcal{K} \subset \{1, \ldots, I\}$, except the empty set,

$$h_G(\mathcal{K}) = \int_{\mathbb{R}} \frac{G(d\theta)}{\prod_{i\in\mathcal{K}}(v_i + 1 + e^{\theta})},$$

where

$$v_i = \frac{\delta_i\delta_{1k} - \delta_1\delta_{ik}}{\delta_0\delta_{ik} - \delta_i\delta_k} \quad \text{for } k \neq 1, i,$$

which are jointly denoted as $\boldsymbol{h}_G$, that is, $\boldsymbol{h}_G = (h_G(\{1\}), h_G\{2\}), \ldots, h_G(\{1, \ldots, I\})$. For details, see San Martín et al. (2013) and Section 4.2.2.

The distribution G generating the person-specific abilities can be identified if an infinite number of items is available. As a matter of fact, using a Bayesian concept of identification, it is possible to obtain the following results:

1. For the semiparametric 1PL model, if $b_1 = 0$ and an infinite quantity of items is available, G is identified by the observations; for a proof, see San Martín et al. (2011) and Theorem 8.6.
2. For the semiparametric 1PL-G model, if $b_1 = 0$, $c_1 = 0$ and an infinite quantity of items is available, G is identified by the observations; for a proof, see San Martín et al. (2013) and Theorem 8.5.

A parameter is said to be Bayesian identified by the observations if it can be written as a measurable function of sampling expectations; for details, see Florens et al. (1990) and Chapter 4, San Martín et al. (2011, Section 3), and San Martín et al. (2013, Section 4.3.1). The identification results obtained in a sampling framework can be embedded in a Bayesian one. It should be mentioned that, although it is possible to compute posterior distribution of unidentified parameters, these do not provide any empirical information about them. Thus, either from a Bayesian point of view, or from a sampling theory framework, identification is a problem that need to be considered. For more discussion on this aspect, see San Martín and González (2010) and San Martín et al. (2013) and Sections 4 and 5.

8.7 Discussion

We have studied the identification problem of several binary IRT models. The problem was considered under three different specifications. The first specification assumes that the person-specific abilities are unknown parameters. The second specification considers the abilities as mutually independent random variables with a common distribution known up to the scale and the location parameters. Finally, the third specification corresponds to semiparametric models, where the distribution generating the person-specific abilities is considered as a parameter of interest. For the fixed-effects specifications, the identification results were obtained for standard IRT models, namely, when the cdf F corresponds to a logistic distribution. For the random-effects specifications, the identification results were obtained for general cdf F, except for the 1PL-G model. For the semiparametric specifications, the results were obtained when the cdf F is a logistic distribution. Table 8.1 summarizes these results, where questions marks are used to indicate that identification of the semiparametric 2PL model is still an open problem. The 3PL model is not mentioned because its identification is still an open problem. However, see San Martín et al. (2015a,b).

From a theoretical perspective, two remarks deserve to be made. First, the proofs of these identification results consist in obtaining the corresponding identified parametrizations of the sampling process. Thereafter, identification restrictions are imposed in such a way that the parameters of interest become identified. This means that the identification restrictions are not only sufficient conditions, but also necessary. Second, in the psychometric literature, there exist heuristic rules establishing that the identification of fixed-effects IRT models implies the identification of the corresponding random-effects IRT models (Adams et al. 1997; Adams and Wu, 2007). However, the identification results discussed in this chapter show that the identification of fixed-effects binary IRT models does not imply the identification of the corresponding random-effects binary IRT models and, by extension, of their semiparametric versions (for more details, see San Martín, 2003; San Martín et al., 2011) and Section 3.2. Consequently, this kind of heuristic rules, although intuitive, need to be considered carefully.

From a practical point of view, these identification results are relevant because doing an identification analysis assumes that the corresponding statistical model is the true model. This means that each binary IRT model should be applied to a set of data which, by design, satisfies the identification restrictions. In order to be more precise, let us consider the identification of the 1PL-G model. Its three different specifications share a common identification restriction, namely, a guessing parameter should be fixed at 0. As it was discussed previously, this identification result imposes a design on the multiple-choice test, namely, to ensure that it include an item that any person will correctly answer by guessing. In practice, this does not mean that any kind of educational data can be analyzed with the 1PL-G model but only those data generated by a multiple-choice test satisfying the design. This last requirement should be ensured by the context of application. Once this is done, the 1PL-G model can be applied.

In the semiparametric case, when a finite number of items is available (which is always the case), the distribution G is not identified by the observations. This lack of

TABLE 8.1
Summary of the Identification of Binary Item Response Models

Specification		1PL-Type Models	2PL-Type Models	1PL-G Models
Fixed-Effects	(1)	$(\theta_{1:P}, \boldsymbol{b}_{1:I}) \in \mathbb{R}^P \times \mathbb{R}^I$	$(\theta_{1:P}, \boldsymbol{a}_{1:I}, \boldsymbol{b}_{1:I}) \in \mathbb{R}^P \times \mathbb{R}_+^I \times \mathbb{R}^I$	$(\theta_{1:P}, \boldsymbol{b}_{1:I}, \boldsymbol{c}_{1:I}) \in \mathbb{R}^P \times \mathbb{R}^I \times [0,1]^I$
	(2)	$b_1 = 0$	$b_1 = 0$ and $a_1 = 0$	$b_1 = 0$ and $c_1 = 0$
$F = \Psi$	(3)	$P \geq 2, 2 \leq I < \infty$	$P \geq 2, 2 \leq I < \infty$	$P \geq 2, 2 \leq I < \infty$
	(4)	$(\theta_{1:P}, \boldsymbol{b}_{2:I}) \in \mathbb{R}^P \times \mathbb{R}^{I-1}$	$(\theta_{1:P}, \boldsymbol{a}_{2:I}, \boldsymbol{b}_{2:I}) \in \mathbb{R}^P \times \mathbb{R}_+^{I-1} \times \mathbb{R}^{I-1}$	$(\theta_{1:P}, \boldsymbol{b}_{2:I}, \boldsymbol{c}_{2:I}) \in \mathbb{R}^P \times \mathbb{R}^{I-1} \times [0,1]^{I-1}$
Random-Effects	(1)	$(\boldsymbol{b}_{1:I}, \sigma) \in \mathbb{R}^I \times \mathbb{R}_+$	$(\boldsymbol{a}_{1:I}, \boldsymbol{b}_{1:I}, \sigma) \in \mathbb{R}_+^I \times \mathbb{R}^I \times \mathbb{R}_+$	$(\boldsymbol{b}_{1:I}, \boldsymbol{c}_{1:I} \times \sigma) \in \mathbb{R}^I \times [0,1]^I \times \mathbb{R}_+$
	(2)	none	$a_1 = 1$	$c_1 = 0$
$\theta_p \overset{\text{iid}}{\sim} G(\cdot;\sigma)$	(3)	$P \geq 1, 2 \leq I < \infty$	$P \geq 1, 3 \leq I < \infty$	$P \geq 1, 3 \leq I < \infty$
	(4)	$(\boldsymbol{b}_{1:I}, \sigma) \in \mathbb{R}^I \times \mathbb{R}_+$	$(\boldsymbol{a}_{2:I}, \boldsymbol{b}_{1:I}, \sigma) \in \mathbb{R}_+^{I-1} \times \mathbb{R}^I \times \mathbb{R}_+$	$(\boldsymbol{b}_{1:I}, \boldsymbol{c}_{2:I} \times \sigma) \in \mathbb{R}^I \times [0,1]^{I-1} \times \mathbb{R}_+$
Random Effects	(1)	$(\boldsymbol{b}_{1:I}, \mu, \sigma) \in \mathbb{R}^I \times \mathbb{R} \times \mathbb{R}_+$	$(\boldsymbol{a}_{1:I}, \boldsymbol{b}_{1:I}, \mu, \sigma) \in \mathbb{R}_+^I \times \mathbb{R}^I \times \mathbb{R} \times \mathbb{R}_+$	$(\boldsymbol{b}_{1:I}, \boldsymbol{c}_{1:I}, \mu, \sigma) \in \mathbb{R}^I \times [0,1]^I \times \mathbb{R} \times \mathbb{R}_+$
	(2)	$b_1 = 0$	$a_1 = 1, b_1 = 0$	$b_1 = 0, c_1 = 0$
$\theta_p \overset{\text{iid}}{\sim} G(\cdot;\mu,\sigma)$	(3)	$P \geq 1, 2 \leq I < \infty$	$P \geq 1, 3 \leq I < \infty$	$P \geq 1, 3 \leq I < \infty$
	(4)	$(\boldsymbol{b}_{2:I}, \mu, \sigma) \in \mathbb{R}^{I-1} \times \mathbb{R} \times \mathbb{R}_+$	$(\boldsymbol{a}_{2:I}, \boldsymbol{b}_{2:I}, \mu, \sigma) \in \mathbb{R}_+^{I-1} \times \mathbb{R}^{I-1} \times \mathbb{R} \times \mathbb{R}_+$	$(\boldsymbol{b}_{2:I}, \boldsymbol{c}_{2:I}, \mu, \sigma) \in \mathbb{R}^{I-1} \times [0,1]^{I-1} \times \mathbb{R} \times \mathbb{R}_+$
Semiparametric	(1)	$(\boldsymbol{b}_{1:I}, G) \in \mathbb{R}^I \times [0,1]^I \times \mathcal{P}(\mathbb{R}, \mathcal{B})$	$(\boldsymbol{a}_{1:I}, \boldsymbol{b}_{1:I}, G) \in \mathbb{R}_+^I \times \mathbb{R}^I \times \mathcal{P}(\mathbb{R}, \mathcal{B})$	$(\boldsymbol{b}_{1:I}, \boldsymbol{c}_{1:I}, G) \in \mathbb{R}^I \times [0,1]^I \times \mathcal{P}(\mathbb{R}, \mathcal{B})$
	(2)	$b_1 = 0$	?	$b_1 = 0, c_1 = 0$
$\theta_p \overset{\text{iid}}{\sim} G, F = \Psi$	(3)	$P \geq 1, 2 \leq I < \infty$	?	$P \geq 1, 3 \leq I < \infty$
	(4)	$(\boldsymbol{b}_{2:I}, \boldsymbol{m}_G) \in \mathbb{R}^{I-1} \times \mathbb{R}_+^{I+1}$	?	$(\boldsymbol{b}_{2:I}, \boldsymbol{c}_{2:I}, \boldsymbol{h}_G) \in \mathbb{R}^{I-1} \times [0,1]^{I-1} \times \mathbb{R}_+^{2^I-1}$
Semiparametric	(1)	$(\boldsymbol{b}_{1:I}, G) \in \mathbb{R}^I \times \mathcal{P}(\mathbb{R}, \mathcal{B})$	$(\boldsymbol{a}_{1:I}, \boldsymbol{b}_{1:I}, G) \in \mathbb{R}_+^I \times \mathbb{R}^I \times \mathcal{P}(\mathbb{R}, \mathcal{B})$	$(\boldsymbol{b}_{1:I}, \boldsymbol{c}_{1:I}, G) \in \mathbb{R}^I \times [0,1]^I \times \mathcal{P}(\mathbb{R}, \mathcal{B})$
	(2)	$b_1 = 0$	?	$b_1 = 0, c_1 = 0$
$\theta_p \overset{\text{iid}}{\sim} G, F = \Psi$	(3)	$P \geq 1, I = \infty$	?	$P \geq 1, I = \infty$
	(4)	$(\boldsymbol{b}_{2:I}, G) \in \mathbb{R}^{I-1} \times \mathcal{P}(\mathbb{R}, \mathcal{B})$	?	$(\boldsymbol{b}_{2:I}, \boldsymbol{c}_{2:I}, G) \in \mathbb{R}^{I-1} \times [0,1]^{I-1} \times \mathcal{P}(\mathbb{R}, \mathcal{B})$

Note: (1) Parameter space; (2) identification restrictions; (3) conditions on P and I; (4) identified parameters.

identifiability jeopardizes the empirical meaning of an estimate for G. This result is of practical relevance, especially considering the large amount of research trying to relax the parametric assumption of G in binary IRT models.

Acknowledgment

This research was partially supported by the ANILLO Project SOC1107 *Statistics for Public Policy in Education: Analysis and Decision Making of Observational Data* from the Chilean government.

References

Adams, R. J., Wilson, M. R., and Wang, W. 1997. The multidimensional random coefficients multinomial logit model. *Applied Psychological Measurement, 21*, 1–23.

Adams, R. J. and Wu, M. L. 2007. The mixed-coefficients multinomial logit model: A generalization for of the Rasch model. In M. von Davier and C. H. Carstensen (Eds.), *Multivariate and Mixture Distribution Rasch Models* (pp. 57–75). Springer, New York, USA.

Andersen, E. B. 1980. *Discrete Statistical Models with Social Science Applications*. North Holland, Amsterdam, The Netherlands.

Bamber, D. and van Santen, J. P. H. 2000. How to asses a model's testability and identifiability. *Journal of Mathematical Psychology, 44*, 20–40.

Barra, J. R. 1981. *Mathematical Basis of Statistics*. Academic Press, New York, USA.

Basu, D. 1975. Statistical information and likelihood (with discussion). *Sankhyā: The Indian Journal of Statistics, 37*, 1–71.

Boomsma, A., van Duijn, M. A. J., and Snijders, T. A. B. 2001. *Essays on Item Response Theory (Lecture Notes in Statistics, 157)*. Springer, New York, USA.

Clifford, P. 1982. Some general comments on nonidentifiability. In L. Le Cam and J. Neyman (Eds.), *Probability Models and Cancer* (pp. 81–83). North Holland, Amsterdam, The Netherlands.

Cox, D. R. and Hinkley, D. V. 1974. *Theoretical Statistics*. Chapman and Hall, London, UK.

De Boeck, P. and Wilson, M. 2004. *Explanatory Item Response Models. A Generalized Linear and Nonlinear Approach*. Springer, New York, USA.

Del Pino, G., San Martín, E., González, J., and De Boeck, P. 2008. On the relationships between sum score based estimation and joint maximum likelihood estimation. *Psychometrika, 73*, 145–151.

Embretson, S. E. and Reise, S. P. 2000. *Item Response Theory for Psychologists*. Lawrence Erlbaum Associates, Publishers, New Jersey, USA.

Engle, R. E., Hendry, D. F., and Richard, J. F. 1983. Exogeneity. *Econometrica, 51*, 277–304.

Fischer, G. H. and Molenaar, I. W. 1995. *Rasch Models Foundations, Recent Developments and Applications*. Springer, New York, USA.

Fisher, R. A. 1922. On the mathematical foundations of theoretical statistics. *Philosophical Transactions of the Royal Society of London, Series A, 222*, 309–368.

Fisher, R. A. 1973. *Statistical Methods for Research Workers*. Hafner Publishing, New York, USA.

Florens, J. P., Mouchart, M., and Rolin, J. M. 1990. *Elements of Bayesian Statistics*. Marcel Dekker, New York, USA.

Hsiao, C. 1983. Identification. In Z. Griliches and M. D. Intriligator (Eds.), *Handbook of Econometrics*, Volume I (pp. 223–283). North Holland, Amsterdam, The Netherlands.

Hurwicz, L. 1950. Generalization of the concept of identification. In T. C. Koopmans (Ed.), *Statistical Inference in Dynamic Economic Models, Cowles Commission Research Monograph 11* (pp. 238–257). Wiley, New York, USA.
Karabatsos, G. and Walker, S. 2009. Coherent psychometric modelling with Bayesian nonparametrics. *British Journal of Mathematical and Statistical Psychology, 62* 1–20.
Koopmans, T. C. 1949. Identification problems in economic model construction. *Econometrica, 17,* 125–144.
Koopmans, T. C. and Reiersøl, O. 1950. The identification of structural characteristics. *The Annals of Mathematical Statistics, 21,* 165–181.
Lancaster, T. 2000. The incidental parameter problem since 1948. *Journal of Econometrics, 95,* 391–413.
Lord, F. M. and Novick, M. R. 1968. *Statistical Theories of Mental Test Scores.* Addison-Wesley Publishing Company, Massachusetts, USA.
Manski, C. F. 1995. *Identification Problems in the Social Sciences.* Harvard University Press, New York, USA.
Maris, G. and Bechger, T. 2009. On interpreting the model parameters for the three parameter logistic model. *Measurement: Interdisciplinary Research and Perspective, 7,* 75–86.
McCullagh, P. 2002. What is a statistical model? (with discussion). *The Annals of Statistics, 30,* 1225–1310.
McHugh, R. B. 1956. Efficient estimation and local identification in latent class analysis. *Psychometrika, 21,* 331–347.
Millsap, R. and Maydeu-Olivares, A. 2009. *Quantitative Methods in Psychology.* Sage, Oaks, USA.
Miyazaki, K. and Hoshino, T. 2009. A Bayesian semiparametric item response model with dirichlet process priors. *Psychometrika, 74,* 375–393.
Molenaar, I. W. 1995. Estimation of item parameters. In G. H. Fischer and I.W. Molenaar (Eds.), *Rasch Models. Foundations, Recent Developments and Applications* (pp. 39–51). Springer, New York, USA.
Mouchart, M. and San Martín, E. 2003. Specification and identification issues in models involving a latent hierarchical structure. *Journal of Statistical Planning and Inference, 111,* 143–163.
Peress, M. 2012. Identification of a semiparametric item response model. *Psychometrika, 77,* 223–243.
Raoult, J. P. 1975. *Structures Statistiques.* Presses Universitaires de France, Paris, France.
Rasch, G. 1960. *Probabilistic Models for Some Intelligence and Attaintment Tests.* The Danish Institute for Educational Research, Copenhagen, Denmark.
Rasch, G. 1966. An individualistic approach to item analysis. In P. F. Lazarsfeld and N. W. Henry (Eds.), *Readings in Mathematical Social Sciences* (pp. 89–107). Chicago Science Reading Association, Chicago, USA.
Roberts, G. O. and Rosenthal, J. 1998. Markov-chain monte carlo: Some practical implications of theoretical results. *The Canadian Journal of Statistics, 26,* 5–20.
Rothenberg, T. J. 1971. Identification in parametric models. *Econometrica, 39,* 577–591.
San Martín, E. 2003. Modeling problems motivated by the specification of latent linear structures. *Journal of Mathematical Psychology, 47,* 572–579.
San Martin, E., Del Pino, G., and De Boeck, P. 2006. IRT models for ability-based guessing. *Applied Psychological Measurement, 30,* 183–203.
San Martín, E. and González, J. 2010. Bayesian identifiability: Contributions to an inconclusive debate. *Chilean Journal of Statistics, 1,* 69–91.
San Martín, E., González, J., and Tuerlinckx, F. 2009. Identified parameters, parameters of interest and their relationships. *Measurement: Interdisciplinary Research and Perspective, 7,* 95–103.
San Martín, E., González, J., and Tuerlinckx, F. 2015a. On the unidentifiability of the fized-effects 3PL model. *Psychometrika, 80,* 450–457.
San Martín, E., González, J., and Tuerlinckx, F. 2015b. Erratum to: On the unidentifiability of the fized-effects 3PL model. *Psychometrika* (DOI: 10.1007/s11336-015-9470-0).
San Martín, E., Jara, A., Rolin, J. M., and Mouchart, M. 2011. On the Bayesian nonparametric generalization of IRT-type models. *Psychometrika, 76,* 385–409.
San Martín, E. and Rolin, J. M. 2013. Identification of parametric Rasch-type models. *Journal of Statistical Planning and Inference, 143,* 116–130.

San Martín, E., Rolin, J. M., and Castro, L. M. 2013. Identification of the 1PL model with guessing parameter: Parametric and semi-parametric results. *Psychometrika, 78*, 341–379.

Thissen, D. 2009. On interpreting the parameters for any item response model. *Measurement: Interdisciplinary Research and Perspective, 7*, 104–108.

Van der Linden, W. and Hambleton, R. K. 1997. *Handbook of Modern Item Response Theory*. Springer, New York, USA.

Woods, C. M. 2006. Ramsay-curve item response theory (RC-IRT) to detect and correct for nonnormal latent variables. *Psychological Methods, 11*, 253–270.

Woods, C. M. 2008. Ramsay-curve item response theory for the three-parameter logistic item response model. *Applied Psychological Measurement, 32*, 447–465.

Woods, C. M. and Thissen, D. 2006. Item response theory with estimation of the latent population distribution using spline-based densities. *Psychometrika, 71*, 281–301.

9

Models with Nuisance and Incidental Parameters

Shelby J. Haberman

CONTENTS

9.1 Introduction

In the statistical and econometric literature, it is common to find discussions of probability models in which some parameters are of direct interest but others are nuisance parameters (Hotelling, 1940) that are not of direct interest but affect the distribution of the observations (Cox and Hinkley, 1974, p. 35). It is also common to encounter cases in which some parameters are deemed structural parameters and some parameters are deemed incidental parameters. Typically, the incidental parameters are treated as nuisance parameters. The structural parameters are regarded as parameters of direct interest. The incidental parameters are regarded as parameters not directly of interest that are associated with individual observations (Neyman and Scott, 1948). The inferential challenge that is relevant to IRT arises when the number of nuisance parameters increases as the number of observations increases.

Incidental parameters are very commonly encountered in IRT, although in many applications it is easy to quarrel with the idea that the incidental parameters are not themselves of interest. The general problem can be described in terms relevant to IRT in terms of a test presented to $P > 1$ examinees p, $1 \leq p \leq P$, which consists of $n > 1$ items i, $1 \leq i \leq n$. For simplicity, matrix sampling is not considered, so that each item is presented to each person.

In addition, the simplifying assumption is made that only response scores are considered. For example, no demographic variables such as gender or race are recorded. It is always convenient to assume that $P \geq n$, so that there are at least as many persons as items.

As is commonly the case in IRT, polytomous item scores are assumed. Thus, each item i has associated integer scores j numbered from 1 to A_i, where the integer A_i is at least 2. The response score of person p on item i is denoted by U_{pi}, and the n-dimensional response vector of person p is denoted by $\mathbf{U}_p$. To each item i corresponds an associated q_i-dimensional vector $\boldsymbol{\beta}_i$ of unknown item parameters β_{ik}, $1 \leq k \leq q_i$. For a positive integer D, to each person p corresponds a D-dimensional vector $\boldsymbol{\theta}_p$ of person parameters θ_{pd}, $1 \leq d \leq D$. The parameters are unrestricted. The item parameters are consistently regarded as structural parameters, while the person parameters will be regarded as nuisance incidental parameters. This description of person parameters as incidental parameters is a bit strange in IRT to the extent that the person parameters are relevant to description of person performance and provide the basis for test scoring, and the test is presumably designed to study performance of individuals. Nonetheless, the usage is helpful in terms of relationships to existing literature. The simplification that a complete array of U_{pi} is available does prevent direct application of results to numerous survey assessments and to many equating applications. In addition, test scoring issues are not considered in which estimates of item parameters are obtained from a sample and then applied to scoring of examinees not in the original sample.

The challenge in models for item responses is that the total number of parameters is very large. Let $v_0 = 0$, and for $1 \leq i \leq n$, let $v_i = v_{i-1} + q_i$. Let $\boldsymbol{\beta}$ be the v_n-dimensional vector with element $v_{i-1} + k$ equal to β_{ik} for $1 \leq k \leq q_i$ and $1 \leq i \leq n$. Let $\boldsymbol{\Theta}$ be the PD-dimensional vector with element $(p-1)D + d$ equal to θ_{pd} for $1 \leq d \leq D$ and $1 \leq p \leq P$. Then $\boldsymbol{\beta}$ and $\boldsymbol{\Theta}$ include a total of $v_n + PD$ elements, a value greater than the number P of persons in the sample. This situation presents basic inferential problems (Neyman and Scott, 1948) that will be discussed in the following sections. This chapter will consider four common approaches to this problem of large numbers of incidental parameters. These approaches are considered in terms of their potential to provide consistent and asymptotically normal estimates of parameters. Computational details are not examined. Before proceeding to these approaches, it is helpful to provide a general description of the models under study.

9.1.1 Models

Two distinct traditions exist concerning model description, a purely conditional approach in which incidental parameters are fixed values and a marginal approach in which incidental parameters are regarded as random variables (Andersen, 1971b). In this chapter, the latter approach is adopted to simplify analysis. Thus it is assumed that the person parameters are independent and identically distributed random vectors with common distribution function F. It is assumed that the pairs $(\mathbf{U}_p, \boldsymbol{\theta}_p)$, $1 \leq p \leq P$, are independent and identically distributed. The standard local independence assumption is imposed that, for each person p, $1 \leq p \leq P$, the item scores U_{pi}, $1 \leq i \leq n$, are conditionally independent given the vector $\boldsymbol{\theta}_p$ of person parameters.

A parametric model is assumed for the conditional distribution of the item response U_{pi} given the person parameters $\boldsymbol{\theta}_p$. Because the model involves both the vector $\boldsymbol{\beta}_i$ of item parameters for item i and the vector $\boldsymbol{\theta}_p$ of person parameters for person p, it is helpful to apply a general concatenation operator to combine two vectors. Let S and T be positive integers, let $\mathbf{x}$ be an S-dimensional vector with elements x_s, $1 \leq s \leq S$, and let $\mathbf{y}$ be a T-dimensional vector with elements y_t, $1 \le t \le T$. Then $\mathbf{x}\#\mathbf{y}$ denotes the concatenated $S+T$-dimensional vector with element s equal to x_s for $1 \le s \le S$ and with element $S+t$ equal to y_t for $1 \le t \le T$. For positive integer response value $a \le A_i$ and item i, $1 \le i \le n$, let the item log probability function g_{ia} be infinitely differentiable real functions on the space B_i of $q_i + D$-dimensional vectors. Because $\exp(g_{ia})$, $1 \le a \le A_i$, will be used for a model for the conditional probability distribution of the item response U_{pi} for each person p given the person parameter vector $\boldsymbol{\theta}_p$, it is assumed that $\sum_{a=1}^{A_i} \exp(g_{ia})$ is the function on B_i with constant value 1. The parametric model then requires that the conditional probability that $U_{pi} = a$ given $\boldsymbol{\theta}_p = \boldsymbol{\theta}$ is

$$P\{U_{pi} = a|\boldsymbol{\theta}_p = \boldsymbol{\theta}\} = \exp[g_{ia}(\boldsymbol{\beta}_i\#\boldsymbol{\theta})] \tag{9.1}$$

for each item response score a, $1 \le a \le A_i$, for each item i, $1 \le i \le n$, for each person p, $1 \le p \le P$, and for $\boldsymbol{\theta}$ a D-dimensional vector with elements θ_d, $1 \le d \le D$.

Some notational conventions based on 9.1 are often useful for formulas. The probability $P\{U_{pi} = a|\boldsymbol{\theta}_p)$ that item response $U_{pi} = a$ given person parameter vector $\boldsymbol{\theta}_p$ is $\exp[g_{ia}(\boldsymbol{\beta}_i\#\boldsymbol{\theta}_p)]$ for each item response score a from 1 to A_i. The expected value of U_{pi} given $\boldsymbol{\theta}_p$ is then

$$E(U_{pi}|\boldsymbol{\theta}_p) = \sum_{a=1}^{A_i} a\, P\{U_{pi} = a|\boldsymbol{\theta}_p\}, \tag{9.2}$$

and the variance of U_{pi} given $\boldsymbol{\theta}_p$ is

$$\sigma^2(U_{pi}|\boldsymbol{\theta}_p) = \sum_{a=1}^{A_i} [a - E(U_{pi}|\boldsymbol{\theta}_p)]^2 P\{U_{pi} = a|\boldsymbol{\theta}_p\}. \tag{9.3}$$

If t is a positive integer and $\mathbf{Y}_i$ is a t-dimensional function on the integers 1 to A_i, then the mean of $\mathbf{Y}(U_{pi})$ given $\boldsymbol{\theta}_p$ is

$$E(\mathbf{Y}(U_{pi})|\boldsymbol{\theta}_p) = \sum_{a=1}^{A_i} \mathbf{Y}(a) P\{U_{pi} = a|\boldsymbol{\theta}_p\}. \tag{9.4}$$

If a prime indicates a transpose and a t-dimensional vector is regarded as a t by 1 matrix, then the conditional covariance matrix $\mathrm{Cov}(\mathbf{Y}(U_{pi})|\boldsymbol{\theta}_p)$ of $\mathbf{Y}(U_{pi})$ given $\boldsymbol{\theta}_p$ is

$$\sum_{a=1}^{A_i} [\mathbf{Y}(a) - E(\mathbf{Y}(U_{pi})|\boldsymbol{\theta}_p)][\mathbf{Y}(a) - E(\mathbf{Y}(U_{pi})|\boldsymbol{\theta}_p)]' P\{U_{pi} = a|\boldsymbol{\theta}_p\}.$$

In the models to be studied, the observations U_{pi}, $1 \le i \le n$, $1 \le p \le P$, do not identify the item vectors $\boldsymbol{\beta}_i$, $1 \le i \le n$ (for a general discussion of identifiability, see Chapter 8). As a consequence, $H \ge 1$ added linear constraints will be imposed. For some H-dimensional vector $\mathbf{z}$ and some H by v_n matrix $\mathbf{Z}$, the requirement is added that $\mathbf{Z}\boldsymbol{\beta} = \mathbf{z}$. The selection of constraints is rather arbitrary, but discussion is simplest if some selection is made. It is assumed that H is the smallest integer that can be used to identify parameters.

The following examples of models will often be considered in this chapter. They are selected because they are relatively common models or relatively simple to analyze.

9.1.2 Partial Credit Model

In the simplest partial credit model (Masters, 1982), the number of item parameters is $q_i = A_i - 1$ for each item i, the dimension of each vector of person parameters is $D = 1$, and, for item response scores a, $1 \le a \le A_i$, the q_i-dimensional vector $\mathbf{y}_{ia}$ is defined so that its first $a - 1$ elements are 1 and its remaining elements are 0. The item log probability function is

$$g_{ia}(\boldsymbol{\beta}_i \# \boldsymbol{\theta}) = (a-1)\theta_1 - \mathbf{y}'_{ia}\boldsymbol{\beta}_i - \log \sum_{a'=1}^{A_i} \exp[(a'-1)\theta_1 - \mathbf{y}'_{ia'}\boldsymbol{\beta}_i] \tag{9.5}$$

for item response scores a, $1 \le a \le A_i$. Here, for q_i -dimensional vectors $\mathbf{x}$ and $\mathbf{y}$ with respective elements x_k and y_k for $1 \le k \le q_i$, $\mathbf{x}'\mathbf{y}$ is the sum $\sum_{k=1}^{q_i} x_k y_k$. Thus, the product $\mathbf{y}'_{ia}\boldsymbol{\beta}_i$ is the sum of the item parameters β_{ik} for all integers k such that $1 \le k < a$. For item response score $a = 1$, this sum is 0. If each A_i is 2, so that the item responses are binary, then the model is a one-parameter logistic (1PL) or Rasch model (Rasch, 1960). To identify parameters, the constraint is imposed that $\mathbf{y}'_{1(A_1)}\boldsymbol{\beta}_1 = 0$, so that $H = 1$.

9.1.3 Generalized Partial Credit Model

In the generalized partial credit model with a one-dimensional person parameter (Muraki, 1992, 1997), for each item i, the vector $\mathbf{y}_{ia}$ is defined as in the partial credit model for each item response a from 1 to A_i, and the number of item parameters is $q_i = A_i$. The dimension of each vector of person parameters is $D = 1$, and the item log probability function

$$g_{ia}(\boldsymbol{\beta}_i \# \boldsymbol{\theta}) = [(-\mathbf{y}_{ia})\#((a-1)\boldsymbol{\theta})]'\boldsymbol{\beta}_i - \log \sum_{a'=1}^{A_i} \exp\{[(-\mathbf{y}_{ia})\#((a'-1)\boldsymbol{\theta})]'\boldsymbol{\beta}_i\} \tag{9.6}$$

for item response scores a from 1 to A_i. The product $[(-\mathbf{y}_{ia})\#(a\boldsymbol{\theta})]'\boldsymbol{\beta}_i$ is $a\theta_i\beta_{iA_i}$ minus the sum of the β_{ik} for all integers k such that $1 \le k < a$. If each A_i is 2, so that the item responses are binary, then the model is a two-parameter logistic (2PL) model (Lord and Novick, 1968, p. 400) with item log probability function

$$g_{ia}(\boldsymbol{\beta}_i \# \boldsymbol{\theta}) = (a-1)(\beta_{i2}\theta_1 - \beta_{i1}) - \log[1 + \exp(\beta_{i2}\theta_1 - \beta_{i1})] \tag{9.7}$$

for item response score a equal 1 or 2. If each β_{iA_i} is 1, then one has the partial credit model. It is common for β_{iA_i} to be termed the item discrimination for item i and to be noted by the symbol a_i, and it is typically the case that a_i is positive; however, this requirement is not employed in this chapter except in the model constraints for parameter identification. The constraints used are that $\sum_{k=1}^{A_i-1} \beta_{1k} = 0$ and $\beta_{A_1 1} = 1$, so that $H = 2$. Note that $\beta_{11} = 0$ if $A_i = 2$.

9.1.4 Additive Model for Nominal Responses

Consider a case in which all the items have the same number of scores, so that $A_i = A$ for $1 \le i \le n$ for some integer $r > 1$. Let the dimension of the person parameter be $D = r - 1$, and let the dimension of the parameters for item i be $q_i = A$ for $1 \le i \le n$. Let the item log probability function

$$g_{ia}(\boldsymbol{\beta}_i\#\boldsymbol{\theta}) = \mathbf{y}'_{1a}(\boldsymbol{\theta} - \boldsymbol{\beta}_i) - \log \sum_{a'=1}^{A} \exp[\mathbf{y}'_{1a'}(\boldsymbol{\theta} - \boldsymbol{\beta}_i)] \tag{9.8}$$

for item response scores a from 1 to A. If $A = 2$, then one again has the 1PL model. The generalized partial credit model corresponds to the case of $\theta_{pd} = \theta_{p1}$ for $1 \le d \le A - 1$. In general, the constraints are imposed that $\boldsymbol{\beta}_1 = \mathbf{0}_{A-1}$, where $\mathbf{0}_{A-1}$ is the vector of dimension $(A - 1)$ with all elements 0. Thus $H = A - 1$. This model is a special case of an additive exponential response model (Haberman, 1977b).

9.1.5 Approaches to Estimation

In this chapter, four common approaches are considered for parameter estimation. First, 9.2, joint estimation is explored. Here, the person parameter $\boldsymbol{\theta}_p$ is regarded as a fixed parameter. Both the item parameter vectors $\boldsymbol{\beta}_i$ and the person parameter vectors $\boldsymbol{\theta}_p$ are estimated simultaneously by joint maximum likelihood. As is well known (Neyman and Scott, 1948; Andersen, 1973), this approach is quite problematic due to inconsistency issues. Nonetheless, it is worth noting that the inconsistency problem diminishes as the number n of items increases (Haberman, 1977b).

Second, conditional estimation is applied. This approach typically applies to a rather restricted class of models that includes the partial credit model and the additive model for nominal responses but does not include the generalized partial credit model. Conditional estimation, which employs conditional maximum likelihood, avoids the inconsistency problems of joint estimation, but there can be a significant computational cost. Conditional estimation applies to the vectors $\boldsymbol{\beta}_i$, $1 \le i \le n$, of item parameters. Estimation of vectors $\boldsymbol{\theta}_p$, $1 \le p \le P$, of person parameters, which must be handled by other methods, is less satisfactory.

Third, unrestricted marginal estimation is considered. Here the vectors $\boldsymbol{\theta}_p$, $1 \le p \le P$, of person parameters are explicitly treated as independent and identically distributed random vectors with a common distribution function F. The distribution function F is assumed to be completely unknown. Estimation is accomplished by maximum marginal likelihood. Results are quite mixed. When conditional estimation is applicable, estimation of item parameters by unrestricted marginal estimation is very similar to estimation of item parameters by conditional estimation. However, estimation of the distribution function F is rather unsatisfactory. When conditional estimation is not available, parameter identification becomes impossible, so that marginal estimation is not applicable. Thus, unrestricted marginal estimation applies to the generalized partial credit model and to the additive model for nominal responses, but it is not useful for the generalized partial credit model.

Finally, consideration is given to restricted marginal estimation in which a parametric model is applied to the distribution function F. When maximum marginal likelihood is used, this case typically provides consistent estimates and permits estimation of the person parameters $\boldsymbol{\theta}_p$. In the end, the frequency with which maximum marginal likelihood is used in IRT reflects the attractive results associated with this estimation approach. Restricted marginal estimation applies to all three models used as examples.

Section 9.6 summarizes the practical implications of the results described in this chapter.

9.2 Joint Estimation

In joint estimation (Lord, 1980, pp. 179–182), maximum likelihood is applied simultaneously to all item and person parameters, and estimation proceeds as if the person parameters are fixed rather than random. The basic procedure is relatively simple; however, the statistical properties are rather complex and not very satisfactory.

To begin, the joint distribution of $\mathbf{U}_p$ given $\boldsymbol{\theta}_p$ is considered. Let $\mathcal{U}$ be the set of possible values of the response vector $\mathbf{U}_p$ of a person p, so that an n-dimensional vector $\mathbf{a}$ is in $\mathcal{U}$ if, and only if, each element a_i, $1 \le i \le n$, is a positive integer no greater than A_i. Conditional on the person parameter vector $\boldsymbol{\theta}_p = \boldsymbol{\theta}$, the joint probability that the vector $\mathbf{U}_p$ of responses for person p is $\mathbf{a}$ in $\mathcal{U}$ is

$$P\{\mathbf{U}_p = \mathbf{a} | \boldsymbol{\theta}_p = \boldsymbol{\theta}\} = \exp[g_{J\mathbf{a}}(\boldsymbol{\beta}\#\boldsymbol{\theta})], \tag{9.9}$$

where the joint log probability function

$$g_{J\mathbf{a}}(\boldsymbol{\beta}\#\boldsymbol{\theta}) = \sum_{i=1}^{n} g_{ia_i}(\boldsymbol{\beta}_i\#\boldsymbol{\theta}) \tag{9.10}$$

for any possible response vector $\mathbf{a}$ in $\mathcal{U}$. The notation $P\{\mathbf{U}_p = \mathbf{a} | \boldsymbol{\theta}_p\}$ is then used for $\exp[g_{J\mathbf{a}}(\boldsymbol{\beta}\#\boldsymbol{\theta}_p)]$.

The joint log-likelihood function is

$$\ell_J(\boldsymbol{\beta}\#\boldsymbol{\Theta}) = \sum_{p=1}^{P} g_{J\mathbf{U}_p}(\boldsymbol{\beta}\#\boldsymbol{\theta}_p). \tag{9.11}$$

In joint estimation, the joint log-likelihood function is then maximized subject to the constraint that $\mathbf{Z}\boldsymbol{\beta} = \mathbf{z}$. To avoid problems of nonexistent or ambiguously defined estimates for such case as $U_{pi} = 1$ for all items i for some person p or $U_{pi} = A_i$ for all items i for some person p, the following convention will be employed. Let Λ be a closed and bounded convex set of $v_n + PD$-dimensional vectors such that $\boldsymbol{\beta}\#\boldsymbol{\Theta}$ is in the interior of Λ. The maximum joint log-likelihood $\hat{\ell}_{J\Lambda}$ is the supremum of $\ell_J(\mathbf{x}\#\mathbf{y})$ subject to the constraints that $\mathbf{x}$ is a v_n-dimensional vector, $\mathbf{y}$ is a PD-dimensional vector, $\mathbf{Zx} = \mathbf{z}$, and $\mathbf{x}\#\mathbf{y}$ is in Λ. The joint maximum-likelihood estimators $\hat{\boldsymbol{\beta}}_{J\Lambda}$ of $\boldsymbol{\beta}$ and $\hat{\boldsymbol{\Theta}}_{J\Lambda}$ of $\boldsymbol{\Theta}$ are defined to be functions of the response vectors $\mathbf{U}_p$, $1 \le p \le P$, such that $\ell_J(\hat{\boldsymbol{\beta}}_{J\Lambda}\#\hat{\boldsymbol{\Theta}}_{J\Lambda}) = \hat{\ell}_{J\Lambda}$, $\mathbf{Z}\hat{\boldsymbol{\beta}}_{J\Lambda} = \mathbf{z}$, and $\hat{\boldsymbol{\beta}}_{J\Lambda}\#\hat{\boldsymbol{\Theta}}_{J\Lambda}$ is in Λ. Given $\hat{\boldsymbol{\beta}}_{J\lambda}$ and $\hat{\boldsymbol{\Theta}}_{J\Lambda}$, obvious estimates $\hat{\boldsymbol{\beta}}_{J\Lambda i}$ of $\boldsymbol{\beta}_i$, $\hat{\beta}_{J\Lambda ik}$ of β_{ik}, $\hat{\boldsymbol{\theta}}_{J\Lambda p}$ of $\boldsymbol{\theta}_p$, and $\hat{\theta}_{J\Lambda pd}$ of θ_{pd} are obtained for $1 \le k \le q_i$, $1 \le i \le n$, $1 \le d \le D$, and $1 \le p \le P$. For example, element $v_{i-1} + k$ of $\hat{\boldsymbol{\beta}}_{J\Lambda}$ is $\hat{\beta}_{J\Lambda ik}$ for $1 \le k \le q_i$ and $1 \le i \le n$. In the partial credit model and the additive model for nominal responses, $\hat{\boldsymbol{\beta}}_{J\Lambda}$ and $\hat{\boldsymbol{\Theta}}_{J\Lambda}$ are uniquely defined.

9.2.1 Joint Estimation for the 1PL Model

The basic problem with joint maximum-likelihood estimation is inconsistency. A large number P of persons does not ensure that estimated item parameters converge to the parameters they estimate. This issue is most thoroughly studied in the 1PL case. For a convenient definition of Λ, for any positive integer t and any t-dimensional vector $\mathbf{x}$ with elements x_s, $1 \leq s \leq t$, define the maximum norm $|\mathbf{x}|$ to be the maximum value of $|x_s|$ for $1 \leq s \leq t$. Let Λ be the set of vectors of dimension $n + P$ with maximum norm no greater than $P + |\boldsymbol{\beta}\#\boldsymbol{\Theta}|$.

The classical example involves two items, so that $n = 2$. In this instance, for fixed n and for P which approaches ∞, $\hat{\beta}_{J\Lambda 21}$ converges to $2\beta_{21}$ with probability 1, and, for each fixed person p, the probability is 1 that $\hat{\theta}_{J\Lambda p1}$ converges to β_{21}, $-\infty$, or ∞ (Andersen, 1973, pp. 66–69). This result is quite consistent with earlier results for joint maximum-likelihood estimation outside of IRT (Neyman and Scott, 1948).

In general, in IRT, inconsistency can be regarded as a problem associated with a fixed number n of items. The situation changes quite substantially if the number n of items becomes large (Haberman, 1977b, 2004, 2008a). In the 1PL case, let P and n both become large, but $|\boldsymbol{\beta}|$ and $\boldsymbol{\Theta}|$ are uniformly bounded by some fixed positive real number. Let $\log(n)/P$ converge to 0. Then $|\hat{\boldsymbol{\beta}}_{J\Lambda} - \boldsymbol{\beta}|$ and $|\hat{\boldsymbol{\Theta}}_{J\Lambda} - \boldsymbol{\Theta}|$ converge in probability to 0, so that all parameters are consistently estimated. In addition, normal approximations apply for person parameters. The conditional variance $\sigma^2(X_p|\theta_p)$ of the person sum $X_p = \sum_{i=1}^{n} U_{pi}$ given θ_p is

$$\sigma^2(X_p|\theta_p) = \sum_{i=1}^{n} \sigma^2(U_{pi}|\theta_p) \tag{9.12}$$

for each person p. The conditional standard deviation of X_p given θ_p is then $\sigma(X_p|\theta_p)$. For any fixed person p, the normalized value $\sigma(X_p|\theta_p)(\hat{\theta}_{J\Lambda p1} - \theta_{p1})$ converges in law to a standard normal distribution.

This result permits construction of approximate confidence intervals for person parameters. For any positive real $\alpha < 1$, let z_α be the real number such that a standard normal random variable has absolute value greater than z_α with probability α. Let

$$\hat{\sigma}^2_{J\Lambda}(U_{pi}|\theta_p) = \frac{\exp(\hat{\theta}_{J\Lambda p1} - \hat{\beta}_{J\Lambda i1})}{[1 + \exp(\hat{\theta}_{J\Lambda p1} - \hat{\beta}_{J\Lambda i1})]^2} \tag{9.13}$$

and let

$$\hat{\sigma}^2_{J\Lambda}(X_p|\theta_p) = \sum_{i=1}^{n} \hat{\sigma}^2_{J\Lambda}(U_{pi}|\theta_p) \tag{9.14}$$

for each person p. For any fixed person p, the probability approaches $1 - \alpha$ that θ_{pi} satisfies

$$\hat{\theta}_{J\Lambda p1} - \frac{z_\alpha}{\hat{\sigma}_{J\Lambda}(X_p|\theta_p)} \leq \theta_{p1} \leq \hat{\theta}_{J\Lambda p1} + \frac{z_\alpha}{\hat{\sigma}_{J\Lambda}(X_p|\theta_p)}. \tag{9.15}$$

Asymptotic independence is also available for estimated person parameters. The normalized values $\sigma(X_p|\theta_p)(\hat{\theta}_{J\Lambda p1} - \theta_{p1})$, $p \geq 1$, are asymptotically independent. If $\mathbf{h}$ is a P-dimensional vector with elements h_p, $1 \leq p \leq P$, if some h_p is not 0, and if the number P' of nonzero h_p is bounded as P and n increase, then

$$\frac{\mathbf{h}'\hat{\Theta}_{J\Lambda} - \mathbf{h}'\boldsymbol{\Theta}}{\left[\sum_{p=1}^{P} \frac{h_p^2}{\sigma_J^2(X_p|\theta_p)}\right]^{1/2}}$$

converges in law to a standard normal distribution. For $0 < \alpha < 1$ the probability approaches $1 - \alpha$ that

$$|\mathbf{h}'\boldsymbol{\Theta} - \mathbf{h}'\hat{\boldsymbol{\Theta}}_{J\Lambda}| \le z_{\alpha} \left[\sum_{p=1}^{P} \frac{h_p^2}{\hat{\sigma}_{J\Lambda}^2(X_p|\boldsymbol{\theta}_p)} \right]^{1/2}. \tag{9.16}$$

It is worth emphasizing that in educational tests with far more persons than items, the confidence interval is not likely to work well unless P is much less than n, so that Equation 9.16 cannot normally be expected to provide useful approximate confidence intervals for the average of the person parameters θ_{p1} for p in a set of 100 persons.

Population information concerning the distribution of the θ_{p1} can be obtained from the estimates $\hat{\theta}_{J\Lambda p1}$ for $1 \le p \le P$. If m is an arbitrary bounded continuous function on the real line, then the average $\bar{m}_{J\Lambda} = P^{-1}\sum_{p=1}^{P} m(\hat{\theta}_{J\Lambda p1})$ converges in probability to the common expected value $E(m(\theta_{p1}))$ of the random variables $m(\theta_{p1})$, $p > 1$. If m is continuously differentiable and P/n^2 approaches 0, then an approximate confidence interval for $E(m(\theta_{p1}))$ is easily obtained. Let $s_{J\Lambda m}$ be the sample variance of $m(\hat{\theta}_{J\Lambda p1})$ for $1 \le p \le P$. Then, the probability approaches $1 - \alpha$ that

$$\bar{m}_{J\Lambda} - z_{\alpha} s_{J\Lambda m}/P^{1/2} \le E(m(\theta_{p1})) \le \bar{m}_{J\Lambda} + z_{\alpha} s_{J\Lambda m}/P^{1/2}. \tag{9.17}$$

In the case of item parameters, normal approximations present a somewhat more complex picture. If P/n^2 converges to 0, then conventional normal approximations apply to item parameters. For each item i from 1 to n, let

$$\sigma^2(U_{+i}|\boldsymbol{\Theta}) = \sum_{p=1}^{P} \sigma^2(U_{pi}|\boldsymbol{\theta}_p) \tag{9.18}$$

be the conditional variance of the item sum $U_{+i} = \sum_{p=1}^{P} U_{pi}$ given $\boldsymbol{\Theta}$. For any fixed item $i > 1$, the normalized value

$$\frac{\hat{\beta}_{J\Lambda i1} - \beta_{i1}}{\left[\frac{1}{\sigma^2(U_{+1}|\boldsymbol{\Theta})} + \frac{1}{\sigma^2(U_{+i}|\boldsymbol{\Theta})}\right]^{1/2}}$$

converges in law to a standard normal distribution. If $\mathbf{h}$ is an n-dimensional vector with elements h_i, $1 \le i \le n$, which are real constants with sum 0, if some h_i is not 0, and if the number of nonzero h_i is bounded as P and n increase, then

$$\frac{\mathbf{h}'\hat{\boldsymbol{\beta}}_{J\Lambda} - \mathbf{h}'\boldsymbol{\beta}}{\left[\sum_{i=1}^{n} \frac{h_i^2}{\sigma^2(U_{+i}|\boldsymbol{\Theta})}\right]^{1/2}}$$

converges in law to a standard normal distribution. Confidence intervals for item parameters can be constructed much as in the case of person parameters.

If P/n^2 does not converge to 0, then a real n-dimensional constant vector $\mathbf{\Delta}$ with elements Δ_i, $1 \le i \le n$, dependent on P and n exists such that $\Delta_1 = 0$ and, for any fixed item $i > 1$, the normalized value

$$\frac{\hat{\beta}_{J\Lambda i} - \beta_i - \Delta_i}{\left[\frac{1}{\sigma^2(U_{+1}|\mathbf{\Theta})} + \frac{1}{\sigma^2(U_{+i}|\mathbf{\Theta})}\right]^{1/2}}$$

converges in law to a standard normal distribution. If $\mathbf{h}$ is an n-dimensional vector with elements h_i, $1 \le i \le n$, with sum 0, if some h_i is not 0, and if the number of nonzero h_i is bounded as P and n increase, then

$$\frac{\mathbf{h}'\hat{\boldsymbol{\beta}}_{J\Lambda} - \mathbf{h}'(\boldsymbol{\beta} + \mathbf{\Delta})}{\left[\sum_{i=1}^{n} \frac{h_i^2}{\sigma^2(U_{+i}|\boldsymbol{\theta})}\right]^{1/2}}$$

converges in law to a standard normal distribution. The products $n\Delta_i$ are uniformly bounded as the number P of persons and the number n of items both become large.

These results presented for the 1PL model should be approached quite cautiously when real applications are considered. The number of items is rarely large enough to assure accuracy of the normal approximations for the estimated person parameters (Haberman, 2004). A basic problem involves discreteness of distributions. This issue must be considered unless $\sigma^2(X_p|\theta_p)$ is relatively large, say more than 5. In the case of item parameters, the difficulty involves the need for the number P of persons to become large at a slower rate than the square n^2 of the number of items. This requirement commonly fails for many common assessments in which hundreds of thousands of examinees receive fewer than 100 items in an assessment.

Some requirements in this section can be relaxed, although some results are affected. For example, the requirement that the θ_{p1} be bounded is only needed to ensure that $|\hat{\Theta}_{J\Lambda} - \Theta|$ converges in probability to 0. Other results continue to apply. The requirement that $\log(P)/n$ approach 0 is also not needed (Haberman, 2004). The specific choice of Λ used here can be changed without affecting results.

9.2.2 Partial Credit Model

Results for the 1PL model apply quite easily to the partial credit model (Haberman, 2006). Inconsistency of parameter estimates is observed unless the number of items becomes large as the number of examinees increases. The basic requirements for simple results are essentially the same as in the 1PL case. The set Λ is defined as the set of vectors of dimension $v_n + P$ with maximum norm not greater than $P + |\boldsymbol{\beta}\#\mathbf{\Theta}|$. For some positive integer A, $A_i \le A$ for all items $i \ge 1$, and $|\boldsymbol{\beta}|$ and $|\mathbf{\Theta}|$ are uniformly bounded by some positive real number as the number n of items and the number P of persons both increase. In the simplest case to consider, $\log(P)/n$ approaches 0. Then $|\hat{\boldsymbol{\beta}}_\Lambda - \boldsymbol{\beta}|$ and $|\mathbf{\Theta}_{J\Lambda} - \mathbf{\Theta}|$ converge in probability to 0. The normal approximation for $\hat{\theta}_{J\Lambda p1}$ is only changed to the extent that the conditional variance $\sigma^2(U_{pi}|\theta_p)$ of U_{pi} given θ_p is different than in the 1PL case if $A_i > 2$.

The normal approximations for the item parameters β_i are changed for items i with $A_i > 2$. Let $\mathbf{Y}_{pi}$ be $\mathbf{y}_{iU_{pi}}$ for $1 \le p \le P$ and $1 \le i \le n$, and let $\mathbf{Y}_{+i}$ be $\sum_{p=1}^{P} \mathbf{Y}_i$ for $1 \le i \le n$. Let $\mathrm{Cov}(\mathbf{Y}_{+i}|\mathbf{\Theta})$ denote the conditional covariance matrix of $\mathbf{Y}_{+i}$ given Θ, and let

$\text{Cov}(\mathbf{Y}_{pi}|\theta_p)$ denote the conditional covariance matrix of $\mathbf{Y}_{pi}$ given θ_p, so that

$$\text{Cov}(\mathbf{Y}_{+i}|\boldsymbol{\Theta}) = \sum_{p=1}^{P} \text{Cov}(\mathbf{Y}_{pi}|\theta_p). \tag{9.19}$$

Let $\mathbf{h}$ be a vector of dimension v_n with elements h_{ik}, $1 \le k \le A_i$, $1 \le i \le n$, let $\mathbf{h}_i$ be a vector of dimension $A_i - 1$ with elements h_{ik}, $1 \le k \le A_i - 1$, for $1 \le i \le n$, and let the summation $\sum_{i=1}^{n} \sum_{k=1}^{A_i - 1} h_{ik} = 0$. Assume that the number of nonzero h_{ik} remains positive and uniformly bounded as P and n become large. Let $\mathbf{h}$ be the vector with dimension v_n with elements h_{ik}, $1 \le k \le A_i - 1, 1 \le i \le n$. If P/n^2 approaches 0, then

$$\frac{\mathbf{h}'\hat{\beta}_{J\Lambda} - \mathbf{h}'\beta}{\left\{\sum_{i=1}^{n} \mathbf{h}_i'[\text{Cov}(\mathbf{Y}_{+i}|\boldsymbol{\Theta})]^{-1}\mathbf{h}_i\right\}^{1/2}}$$

converges in law to a standard normal distribution. In the case that P/n^2 does not converge to 0, then a vector $\boldsymbol{\Delta}$ of dimension v_n with elements Δ_{ik}, $1 \le k \le A_i$, $1 \le i \le n$, exists such that $\Delta_{1k} = 0$ for $1 \le k \le A_1 - 1$. The vector $\boldsymbol{\Delta}$ depends on P and n. In this case,

$$\frac{\mathbf{h}'\hat{\beta}_{J\Lambda} - \mathbf{h}'(\beta + \boldsymbol{\Delta})}{\left\{\sum_{i=1}^{n} \mathbf{h}_i'[\text{Cov}(\mathbf{Y}_{+i}|\boldsymbol{\Theta})]^{-1}\mathbf{h}_i\right\}^{1/2}}$$

converges in law to a standard normal distribution. The product $n|\boldsymbol{\Delta}|$ is bounded as the number P of persons and the number n of items both become large.

Results for estimation of expectations $E(m(\theta_{p1}))$ are essentially the same as in the 1PL case.

9.2.3 Additive Model for Nominal Responses

The results for the 1PL case apply readily to the additive model for nominal responses (Haberman, 1977b). The requirement that $\log(n)/P$ approaches 0 is maintained, and $|\beta|$ and $|\boldsymbol{\Theta}|$ are assumed uniformly bounded. The set Λ may be defined as the set of vectors of dimension $v_n + PD$ with maximum norm not greater than $P + |\beta\#\boldsymbol{\Theta}|$. Once again, $|\hat{\beta}_{J\Lambda} - \beta|$ and $|\hat{\boldsymbol{\Theta}}_{J\Lambda} - \boldsymbol{\Theta}|$ converge in probability to 0. The results for the normal approximation for β are the same as in the partial credit model for $A_i = A$ for all items. In the case of $\boldsymbol{\Theta}$, let $\mathbf{Y}_p$ be the sum of the $\mathbf{Y}_{pi}$, $1 \le i \le n$, so that the conditional covariance matrix of $\mathbf{Y}_{p+}$ given θ_p is

$$\text{Cov}(\mathbf{Y}_{p+}|\theta_p) = \sum_{i=1}^{n} \text{Cov}(\mathbf{Y}_{pi}|\theta_p). \tag{9.20}$$

For a given person p, if $\mathbf{h}_p$ is an $(A-1)$-dimensional vector that has some nonzero element, then

$$\frac{\mathbf{h}_p'\hat{\theta}_{J\Lambda p} - \mathbf{h}_p'\theta_p}{\{\mathbf{h}_p'[\text{Cov}(\mathbf{Y}_{p+}|\theta_p)]^{-1}\mathbf{h}_p\}^{1/2}}$$

converges in law to a standard normal distribution. If $\mathbf{h}_p$, $1 \le p \le P$, are $(A-1)$-dimensional vectors such that some $\mathbf{h}_p$ is nonzero and the number of nonzero $\mathbf{h}_p$ is bounded as P and n become large and if $\mathbf{h}$ is a vector of dimension PD such that element $D(p-1)+d$ of $\mathbf{h}$ is element d of $\mathbf{h}_p$, then

$$\frac{\mathbf{h}'\hat{\boldsymbol{\Theta}}_{J\Lambda} - \mathbf{h}'\boldsymbol{\Theta}}{\left\{\sum_{p=1}^{P} \mathbf{h}'_p[\mathrm{Cov}(\mathbf{Y}_{p+}|\boldsymbol{\theta}_p)]^{-1}\mathbf{h}_p\right\}^{1/2}}$$

converges in law to a standard normal distribution.

Results for estimation of expectations $E(m(\boldsymbol{\theta}_p))$ for bounded continuous real functions m on the set of $(A-1)$-dimensional vectors are essentially the same as results for the 1PL model. The average $\bar{m}_{J\Lambda} = P^{-1}\sum_{p=1}^{P} m(\hat{\boldsymbol{\theta}}_{J\Lambda p})$ converges in probability to the common expected value $E(m(\boldsymbol{\theta}_p))$ of the random variables $m(\boldsymbol{\theta}_p)$, $p > 1$. If m is continuously differentiable, then an approximate confidence interval for $E(m(\boldsymbol{\theta}_p))$ is easily obtained. Let $s_{J\Lambda m}$ be the sample variance of $m(\hat{\boldsymbol{\theta}}_{J\Lambda p})$ for $1 \le p \le P$. Then, the probability approaches $1-\alpha$ that

$$\bar{m}_{J\Lambda} - z_{\alpha}s_{J\Lambda m}/P^{1/2} \le E(m(\boldsymbol{\theta}_p)) \le \bar{m}_{J\Lambda} + z_{\alpha}s_{J\Lambda m}/P^{1/2}. \tag{9.21}$$

9.2.4 Generalized Partial Credit Model

The generalized partial credit model has received less attention in terms of verification of consistency properties as the number P of persons and the number n of items both approach ∞. Unlike the previously considered cases, the difference $g_{ia}(\boldsymbol{\beta}_i\#\boldsymbol{\theta}) - g_{i1}(\boldsymbol{\beta}_i\#\boldsymbol{\theta})$ is not a linear function for each item response score a, $1 \le a \le A_i$ and item i, $1 \le i \le n$. As a consequence, the approach in Haberman (1977b, 2004, 2006, 2008a) does not apply. Modification can be attempted in a manner similar to that used in Haberman (1988); however, complications still exist, for the Hessian matrices $\nabla^2 g_{ia}$ of the item log probability functions g_{ia} need not be negative semidefinite. Thus, no straightforward method exists to verify that a local maximum of the joint log-likelihood is also a global maximum. Consistency and asymptotic normality results similar to those in the partial credit model appear to hold if the set Λ of permitted parameter values in somewhat different than in previous cases. Let $\log(P)/n$ approach 0, let $|\boldsymbol{\beta}|$ and $|\boldsymbol{\Theta}|$ be uniformly bounded by a positive constant c as P and n become large, let θ_{p1} have a positive variance, and let all item discriminations β_{iA_i} be bounded below by a positive constant d for all items $i \ge 1$. Let c' be a positive constant greater than c, and let d' be a positive constant less than d. Let Λ consist of all vectors $\mathbf{x}$ of dimension $v_n + P$ with maximum norm not greater than c' and with each element v_i of $\mathbf{x}$, $1 \le i \le n$, not less than d'. Then $|\hat{\boldsymbol{\beta}}_{J\Lambda} - \boldsymbol{\beta}|$ and $|\hat{\boldsymbol{\Theta}}_{J\Lambda} - \boldsymbol{\Theta}|$ both converge in probability to 0. Normal approximations are then available for person parameters. If P/n^2 approaches 0, then normal approximations are available for item parameters. The approximations are quite similar to those previously described for other models.

9.3 Conditional Estimation

In conditional estimation (Andersen, 1970, 1973), incidental parameters are eliminated, so that the conditional maximum-likelihood estimate involves only a relatively limited number of parameters. Whether incidental parameters are regarded as fixed or random does not affect the basic estimation procedures. Conditional estimation can be applied in the partial credit model or in the additive model for nominal responses. On the other hand, conditional estimation does not apply to the generalized partial credit model. When conditional estimation applies, item parameters are consistently estimated without a requirement that the number of items becomes large; however, estimation of person parameters is somewhat less satisfactory than estimation of item parameters unless the number of items is large.

To apply conditional estimation, let the item log probability function g_{ia}, $1 \le a \le A_i$, $1 \le i \le n$, have the additive form

$$g_{ia}(\boldsymbol{\beta}_i \# \boldsymbol{\theta}) = g_{i1}(\boldsymbol{\beta}_i \# \boldsymbol{\theta}) + \mathbf{t}'_{1ia}\boldsymbol{\beta}_i + \mathbf{t}'_{2ia}\boldsymbol{\theta} \tag{9.22}$$

for q_i-dimensional vectors $\mathbf{t}_{1ia}$ and D-dimensional vectors $\mathbf{t}_{2ia}$. For $a = 1$, all elements of $\mathbf{t}_{i1a}$ and $\mathbf{t}_{i2a}$ are 0. For each person p, $1 \le p \le P$, define the summation

$$\mathbf{T}_{2p} = \sum_{i=1}^{n} \mathbf{t}_{2iU_{ip}}, \tag{9.23}$$

let $\mathbf{t}_{1\mathbf{a}}$ be the v_n-dimensional vector with element $v_{i-1} + k$ equal to element k of t_{1ia_i} for $1 \le k \le q_i$ and $1 \le i \le n$, and let $\mathbf{T}_{1p} = \mathbf{t}_{1\mathbf{U}_p}$. Let the conditional information matrix $\mathbf{I}_C$ be $E(\mathrm{Cov}(\mathbf{T}_{1p}|\mathbf{T}_{2p})$, the expected conditional covariance matrix of $\mathbf{T}_{1p}$ given $\mathbf{T}_{2p}$. The key to conditional estimation involves inference conditional on the sums $\mathbf{T}_{2p}$. The approach is only appropriate if $\mathbf{I}_C + \mathbf{Z}'\mathbf{Z}$ is positive definite. Let $\mathcal{T}_2$ be the possible values of $\mathbf{T}_{2p}$, and let $\mathcal{U}(\mathbf{t})$, $\mathbf{t}$ in $\mathcal{T}_2$, be the members $\mathbf{a}$ of $\mathcal{U}$ such that $\sum_{i=1}^{n} \mathbf{t}_{2ia_i} = \mathbf{t}$. For any $\mathbf{t}$ in $\mathcal{T}_2$, the probability that $\mathbf{T}_{2p} = \mathbf{t}$ given $\boldsymbol{\theta}_p = \boldsymbol{\theta}$ is

$$P\{\mathbf{T}_{2p} = \mathbf{t}|\boldsymbol{\theta}_p = \boldsymbol{\theta}) = \exp[g_{T\mathbf{t}}(\boldsymbol{\beta}\#\boldsymbol{\theta})], \tag{9.24}$$

where the marginal log probability function

$$g_{T\mathbf{t}}(\boldsymbol{\beta}\#\boldsymbol{\theta}) = \log \sum_{\mathbf{a}\in\mathcal{U}(\mathbf{t})} \exp[g_{J\mathbf{a}}(\boldsymbol{\beta}\#\boldsymbol{\theta})]. \tag{9.25}$$

For any $\mathbf{a}$ in $\mathcal{U}(\mathbf{t})$ and $\mathbf{t}$ in $\mathcal{T}_2$, the difference $g_{J\mathbf{a}}(\boldsymbol{\beta}\#\boldsymbol{\theta}) - g_{T\mathbf{t}}(\boldsymbol{\beta}\#\boldsymbol{\theta})$ has the same value $g_{C\mathbf{a}}(\boldsymbol{\beta})$ for all D-dimensional vectors $\boldsymbol{\theta}$. Therefore, the conditional probability that $\mathbf{U}_p = \mathbf{a}$ given that $\mathbf{T}_{2p} = \mathbf{t}$ is

$$P\{\mathbf{U}_p = \mathbf{a}|\mathbf{T}_{2p} = \mathbf{t}) = \exp(g_{C\mathbf{a}}(\boldsymbol{\beta})], \tag{9.26}$$

where the conditional joint log probability function is

$$g_{C\mathbf{a}}(\boldsymbol{\beta}) = \sum_{i=1}^{n} \mathbf{t}'_{1a_i i}\boldsymbol{\beta}_i - \log\left[\sum_{\mathbf{u}\in\mathcal{U}(\mathbf{t})} \exp\left(\sum_{i=1}^{n} \mathbf{t}'_{1u_i i}/\boldsymbol{\beta}_i\right)\right]. \tag{9.27}$$

The conditional log-likelihood ℓ_C then satisfies

$$\ell_C(\boldsymbol{\beta}) = \sum_{p=1}^{P} g_{C\mathbf{U}_p}(\boldsymbol{\beta}). \tag{9.28}$$

In conditional estimation, the conditional log-likelihood function is then maximized subject to the constraint that $\mathbf{Z}\boldsymbol{\beta} = \mathbf{z}$. To avoid problems of nonexistent or ambiguously defined estimates, let Λ be a closed and bounded convex set of v_n-dimensional vectors such that $\boldsymbol{\beta}$ is in the interior of Λ. The maximum conditional log-likelihood $\hat{\ell}_{C\Lambda}$ is the supremum of $\ell_C(\mathbf{x})$ subject to the constraints that $\mathbf{Z}\mathbf{x} = \mathbf{z}$ and $\mathbf{x}$ is in Λ. The conditional maximum-likelihood estimator $\hat{\boldsymbol{\beta}}_{C\Lambda}$ of $\boldsymbol{\beta}$ is defined to be a function of the response vectors $\mathbf{U}_p$, $1 \le p \le P$, such that $\ell_C(\hat{\boldsymbol{\beta}}_{C\Lambda}) = \hat{\ell}_{C\Lambda}$, $\mathbf{Z}\hat{\boldsymbol{\beta}}_{C\Lambda} = \mathbf{z}$, and $\hat{\boldsymbol{\beta}}_{C\Lambda}$ is in Λ. As in joint maximum likelihood, given $\hat{\boldsymbol{\beta}}_{C\Lambda}$, estimates $\hat{\boldsymbol{\beta}}_{C\Lambda i}$ of $\boldsymbol{\beta}_i$ and $\hat{\beta}_{C\Lambda ik}$ of β_{ik} are defined as in the case of joint estimation.

Conditional estimation does not lead directly to estimation of person parameters. One simple approach is to estimate $\boldsymbol{\Theta}$ by maximization of the joint likelihood function $\ell_J(\hat{\boldsymbol{\beta}}_{C\Lambda}\#\boldsymbol{\Theta})$ over possible values of $\boldsymbol{\Theta}$. Let Γ be a convex subset of the space of PD-dimensional vectors such that $\boldsymbol{\Theta}$ is in its interior. Let $\hat{\ell}_{C\Lambda\Gamma}$ be the maximum of $\ell_J(\hat{\boldsymbol{\beta}}_{C\Lambda}\#\mathbf{y})$ for $\mathbf{y}$ in Γ. Then $\hat{\boldsymbol{\Theta}}_{C\Lambda\Gamma}$ is a function of the observed item response scores U_{pi}, $1 \le i \le n$, $1 \le p \le P$, such that $\ell_J(\hat{\boldsymbol{\beta}}_{C\Lambda}\#\hat{\boldsymbol{\Theta}}_{C\Lambda\Gamma}) = \hat{\ell}_{C\Lambda\Gamma}$ and $\hat{\boldsymbol{\Theta}}_{C\Lambda\Gamma}$ is in Γ. One may then define corresponding estimates $\hat{\boldsymbol{\theta}}_{C\Lambda\Gamma p}$ and $\hat{\theta}_{C\Lambda\Gamma pd}$ for $1 \le d \le D$ and $1 \le p \le P$.

Unlike the case of joint estimation, conditional estimation normally leads to very satisfactory consistency properties for estimation of item parameters even for a fixed number of items (Andersen, 1973). To begin, let the number of items be fixed, and let the number of examinees approach infinity. Let $\mathbf{I}_C + \mathbf{Z}'\mathbf{Z}$ be positive definite. Then $\hat{\boldsymbol{\beta}}_{C\Lambda\Gamma}$ converges to $\boldsymbol{\beta}$ with probability 1, and $P^{1/2}(\hat{\boldsymbol{\beta}}_{C\Lambda\Gamma} - \boldsymbol{\beta})$ converges in law to a multivariate normal distribution with zero mean and with covariance matrix

$$(\mathbf{I}_C + \mathbf{Z}'\mathbf{Z})^{-1}\mathbf{I}_C(\mathbf{I}_C + \mathbf{Z}'\mathbf{Z})^{-1}.$$

The matrix $\mathbf{I}_C$ may be estimated by $\hat{I}_{C\Lambda} = -P^{-1}\nabla^2\ell_C(\hat{\boldsymbol{\beta}}_{C\Lambda})$, where $\nabla^2\ell_C$ is the Hessian matrix of the conditional log-likelihood ℓ_C. Conditional estimation of $\boldsymbol{\beta}$ is not typically as efficient as estimation of $\boldsymbol{\beta}$ via maximum marginal likelihood if a parametric model for the $\boldsymbol{\theta}_p$ is used, although the loss of efficiency is normally quite small (Eggen, 2000). If the number n of items is fixed, then for any fixed person p, the difference $|\hat{\boldsymbol{\theta}}_{C\Lambda\Gamma p} - \boldsymbol{\theta}_p|$ does not converge in probability to 0 as the number P of persons becomes large, a predictable result given that the number of items providing information on person p does not increase for $P \ge p$.

Conditional estimation of item parameters remains satisfactory when the number n of items becomes large but the number P of persons remains at least as large as n. In addition, estimation of person parameters becomes much more satisfactory, and estimation of the distribution of person parameters becomes feasible (Haberman, 2004, 2006). The 1PL model provides a typical example. As in the case of joint estimation, assume that the $|\boldsymbol{\beta}|$ is uniformly bounded for all items i. Let Λ consist of all n-dimensional vectors with maximum norm no greater than $P + |\boldsymbol{\beta}|$. Let Γ consist of all n-dimensional vectors with maximum norm no greater than $P + |\boldsymbol{\Theta}|$. Then, $|\hat{\boldsymbol{\beta}}_{C\Lambda\Gamma} - \boldsymbol{\beta}|$ converge in probability to 0. If $\log(P)/n$ converges to 0, then $|\hat{\boldsymbol{\Theta}}_{C\Lambda\Gamma} - \boldsymbol{\Theta}|$ also converges in probability to 0.

For any fixed item $i > 1$, the normalized value

$$\frac{\hat{\beta}_{C\Lambda\Gamma i} - \beta_i}{\left[\frac{1}{\sigma^2(U_{+1}|\boldsymbol{\Theta})} + \frac{1}{\sigma^2(U_{+i}|\boldsymbol{\Theta})}\right]^{1/2}}$$

converges in law to a standard normal distribution. If h_i, $1 \le i \le n$, are real constants with sum 0, if some h_i is not 0, and if the number of nonzero h_i is bounded as P and n increase, then

$$\frac{\mathbf{h}'\hat{\boldsymbol{\beta}}_{C\Lambda\Gamma} - \mathbf{h}'\boldsymbol{\beta}}{\left[\sum_{i=1}^{n} \frac{h_i^2}{\sigma^2(U_{+i}|\boldsymbol{\Theta})}\right]^{1/2}}$$

converges in law to a standard normal distribution. If n^2/P approaches 0, then for any nonzero n-dimensional vector $\mathbf{h}$,

$$\frac{P^{1/2}(\mathbf{h}'\hat{\boldsymbol{\beta}}_{C\Lambda\Gamma} - \mathbf{h}'\boldsymbol{\beta})}{(\mathbf{h}'\mathbf{I}_C\mathbf{h})^{1/2}}$$

converges in law to a standard normal distribution. When conditions for large-sample properties apply to both joint estimation and conditional estimation, $\mathbf{h}'\boldsymbol{\beta}$ is estimated, and elements h_i of $\mathbf{h}$ are 0 for $i > n'$ for some fixed $n' \ge 0$, then the same normal approximation applies to both conditional and joint estimation.

For any fixed person p, the normalized value $\sigma(U_{p+}|\theta_p)(\hat{\theta}_{C\Lambda\Gamma p1} - \theta_{p1})$ converges in law to a standard normal distribution. If m is an arbitrary bounded continuous function on the real line, then the average $\bar{m}_C = P^{-1}\sum_{p=1}^{P} m(\hat{\theta}_{Cp1})$ converges in probability to the common expected value $E(m(\theta_{p1}))$ of the random variables $m(\theta_{p1})$, $p > 1$. Conditions and results for a normal approximation are essentially the same as in joint estimation.

A very important gain in conditional estimation involves the ability to test a model against an alternative hypothesis (Andersen, 1971a; Haberman, 1977a; Glas and Verhelst, 1995; Haberman, 2007). Conditional likelihood-ratio tests, conditional Wald tests, generalized residuals, and conditional efficient scores tests are all possible. The resulting test statistics have approximate normal or chi-square distributions when the number P of persons is large.

9.4 Unrestricted Marginal Estimation

In unrestricted marginal estimation, inferences are based on the marginal distribution of the response vectors $\mathbf{U}_p$. The marginal probability

$$P\{\mathbf{U}_p = \mathbf{a}\} = \exp[g_{M\mathbf{a}}(\boldsymbol{\beta}, F)], \tag{9.29}$$

where $\exp[g_{M\mathbf{a}}(\boldsymbol{\beta}, F)]$ is the expectation of $\exp[g_{J\mathbf{a}}(\boldsymbol{\beta}\#\boldsymbol{\omega})]$ for $\boldsymbol{\omega}$ a D-dimensional random vector with distribution function F. Thus $g_{M\mathbf{a}}(\boldsymbol{\beta}, F)$ is the unrestricted marginal log probability function for the vector $\mathbf{a}$ of response scores. The restraint is imposed on $\boldsymbol{\beta}$ that $\mathbf{Z}\boldsymbol{\beta} = \mathbf{z}$, but no restriction is placed on F (Kiefer and Wolfowitz, 1956; Tjur, 1982; Cressie and Holland, 1983). The marginal log-likelihood is then

$$\ell_M(\boldsymbol{\beta}, F) = \sum_{p=1}^{P} g_{M\mathbf{U}_p}(\boldsymbol{\beta}, F). \tag{9.30}$$

Maximization of the marginal log-likelihood involves far more basic problems than those encountered in previous sections of this chapter. To avoid problems of nonexistent or multiple maxima, it is a simple matter to consider a closed convex set Λ of v_n-dimensional vectors such that $\boldsymbol{\beta}$ is in the interior of Λ; however, the set $\mathcal{F}$ of distribution functions has infinite dimension. Identifiability of F is a major issue. To any distribution function G in $\mathcal{F}$ and any v_n-dimensional vector $\mathbf{x}$ such that $\mathbf{Zx} = \mathbf{z}$ corresponds a distribution function G^* of a polytomous D-dimensional random vector $\boldsymbol{\theta}^*$ with the following characteristics (Rockafellar, 1970, Section 17):

1. The variable $\boldsymbol{\theta}^*$ assumes values in a finite set W.
2. To each $\boldsymbol{\theta}$ in W corresponds a positive probability $P\{\boldsymbol{\theta}^* = \boldsymbol{\theta}\}$ that $\boldsymbol{\theta}^* = \boldsymbol{\theta}$.
3. If $\alpha(\boldsymbol{\theta}) \geq 0$ for $\boldsymbol{\theta}$ in W, $\sum_{\boldsymbol{\theta}\in W} \alpha(\boldsymbol{\theta}) = 1$, and G^+ is the distribution function of a polytomous random variable $\boldsymbol{\theta}^+$ such that $P\{\boldsymbol{\theta}^+ = \boldsymbol{\theta}\} = \alpha(\boldsymbol{\theta})$ for all $\boldsymbol{\theta}$ in W, then $g_{M\mathbf{a}}(\mathbf{x}, G) = g_{M\mathbf{a}}(\mathbf{x}, G^+)$ for all $\mathbf{a}$ in $\mathcal{U}$ if, and only if, $G^+ = G^*$.

One may say that G^* is a distribution function of a polytomous random vector that corresponds to G for the v_n-dimensional vector $\mathbf{x}$. The condition $G^+ = G^*$ implies that $\alpha(\boldsymbol{\theta}) = P\{\boldsymbol{\theta}^* = \boldsymbol{\theta}\}$ for all $\boldsymbol{\theta}$ in W, and it must be the case that W has no more than $R = \prod_{i=1}^{n} A_i$ values. It should be emphasized that, although G and G^* are not distinguished by the observed marginal distribution of the response vector $\mathbf{U}_p$, the two distribution functions need not have any intuitive relationship or resemblance.

Let V be a closed, convex, and bounded nonempty set of D-dimensional vectors, and let $\mathcal{G}$ be the set of distribution functions of polytomous random variables with all values in V. Assume that a distribution function F^* of a polytomous random variable $\boldsymbol{\theta}^*$ that corresponds to F for $\boldsymbol{\beta}$ has all values of $\boldsymbol{\theta}^*$ in the interior of V. Let $\hat{\ell}_{M\Lambda\mathcal{G}}$ be the supremum of $\ell_M(\mathbf{x}, G)$ for $\mathbf{x}$ in λ such that $\mathbf{Zx} = \mathbf{z}$ and G in $\mathcal{G}$. Then, the unrestricted maximum-likelihood estimates $\hat{\boldsymbol{\beta}}_{M\Lambda\mathcal{G}}$ and $\hat{F}_{M\Lambda\mathcal{G}}$ are defined as functions on the responses $U_{pi}, 1 \leq i \leq n, 1 \leq p \leq P$, such that $\mathbf{Z}\hat{\boldsymbol{\beta}}_{M\Lambda V} = \mathbf{z}$, $\ell_M(\hat{\boldsymbol{\beta}}_{M\Lambda\mathcal{G}}, \hat{F}_{M\Lambda\mathcal{G}})$ is $\hat{\ell}_{M\Lambda\mathcal{G}}$, $\hat{\boldsymbol{\beta}}_{M\Lambda\mathcal{G}}$ is in Λ, and $\hat{F}_{M\Lambda\mathcal{G}})$ is in $\mathcal{G}$.

The estimates $\hat{\boldsymbol{\beta}}_{M\Lambda\mathcal{G}}$ and $\hat{F}_{M\Lambda\mathcal{G}}$ often have no useful properties (Haberman, 2005). Let $\mathcal{P}$ be the set of real functions on $\mathcal{U}$ such that each function P in $\mathcal{P}$ satisfies $P(\mathbf{a}) = \exp[g_{M\mathbf{u}}(\mathbf{x}, G)]$, $\mathbf{a}$ in $\mathcal{U}$, for some vector $\mathbf{x}$ of dimension v_n such that $\mathbf{Zx} = \mathbf{z}$ and some distribution function G in $\mathcal{F}$. For each such vector $\mathbf{x}$, let $\mathcal{P}(\mathbf{x})$ be the set of functions P in $\mathcal{P}$ such that $P(\mathbf{u}) = \exp[g_{M\mathbf{u}}(\mathbf{x}, G)]$, $\mathbf{u}$ in $\mathcal{U}$, for some distribution function G in $\mathcal{F}$. Then $\mathcal{P}(\mathbf{x})$ is a nonempty convex set of dimension $S(\mathbf{x}) \leq R - 1$. It is the smallest convex set of real functions on $\mathcal{U}$ that for each D -dimensional vector $\boldsymbol{\theta}$ includes the function $g_J(\,\mathbf{x}\#\boldsymbol{\theta})$ with value $g_{J\mathbf{a}}(\mathbf{x}\#\boldsymbol{\theta})$ for $\mathbf{a}$ in $\mathcal{U}$.

The worst possible situation occurs if $S(\boldsymbol{\beta}) = R - 1$ (Haberman, 2005). In this case, some positive constant $\delta > 0$ exists such that $S(\mathbf{x}) = R - 1$ for any v_n-dimensional vector $\mathbf{x}$ such that $\mathbf{Zx} = \mathbf{z}$ and $|\mathbf{x} - \boldsymbol{\beta}| < \delta$. It follows in such a case that the unrestricted marginal log-likelihood cannot provide consistent estimation of $\boldsymbol{\beta}$. Let $\bar{P}(\mathbf{a})$, $\mathbf{a}$ in $\mathcal{U}$, be the fraction of persons p, $1 \leq p \leq P$, such that $\mathbf{U}_p = \mathbf{a}$. There exists some real $\epsilon > 0$ such that if $|\bar{P}(\mathbf{a}) - P\{\mathbf{U}_p = \mathbf{a}\}| < \epsilon$ for all $\mathbf{a}$ in $\mathcal{U}$, then $\ell_M(\mathbf{x}, G)$ is equal to the supremum of ℓ_M whenever $\exp[g_{M\mathbf{a}}(\mathbf{x}, G)] = \bar{P}(\mathbf{a})$ for all $\mathbf{a}$ in $\mathcal{U}$. In addition, for some real $\epsilon' > 0$, to any v_n-dimensional vector $\mathbf{x}$ such that $\mathbf{Zx} = \mathbf{z}$ and $|\mathbf{x} - \boldsymbol{\beta}| < \epsilon'$ corresponds a G in $\mathcal{F}$ such that $\bar{P}(\mathbf{a}) = \exp[g_{M\mathbf{a}}(\mathbf{x}, G)]$ for all $\mathbf{a}$ in $\mathcal{U}$. It follows that unrestricted maximum marginal likelihood cannot provide consistent estimation of the vector $\boldsymbol{\beta}$ of item parameters. This worst possible situation is not a theoretical threat. It applies to the generalized partial credit model.

In practice, the only cases in which useful results are obtained are cases in which conditional estimation can be applied. Let Equation 9.22 hold, and let $\mathbf{I}_C + \mathbf{Z}'\mathbf{Z}$ be positive definite. Given notation in Section 9.3, for $\mathbf{a}$ in $\mathcal{U}(\mathbf{t})$ and $\mathbf{t}$ in $\mathcal{T}_2$,

$$g_{M\mathbf{a}}(\boldsymbol{\beta}, F) = g_{C\mathbf{a}}(\boldsymbol{\beta}) + \log E(\exp[g_{T\mathbf{t}}(\boldsymbol{\beta}\#\boldsymbol{\omega})]), \tag{9.31}$$

where $\boldsymbol{\omega}$ is again a D-dimensional random vector with distribution function F. For each $\boldsymbol{\beta}$, the basic requirement for simple results is that for some distinct D-dimensional vectors $\boldsymbol{\upsilon}(\mathbf{t})$, $\mathbf{t}$ in $\mathcal{T}_2$, unique positive weights $\alpha(\mathbf{t})$, $\mathbf{t}$ in $\mathcal{T}_2$, exist such that the sum of the $\alpha(\mathbf{t})$ is 1 and

$$P\{\mathbf{t}_{2p} = \mathbf{t}\} = \sum_{\mathbf{w}\in\mathcal{T}_2} \alpha(\mathbf{w}) g_{T\mathbf{t}}(\boldsymbol{\beta}\#\boldsymbol{\upsilon}(\mathbf{w})) \tag{9.32}$$

for all $\mathbf{t}$ in $\mathcal{T}_2$. In such a case, for Λ defined as in Section 9.3, the probability is 1 that $\hat{\boldsymbol{\beta}}_{M\Lambda\mathcal{G}} = \hat{\boldsymbol{\beta}}_{C\Lambda}$ for P sufficiently large. Large-sample results are then the same for $\hat{\boldsymbol{\beta}}_{M\Lambda\mathcal{G}}$ and $\hat{\boldsymbol{\beta}}_{C\Lambda\Gamma}$ (Tjur, 1982; Cressie and Holland, 1983).

The 1PL model has been examined in some detail (Cressie and Holland, 1983). In this case, $\mathcal{T}_2$ consists of vectors of dimension 1 with integer elements from 0 to n, and Equation 9.32 reduces to conditions on the first n moments of the person sum $X_p = \sum_{i=1}^{n} U_{pi}$. In the 1PL case and in other cases in which conditional estimation applies, the large-sample results for a fixed number of items can be deceptive for educational tests in which the number of items is relatively large, say greater than 50. In typical cases, some probabilities $P\{\mathbf{T}_{2p} = \mathbf{t}\}$ are so small that the probability that no $\mathbf{T}_{2p}$ is $\mathbf{t}$ is quite high for realistic values of P. In such a case, $\hat{\boldsymbol{\beta}}_{C\Lambda}$ is normally not the same as $\hat{\boldsymbol{\beta}}_{M\Lambda\mathcal{G}}$. This problem is obviously also present if the number of items increases as the number of persons increases. In summary, unrestricted maximum marginal likelihood is not very satisfactory for the type of models under study.

9.5 Restricted Marginal Estimation

Marginal maximum likelihood is most commonly employed in IRT when the common distribution function F of the person parameter vector $\boldsymbol{\theta}_p$ for each person p, $1 \le p \le P$ is assumed to satisfy $F = G(\boldsymbol{\gamma})$ for some member $\boldsymbol{\gamma}$ of an open subset O of the space of ψ-dimensional vectors, where ψ is a positive integer and G is a function from O to the set of D-dimensional distribution functions (Bock and Lieberman, 1970; Bock and Aitkin, 1981). In the most common case, $\boldsymbol{\theta}_p$ is assumed to have a normal distribution with some mean $\boldsymbol{\mu}$ and some positive-definite covariance matrix $\boldsymbol{\Sigma}$. In this case ψ is $D(D+3)/2$. Cases have also been considered in which for some given finite set $\mathcal{V}$, F is assumed to be a distribution function of a random vector that only has values in $\mathcal{V}$ and assumes each such value with positive probability (Heinen, 1996). The important difference from the case of unrestricted marginal estimation is the restricted number of elements of $\mathcal{V}$. Here, ψ is one less than the number of elements of $\mathcal{V}$. A virtue of restricted marginal estimation is that it is applicable to all models for item responses that have been considered in this chapter.

For response vectors $\mathbf{a}$ in $\mathcal{U}$, the marginal probability

$$P\{\mathbf{U}_p = \mathbf{a}\} = \exp[g_{R\mathbf{a}}(\boldsymbol{\beta}\#\boldsymbol{\gamma})], \tag{9.33}$$

where $\exp[g_{R\mathbf{a}}(\boldsymbol{\beta}\#\boldsymbol{\gamma})]$ is the expectation of $\exp[g_{J\mathbf{a}}(\boldsymbol{\beta}\#\boldsymbol{\omega})]$ for $\boldsymbol{\omega}$ a D-dimensional random vector with distribution function $G(\psi)$. The restraint is still imposed on $\boldsymbol{\beta}$ that $\mathbf{Z}\boldsymbol{\beta} = \mathbf{z}$. The marginal log-likelihood is then

$$\ell_R(\boldsymbol{\beta}\#\boldsymbol{\gamma}) = \sum_{p=1}^{P} g_{R\mathbf{U}_p}(\boldsymbol{\beta}\#\boldsymbol{\gamma}). \tag{9.34}$$

Let Λ be a closed convex set of v_n-dimensional vectors such that $\boldsymbol{\beta}$ is in the interior of Λ, and let Γ be a closed convex subset of O such that $\boldsymbol{\gamma}$ is in the interior of Γ. Let $\hat{\ell}_{R\Lambda\Gamma}$ be the supremum of $\ell_R(\mathbf{x}\#\mathbf{y})$ for $\mathbf{x}$ in Λ and $\mathbf{y}$ in Γ such that $\mathbf{Z}\mathbf{x} = \mathbf{z}$. In a manner similar to that used in other sections, one may define the restricted maximum-likelihood estimates $\hat{\boldsymbol{\beta}}_{R\Lambda\Gamma}$ and $\hat{\boldsymbol{\gamma}}_{R\Lambda\Gamma}$ to be functions of the observed responses U_{pi}, $1 \le i \le n$, $1 \le p \le P$, such that $\mathbf{Z}\hat{\boldsymbol{\beta}}_{R\Lambda\Gamma} = \mathbf{z}$, $\hat{\boldsymbol{\beta}}_{R\Lambda\Gamma}$ is in Λ, $\hat{\boldsymbol{\gamma}}_{R\Lambda\Gamma}$ is in Γ, and $\ell_R(\hat{\boldsymbol{\beta}}_{R\Lambda\Gamma}\#\hat{\boldsymbol{\gamma}}_{R\Lambda\Gamma}) = \hat{\ell}_{R\Lambda\Gamma}$.

Standard results at least require that the log probability function $g_{R\mathbf{a}}$ be twice continuously differentiable. Let $\nabla^2 g_{R\mathbf{a}}(\boldsymbol{\beta}\#\boldsymbol{\gamma})$ be the Hessian matrix of $g_{R\mathbf{a}}$ at $\boldsymbol{\beta}\#\boldsymbol{\gamma}$, and let the information matrix $\mathbf{I}_R$ be the expectation of $-\nabla^2 g_{R\mathbf{U}_p}(\boldsymbol{\beta}\#\boldsymbol{\gamma})$. Let $\mathbf{Z}_*$ be the $v_n + \psi$ by H matrix with the first v_n columns equal to $\mathbf{Z}$ and with the remaining ψ columns with all elements 0. Let $\mathbf{I}_R + \mathbf{Z}_*'\mathbf{Z}_*$ be positive definite. The basic approximation for a fixed number of items and a large number of persons is somewhat similar to a previous result for the joint estimation for a generalized partial credit model to the extent that the sets Λ and Γ often need to be restricted even for large samples. The sets Λ and Γ may be defined so that $\hat{\boldsymbol{\beta}}_{R\Lambda\Gamma}$ converges with probability 1 to $\boldsymbol{\beta}$ and $\hat{\boldsymbol{\gamma}}_{R\Lambda\Gamma}$ converges to $\boldsymbol{\gamma}$ with probability 1 as the number P of persons becomes arbitrarily large. In addition, the distribution of $P^{1/2}(\hat{\boldsymbol{\beta}}_{R\Lambda\Gamma}\#\hat{\boldsymbol{\gamma}}_{R\Lambda\Gamma} - \boldsymbol{\beta}\#\boldsymbol{\gamma})$ converges in law to a multivariate normal distribution with zero mean and with covariance matrix

$$(\mathbf{I}_R + \mathbf{Z}_*'\mathbf{Z}_*)^{-1}\mathbf{I}_R(\mathbf{I}_R + \mathbf{Z}_*'\mathbf{Z}_*)^{-1}.$$

For construction of asymptotic confidence intervals, $\mathbf{I}_R$ may be approximated by $-P^{-1}\nabla^2\ell_R(\hat{\boldsymbol{\beta}}_{R\Lambda\Gamma}\#\hat{\boldsymbol{\gamma}}_{R\Lambda\Gamma})$.

In the case of a bounded and continuous real function m on the set of D-dimensional vectors, the expectation $E(m(\boldsymbol{\theta}_p))$ can be written as a Lebesgue–Stieltjes integral $\int m\, dG(\boldsymbol{\beta})$. If $\boldsymbol{\theta}_p$ has a continuous distribution with continuous density f, then the integral is the conventional Riemann integral $\int m(\boldsymbol{\theta}) f(\boldsymbol{\theta}) d\boldsymbol{\theta}$ over the entire set of D-dimensional vectors. If $\boldsymbol{\theta}_p$ is polytomous with all values in $\mathcal{V}$, then the integral is $\sum_{\mathbf{y}\in\mathcal{V}} m(\boldsymbol{\theta}) P\{\boldsymbol{\theta}_p = \boldsymbol{\theta}\}$. If e is the real function e on O such that $e(\mathbf{y}) = \int m\, dG(\mathbf{y})$ for $\mathbf{y}$ in O, then $e(\hat{\boldsymbol{\gamma}}_{R\Lambda\Gamma}) = \int m\, dG(\hat{\boldsymbol{\gamma}}_{R\Lambda\Gamma})$ converges to $e(\boldsymbol{\gamma}) = E(m(\boldsymbol{\theta}_p))$ with probability 1. If e is continuously differentiable, then a normal approximation and corresponding approximate confidence bounds can be readily obtained for $P^{1/2}[e(\hat{\boldsymbol{\gamma}}_{R\Lambda\Gamma}) - e(\boldsymbol{\gamma})]$.

Estimation of $\boldsymbol{\theta}_p$ can be accomplished by customary maximum likelihood (ML), maximum *a posteriori* (MAP), or expected *a posteriori* (EAP) estimation (Bock and Aitkin, 1981).

Although these results are relatively routine, there appear to be a number of issues of concern. It is obviously desirable that Λ and Γ clearly include $\boldsymbol{\beta}$ and $\boldsymbol{\gamma}$. Because the Hessian of the restricted log marginal probability function $g_{R\mathbf{u}}$ is not necessarily negative semidefinite, it is often unclear how the requirement can be met. In addition, substantial practical limitations on the quality of large-sample approximations arise unless the dimension ψ of

γ is small (Heinen, 1996; Haberman, 2005). For example, in the 1PL case, ψ cannot exceed the number of items, and large-sample results provide poor approximation even for much smaller ψ and for a quite large number P of persons (Haberman, 2004). A basic problem involves the limited dependence of the posterior distribution of θ_p given $\mathbf{U}_p$ on the distribution function F of θ_p when θ_p has a continuous distribution (Holland, 1990). This limited dependence is especially pronounced when the number of items is relatively large.

As in earlier sections, asymptotic approximations do apply when both the number of items and the number of persons both become large. Conditions and results are quite similar to those previously encountered, and approaches for incomplete observation in log-linear models may be applied without unusual difficulty (Haberman, 1988, 2008b). In summary, restricted marginal estimation is a practical approach. It avoids the consistency problems of joint estimation, and it has wider application than either conditional estimation or unrestricted marginal estimation.

9.6 Conclusions

In this review of methods for the treatment of incidental parameters within the context of IRT, some basic conclusions can be reached. Joint estimation does not appear to be an appropriate method to use, although its properties are less undesirable in large samples with large numbers of items. Conditional estimation appears quite satisfactory when applicable for estimation of item parameters. It is less satisfactory for estimation of person parameters, although results improve as the number of items increases. Unrestricted marginal estimation does not really help in practice. Restricted marginal estimation does have attractive properties; however, significant limitations exist in terms of the number of parameters ψ that can be used without compromising the quality of large-sample approximations. In the end, a rather conventional approach to estimation, restricted marginal maximum likelihood, may well be the most appropriate approach.

The analysis in this chapter has not treated Bayesian inference within the context of incidental parameters encountered within IRT (Patz and Junker, 1999). On the whole, Bayesian inference does not appear to change the basic situation with respect to consistency and asymptotic normality in the presence of incidental parameters, but results are differently expressed and, as usual in Bayesian inference, reasonableness of prior distributions must be considered. The simplest case involves a fixed number of items and inferences for the model used in restricted marginal estimation. Here a continuous prior distribution can be applied for the parameter vector $\beta\#\gamma$. Large-sample results based on normal approximations can then be obtained for the posterior distribution of the parameter vector (Ghosh et al., 2006, Chapter 4). In the case of Bayesian inference, literature for the case corresponding to joint estimation appears difficult to find.

References

Andersen, E. B. 1970. Asymptotic properties of conditional maximum-likelihood estimators. *Journal of the Royal Statistical Society, Series B*, 32, 283–301.

Andersen, E. B. 1971a. The asymptotic distribution of conditional likelihood ratio tests. *Journal of the American Statistical Association, 66*, 630–633.

Andersen, E. B. 1971b. A strictly conditional approach to estimation theory. *Scandinavian Actuarial Journal, 1971*, 39–49.

Andersen, E. B. 1973. *Conditional Inference and Models for Measuring*. Copenhagen: Mentalhygiejnisk Forlag.

Bock, R. D. and Aitkin, M. 1981. Marginal maximum likelihood estimation of item parameters: Application of an EM algorithm. *Psychometrika, 46*, 443–459.

Bock, R. D. and Lieberman, M. 1970. Fitting a response curve model for dichotomously scored items. *Psychometrika, 35*, 179–198.

Cox, D. R. and Hinkley, D. V. 1974. *Theoretical Statistics*. London: Chapman and Hall.

Cressie, N. and Holland, P. W. 1983. Characterizing the manifest probabilities of latent trait models. *Psychometrika, 48*, 129–141.

Eggen, T. J. H. M. 2000. On the loss of information in conditional maximum likelihood estimation of item parameters. *Psychometrika, 65*, 337–362.

Ghosh, J. K., Delampady, M., and Samanta, T. 2006. *An Introduction to Bayesian Analysis: Theory and Methods*. New York: Springer.

Glas, C. A. W. and Verhelst, N. D. 1995. Testing the Rasch model. In G. H. Fischer and I. W. Molenaar (Eds.), *Rasch Models: Foundations, Recent Developments, and Applications* (pp. 69–95). New York: Springer.

Haberman, S. J. 1977a. Log-linear models and frequency tables with small expected cell counts. *The Annals of Statistics, 5*, 1148–1169.

Haberman, S. J. 1977b. Maximum likelihood estimates in exponential response models. *The Annals of Statistics, 5*, 815–841.

Haberman, S. J. 1988. A stabilized Newton-Raphson algorithm for log-linear models for frequency tables derived by indirect observation. *Sociological Methodology, 18*, 193–211.

Haberman, S. J. 2004. *Joint and Conditional Maximum-Likelihood Estimation for the Rasch Model for Binary Responses* (Research Rep. No. RR-04-20). Princeton, NJ: ETS.

Haberman, S. J. 2005. *Identifiability of Parameters in Item Response Models with Unconstrained Ability Distributions* (Research Rep. No. RR-05-24). Princeton, NJ: ETS.

Haberman, S. J. 2006. *Joint and Conditional Estimation for Implicit Models for Tests with Polytomous Item Scores* (Research Rep. No. RR-06-03). Princeton, NJ: ETS.

Haberman, S. J. 2007. The interaction model. In M. von Davier and C. H. Carstensen (Eds.), *Multivariate and Mixture Distribution Rasch Models* (pp. 201–216). New York: Springer Science.

Haberman, S. J. 2008a. *Asymptotic Limits of Item Parameters in Joint Maximum-Likelihood Estimation for the Rasch Model* (Research Rep. No. RR-08-04). Princeton, NJ: ETS.

Haberman, S. J. 2008b. *A General Program for Item-Response Analysis That Employs the Stabilized Newton-Raphson Algorithm* (Research Rep. No. RR-13-xx). Princeton, NJ: ETS.

Heinen, T. 1996. *Latent Class and Discrete Latent Trait Models*. Thousand Oaks, CA: Sage.

Holland, P. W. 1990. The Dutch identity: A new tool for the study of item response models. *Psychometrika, 55*, 5–18.

Hotelling, H. 1940. The selection of variates for use in prediction with some comments on the general problem of nuisance parameters. *The Annals of Mathematical Statistics, 11*, 271–283.

Kiefer, J. and Wolfowitz, J. 1956. Consistency of the maximum likelihood estimator in the presence of infinitely many incidental parameters. *Annals of Mathematical Statistics, 27*, 887–906.

Lord, F. M. 1980. *Applications of Item Response Theory to Practical Testing Problems*. Mahwah, NJ: Lawrence Erlbaum.

Lord, F. M. and Novick, M. R. 1968. *Statistical Theories of Mental Test Scores*. Reading, MA: Addison-Wesley.

Masters, G. N. 1982. A Rasch model for partial credit scoring. *Psychometrika, 47*, 149–174.

Muraki, E. 1992. A generalized partial credit model: Application of an EM algorithm. *Applied Psychological Measurement, 16*, 159–176.

Muraki, E. 1997. A generalized partial credit model. In W. J. van der Linden and R. K. Hambleton (Eds.), *Handbook of Modern Item Response Theory*. New York: Springer-Verlag.

Neyman, J. and Scott, E. L. 1948. Consistent estimation from partially consistent observations. *Econometrica, 16*, 1–32.

Patz, R. J. and Junker, B. W. 1999. A straightforward approach to Markov chain Monte Carlo methods for item response models. *Journal of Educational and Behavioral Statistics, 24*, 146–178.

Rasch, G. 1960. *Probabilistic Models for Some Intelligence and Attainment Tests.* Copenhagen: Danish Institute for Educational Research.

Rockafellar, R. T. 1970. *Convex Analysis*. Princeton, NJ: Princeton University Press.

Tjur, T. 1982. A connection between Rasch's item analysis model and a multiplicative Poisson model. *Scandinavian Journal of Statistics, 9*, 23–30.

10

Missing Responses in Item Response Modeling

Robert J. Mislevy

CONTENTS

10.1 Missing Responses in Item Response Theory

Missing responses occur in item response theory (IRT) applications in ways that may or may not have been intended by the examiner and may or may not be related to people's ability θ. The mechanisms that produce missingness must be taken into account to draw correct inferences. This chapter addresses missingness in IRT through the lens of Rubin's (1976, 1977) missing-data framework. Implications for Bayesian and likelihood estimation are discussed for (a) random assignment of items, including alternate test forms and matrix sampling, (b) targeted testing, including vertical linking, (c) adaptive testing, (d) not-reached items, (e) intentional omits, and (f) examinee choice of items.

We first review notation, definitions, and key results as they apply to missingness in IRT, including missing at random (MAR) and ignorability. We next discuss how types of missingness affect inference about θ when item parameters ($\boldsymbol{\beta}$) are known. We then address inference about $\boldsymbol{\beta}$ when some responses are missing.

10.2 Notation, Concepts, and General Results

10.2.1 Notation

Under an IRT model with person ability θ and local independence,

$$P\{U_1 = u_1, \ldots, U_I = u_I;\ \theta, \beta_1, \ldots, \beta_I | y\} = \prod_{i=1}^{I} P\{U_i = u_i; \theta, \beta_i\},$$

or

$$P\{\mathbf{U} = \mathbf{u}; \theta, \boldsymbol{\beta} | y\} = \prod_{i=1}^{I} f_\theta(u_i; \beta_i) \equiv f_\theta(\mathbf{u}; \boldsymbol{\beta}), \tag{10.1}$$

where U_i is the response for Item i and u_i is a possible value, $f_\theta(u_i; \beta_i)$ is the response function for Item i with possibly vector-valued parameter β_i, and y is a person covariate such as grade. We write $f_\theta(u_i; \boldsymbol{\beta}_i)$ as $f_\theta(u_i)$ and $f_\theta\ (\mathbf{u};\ \beta)$ as $f_\theta(\mathbf{u})$ when item parameters are known and $f_{\theta\beta}(u_i)$ and $f_{\theta\beta}(\mathbf{u})$ when they are not. After $\mathbf{u}$ is observed, $f_\theta(\mathbf{u})$ is interpreted as a likelihood function $L(\theta|\mathbf{u})$. The maximizing value is the MLE $\hat{\theta}$. Bayesian inference about θ is based on the posterior distribution $p(\theta|\mathbf{u}) \propto f_\theta(\mathbf{u})p(\theta)$, or when there is a covariate, $p(\theta|\mathbf{u}, y) \propto f_\theta(\mathbf{u})p(\theta|y)$.

With suitable specification of $f_\theta(\mathbf{u})$, θ, β_i, and u_i, Equation 10.1 encompasses the IRT models described in *Handbook of Item Response Theory, Volume One: Models* (Chapters 1 through 17) of this handbook. The results for cases (a)–(d) apply generally. For cases (e) and (f), we supplement the general results with specifics for unidimensional dichotomous IRT models such as the Rasch model (Volume One, Chapter 3) and three-parameter logistic (3PL) (Volume One, Chapter 2).

If we observe responses to only a subset of items that have been or could have been administered to a person, the data then indicate which responses are observed and their values (Little and Rubin, 1987; Rubin, 1976). Define $\mathbf{U} = (U_1, \ldots, U_I)$, the possibly hypothetical vector of responses to all items, and $\mathbf{M} = (M_1, \ldots, M_I)$, a missingness indicator with $m_i = 1$ if the value of U_i is observed and 0 if it is missing. $\mathbf{M}$ induces the partition $(\mathbf{U}_{\text{mis}}, \mathbf{U}_{\text{obs}})$ and a realized value $\mathbf{m}$ induces $(\mathbf{u}_{\text{mis}}, \mathbf{u}_{\text{obs}})$. In IRT, $f_\theta(\mathbf{u}) = f_\theta(\mathbf{u}_{\text{obs}}) f_\theta(\mathbf{u}_{\text{mis}})$ under local independence.

Inferences about θ should be based on the data that are actually observed, $(\mathbf{U}_{\text{obs}}, \mathbf{M})$, and our beliefs about the processes that caused the missingness. There are three basic approaches. The first two approaches model the hypothetical complete-data vector $(\mathbf{U}, \mathbf{M})$.

1. A model can be constructed for missingness given all responses, $g_\phi(\mathbf{m}|\mathbf{u})$, with parameter ϕ (Rubin, 1976). Its properties determine whether missingness is ignorable.
2. A predictive distribution can be constructed for the missing data given the observed data and missingness pattern, $h_{\phi\theta}(\mathbf{u}_{\text{mis}}|\mathbf{u}_{\text{obs}}, \mathbf{m})$ (Little and Rubin, 1987; Rubin, 1977, 1987). When missingness is not ignorable, $h_{\phi\theta}$ is pivotal in imputation methods and in Markov chain Monte Carlo estimation (MCMC; Gelman et al., 2004; see Chapter 15).
3. One can model $(\mathbf{U}_{\text{obs}}, \mathbf{M})$ directly, in ways that do not address $\mathbf{U}_{\text{mis}}$ (e.g., Moustaki, 1996).

The following sections explain general results of the first two approaches.

10.2.2 Inference under Ignorability

Definition 10.1

Define the missingness process $g_\phi(\mathbf{m}|\mathbf{u})$ as $P\{\mathbf{M} = \mathbf{m}; \phi|\mathbf{U} = \mathbf{u}\}$, with ϕ the possibly vector-valued parameter of the missingness process. ϕ may include θ. When this is central to a discussion, we make this explicit by writing $g_{\phi\theta}(\mathbf{m}|\mathbf{u})$.

The form of g depends on the process. Elements of ϕ can characterize the person, the item, the administration scheme, or some combination. In general,

$$\begin{aligned} P\{\mathbf{U} = \mathbf{u}, \mathbf{M} = \mathbf{m}; \theta, \phi\} &= P\{\mathbf{U} = \mathbf{u}; \theta\} P\{\mathbf{M} = \mathbf{m}|\mathbf{U} = \mathbf{u}, \theta, \phi\} \\ &= f_\theta(\mathbf{u}) g_\phi(\mathbf{m}|\mathbf{u}). \end{aligned} \tag{10.2}$$

The likelihood function for (θ, ϕ) induced by $(\mathbf{m}, \mathbf{u}_{\text{obs}})$ is obtained from $P\{\mathbf{m}, \mathbf{u}_{\text{obs}}; \theta, \phi\}$ as

$$L(\theta, \phi; \mathbf{m}, \mathbf{u}_{\text{obs}}) = \delta_{\theta\phi} \int f_\theta(\mathbf{u}_{\text{obs}}, \mathbf{u}_{\text{mis}}) g_\phi(\mathbf{m}|\mathbf{u}_{\text{obs}}, \mathbf{u}_{\text{mis}}) \partial \mathbf{u}_{\text{mis}}, \tag{10.3}$$

with $\delta_{\theta\phi} = 1$ if a value of (θ, ϕ) is in the parameter space and 0 if not.* The observed-data likelihood is thus an average over the likelihoods of all the complete-data response patterns that accord with $\mathbf{u}_{\text{obs}}$, each weighted by the probability of the observed missingness pattern given the observed and the possible missing responses. Under local independence in IRT,

$$L(\theta, \phi; \mathbf{m}, \mathbf{u}_{\text{obs}}) = \delta_{\theta\phi} f_\theta(\mathbf{u}_{\text{obs}}) \int f_\theta(\mathbf{u}_{\text{mis}}) g_\phi(\mathbf{m}|\mathbf{u}_{\text{obs}}, \mathbf{u}_{\text{mis}}) \partial \mathbf{u}_{\text{mis}}. \tag{10.4}$$

Likelihood inferences are based on relative values of $L(\theta, \phi; \mathbf{m}, \mathbf{u}_{\text{obs}})$ over (θ, ϕ). Bayesian inference is based on the posterior

$$p(\theta, \phi|\mathbf{m}, \mathbf{u}_{\text{obs}}) \propto L(\theta, \phi; \mathbf{m}, \mathbf{u}_{\text{obs}}) p(\theta, \phi), \tag{10.5}$$

* Integration notation indicates marginalizing over the variable of interest, whether continuous or discrete.

with prior $p(\theta, \phi)$, or $p(\theta, \phi|y)$ if there is a covariate. We omit y from discussion except where it is germane to a point.

Ignoring the missingness process when drawing inferences about θ means instead of using the likelihood $L(\theta, \phi; \mathbf{m}, \mathbf{u}_{\text{obs}})$, using a facsimile of Equation 10.1 with $\mathbf{u}_{\text{obs}}$ alone,

$$L^*(\theta; \mathbf{u}_{\text{obs}}) = \delta_\theta f_\theta(\mathbf{u}_{\text{obs}}). \tag{10.6}$$

(The asterisk superscript denotes a simpler function that is related to its unmarked counterpart and can be used for inference under certain conditions. Similar notation will appear throughout the chapter.) Likelihood inferences about θ that ignore the missingness compare values of L^* at various values of θ. Bayesian inferences that ignore the missingness use

$$p^*(\theta|\mathbf{u}_{\text{obs}}) \propto L^*(\theta; \mathbf{u}_{\text{obs}})p(\theta). \tag{10.7}$$

Definition 10.2

A missingness process in IRT is ignorable with respect to likelihood inference if inferences about θ through Equation 10.6 using $\mathbf{u}_{\text{obs}}$ are equivalent to inferences through Equation 10.4 using $(\mathbf{u}_{\text{obs}}, \mathbf{m})$.

Definition 10.3

A missingness process in IRT is ignorable with respect to Bayesian inference if inferences about θ through Equation 10.7 using $\mathbf{u}_{\text{obs}}$ are equivalent to inferences through Equation 10.5 using $(\mathbf{u}_{\text{obs}}, \mathbf{m})$.

Rubin (1976) derived conditions for ignorability under sampling distribution, direct likelihood, and Bayesian inference. This chapter focuses on sufficient conditions for the latter two, which dominate IRT applications. The following concepts are required.

Definition 10.4

Missing responses are missing completely at random (MCAR) if for each value of ϕ and each fixed value of $\mathbf{m}$, $g_\phi(\mathbf{m}|\mathbf{u})$ takes the same value for all $\mathbf{u}$. That is, $g_\phi(\mathbf{m}|\mathbf{u}) = g_\phi(\mathbf{m})$, and if there are covariates, $g_\phi(\mathbf{m}|\mathbf{u}, y) = g_\phi(\mathbf{m})$. The probability of a missingness pattern does not depend on the missing responses, the observed responses, or covariates.

Definition 10.5

Missing responses are missing at random (MAR) if for each value of ϕ and each fixed value of $\mathbf{m}$ and $\mathbf{u}_{\text{obs}}$, $g_\phi(\mathbf{m}|\mathbf{u}_{\text{obs}}, \mathbf{u}_{\text{mis}})$ takes the same value for all $\mathbf{u}_{\text{mis}}$. That is, $g_\phi(\mathbf{m}|\mathbf{u}) = g_\phi(\mathbf{m}|\mathbf{u}_{\text{obs}})$, or if there are covariates, $g_\phi(\mathbf{m}|\mathbf{u}, y) = g_\phi(\mathbf{m}|\mathbf{u}_{\text{obs}}, y)$. The probability of a missingness pattern does not depend on the missing responses, but may depend on observed responses and covariates. MCAR implies MAR.

Definition 10.6

θ is distinct (D) from ϕ if their joint parameter space factors into a θ space and a ϕ space, and when prior distributions are specified, they are independent.

The key result is a sufficient condition for ignorability:

Theorem 10.1 (Rubin, 1976, p. 581)

For direct likelihood or Bayesian inferences about θ, it is appropriate to ignore a process that causes missing data if MAR and D are satisfied.*

10.2.3 Inference When Missingness Is Not Ignorable

Three ways to handle nonignorable missingness in IRT are (a) inference using Equation 10.4, (b) imputing values for $\mathbf{u}_{\text{mis}}$, and (c) modeling $(\mathbf{U}_{\text{obs}}, \mathbf{M})$ directly. The first two involve predictive distributions for the missing responses.

10.2.3.1 Predictive Distributions for Missing Responses

Definition 10.7

Define $h_{\phi\theta}(\mathbf{u}_{\text{mis}}|\mathbf{u}_{\text{obs}}, \mathbf{m})$ as $P\{\mathbf{U}_{\text{mis}} = \mathbf{u}_{\text{mis}};\ \phi,\ \theta|\mathbf{U}_{\text{obs}} = \mathbf{u}_{\text{obs}},\ \mathbf{M} = \mathbf{m}\}$, the predictive distribution of $\mathbf{u}_{\text{mis}}$ given the observed responses, the missingness pattern, ϕ, and θ.

For example, if we believe that a person's response to a not-reached Item i would follow the IRT model, $h_{\phi\theta}(u_{\text{mis},i}|\mathbf{u}_{\text{obs}}, \mathbf{m}) = f_\theta(u_i)$. If we believe an omitted response is a guess among A alternatives, $h_{\phi\theta}(u_{\text{mis},i}|\mathbf{u}_{\text{obs}}, \mathbf{m})$ is Bernoulli (A^{-1}). Neither example involves ϕ, but it is possible. We write $h_\theta(\mathbf{u}_{\text{mis}}|\mathbf{u}_{\text{obs}}, \mathbf{m})$ for when ϕ is not involved or has been integrated out, as $\int h_{\phi\theta}(\mathbf{u}_{\text{mis}}|\mathbf{u}_{\text{obs}}, \mathbf{m})p(\phi|\mathbf{u}_{\text{obs}}, \mathbf{m})\partial\phi$.

10.2.3.2 Markov Chain Monte Carlo Estimation

MCMC estimation can be used to carry out inference through the missing-data likelihood (10.4). Bayesian MCMC provides a numerical approximation of the posterior distribution of parameters and missing observations by sampling from the sequences of simpler distributions designed to produce this result. The draws for a given variable approximate its posterior distribution, providing estimates of posterior means, variances, etc. This section reviews key ideas of MCMC estimation for a sample of persons and unknown item parameters, highlighting the role of $h_{\phi\theta}(\mathbf{u}_{\text{mis}}|\mathbf{u}_{\text{obs}}, \mathbf{m})$; for a general introduction to MCMC estimation (see Chapter 15).

Bayesian inference begins with the joint distribution for all variables in the model—in IRT, this is item responses $\mathbf{u}$, missingness indicators $\mathbf{m}$, person parameters θ and item parameters β, and missingness parameters ϕ. For IRT with missingness that is independent across persons conditional on θ, β, and ϕ, the joint distribution over P persons

---

* Proofs are omitted as they are straightforward or appear in the citations.

and I items is

$$p((\mathbf{m}, \mathbf{u}_{\text{obs}}, \mathbf{u}_{\text{mis}}, \theta, \phi^{(p)}), \beta, \phi^{(i)})$$
$$= \prod_p f_{\theta\beta}(\mathbf{u}_{p,\text{obs}}, \mathbf{u}_{p,\text{mis}}; \beta) g_{\phi}(\mathbf{m}_p; \beta, \phi^{(i)} | \mathbf{u}_{p,\text{obs}}, \mathbf{u}_{p,\text{mis}}) p(\theta, \phi^{(p)}) p(\beta) p(\phi^{(i)}), \quad (10.8)$$

where $(\mathbf{m}, \mathbf{u}_{\text{obs}}, \mathbf{u}_{\text{mis}}, \theta, \phi^{(p)})$ denotes missingness patterns, item responses, proficiencies, and person missingness parameters for Ppersons, and $\phi^{(i)}$ indicates item-associated missingness parameters. Writing $f_{\theta\beta}$ emphasizes that β is not treated as known here.

Once $(\mathbf{m}, \mathbf{u}_{\text{obs}})$ is observed, the posterior for $(\mathbf{u}_{\text{mis}}, \theta, \phi^{(p)})$, β and $\phi^{(i)}$ is

$$p((\mathbf{u}_{\text{mis}}, \theta, \phi^{(p)}), \beta, \phi^{(i)} | (\mathbf{m}, \mathbf{u}_{\text{obs}}))$$
$$\propto \prod_p f_{\theta\beta}(\mathbf{u}_{p,\text{obs}}, \mathbf{u}_{p,\text{mis}}; \beta) g_{\phi_p}(\mathbf{m}_p; \beta, \phi^{(i)} | \mathbf{u}_{p,\text{obs}}, \mathbf{u}_{p,\text{mis}}) p(\theta, \phi^{(p)}) p(\beta) p(\phi^{(i)}), \quad (10.9)$$

with $(\mathbf{m}, \mathbf{u}_{\text{obs}})$ fixed at the observed values. The Gibbs sampling variant of MCMC proceeds in each cycle by drawing a value for each unknown variable from its so-called full-conditional distribution, that is, its distribution conditional on the observed data and the previous cycle's draws for all other variables. The full-conditional value for the Person p's missing responses is $h_{\phi\theta}(\mathbf{u}_{p,\text{mis}} | \mathbf{u}_{p,\text{obs}}, \mathbf{m}_p)$ with θ_p, $\phi_p^{(p)}$, $\phi^{(i)}$, and β fixed at their previous draws. The distribution of draws for $\mathbf{u}_{p,\text{mis}}$ across cycles approximates the predictive distribution for the missing responses marginalized over θ_p, $\phi_p^{(p)}$, $\phi^{(i)}$, and β.

10.2.3.3 Imputation for Missing Responses

Imputation fills in missing responses in some way and carries out inference as if the imputations were observed values. From a Bayesian perspective, the predictive distribution h fully characterizes knowledge about missing responses given the observed data (Rubin, 1977). Several imputation methods for missing responses in IRT appear in the literature. We can see how they can be understood in terms of h and consider when they support reasonable inferences.

Applied to missingness in IRT, Rubin's (1977, 1987) imputation approach for some function $S(\mathbf{u})$ such as $\hat{\theta}$ when responses are missing is to base inferences on its conditional expectation given the observed data $(\mathbf{m}, \mathbf{u}_{\text{obs}})$:

$$E[S(\mathbf{u}) | \mathbf{m}, \mathbf{u}_{\text{obs}}] = \iiint S(\mathbf{u}_{\text{obs}}, \mathbf{u}_{\text{mis}})\, h_{\phi\theta}(\mathbf{u}_{\text{mis}} | \mathbf{u}_{\text{obs}}, \mathbf{m}) p(\theta, \phi | \mathbf{u}_{\text{obs}}, \mathbf{m}) \partial \mathbf{u}_{\text{mis}} \partial\theta\, \partial\phi. \quad (10.10)$$

Note that Equation 10.10 is conditional on $\mathbf{m}$. Under some missingness processes, $\mathbf{M}$ depends on θ, such as with intentional omits. Not including a term for $P\{\mathbf{M} = \mathbf{m} | \mathbf{u}_{\text{obs}}, \theta, \phi\}$ means forfeiting information that $\mathbf{m}$ may convey about θ.

The idea can be adapted to likelihood functions:

Theorem 10.2

A conditional-on-$\mathbf{m}$ expected likelihood for θ given $(\mathbf{m}, \mathbf{u}_{\text{obs}})$ is

$$\begin{aligned} L^{(c)}(\theta; \mathbf{u}_{\text{obs}}|\mathbf{m}) &= E[L(\theta; \mathbf{u})|\mathbf{m}, \mathbf{u}_{\text{obs}}] \\ &= \int L(\theta; \mathbf{u}_{\text{obs}}, \mathbf{u}_{\text{mis}}) h_\theta(\mathbf{u}_{\text{mis}}|\mathbf{u}_{\text{obs}}, \mathbf{m}) \partial \mathbf{u}_{\text{mis}} \\ &= f_\theta(\mathbf{u}_{\text{obs}}) \int f_\theta(\mathbf{u}_{\text{mis}}) h_\theta(\mathbf{u}_{\text{mis}}|\mathbf{u}_{\text{obs}}, \mathbf{m}) \partial \mathbf{u}_{\text{mis}}. \end{aligned} \tag{10.11}$$

We will use Equation 10.11 to evaluate imputation pseudolikelihoods, which take the form of IRT likelihoods with observed responses and with missing responses filled in with values $\mathbf{u}_{\text{mis}}^{(\text{imp})}$, where $u_{\text{mis,k}}^{(\text{imp})}$ is an imputation for kth item response if $m_k = 0$.

Definition 10.8

The imputation pseudolikelihood is

$$L^{(\text{imp})}(\theta; \mathbf{m}, \mathbf{u}_{\text{obs}}) \equiv f_\theta(\mathbf{u}_{\text{obs}}, \mathbf{u}_{\text{mis}}^{(\text{imp})}) = f_\theta(\mathbf{u}_{\text{obs}}) f_\theta(\mathbf{u}_{\text{mis}}^{(\text{imp})}) = f_\theta(\mathbf{u}_{\text{obs}}) \prod_{k:m_k=0} f_\theta(u_{\text{mis,k}}^{(\text{imp})}), \tag{10.12}$$

where the product operator is over all items for which responses are missing, that is, $\{k : m_k = 0\}$.

The conditional independence form in Equation 10.12 evaluated with imputations for the missing values, or $\prod_{k:m_k=0} f_\theta(u_{\text{mis,k}}^{(\text{imp})})$, replaces the marginalization of $f_\theta(\mathbf{u}_{\text{mis}})$ over the predictive distribution $h_\theta(\mathbf{u}_{\text{mis}}|\mathbf{u}_{\text{obs}}, \mathbf{m})$. We analyze an imputation approach with respect to two attributes. The first is the conditional independence form.

Definition 10.9

A predictive distribution for $\mathbf{u}_{\text{mis}}$ is conditionally independent predictively (CIP) if the missing responses are conditionally independent given $\mathbf{m}$, $\mathbf{u}_{\text{obs}}$, and θ and ϕ:

$$h_\theta(\mathbf{u}_{\text{mis}}|\mathbf{u}_{\text{obs}}, \mathbf{m}) = \prod_{k:m_k=0} h_\theta(u_{\text{mis},k}|\mathbf{u}_{\text{obs}}, \mathbf{m}). \tag{10.13}$$

CIP is practically important when different responses are missing for different reasons; some items not presented, others not reached, still others intentionally omitted. CIP allows different forms and parameters to be used independently for missing items' predictive distributions. CIP will be most seriously violated if missingness for the blocks of items depends on the same unknown person-related missingness parameter $\phi^{(p)}$. Examples are a propensity to omit items, and working speed that determines not reaching items and all items after the first not-reached item are necessarily not reached. If there are few such missing responses, the effect is apt to be negligible.

The second, more important, attribute of imputations is their relation to belief about what the missing response might be, expressed by $h_\theta(u_{\text{mis},k}|\mathbf{u}_{\text{obs}}, \mathbf{m})$. In some cases, the value of a missing response is known with certainty, and $h_{k\theta\phi}(u_{\text{mis},k};\ \theta,\ \phi|\mathbf{u}_{\text{obs}}, \mathbf{m})$ has support on that single point, $\dot{u}_{\text{mis},k}$.

Definition 10.10 (Known-value imputation)

When the value of each missing response is known to be $\dot{u}_{\text{mis},k}$, define $\dot{\mathbf{u}}$ by augmenting $\mathbf{u}_{\text{obs}}$ with $\dot{u}_{\text{mis},k}$ for each $u_{\text{mis},k}$. Then, $L^{(c)}(\theta; \mathbf{u}_{\text{obs}}|\mathbf{m}) = f_\theta(\dot{\mathbf{u}}) = L^{(\text{imp})}(\theta; \mathbf{m}, \mathbf{u}_{\text{obs}})$, and it provides correct likelihood and Bayesian inference conditional on $\mathbf{m}$.

Definition 10.11 (Expected-value imputation (or corrected mean imputation; Sijtsma and van der Ark, 2003))

Fill in each missing response with the expected value of its predictive distribution, $\bar{u}_{\text{mis},k} \equiv E[h_{k\theta}(u_{\text{mis},k}|\mathbf{u}_{\text{obs}}, \mathbf{m})]$, to obtain $\bar{\mathbf{u}}$. Use $\bar{L}(\theta; \mathbf{m}, \mathbf{u}_{\text{obs}}) \equiv f_\theta(\bar{\mathbf{u}})$ for likelihood and Bayesian inference conditional on $\mathbf{m}$.

Lord (1974) proposes expected-value imputation of A^{-1} for omitted responses in dichotomous multiple-choice items with A alternatives, under the assumption that the response would have been a random guess. The standard error from $\bar{L}$ overstates precision because treating the expectations as observed values ignores the variation in h.

Definition 10.12 (Single imputation)

Define $\tilde{\mathbf{u}}$ by filling in each missing response with a draw $\tilde{u}_{\text{mis},k}$ from $h_{k\theta}(u_{\text{mis},k}|\mathbf{u}_{\text{obs}}, \mathbf{m})$. Define $\tilde{L}(\theta; \mathbf{m}, \mathbf{u}_{\text{obs}}) \equiv f_\theta(\tilde{\mathbf{u}})$ and use it for likelihood and Bayesian inference conditional on $\mathbf{m}$.

$\tilde{L}$ is a crude numerical approximation of Equation 10.13 that provides the right expectation of $L^{(c)}$. But different draws have different maxima, and the curvature of each $\tilde{L}$ overstates precision by ignoring the variation in h. Multiple imputation corrects these deficiencies (Finch, 2008).

Definition 10.13 (Multiple imputation (Rubin, 1987))

Define *Rep* replicate response vectors $\tilde{\mathbf{u}}^{(r)}$, each filling in $u_{\text{mis},k}$ with a random draw $\tilde{u}^{(r)}_{\text{mis},k}$ from $h_{k\theta}(u_{\text{mis},k}|\mathbf{u}_{\text{obs}}, \mathbf{m})$.

Each $\tilde{\mathbf{u}}^{(r)}$ is produced by drawing parameter values from their predictive distributions, then drawing item responses conditional on the drawn parameter values. If h does not depend on θ or ϕ, they can be omitted from the procedure.

Step 1: Draw $(\tilde{\theta}^{(r)}, \tilde{\phi}^{(r)})$ from

$$\begin{aligned} p(\theta, \phi|\mathbf{u}_{\text{obs}}, \mathbf{m}) &\propto p(\mathbf{u}_{\text{obs}}, \mathbf{m}|\theta, \phi)p(\theta, \phi) \\ &= f_\theta(\mathbf{u}_{\text{obs}}) \int g_{\phi\theta}(\mathbf{m}|\mathbf{u}_{\text{obs}}, \mathbf{u}_{\text{mis}})\partial \mathbf{u}_{\text{mis}}\, p(\theta, \phi). \end{aligned} \tag{10.14}$$

Step 2: Draw each $\tilde{u}^{(r)}_{\text{mis},k}$ from

$$h_{k\theta\phi}(u_{\text{mis},k}|\mathbf{u}_{\text{obs}}, \mathbf{m}, \theta = \tilde{\theta}^{(r)}, \phi = \tilde{\phi}^{(r)}). \tag{10.15}$$

Theorem 10.3

If the missingness process is ignorable, Equation 10.14 simplifies to $p(\theta, \phi|\mathbf{u}_{obs}, \mathbf{m}) = p(\theta|\mathbf{u}_{obs}) = f_\theta(\mathbf{u}_{obs})p(\theta)$. Draw $\tilde{\theta}^{(r)}$ from $p(\theta|\mathbf{u}_{obs})$ and $\mathbf{u}_{mis}^{(r)}$ from $f_\theta(\mathbf{U}_{mis})$ with $\theta = \tilde{\theta}^{(r)}$.

Theorem 10.4 (Inference from multiple imputations (Rubin, 1987))

Let $S(\mathbf{u})$ be a statistic of the full-response vector and $V(\mathbf{u})$ be an estimate of its variance, such as SEM^2 or posterior variance. Estimates of S and its precision conditional on $\mathbf{m}$ are

$$\bar{S}(\mathbf{m}, \mathbf{u}_{obs}) = Rep^{-1} \sum_{r=1}^{Rep} S(\tilde{\mathbf{u}}^{(r)}) \quad \text{and} \quad T(\mathbf{m}, \mathbf{u}_{obs}) = \bar{V} + (1 + Rep^{-1})B, \tag{10.16}$$

where

$$\bar{V} = Rep^{-1} \sum_r V(\tilde{\mathbf{u}}^{(r)}) \quad \text{and} \quad B = (Rep - 1)^{-1} \sum_r (S(\tilde{\mathbf{u}}^{(r)}) - \bar{S})^2. \tag{10.17}$$

Multiple imputation provides more accurate estimates of S than single imputation. The variance among the $S(\tilde{\mathbf{u}}^{(r)})$s, namely, B in Equation 10.17, picks up variance in the hs as it affects inference about S.

10.3 Common Types of Missingness

10.3.1 Random Assignment

Consider I items with known item parameters, and a testing procedure under which a person will be administered a subset from the K allowable patterns $\{\mathbf{m}^{(1)}, \ldots, \mathbf{m}^{(K)}\}$. An examinee is administered a set according to the administrator-determined probabilities $(\phi_1, \ldots, \phi_K)$. The probabilities do not depend on $\mathbf{u}_{obs}$, $\mathbf{u}_{mis}$, θ, or y. Examples include alternate test forms, matrix sampling, and linear-on-the-fly tests (LOFT), where each item is selected with equal probability from those not yet administered. Thus

$$g_\phi(\mathbf{m}|\mathbf{u}) = \begin{cases} \phi_k & \text{for all } \mathbf{u} \text{ if } \mathbf{m} = \mathbf{m}^{(k)} \\ 0 & \text{otherwise} \end{cases} \tag{10.18}$$

Theorem 10.5

Random assignment satisfies MCAR.

Theorem 10.6

The missingness induced by random assignment of items is ignorable under likelihood and Bayesian inference.

10.3.2 Targeted Testing

Targeted testing uses person covariates y related to θ in item selection to make estimates of θ more precise. An easy test form can be administered to fourth graders and a hard form to sixth graders. Again consider I items with known parameters, and a procedure that administers a person an allowable item subset corresponding to $\{\mathbf{m}^{(1)}, \ldots, \mathbf{m}^{(K)}\}$, according to administrator-determined probabilities $(\phi_1(y), \ldots, \phi_K(y))$.

Theorem 10.7

Targeted testing assignment satisfies MCAR.

Theorem 10.8

The missingness process ϕ is distinct from θ as required for likelihood inference if all values of θ can occur at all values of y.

Theorem 10.9

The missingness process ϕ is distinct from θ as required for Bayesian inference only if θ and y are independent *a priori.*

Theorem 10.10

The missingness process ϕ is distinct from θ as required for Bayesian inference conditional on y.

Theorem 10.11

The missingness induced by targeted testing is ignorable under Bayesian inference conditional on y, that is, the correct posterior is proportional to

$$p^*(\theta|\mathbf{u}_{\text{obs}}, y) \propto L^*(\theta; \mathbf{u}_{\text{obs}}) p(\theta|y). \tag{10.19}$$

10.3.3 Adaptive Testing

Adaptive testing uses a person's preceding responses to select items, to increase information about each person (van der Linden and Glas, 2010; Volume Three, Chapter 10). The data in adaptive testing are a sequence of n ordered pairs $\mathbf{s}_n = ((I_1, u_{\text{obs},1}), \ldots, (I_n, u_{\text{obs},n}))$, where I_k identifies the kth item administered and $u_{\text{obs},k}$ is the response. Define the partial response sequence $\mathbf{S}_k$ as the first k pairs, with the null sequence $\mathbf{S}_0$ representing the start of the test. Testing proceeds until a desired precision or test length is reached. Selecting the fictitious Item 0 corresponds to stopping the test. It is the $n+1$st item, but has no response.

A test administrator specifies, explicitly or implicitly, for all items i and all realizable partial response sequences $\mathbf{s}_k$, the probabilities $\phi(i, \mathbf{s}_k)$ that Item i will be selected as the $k+1$st item after $\mathbf{s}_k$ has been observed (e.g., present the unadministered item with the most information at the provisional MLE $\hat{\theta}_k$). Multistage testing is expressed by constrained $\phi(i, \mathbf{s}_k)$s that effect selecting among specified blocks of items at specified branching points.

$\mathbf{s}_n$ conveys the value of $(\mathbf{m}, \mathbf{u}_{\text{obs}})$, because $m_i = 1$ if $I_k = i$ for some k and 0 if not, and the responses to the administered items constitute $\mathbf{u}_{\text{obs}}$.

Theorem 10.12

The probability of $\mathbf{s}_n$ given θ is

$$P\{\mathbf{S}_n = \mathbf{s}_n; \theta\} = f_\theta(\mathbf{u}_{\text{obs}}) \prod_{k=1}^{n+1} \phi(i_k, \mathbf{s}_{k-1}). \tag{10.20}$$

Theorem 10.13 (Mislevy and Chang, 2000; Mislevy and Wu, 1996)

The missingness mechanism for adaptive testing is

$$g_{\theta\phi}(\mathbf{m}|\mathbf{u}) = g_{\theta\phi}(\mathbf{m}|\mathbf{u}_{\text{obs}}) = \sum_{\mathbf{s}_n^* \in T} \prod_{k=1}^{n+1} \phi(i_k, \mathbf{s}_{k-1}^*) \tag{10.21}$$

where T is the set of response sequences with $\mathbf{m}$ and $\mathbf{u}_{\text{obs}}$ that match those of $\mathbf{s}_n$.

Theorem 10.14

Adaptive testing missingness is MAR.

Theorem 10.15

Adaptive testing missingness satisfies D as it applies to likelihood inference.

Theorem 10.16

The missingness induced by adaptive testing is ignorable under likelihood inference. Correct inferences are thus obtained from $L^*(\theta; \mathbf{u}_{\text{obs}})$.

Theorem 10.17

The missingness induced by adaptive testing is ignorable under Bayesian inference. Correct inferences are thus obtained from $p^*(\theta|\mathbf{u}_{\text{obs}}) \propto L^*(\theta; \mathbf{u}_{\text{obs}})p(\theta)$.

Theorem 10.18

If any item selection probabilities depend on a person covariate y that is not independent of θ, the missingness induced by adaptive testing is ignorable under Bayesian inference conditional on y. Correct inferences are obtained from

$$p^*(\theta|\mathbf{u}_{\text{obs}}, y) \propto L^*(\theta; \mathbf{u}_{\text{obs}})p(\theta|y). \tag{10.22}$$

10.3.4 Not-Reached Items

IRT is intended for "power" tests, in which persons' chances of responding correctly would not differ appreciably if the time limit were more generous. In practice, some persons may not reach all items.

Consider first the case in which the following conditions are met:

1. $f_\theta(u)$ governs responses to both reached and unreached items.
2. Missingness of not-reached items is not based on the person's examination of their content.
3. The administrator knows which items have not been reached (known in computer-based tests, plausible in paper and pencil testing when a person starts at the beginning and proceeds linearly until time ends).

Under Condition 3, $I+1$ patterns of missingness can occur: Let $\mathbf{m}^{(k)}$ for $k=0,\ldots,n$, denote not reaching the last k items: I–k 1's followed by k 0's. Missingness is characterized by the person speed parameter $\phi = (\phi_0,\ldots,\phi_n)$ where $\phi_k = P\{\mathbf{M} = \mathbf{m}^{(k)}\}$, and

$$g_\phi(\mathbf{m}|\mathbf{u}) = \begin{cases} \phi_k & \text{for all } \mathbf{u} \text{ if } \mathbf{m} = \mathbf{m}^{(k)} \\ 0 & \text{otherwise} \end{cases} \tag{10.23}$$

A more parsimonious parameterization such as a log-linear model may be employed for ϕ.

Theorem 10.19

Under Conditions 1 through 3, not-reached items are MCAR.

Theorem 10.20

Under Conditions 1 through 3, not-reached items are ignorable under direct likelihood inference if all values of θ can occur at all values of ϕ. This is generally the case.

Theorem 10.21

Under Conditions 1 through 3, not-reached items are ignorable under Bayesian inference if θ and ϕ are independent, that is, $p(\theta,\phi) = p(\theta)p(\phi)$.

Theorem 10.20 says that the likelihood based on only responses to the reached items, $L^*(\theta;\mathbf{u}_{\text{obs}}) = f_\theta(\mathbf{u}_{\text{obs}})$ is appropriate for likelihood inference about θ. Lord (1974) endorses this practice as long as a person has answered most items. Theorem 10.21 says that ignorability under Bayesian inference requires prior independence of θ and ϕ. Studies suggest this need not be so. For example, van den Wollenberg (1979) reported positive correlations between percent-correct scores on the first 11 items (which were reached by all persons) and the total number of items reached in four of six intelligence tests in the ISI battery.

Theorem 10.22 (Mislevy and Wu, 1996, p. 14)

Assume Conditions 1 through 3. If θ and ϕ are not independent *a priori*, then

a. Bayesian inference about θ requires accounting for the relationship between θ and ϕ, and thus, the information about θ conveyed by $\mathbf{m}$, through

$$p(\theta,\phi|\mathbf{m},\mathbf{u}_{\text{obs}}) \propto L^*(\theta;\mathbf{u}_{\text{obs}})g_\phi(\mathbf{m})p(\theta,\phi) = L^*(\theta;\mathbf{u}_{\text{obs}})p(\theta|\phi)g_\phi(\mathbf{m})p(\phi). \tag{10.24}$$

The full-information marginal posterior for θ is

$$p(\theta|\mathbf{m}, \mathbf{u}_{\text{obs}}) \propto L^*(\theta; \mathbf{u}_{\text{obs}}) \int g_\phi(\mathbf{m}) p(\theta|\phi) p(\phi) \partial\phi$$
$$= L^*(\theta; \mathbf{u}_{\text{obs}}) p(\theta|\mathbf{m}). \tag{10.25}$$

b. $p^*(\theta|\mathbf{u}_{\text{obs}}) \propto L^*(\theta; \mathbf{u}_{\text{obs}}) p(\theta)$ provides consistent but possibly biased partial-information inference about θ.

$L^*(\theta; \mathbf{u}_{\text{obs}})$ is the correct likelihood because MAR holds, so $\hat{\theta}$ is consistent as the number of reached items increases. However, $p^*(\theta|\mathbf{u}_{\text{obs}})$ differs from $p(\theta, \phi|\mathbf{m}, \mathbf{u}_{\text{obs}})$ to the extent that $p(\theta|\mathbf{m})$ differs from $p(\theta)$. Estimates of θ from $p^*(\theta|\mathbf{u}_{\text{obs}})$, when items are not reached, are biased upwards if θ and ϕ are positively correlated, increasingly so as fewer are reached (Glas and Pimentel, 2008).

There is not enough information in a single administration with some not-reached items to estimate $p(\theta|\mathbf{m})$ in much detail. Pimentel (2005) and Glas and Pimentel (2008) used a simple "steps" model (Verhelst et al., 1997) in such cases. Special studies where $\mathbf{u}_{\text{mis}}$ is subsequently obtained could be used to estimate, say, a smoothed family of beta-binomial distributions for $\mathbf{m}$ given θ, and reversed with Bayes Theorem to provide $p(\theta|\mathbf{m})$.

Condition 1, assuming a constant IRT model, can be problematic. Persons who know their not-reached responses will be ignored may spend more time on early items and enjoy a higher effective θ than they would have had on the not-reached items. Therefore, most operational adaptive tests do not report a score unless a person reaches a certain proportion of items, say 90%.

To impute responses for not-reached items, one first draws $\tilde{\theta}$ from Equation 10.25, then each $u_{\text{mis},k}$ from $f_\theta(u_{\text{mis},k})$ with $\theta = \tilde{\theta}$. Treating not-reached items as wrong is known-value imputation assuming that the missing responses would have certainly been wrong. This is dubious for multiple-choice items, and biases θ estimates downward.

10.3.5 Intentionally Omitted Items

A missing response is an intentional omission when a person is administered an item, appraises it, and decides not to respond. After arguing that such omissions cannot generally be considered ignorable, we discuss ways to deal with them. We focus on dichotomous responses and widely used IRT models, but the ideas extend to ordered-category responses and testlets.

Definition 10.14 (Zwick, 1990)

IRT models that satisfy the following conditions are said to be SMURFLI$^{(2)}$:

- Strict monotonicity, that is, $\theta' > \theta'' \Rightarrow P\{U_i = 1|\theta'\} > P\{U_i = 1|\theta''\}$
- Unidimensional response functions
- Local independence
- Two possible responses, $u_i = l$ correct and 0 incorrect

The Rasch model is SMURFLI$^{(2)}$, as are two- and three-parameter normal and logistic models with positive slope parameters and lower asymptotes less than one.

10.3.5.1 Scoring Rules, Omitting Behavior, and Ignorability

Let $\mathbf{v}$ be a pattern of right, wrong, and omitted responses, where $v_i = u_i$ if $m_i = 1$ and $v_i = *$ if $m_i = 0$. A formula score $T(\mathbf{v})$ takes the form

$$T(\mathbf{v}) = R^+(\mathbf{v}) - C\ W^+(\mathbf{v}), \tag{10.26}$$

where $R^+(\mathbf{v})$ and $W^+(\mathbf{v})$ are counts of right and wrong responses and C, with $0 \leq C \leq 1$, is an administrator-selected constant. $C = 0$ gives number-right scores; $C = 1$ gives right-minus-wrong scores; for multiple-choice items with A alternatives, $C = 1/(A-1)$ gives "corrected-for-guessing" scores.

Since a correct response to an item always gives a higher score than an incorrect response, people will make responses they think are correct. They chose whether to respond to an item based on the contribution they expect from their response. They maximize their expected scores by answering items for which their probability of being correct is at least $C/(1+C)$. They do not this know probability, but may have a subjective estimate w_{pi} and should answer if $w_{pi} \geq C/(1+C)$. They should answer every item under number-right scoring, and whenever $w_{pi} \geq 1/A$ under corrected-for-guessing scoring.

People may differ in the accuracy of their w_{pi}s and their propensities to omit when they are uncertain (Matters and Burnett, 2003). Such characteristics are person components of ϕ in intentional omission, as are the w_{pi}s. Analyzing responses that examinees originally omitted under right-minus-wrong scoring, Sherriffs and Boomer (1954) found about half of the omitted responses would have been correct among low-risk aversion examinees, but two-thirds would have been correct among high-risk aversion examinees (also see Cross and Frary, 1977). The following conditions typify omitting behavior in educational tests:

4. People are more likely to omit items when they think their answers would be incorrect than when they think their answers would be correct.
5. Peoples' w_{pi}s tend to better calibrated to u_{pi}s as θ increases, that is, they are more likely to know when their response would be correct or incorrect.

Theorem 10.23

For a given $\mathbf{u}_{\text{obs}}$ and $\mathbf{m}$ with omits, the missingness is MAR only if this $\mathbf{m}$ is equally likely for all values of $\mathbf{u}_{\text{mis}}$.

This theorem follows from the definition of MAR. Its condition does not jibe well with Conditions 4 and 5, particularly for mostly correct $\mathbf{u}_{\text{obs}}$. These patterns are more likely from high θ persons, who are more likely to perceive correctly they do not know the answers to items that they omit. Further, Rubin's (1976) necessary and sufficient condition for ignorability under likelihood inference would require high-ability persons to be just as likely to produce any given missingness pattern as low-ability persons, given $\mathbf{u}_{\text{obs}}$. It is hard to reconcile ignorability of omitted responses with the conditions. We must therefore base inference on $L(\theta, \phi; \mathbf{m}, \mathbf{u}_{\text{obs}})$. Accepting that omits are not ignorable also means $g_\phi(\mathbf{m}|\mathbf{u})$ depends on $\mathbf{u}_{\text{mis}}$ and cannot be estimated from $(\mathbf{m}, \mathbf{u}_{\text{obs}})$. Modeling intentional omissions requires the analyst to either specify the mechanism of the omitting process or estimate it from data in which omitted responses are subsequently obtained.

10.3.5.2 Lord's 1974 Recommendation

Lord (1974) argues against both ignoring omitted responses to multiple-choice items under guessing-corrected scoring and treating them as wrong. The information that they provide can be handled with standard IRT estimation routines, he notes, if they are treated as fractionally correct, with the value $c = 1/(\#\text{ alternatives})$. Lord assumed "rational" omitting behavior: Examinees omit only if their chances of responding correctly would have been c, so $h_{k\theta}(u_{\text{mis},k}|\mathbf{u}_{\text{obs}}, \mathbf{m})$ is Bernoulli (c). Lord's recommendation is expected-value imputation conditional on $\mathbf{m}$. It is computationally convenient and seems defensible if omits are infrequent and persons have been informed of the optimal strategy.

Treating omitted responses to multiple-choice items as wrong is known-value imputation under the strong assumption that each one must be wrong. Logic and empirical evidence argue to the contrary (De Ayala et al., 2001; Hohensinn and Kubinger, 2011), and estimates of θ are biased downward.

10.3.6 Intentional Omits as an Additional Response Category

Multiple-category IRT models (Bock, 1972) can be applied with "omit" treated as an additional response category (Moustaki and O'Muircheartaigh, 2000). A SMURFLI$^{(2)}$ IRT model with omits would be extended to three responses v_i to each item, 0, 1, and *; three response functions, $f^*_{\theta i}(0)$, $f^*_{\theta i}(1)$, and $f^*_{\theta i}(*)$ and the conditional probability function would be $\prod f^*_{\theta i}(v_i)$. This approach captures the relationship between θ and the tendency to omit Item i. Lord (1983) expresses reservations, "... since it treats probability of omitting as dependent only on the examinee's ability, whereas it actually depends on a dimension of temperament. It seems likely that local unidimensional independence may not hold" (p. 477). Mislevy and Wu (1996, Theorem 7.2.5) show that local independence will hold in this approach only if all persons with the same θ have the same omitting parameter(s) ϕ.

10.3.7 Model v with Correctness and Missingness Processes

Alternatively, one can model $\mathbf{v}$ directly with a combined structure for correctness and omission, where persons can vary as to missingness propensities ($\phi^{(p)}$) and the missingness process may depend on θ (Goegebeur et al., 2006; Lord, 1983). Recent work builds on Albanese and Knott (1992) and Moustaki (1996). It distinguishes response and missing processes, with responses depending on only θ. Models can be formulated at the test level and the item level and can incorporate covariates y (Moustaki and Knott, 2000; see Rose et al., 2010, for an application).

A test-level formulation introduces person-omitting propensities $\phi^{(p)}$ that may be related to θ, along with item-omission parameters $\phi^{(i)}$ that allow items to differ as to their tendency to be omitted. The response and omission models have separate parameters:

$$P_i(\theta) \equiv \Pr\{V_i = 1|\theta, \beta_i, V_i \neq *\} \equiv \Pr\{U_i = 1|\theta, \beta_i, m_i = 1\} \tag{10.27}$$

$$R_i(\phi^{(p)}) \equiv \Pr\{V_i \neq *; \phi^{(i)}, \phi^{(p)}\} \equiv \Pr\{M_i = 1; \phi^{(i)}, \phi^{(p)}\}, \tag{10.28}$$

but $p(\theta, \phi) \neq p(\theta)p(\phi)$. Standard IRT forms, such as the 2PL, can be used for the P_is and R_is, and ys can be incorporated into the R_is. This model has the same ignorability status as the not-reached process: Correct-likelihood inferences for θ given β are obtained from $L^*(\theta; \mathbf{u}_{\text{obs}}) = f_\theta(\mathbf{u}_{\text{obs}})$ but Bayesian inference requires $p(\theta|\mathbf{m}, \mathbf{u}_{\text{obs}}) \propto L^*(\theta; \mathbf{u}_{\text{obs}})p(\theta|\mathbf{m})$.

An item-level formulation takes the same form for $P_i(\theta)$ but allows θ to appear in the missingness processes:

$$R_i(\phi^{(p)}, \theta) \equiv \Pr\{V_i \neq *; \theta, \phi^{(i)}, \phi^{(p)}\} \equiv \Pr\{M_i = 1; \theta, \phi^{(i)}, \phi^{(p)}\} \quad (10.29)$$

(Albanese and Knott, 1992; O'Muircheartaigh and Moustaki, 1999). Items may thus differ in their tendency be omitted by people with different abilities. Inference about θ, given both kinds of item parameters β and $(\phi^{(i)})$, are based on the full likelihood,

$$L(\theta, \phi^{(p)}; \mathbf{v}|\beta, (\phi^{(i)})) = \delta_{\theta\phi} \prod_i \Pr\{V_i = v_i; \theta, \phi^{(p)}, \beta_i, \phi_i^{(i)}\}, \quad (10.30)$$

with

$$P\{V_i = v; \theta, \phi^{(p)}\} = \begin{cases} 1 - R_i(\phi^{(p)}, \theta) & \text{if } v = * \\ R_i(\phi^{(p)}, \theta) P_i(\theta) & \text{if } v = 1 \\ R_i(\phi^{(p)}, \theta)[1 - P_i(\theta)] & \text{if } v = 0 \end{cases} \quad (10.31)$$

Note in Equation 10.27 that these models do not address $\mathbf{u}_{\text{mis}}$; θ concerns only $\mathbf{u}_{\text{obs}}$. The θ in Equation 10.27 is not necessarily the same as the θ in Equation 10.1. Although there are model-checking tools (e.g., Pimentel's, 2005, splitter item technique), we cannot check whether a common interpretation holds unless we ascertain $\mathbf{u}_{\text{mis}}$ in a follow-up study, then test whether the same θ holds across $\mathbf{u}_{\text{obs}}$ and $\mathbf{u}_{\text{mis}}$.

10.3.8 Examinee Choice

The College Entrance Examination Board's 1905 Botany test included a 10-item section directing a person to answer seven items (Wainer and Thissen, 1994). We now consider tests that incorporate "choose n_c items from N_c." The unchosen items are intentional omissions, but the administrator has forced omission and constrained the permissible missingness patterns.

10.3.8.1 Are Missing Responses to Unchosen Items Ignorable?

Motivated persons respond to the n_c items with the highest w_{pi} values in the set. Assume $f_\theta(\mathbf{U})$ is SMURFLI$^{(2)}$ and consider scores that count correct responses in the n_c chosen items, $R^+(\mathbf{v}) = R^+(\mathbf{u}_{\text{obs}})$. Again persons may differ as to how well their w_{pi}s are calibrated; such characteristics constitute person components of ϕ. Research suggests that high-ability persons are generally better at predicting their scores under different choice patterns and selecting patterns that produce higher scores (Wang et al., 1995). If so, ignorability is not plausible.

Theorem 10.24 (Mislevy and Wu, 1996, Theorem 8.1.2)

For a given $\mathbf{u}_{\text{obs}}$ and $\mathbf{m}$ in which missing responses are due to forced examinee choice, the missingness is MAR only if $\mathbf{m}$ is equally likely for all values of $\mathbf{u}_{\text{mis}}$.

For example, if all three items chosen by a person are wrong, for MAR to hold the items not chosen are just as likely to be all right as to be all wrong. This is not plausible, so likelihood and Bayesian inference based on $L^*(\theta; \mathbf{u}_{\text{obs}})$ will not generally be correct.

10.3.8.2 Inference under Choice Processes That Depend on Subjective Probabilities

Mislevy and Wu (1996) detail missingness mechanisms, likelihoods, and numerical illustrations for three cases of examinee choice: No information from w_{pi}, perfect information from w_{pi}, and a mixture of these. Highlights follow.

Case 1: Uninformed choice. A person's choice of items is independent of $\mathbf{u}$; for example, she chooses items with the highest w_{pi} but they are unrelated to actual correctness u_{pi}. This missingness is MCAR and inference can ignore the choice mechanism.

Case 2: Fully efficient choice. A person has perfect knowledge of $\mathbf{u}$, that is, $w_{pi} = u_{pi}$. Let $R^{\text{max}}(\mathbf{u})$ be the highest score obtainable from a given $\mathbf{u}$ with a permissible $\mathbf{m}$. This person will chose an $\mathbf{m}$ so that $R^+(\mathbf{v}) = R^{\text{max}}(\mathbf{u}) = \min(R^+(\mathbf{u}), n_c)$. Mislevy and Wu (1996) show with an illustrative g that this missingness mechanism is not apt to be MAR because the condition $R^+(\mathbf{u}_{\text{obs}}) = R^{\text{max}}(\mathbf{u})$ depends on both $\mathbf{u}_{\text{mis}}$ and $\mathbf{u}_{\text{obs}}$. $L(\theta, \phi; \mathbf{m}, \mathbf{u}_{\text{obs}})$ is then the average over all $\mathbf{u}$ that are consistent with $\mathbf{u}_{\text{obs}}$.

- When $R^+(\mathbf{u}) \geq n_c$, the person chooses a permissible $\mathbf{m}$ with all correct responses so that $R^+(\mathbf{u}_{\text{obs}}) = n_c$, and the remaining items have $m_i = 0$. $L(\theta, \phi; \mathbf{m}, \mathbf{u}_{\text{obs}})$ is the average of likelihoods over all $\mathbf{u}$ with these $R^{\text{max}}(\mathbf{u})$ right answers, from all $u_{\text{mis}} = 1$ to all $u_{\text{mis}} = 0$. Estimates of θ are biased downward for more able persons who tend to have higher $R^+(\mathbf{u}_{\text{mis}})$ and similarly biased upward for more less able examinees with lower $R^+(\mathbf{u}_{\text{mis}})$.
- When $R^+(\mathbf{u}) = R^{\text{max}}(\mathbf{u}) < n_c$, the person chooses an $\mathbf{m}$ that includes the $R^+(\mathbf{u})$ correct responses and $n_c - R^+(\mathbf{u})$ incorrect responses. $L(\theta, \phi; \mathbf{m}, \mathbf{u}_{\text{obs}})$ simplifies to likelihood with this $\mathbf{u}_{\text{obs}}$ and $\mathbf{u}_{\text{mis}} = 0$ for all unobserved items. Known-value imputation of 0 for each u_{mis} is appropriate in this special case, correct likelihood and Bayesian inferences result, and examinee choice actually provides additional information beyond the observed responses. Fully efficient choice, however, is a very strong assumption.

Case 3: Partially efficient choice. Suppose w_{pi} contains less-than-perfect information about u_{pi}. An illustrative g is a mixture of uniformed choice and fully efficient choice, with mixing proportion ϕ a person parameter of the missingness process. $L(\theta, \phi; \mathbf{m}, \mathbf{u}_{\text{obs}})$ is now the weighted average of likelihoods over all response patterns $\mathbf{m}$ that are compatible with $\mathbf{u}_{\text{obs}}$—including, with weights that depend on $1 - \phi$, ones that could have produced higher scores than the observed pattern. Low scores can now be caused by both low ability and poor choice (Bradlow and Thomas, 1998).

These results indicate that "choose n_c of N_c items" format typically reduces information about θ when an underlying SMURFLI$^{(2)}$ is assumed and the target of interest is θ. They support Wainer and Thissen's (1994) conclusion that examinee choice is ill-suited to standardized testing when a test is unidimensional and the target of inference is average performance in the domain.

10.4 Item-Parameter Estimation

The preceding sections addressed inference about θ when responses are missing, with β known. This section addresses inference about β. The focus is on Bayesian estimation (see Chapter 13) and marginal maximum-likelihood (MML) estimation (see Chapter 11). Attention is restricted to the missingness mechanisms discussed previously, which are functionally independent of β with noted exceptions.

Ignorability conditions are discussed first. The general form for Bayesian inference is then reviewed. Handling nonignorable missing with conditional-on-**m** expected information estimation is addressed next, with comments on imputation approaches.

Underscores will be used to indicate response vectors, missingness vectors, parameter vectors, etc. for a set of persons. For example, the observed response vectors of P persons are denoted $\underline{\mathbf{u}}_{\text{obs}} \equiv (\mathbf{u}_{\text{obs},1}, \ldots, \mathbf{u}_{\text{obs},p}, \ldots, \mathbf{u}_{\text{obs},P})$.

10.4.1 Ignorability Results

Definition 10.15

An IRT missingness process $g_\phi(\mathbf{m}|\mathbf{u}_{\text{obs}}, \mathbf{u}_{\text{mis}})$ is functionally independent (FI) of β if it does not depend on β, that is, β is not a component of ϕ. A process can be functionally independent of β yet depend on other parameters that are not independent of β, such as experts' categorizations E_i of item difficulties.

If $g_\phi(\mathbf{m}|\mathbf{u}_{\text{obs}}, \mathbf{u}_{\text{mis}})$ is functionally independent of β, then by Equation 10.3 the likelihood function for $(\underline{\theta}, \underline{\phi}, \beta)$ induced by the observation of P persons' missingness patterns and observed responses, $(\underline{\mathbf{m}}, \underline{\mathbf{u}}_{\text{obs}})$, is obtained from $P\{\underline{\mathbf{M}} = \underline{\mathbf{m}}, \underline{\mathbf{U}}_{\text{obs}} = \underline{\mathbf{u}}_{\text{obs}}|\underline{\theta}, \underline{\phi}, \beta\}$ as

$$L(\underline{\theta}, \underline{\phi}, \beta; \underline{\mathbf{m}}, \underline{\mathbf{u}}_{\text{obs}}) = \delta_{\theta\phi\beta} \prod_p \int f_{\theta\beta}(\mathbf{u}_{\text{obs},p}, \mathbf{u}_{\text{mis},p})\, g_\phi(\mathbf{m}_p|\mathbf{u}_{\text{obs},p}, \mathbf{u}_{\text{mis},p})\partial \mathbf{u}_{\text{mis},p}. \quad (10.32)$$

Theorem 10.25

Let g be FI of β. Missingness is ignorable for Bayesian inference about β if g is ignorable for Bayesian inference about θ given β. That is, correct Bayesian inferences about β are obtained from the following equation:

$$p^*(\theta, \beta|\mathbf{u}_{\text{obs}}) \propto L^*(\theta, \beta; \mathbf{u}_{\text{obs}})\, p(\theta, \beta) \quad (10.33)$$

where

$$L^*(\theta, \beta; \mathbf{u}_{\text{obs}}) = \prod_p L^*(\theta_p, \beta; \mathbf{u}_{\text{obs},p}).$$

This result is obtained as follows:

$$\begin{aligned} &p(\theta, \phi, \beta|\underline{\mathbf{m}}, \underline{\mathbf{u}}_{\text{obs}}) \\ &\quad = L(\theta, \phi, \beta; \underline{\mathbf{m}}, \underline{\mathbf{u}}_{\text{obs}})\, p(\theta, \phi)\, p(\beta) \end{aligned}$$

$$= \delta_{\theta\phi\beta} \prod_p \left[\int f_{\theta\beta}(\mathbf{u}_{\text{obs},p}, \mathbf{u}_{\text{mis},p})\, g_\phi(\mathbf{m}_p | \mathbf{u}_{\text{obs},p}, \mathbf{u}_{\text{mis},p}) \partial \mathbf{u}_{\text{mis},p}\, p(\theta_p, \phi_p) \right] p(\boldsymbol{\beta})$$

$$= \delta_{\theta\beta} \prod_p f_{\theta\beta}(\mathbf{u}_{\text{obs},p})\, p(\theta_p) p(\boldsymbol{\beta})$$

$$= L^*(\underline{\theta}, \boldsymbol{\beta}; \underline{\mathbf{u}}_{\text{obs}}) p(\underline{\theta}, \boldsymbol{\beta}),$$

as person-level Bayesian ignorability of missingness gives the key simplification.

Theorem 10.26

Let g be FI of β and the prior distributions for θs be independent. Missingness is ignorable for MML inference about β if g is ignorable for Bayesian inference about θ given β. That is, correct MML inferences about β are obtained from the following:

$$L^*_{\text{MML}}(\boldsymbol{\beta}; \underline{\mathbf{u}}_{\text{obs}}) = \delta_\beta \prod_p \int f_{\theta\beta}(\mathbf{u}_{\text{obs},p}) p(\theta)\, \partial\theta. \tag{10.34}$$

Bayesian ignorability for θ provides independent priors for θ and person components $\phi^{(p)}$ of ϕ, so the marginalization over θ in MML estimation of β does not bring in $\phi^{(p)}$—they have been factored out as in Theorem 10.1.

Theorems 10.25 and 10.26 allow MML and Bayesian item-parameter estimation to ignore missingness due to random item assignment, such as matrix sampling and alternate test forms.

Appropriate solutions are thus provided by IRT software packages such as BILOG-MG (Zimowski et al., 2003) and IRTPRO (Cai et al., 2011) using a "not-presented" code for missing responses. The MCMC program WinBUGS (Lunn et al., 2000) uses an analogous code "NA." These codes indicate missing responses that are to be treated as MAR.

Not-reached items are MAR but not generally ignorable under Bayesian inference because the probability of reaching an item may be correlated with θ (Theorem 10.22a). MML and Bayesian inference about β can therefore be biased when $p(\theta|\mathbf{m})$ is replaced by $p(\theta)$ in MML (Holman and Glas, 2005). However, by Theorem 10.22b, ignoring not-reached items yields consistent partial-information inference about θ. Treating not reached as not presented in item-parameter estimation can be defended for long tests with few not-reached items. In contrast, coding them as wrong usually cannot be defended, since the biases it causes do not decline with increasing numbers of observed responses.

Targeted testing employs item assignment probabilities that depend on both beliefs about item difficulties and person covariates y related to θ, as when we administer younger students items we expect to be easier. That is, g depends on both person covariates y_p and item covariates E_i. This situation arises in concurrent calibration in vertical scaling applications. By Theorem 10.11, it is necessary to condition on y for targeted–testing missingness to be ignorable. The appropriate marginal likelihood and Bayesian expressions are as follows:

$$L^*(\boldsymbol{\beta}; \underline{\mathbf{u}}_{\text{obs}}, \underline{\mathbf{y}}) = \delta_\beta \prod_p \int f_{\theta\beta}(\mathbf{u}_{\text{obs},p}) p(\theta | y_p) \partial\theta \tag{10.35}$$

and

$$p(\underline{\theta}, \boldsymbol{\beta} | \underline{\mathbf{u}}_{\text{obs}}, \underline{\mathbf{y}}, \mathbf{E}) \propto \prod_p f_{\theta\beta}(\mathbf{u}_{\text{obs},p}) p(\theta | y_p) p(\boldsymbol{\beta} | \mathbf{E}), \tag{10.36}$$

where δ_β is an indicator which takes the value 1 if β is in the parameter space and 0 if not. Equations 10.35 and 10.36 are multiple-group calibration. Using Equation 10.33 or 10.34 instead under targeted testing produces biased estimates of β (DeMars, 2002; Mislevy and Sheehan, 1989).

By Theorems 10.16 through 10.18, missing responses in adaptive testing are ignorable with respect to θ given β (and y if item selection probabilities use it). Items are typically administered with estimates of β treated as known, say β_{old}. Online calibration brings new items into the item pool by seeding them randomly into adaptive tests. Unobserved responses to new items by persons who did not receive them are MCAR. Modified versions Equations 10.33 through 10.36 can be used to estimate β_{new} via MML; for example, maximize

$$L^*(\beta_{\text{new}}; \underline{\mathbf{u}}_{\text{obs}}, \beta_{\text{old}}) = \prod_p \int f_\theta(\mathbf{u}_{p,\text{obs}}; \beta_{\text{new}}, \beta_{\text{old}}) p(\theta) \partial\theta \tag{10.37}$$

with respect to β_{new}, with β_{old} fixed.

10.4.2 Full Bayesian Model

Equation 10.8 in the section on MCMC inference gave the full Bayesian model for IRT when some responses are missing, assuming FI:

$$\begin{aligned} &p(\underline{\mathbf{m}}, \underline{\mathbf{u}}_{\text{obs}}, \underline{\mathbf{u}}_{\text{mis}}, \underline{\theta}, \underline{\phi}^{(p)}, \beta, \phi^{(i)}) \\ &\quad = \prod_p f_{\theta\beta}(\mathbf{u}_{p,\text{obs}}, \mathbf{u}_{p,\text{mis}}; \beta) g_\phi(\mathbf{m}_p; \beta, \phi^{(i)} | \mathbf{u}_{p,\text{obs}}, \mathbf{u}_{p,\text{mis}}) p(\theta, \phi^{(p)}) p(\beta) p(\phi^{(i)}). \end{aligned} \tag{10.38}$$

The posterior for $(\underline{\mathbf{u}}_{\text{mis}}, \underline{\theta}, \underline{\phi}^{(p)})$, β and $\phi^{(i)}$ after observing $(\underline{\mathbf{m}}, \underline{\mathbf{u}}_{\text{obs}})$ is

$$\begin{aligned} &p(\underline{\mathbf{u}}_{\text{mis}}, \underline{\theta}, \underline{\phi}^{(p)}, \beta, \phi^{(i)} | \underline{\mathbf{m}}, \underline{\mathbf{u}}_{\text{obs}}) \\ &\quad \propto \prod_p f_\theta(\mathbf{u}_{\text{obs},p}, \mathbf{u}_{\text{mis},p}; \beta) g_{\phi_p}(\mathbf{m}_p; \beta, \phi^{(i)} | \mathbf{u}_{\text{obs},p}, \mathbf{u}_{\text{mis},p}) p(\theta, \phi^{(p)}) p(\beta) p(\phi^{(i)}). \end{aligned} \tag{10.39}$$

As noted, Bayesian inference for the unobserved parameters and variables can be obtained using software such as WinBUGS by providing functional forms and parameterizations for the missingness and response processes.

10.4.3 Inference from Conditional-on-m Expected Likelihoods

Theorem 10.2 gave a conditional-on-**m** expected likelihood for inference about θ given β. It built around $h_\theta(\mathbf{u}_{\text{mis}} | \mathbf{u}_{\text{obs}}, \mathbf{m})$, the predictive distribution of $\mathbf{u}_{\text{mis}}$ given $\mathbf{u}_{\text{obs}}$ and $\mathbf{m}$. The approach can be extended to inference about β and be viewed as the foundation for imputation methods for estimating β with standard IRT software.

Equation 10.11 gives the expectation of the likelihood for θ given β induced by $\mathbf{u}$, conditional on $\mathbf{m}$, when $(\mathbf{m}, \mathbf{u}_{\text{obs}})$ is observed. From the observed responses and missingness patterns of P persons $(\underline{\mathbf{u}}_{\text{obs}}, \underline{\mathbf{m}})$, we obtain a pseudo expected joint likelihood for $\underline{\theta}$ and β

conditional on $\underline{\mathbf{m}}$ by taking the product of these terms and considering β unknown:

$$\prod_p \delta_{\theta\beta} E[L(\theta_p, \boldsymbol{\beta}; \mathbf{u}_p)|\mathbf{m}_p, \mathbf{u}_{\text{obs},p}]$$

$$= \prod_p \delta_{\theta\beta} \int L(\theta_p, \boldsymbol{\beta}; \mathbf{u}_{\text{obs},p}, \mathbf{u}_{\text{mis},p}) h_\theta(\mathbf{u}_{\text{mis},p}|\mathbf{u}_{\text{obs},p}, \mathbf{m}_p) \partial \mathbf{u}_{\text{mis},p}$$

$$= \prod_p \delta_{\theta\beta} f_{\theta\beta}(\mathbf{u}_{\text{obs},p}) \int f_{\theta\beta}(\mathbf{u}_{\text{mis},p}) h_\theta(\mathbf{u}_{\text{mis},p}|\mathbf{u}_{\text{obs},p}, \mathbf{m}_p) \partial \mathbf{u}_{\text{mis},p}. \quad (10.40)$$

Again, $\delta_{\theta\beta}$ is the 1/0 indicator as to whether given (θ, β) values are in the parameter space.

Theorem 10.27

Assuming CIP, a conditional-on-**m** expected joint likelihood for $\underline{\theta}$ and β is

$$L^{(c)}(\theta, \boldsymbol{\beta}; \underline{\mathbf{u}}_{\text{obs}}|\underline{\mathbf{m}}) = \prod_p \delta_{\theta\beta} f_{\theta\beta}(\mathbf{u}_{\text{obs},p}) \int f_{\theta\beta}(\mathbf{u}_{\text{mis},p}) h_\theta(\mathbf{u}_{\text{mis},p}|\mathbf{u}_{\text{obs},p}, \mathbf{m}_p) \partial \mathbf{u}_{\text{mis},p}$$

$$= \prod_p \delta_{\theta\beta} f_{\theta\beta}(\mathbf{u}_{\text{obs},p}) \prod_{k:m_{pk}=0} \int f_{\theta\beta}(u_{\text{mis},pk}) h_\theta(u_{\text{mis},pk}|\mathbf{u}_{\text{obs},p}, \mathbf{m}_p) \partial u_{\text{mis},pk}. \quad (10.41)$$

$L^{(c)}$ has the form of the IRT model for the observed responses and item-by-item predictive distributions for the missing responses. By the same arguments provided for Theorem 10.2, Equation 10.41 is an approximation of Equation 10.40 that yields conditional-on-**m** expected likelihood inferences for β. It foregoes the information each $\mathbf{m}_p$ may provide about θ_p.

Equation 10.41 can provide a starting point for MML and Bayesian inference. Bayesian inference multiplies Equation 10.41 by $\Pi p(\theta_p)$ and $p(\beta)$, assuming independent priors. MCMC estimation can be operationalized by crafting the predictive distributions for each missing response in accordance with the missing mechanism—for example, $f_{\theta\beta}(u_{\text{mis},pk})$ for responses that are MAR and Bernoulli (1/*A*) for omitted responses to multiple-choice items with *A* options.

Marginal maximum-likelihood estimation is obtained by integrating Equation 10.41 over $\Pi p(\theta_p)$ or $\Pi p(\theta_p\,|y)$ as needed. Standard software such as BILOG-MG can be used by filling in the $\mathbf{u}_{\text{mis}}$ with draws from each $h_\theta(u_{\text{mis},pk}|\mathbf{u}_{\text{obs},p}, \mathbf{m}_p)$. By generating multiple filled-in datasets and estimating β in each, one can use Rubin's (1987) multiple-imputation formulas to account for the uncertainty about β associated with the *h*s.

10.4.4 Specialized Models

The earlier discussion of not-reached items and intentional omits cited examples of particular models for these kinds of nonignorable response. The references cited there discuss their estimation procedures, all MML or Bayesian MCMC.

10.5 Conclusion

In practical applications of IRT, item responses may not be observed from all examinees to all items for many reasons, such as random assignment as in alternate forms and matrix sampling; targeted testing, including vertical linking, where forms are administered based on person covariate; adaptive and multistage testing; not-reached items, intentional omissions, and examinee choice.

Rubin's (1976, 1977) missing-data framework provides conditions for determining whether missing data can be ignored. Sufficient conditions for ignorability under likelihood and Bayesian inference are missingness at random (MAR) and distinctness parameters for the response and the missingness processes. Table 10.1 summarizes ignorability results for person parameters θ given item parameters β, and Table 10.2 for β under Bayesian and MML estimation.

When ignorability does not hold, one must address the response process and missingness process jointly. Rubin's framework provides general forms for missingness processes and predictive distributions for missing responses. Examples included Glas and Pimentel (2008) for not-reached items and Moustaki and Knott (2000) for omits.

TABLE 10.1
Ignorability Results for θ Given β

	Type of Inference	
Missingness	**Direct Likelihood**	**Bayesian**
Random assignment	Yes	Yes
Targeted forms	Yes	Yes, conditional on person covariates
Adaptive testing	Yes	Yes, conditional on person covariates if used in item selection
Not reached	Yes	No, unless speed and proficiency are independent
Intentional omission	No	No
Examine choice	No	No

TABLE 10.2
Ignorability Results for β Marginalizing over θ

	Type of Inference	
Missingness	**Direct Likelihood**	**Bayesian**
Random assignment	Yes	Yes
Targeted forms	Yes, conditional on person covariates	Yes, conditional on person and item covariates
Adaptive testing	Yes, conditional on person covariates if used in item selection	Yes, conditional on person and item covariates if used in item selection
Not reached	No, unless speed and proficiency are independent	No, unless speed and proficiency are independent
Intentional omission	No	No
Examinee choice	No	No

Earlier techniques for missingness in IRT can be examined through Rubin's lens. Conditions were presented under which some techniques can be justified as approximations of expected likelihoods, conditional on the observed pattern of missingness. Others, such as treating not-reached items and intentional omits to multiple-choice items as wrong, can be understood as bad ideas.

References

Albanese, M. T. and Knott, M. 1992. *TWOMISS: A Computer Program for Fitting a One- or Two-Factor Logit-Probit Latent Variable Model to Binary Data When Observations May Be Missing*. London: London School of Economics and Political Science.

Bock, R. D. 1972. Estimating item parameters and latent ability when responses are scored in two or more nominal categories. *Psychometrika, 37*, 29–51.

Bradlow, E. T. and Thomas, N. 1998. Item response theory models applied to data allowing examinee choice. *Journal of Educational and Behavioral Statistics, 23*, 236–243.

Cai, L., Thissen, D., and du Toit, S. H. C. 2011. *IRTPRO: Flexible, Multidimensional, Multiple Categorical IRT Modeling* [Computer software]. Lincolnwood, IL: Scientific Software International.

Cross, L. H. and Frary, R. B. 1977. An empirical test of Lord's theoretical results regarding formula scoring of multiple choice tests. *Journal of Educational Measurement, 14*, 313–321.

De Ayala, R. J., Plake, B. S., and Impara, J. C. 2001. The impact of omitted responses on the accuracy of ability estimation in item response theory. *Journal of Educational Measurement, 38*, 213–234.

DeMars, C. 2002. Incomplete data and item parameter estimates under JMLE and MML estimation. *Applied Measurement in Education, 15*, 15–31.

Finch, H. 2008. Estimation of item response theory parameters in the presence of missing data. *Journal of Educational Measurement, 45*, 225–245.

Gelman, A., Carlin, J. B., Stern, H. S., and Rubin, D. B. 2004. *Bayesian Data Analysis* (second edition). London: Chapman & Hall.

Glas, C. A. W. and Pimentel, J. 2008. Modeling nonignorable missing data in speeded tests. *Educational and Psychological Measurement, 68*, 907–922.

Goegebeur, Y., De Boeck, P., Molenberghs, G., and del Pino, G. 2006. A local-influence-based diagnostic approach to a speeded item response theory model. *Applied Statistics, 55*, 647–676.

Hohensinn, C. and Kubinger, K. D. 2011. On the impact of missing values on item fit and the model validness of the Rasch model. *Psychological Test and Assessment Modeling, 53*, 380–393.

Holman, R. and Glas, C. A. W. 2005. Modelling non-ignorable missing data mechanism with item response theory models. *British Journal of Mathematical and Statistical Psychology, 58*, 1–18.

Little, R. J. A. and Rubin, D. B. 1987. *Statistical Analysis with Missing Data*. New York, NY: Wiley.

Lord, F. M. 1974. Estimation of latent ability and item parameters when there are omitted responses. *Psychometrika, 39*, 247–264.

Lord, F. M. 1983. Maximum likelihood estimation of item response parameters when some responses are omitted. *Psychometrika, 48*, 477–482.

Lunn, D. J., Thomas, A., Best, N., and Spiegelhalter, D. 2000. WinBUGS—A Bayesian modelling framework: Concepts, structure, and extensibility. *Statistics and Computing, 10*, 325–337.

Matters, G. and Burnett, P. C. 2003. Psychological predictors of the propensity to omit short response items on a high-stakes achievement test. *Educational and Psychological Measurement, 63*, 239–256.

Mislevy, R. J. and Chang, H. H. 2000. Does adaptive testing violate local independence? *Psychometrika, 65*, 149–165.

Mislevy, R. J. and Sheehan, K. M. 1989. The role of collateral information about examinees in item parameter estimation. *Psychometrika, 54*, 661–679.

Mislevy, R. J. and Wu, P. K. 1996. Missing responses and Bayesian IRT ability estimation: Omits, choice, time limits, and adaptive testing. Research Report RR-96-30-ONR. Princeton: Educational Testing Service.

Moustaki, I. 1996. *Latent Variable Models for Mixed Manifest Variables.* PhD thesis. London: The London School of Economics and Political Science.

Moustaki, I. and Knott, M. 2000. Weighting for item non-response in attitude scales using latent variable models with covariates. *Journal of the Royal Statistical Society, Series A, 163,* 445–459.

Moustaki, I. and O'Muircheartaigh, C. 2000. A one dimension latent trait model to infer attitude from nonresponse for nominal data. *Statistica, 60* 259–276.

O'Muircheartaigh, C. and Moustaki, I. 1999. Symmetric pattern models: A latent variable approach to item non-response in attitude scales. *Journal of the Royal Statistical Society, Series A, 162* 177–194.

Pimentel, J. L. 2005. *Item Response Theory Modeling with Nonignorable Missing Data.* PhD thesis. University of Twente, the Netherlands.

Rose, N., von Davier, M., and Xu, X. 2010. *Modeling Non-Ignorable Missing Data with IRT. ETS-RR-10-10.* Princeton: Educational Testing Service.

Rubin, D. B. 1976. Inference and missing data. *Biometrika, 63,* 581–592.

Rubin, D. B. 1977. Formalizing subjective notions about the effect of nonrespondents in sample surveys. *Journal of the American Statistical Association, 72,* 538–543.

Rubin, D. B. 1987. *Multiple Imputation for Nonresponse in Surveys.* New York, NY: Wiley.

Sherriffs, A. C. and Boomer, D. S. 1954. Who is penalized by the penalty for guessing? *Journal of Educational Psychology, 45,* 81–90.

Sijtsma, K. and van der Ark, L. A. 2003. Investigation and treatment of missing item scores in test and questionnaire data. *Multivariate Behavioral Research, 38* 505–528.

van den Wollenberg, A. L. 1979. *The Rasch Model and Time-Limit Tests.* Doctoral dissertation, University of Nijmegen.

van der Linden, W. J. and Glas, C. A. W. (Eds.). 2010. *Elements of Adaptive Testing.* New York: Springer.

Verhelst, N. D., Glas, C. A. W., and de Vries, H. H. 1997. A steps model to analyze partial credit. In van der Linden W. J. and Hambleton R. K. (Eds.), *Handbook of Modern Item Response Theory.* New York: Springer, pp. 123–138.

Wainer, H. and Thissen, D. 1994. On examinee choice in educational testing. *Review of Educational Research, 64,* 159–195.

Wang, X. B., Wainer, H., and Thissen, D. 1995. On the viability of some untestable assumptions in equating exams that allow examinee choice. *Applied Measurement in Education, 8,* 211–225.

Zimowski, M. F., Muraki, E., Mislevy, R. J., and Bock, R. D. 2003. *BILOG-MG 3 for Windows: Multiple-Group IRT Analysis and Test Maintenance for Binary Items* [Computer software]. Skokie, IL: Scientific Software International, Inc.

Zwick, R. 1990. When do item response function and Mantel-Haenzel definitions of differential item functioning coincide? *Journal of Educational Statistics, 15,* 185–197.

Section III

Parameter Estimation

11

Maximum-Likelihood Estimation

Cees A. W. Glas

CONTENTS

11.1 Introduction

Item response theory (IRT) models are used to explain $P \times I$ matrices of responses to test items, $\mathbf{U}$, using person parameters, θ, and item parameters, β. In general, every person and every item has at least one parameter. A naive approach to maximum-likelihood (ML) estimation would be to maximize the likelihood function jointly with respect to θ and β, using a standard technique such as a Newton–Raphson algorithm. But there are two problems with this approach. The first problem is a practical one arising from the large number of parameters. Application of a Newton–Raphson algorithm needs the inversion of the matrix of second-order derivatives of the likelihood function with respect to all parameters, so for larger numbers of persons this approach quickly becomes infeasible. Still, this practical problem might be solvable. A second, more fundamental problem is related to the consistency of the parameter estimates. Neyman and Scott (1948) showed that if the number of parameters grows proportional with the number of observations, this can lead to inconsistent parameter estimates (see Chapter 9). In a basic application of an IRT model, the problem applies to the person parameters. If we try to increase the precision of the item parameter estimates by sampling more persons, we automatically increase the number of person parameters. Simulation studies by Wright and Panchapakesan (1969) and Fischer and Scheiblechner (1970) showed that inconsistencies in the estimates for IRT models can occur indeed.

To obtain consistent item parameter estimates, the person parameters must, in some sense, be removed from the model. One way of accomplishing this is to introduce the assumption of common distributions of the person parameters (often called "population distributions"). The person parameters can then be integrated out of the likelihood to obtain a so-called marginal likelihood. Maximization of this marginal likelihood is known as maximum marginal-likelihood (MML) estimation. The three-parameter logistic (3PL) model was developed by Bock and Aitkin (1981) (also see, Thissen, 1982; Rigdon and Tsutakawa, 1983; Mislevy, 1984, 1986). This approach will be outlined first. The second approach uses conditioning on sufficient statistics for the person parameters. It can only be applied to exponential-family IRT models, examples of which are described in Chapters 3, 7, and 32 of *Handbook of Item Response Theory, Volume One: Models*. The method of conditional maximum likelihood (CML) was developed by Rasch (1960) (also see, Andersen, 1973; Fischer, 1981; Kelderman, 1984, 1995). Both MML and CML estimation will be treated more generally first, and then, the results will be applied to various specific models.

Once the item parameters and, when applicable, the population distribution parameters are estimated, the person parameters can be estimated by treating all other parameters as known constants. This chapter ends with the description of two classes of approaches for estimating person parameters: Frequentist approaches (maximum likelihood and weighted maximum likelihood) and Bayesian approaches (based either on the expectation or the mode of the posterior distributions of the person parameters).

11.2 Estimation of Item and Population Parameters

The derivation of MML estimation equations and expressions for the covariance matrix of the estimates can be greatly simplified by making use of a device called Fisher's identity (Efron, 1977; Louis, 1982; see also, Glas, 1999). In this section, the identity is used to derive the MML equations for a very broad class of IRT models, and then applied to specific IRT models. The identity plays an important role in the framework of the EM algorithm, which is an algorithm for finding the maximum of a likelihood marginalized over unobserved data. The principle can be summarized as follows. Let $L_o(\lambda)$ be the log-likelihood function of parameters λ given observed data x_o, and let $L_c(\lambda)$ be the log-likelihood function given both observed data x_o and unobserved missing data x_m. The latter is called the complete-data log-likelihood. The interest is in finding expressions for the first-order derivatives of $L_o(\lambda)$, that is, expressions for $L'_o(\lambda) = \partial L_o(\lambda)/\partial\lambda$. Define the first-order derivatives with respect to the complete-data log-likelihood as $L'_c(\lambda) = \partial L_c(\lambda)/\partial\lambda$. Then, Fisher's identity states that $L'_o(\lambda)$ is equal to the expectation of $L'_c(\lambda)$ with respect to the posterior distribution of the missing data given the observed data, $p(x_m|x_o;\lambda)$, that is,

$$L'_o(\lambda) = E\left[L'_c(\lambda)\middle| x_o\right] = \int L'_c(\lambda)p(x_m|x_o;\lambda)dx_m. \tag{11.1}$$

If independence between persons is assumed, Fisher's identity can also be expressed in terms of the person-level derivatives of the observed and complete-data log-likelihood, $L_o^{(p)\prime}(\lambda)$ and $L_c^{(p)\prime}(\lambda)$, using the person-level posterior densities $p(x_m^{(p)}|x_o^{(p)};\lambda)$. Fisher's identity then becomes

$$L'_o(\lambda) = \sum_{p=1}^{P} L_o^{(p)'}(\lambda) = \sum_{p=1}^{P} E\left[L_c^{(p)'}(\lambda)\middle| x_o^{(p)}\right] = \sum_{p=1}^{P} \int L_c^{(p)'}(\lambda) p(x_m^{(p)}|x_o^{(p)};\lambda) dx_m. \quad (11.2)$$

Also the standard errors for the estimates of λ are easily derived in this framework: the information matrix $I(\lambda)$ (the negative of the matrix of second-order derivatives) can be approximated as

$$I(\lambda) \approx \sum_{p=1}^{P} E\left[L_c^{(p)'}(\lambda)\middle| x_o^{(p)}\right] E\left[L_c^{(p)'}(\lambda)\middle| x_o^{(p)}\right]^t, \quad (11.3)$$

and the standard errors are the square roots of the diagonal elements of the inverse of this matrix. Note that this approximation does not need the actual computation of any second-order derivatives, since it is based on the outer product of the vectors of first-order derivatives at the person-level.

To apply this framework to IRT, a very general definition of an IRT model is adopted. Assume an IRT model is defined by the probability of a response vector $\mathbf{u}_p$, which is a function of, possibly vector-valued, person parameters θ_p and item parameters $\boldsymbol{\beta}$. So the IRT model is given by $p(\mathbf{u}_p|\theta_p, \beta)$. Note that this density is not further factored to the level of the individual responses, and the usual assumption of local independence is not made. Assume further that the person parameter θ_p has a density $g(\theta_p; \delta, \mathbf{x}_p)$, with parameters δ and, possibly, covariates $\mathbf{x}_p$. The key idea is to view the person parameters $\boldsymbol{\theta}_p$ as missing data and the item and population parameters $\boldsymbol{\beta}$ and δ as structural parameters to be estimated. To apply Equation 11.2, write the complete-data log-likelihood as

$$L_c^{(p)}(\beta, \delta) = \log p(\mathbf{u}_p|\theta_p, \beta) + \log g(\theta_p; \delta, \mathbf{x}_p) \quad (11.4)$$

and so

$$L_c^{(p)'}(\boldsymbol{\beta}) = \partial \log p(\mathbf{u}_p|\boldsymbol{\theta}_p, \beta)/\partial\beta, \quad (11.5)$$

$$L_c^{(p)'}(\boldsymbol{\delta}) = \partial \log g(\boldsymbol{\theta}_p; \boldsymbol{\delta}, \mathbf{x}_p)/\partial\delta. \quad (11.6)$$

Fisher's identity says that the MML equations are the expectations of Equations 11.5 and 11.6 relative to the posterior distribution of the missing data $\boldsymbol{\theta}_p$ given the observations $\mathbf{u}_p$ and, if applicable, $\mathbf{x}_p$, that is, expectations are taken with respect to

$$p(\theta_p|\mathbf{u}_p, \mathbf{x}_p; \beta, \delta) = \frac{p(\mathbf{u}_p|\theta_p, \beta) g(\theta_p; \delta, \mathbf{x}_p)}{\int p(\mathbf{u}_p|\theta_p, \beta) g(\theta_p; \delta, \mathbf{x}_p) d\theta_p}. \quad (11.7)$$

In the next subsections, this will be worked out in more detail for a number of specific IRT models.

11.2.1 MML Estimation for the 1PLM

In this example, it is assumed that all P persons respond to all I items. Consider dichotomously scored items; the responses will be represented by stochastic variables U_{pi}, with realizations $u_{pi} = 1$ for a correct response and $u_{pi} = 0$ for an incorrect response. Define θ_p and b_i as the person and item parameter, respectively. In the one-parameter logistic model (1PLM), also known as the Rasch model (Rasch, 1960), the probability of a correct response is given by

$$P(U_{pi} = 1; \theta_p, b_i) = P_{pi} = \frac{\exp(\theta_p - b_i)}{1 + \exp(\theta_p - b_i)} \tag{11.8}$$

(Volume One, Chapter 3). The logarithm of the probability of I-dimensional response vector $\mathbf{u}_p$ with entries u_{pi} given person parameters θ_p and item difficulty parameters $\mathbf{b}$ can be written as

$$\begin{aligned}
\log \prod_{p=1}^{P} p(\mathbf{u}_p; \theta_p, \mathbf{b}) &= \log \prod_{p=1}^{P} \prod_{i=1}^{I} P_{pi}^{u_{pi}} (1 - P_{ip})^{1-u_{pi}} \\
&= \sum_{p=1}^{P} \sum_{i=1}^{I} \log \frac{\exp(u_{pi}(\theta_p - b_i))}{1 + \exp(\theta_p - b_i)} \\
&= \sum_{p=1}^{P} \left(\sum_{i=1}^{I} u_{pi} \right) \theta_p - \sum_{i=1}^{I} \left(\sum_{p=1}^{P} u_{pi} \right) b_i \\
&\quad - \sum_{p=1}^{P} \sum_{i=1}^{I} \log(1 + \exp(\theta_p - b_i)) \\
&= \sum_{p=1}^{P} x_p \theta_p - \sum_{i=1}^{I} s_i b_i - H(\theta_p, \mathbf{b}),
\end{aligned} \tag{11.9}$$

where x_p is the number of correct responses given by person p, and s_i is the number correct responses given to item i. X_p and S_i are sufficient statistics for θ_p and b_i, respectively. Notice that $H(\theta_p, \mathbf{b})$ is implicitly defined as $\sum_p \sum_i \log(1 + \exp(\theta_p - b_i))$.

The first-order derivatives of the complete-data likelihood can be obtained in two ways. The first option is to take the first-order derivatives of the expression in Equation 11.9. The second option is to use the fact that the model defined by this expression is an exponential-family model, because its log-likelihood can be expressed as a linear function of (i) products of sufficient statistics and parameters and (ii) a function of the parameters $H(\theta_p, \mathbf{b})$, which does not depend on the data. In an exponential-family model, these first-order derivatives are the differences between the sufficient statistics and their expected values (see, for instance, Andersen, 1980). So in the present case, for instance, the estimation equation for item parameter b_i is given by $-s_i + E(S_i)$, that is

$$-s_i + \sum_{p=1}^{P} P_{pi}. \tag{11.10}$$

It is easily verified, that this expression is the same as the one obtained when actually deriving the first-order derivatives.

In MML estimation, it is commonly assumed that the person parameters are independently drawn from a normal distribution. For the Rasch model, the mean of this normal distribution is set equal to zero to make the model identifiable. So the variance σ^2 is the only parameter of this distribution that must be estimated. Also, the normal distribution is an exponential-family model, and if the parameters θ_p are considered missing data, the first-order derivative of the complete-data log-likelihood is

$$\frac{\sum_{p=1}^{P}\theta_p^2 - \sigma^2}{\sigma^3}. \tag{11.11}$$

Using Fisher's identity and setting the first-order derivatives to zero results in the MML estimation equations,

$$s_i = \sum_{p=1}^{P} E\left[\, P_{pi} \,|\, \mathbf{u}_p; \mathbf{b}, \sigma^2\right] \tag{11.12}$$

for $i = 1, \ldots, I$, and

$$\sigma^2 = \frac{1}{P}\sum_{p=1}^{P} E\left[\theta_p^2 \,\middle|\, \mathbf{u}_p; \mathbf{b}, \sigma^2\right]. \tag{11.13}$$

The expectations are expectations with respect to the posterior distribution defined in Equation 11.7. For the Rasch model, by substituting $\beta = \mathbf{b}$ and $\delta = \sigma^2$ in Equation 11.7, the density specializes to

$$\begin{aligned} p(\theta_p|\mathbf{u}_p;\mathbf{b},\sigma^2) &= \frac{p(\mathbf{u}_p|\theta_p,\mathbf{b})g(\theta_p;\sigma^2)}{\int p(\mathbf{u}_p|\theta_p,\mathbf{b})g(\theta_p;\sigma^2)d\theta_p} \\ &= \frac{\left[\prod_{i=1}^{I}\exp(-u_{pi}b_i)\right]\exp(x_p\theta_p)H(\theta,\mathbf{b})g(\theta_p;\sigma^2)}{\left[\prod_{i=1}^{I}\exp(-u_{pi}b_i)\right]\int\exp(x_p\theta_p)H(\theta,\mathbf{b})g(\theta_p;\sigma^2)d\theta_p} \\ &= \frac{\exp(x_p\theta_p)H(\theta,\mathbf{b})g(\theta_p;\sigma^2)}{\int\exp(x_p\theta_p)H(\theta,\mathbf{b})g(\theta_p;\sigma^2)d\theta_p}. \end{aligned} \tag{11.14}$$

Note that the density depends only on the total score x_p and can be written as $p(\theta_p|x_p;\mathbf{b},\sigma^2)$. So the estimation equations can be further simplified to

$$s_i = \sum_{x=1}^{P} n_x E\left[\, P_i \,|\, x; \mathbf{b}, \sigma^2\right], \quad \text{for } i = 1, \ldots, I \quad and \quad \sigma^2 = \sum_{x=1}^{P} n_x E\left[\theta^2 \,\middle|\, x; \mathbf{b}, \sigma^2\right],$$

where n_x is the number of persons obtaining sum score x.

11.2.2 MML Estimation for Dichotomously Scored Items

In the previous section, the Rasch model was used as an example to introduce MML estimation. The example was chosen for explanatory purposes only, because the Rasch is usually estimated using CML. Further, in many applications, a much more flexible model than the Rasch model is needed. In this section, the procedure outlined above will be generalized to models with discrimination and guessing parameters, multidimensional models, applications for incomplete designs, multigroup IRT models, and Bayes modal estimation.

Define the logistic function $\Psi(.)$ as

$$\Psi(x) = \frac{\exp(x)}{1 + \exp(x)}. \tag{11.15}$$

In the 3PLM, the probability of a correct response is given by

$$\Pr\left(U_{pi} = 1\right) = P_{pi} = c_i + (1 - c_i)\Psi(\tau_{pi}) \tag{11.16}$$

with

$$\tau_{pi} = a_i(\theta_p - b_i), \tag{11.17}$$

where $b_i \in \mathbb{R}$, $a_i \in \mathbb{R}^+$, and $c_i \in [0, 1]$ are the location, discrimination, and guessing parameter of item i, respectively (Volume One, Chapter 3). If all guessing parameters are set to zero, the model in Equation 11.16 specializes to the two-parameter logistic model (2PLM). And if, in addition, all discrimination parameters are set equal to one the model specialized to the 1PLM, which is the Rasch model.

The same model can be generalized further to a multidimensional model, where the responses no longer depend on a scalar-valued person parameters θ_p but on D-dimensional vectors of person parameters, $\theta_p = (\theta_{p1}, \ldots, \theta_{pd}, \ldots, \theta_{pD})$. This model generalizes Equation 11.17 to

$$\tau_{pi} = \left(\sum_{d=1}^{D} a_{id}\theta_{pd}\right) - b_i. \tag{11.18}$$

Discrimination parameters a_{id} $(d = 1, \ldots, D)$ represent the relative importance of the various dimensions for a response on item i. In addition, it will be assumed that the person parameters $\theta_{p1}, \ldots, \theta_{pq}, \ldots, \theta_{pQ}$ have a D-variate normal distribution with parameters μ and $\mathbf{\Sigma}$. Takane and de Leeuw (1987) show that if the logistic function $\Psi(\tau_{pi})$ is replaced by the normal cumulative distribution function, $\Phi(\tau_{pi})$, and all guessing parameters are omitted, the resulting model is equivalent with a factor analysis model. Therefore, parameters $a_{i1}, \ldots, a_{iD}$ are often called factor-loadings and the proficiency parameters $\theta_{p1}, \ldots, \theta_{nd}, \ldots, \theta_{nD}$ are viewed as factor scores. As the model is overparameterized, allowing for the rotation of the parameters $a_{i1}, \ldots, a_{iD}$, the model has to be made identifiable. One of many possibilities is setting the mean and the covariance matrix equal to zero and the identity matrix, respectively, and introducing the constraints $a_{id} = 0$, for $i = 1, \ldots, D - 1$ and $d = i + 1, \ldots, D$. Another possibility is treating the parameters of the ability distribution as unknown and setting D item parameters b_i equal to zero while, for $i = 1, \ldots, D$ and $d = 1, \ldots, D$, imposing the restrictions $a_{id} = 1$, if $i = d$, and $a_{id} = 0$, if $i \neq d$.

Another generalization deals with the fact that in many IRT applications not all items are administered to all persons. For instance, in large-scale educational surveys, such as PISA and TIMSS, samples of students respond to subsets of items, often called booklets. Also, calibration designs for item banks used for computerized adaptive testing and equating designs for tests and examinations are examples of so-called incomplete designs. In the following sections, we will show how IRT modeling accommodates to missing data. In doing so, it will be assumed that the mechanism which has caused the missing data is ignorable, that is, this mechanism does not need to be modeled concurrently with the responses model. For more details on ignorability, refer to Mislevey (see Chapter 10) and

Rubin (1976). To account for missing data in the formulas below, we use missing data indicators d_{pi}, which are equal to one if person p responded to item i, and zero otherwise.

The last generalization introduced in this section is replacement of the assumption that all persons are drawn from one population by the assumption of different samples of persons drawn from distinct populations. Usually, it is assumed that each population has a normal ability distribution indexed by a unique parameters. Bock and Zimowski (1997) point out that this generalization, together with the possibility of analyzing incomplete item-administration designs, is required to provide a unified approach to analyzing such problems as differential item functioning, item parameter drift, nonequivalent groups equating, vertical equating, and matrix-sampled educational assessment.

Application of the framework for deriving the MML estimation equations outlined in the previous sections proceeds as follows. The equations for the item parameters are obtained upon setting Equation 11.2 to zero; so explicit expressions are needed for Equation 11.5. The logarithm of the probability of response vector $\mathbf{u}_p$, given the design variables d_{pi}, person parameter $\boldsymbol{\theta}_p$, and item difficulty parameters $\mathbf{a}$, $\mathbf{b}$, and $\mathbf{c}$, can be written as

$$L_c^{(p)}(\boldsymbol{\theta}_p, \mathbf{a}, \mathbf{b}, \mathbf{c}) = \sum_{i=1}^{I} d_{pi}\left(u_{pi} \log P_{ip} + (1 - u_{pi}) \log(1 - P_{pi})\right).$$

In the previous section, we could use the fact that the Rasch model is an exponential-family model to easily derive the complete-data estimation equations. However, the present model does not belong to an exponential family, so we must proceed is a different manner. For any parameter appearing only in the response probability of item i, the first-order derivative is

$$L_c^{(p)\prime} = \frac{d_{pi}(u_{pi} - P_{pi})P'_{pi}}{P_{pi}(1 - P_{pi})}. \tag{11.19}$$

Applying this result to the discrimination, difficulty and guessing parameters gives

$$L_c^{(p)\prime}(a_{id}) = \frac{d_{pi}(u_{pi} - P_{ip})(1 - c_i)\Psi_{pi}(1 - \Psi_{pi})(\theta_{pd} - b_i)}{P_{pi}(1 - P_{pi})}, \tag{11.20}$$

$$L_c^{(p)\prime}(b_i) = \frac{d_{pi}(P_{ip} - u_{pi})(1 - c_i)\Psi_{pi}(1 - \Psi_{pi})a_i}{P_{pi}(1 - P_{pi})}, \tag{11.21}$$

and

$$L_c^{(p)\prime}(c_i) = \frac{d_{pi}(u_{pi} - P_{pi})(1 - \Psi_{pi})}{P_{pi}(1 - P_{pi})}, \tag{11.22}$$

respectively, where Ψ_{pi} is a short-hand notation for $\Psi(\tau_{pi})$. The likelihood equations for the item parameters are found upon inserting these expressions into Equation 11.2 and setting the resulting expressions to zero. Note that as a result of the presence of factor d_{pi} in all three expressions 11.20 through 11.22, the MML equations only involve responses to items that are actually observed.

Unlike the derivation of the likelihood equations for the item parameters, the derivation of the equations for the population parameters can be done exploiting the fact that multivariate normal distributions belong to an exponential family. If the person parameters were observed rather than missing, their mean in population $g = 1, \ldots, G$ would have been sufficient statistics for the parameters μ_g. Taking posterior expectations result in the MML equations

$$\mu_g = \frac{1}{P_g} \sum_{p|g} E\left[\theta_p | \mathbf{u}_p; \mathbf{a}, \mathbf{b}, \mathbf{c}, \mu_g, \boldsymbol{\Sigma}_g\right] \tag{11.23}$$

for $g = 1, \ldots, G$, where P_g is the number of persons sampled from population g and the sum is over all persons sampled from population g.

The MML equation for covariance matrix $\boldsymbol{\Sigma}_g$ can be easily derived following an analogous argument; the equation involves a posterior expectation of the equation that would have been used if the person parameters were observed, that is,

$$\boldsymbol{\Sigma}_g = \frac{1}{P_g} \sum_{p|g} E\left[\theta_p \theta_p^t | \mathbf{u}_p; \mathbf{a}, \mathbf{b}, \mathbf{c}, \mu_g, \boldsymbol{\Sigma}_g\right] - \mu_g \mu_g^t \tag{11.24}$$

for $g = 1, \ldots, G$.

Models with multiple population distributions can be made identifiable by restricting the parameters of one the distributions. The parameters of the other population distributes remain free.

11.2.3 MML Estimation for Polytomously Scored Items

The family of IRT models for polytomously scored items is quite large, and it is beyond the scope of this section to cover every possible model in detail. However, the steps remain the same as for the case of IRT models for dichotomously scored items: (i) derive the complete-data estimation equations treating the person parameters as missing data and (ii) apply Fisher's identity to derive MML estimation equations and standard errors. For some models, the generalization of the procedure is quite trivial. For instance, the sequential model (Tutz, 1990; Chapter 9 in *Handbook of Item Response Theory, Volume One: Models*) can be viewed as an IRT model for dichotomously scored virtual items with missing responses, so the presentation in the previous section directly applies. Likewise, graded response models (Samejima, 1969; Chapter 6 in *Handbook of Item Response Theory, Volume One: Models*) can be viewed as differences between response functions of virtual dichotomously scored items. So the generalization here is also minor, except for the fact that these models require order restrictions on item parameters to assure that all response probabilities are positive. These restrictions, however, do not create fundamental problems.

A third often-used class of IRT models contains the partial credit model (PCM) (Masters, 1982; Chapter 7 in *Handbook of Item Response Theory, Volume One: Models*), the generalized partial credit model (GPCM) (Muraki, 1992; Chapter 8 in *Handbook of Item Response Theory, Volume One: Models*), and the nominal response model (Bock, 1972; Thissen and Cai, Chapter 4 in *Handbook of Item Response Theory, Volume One: Models*). We briefly outline the approach for the GPCM, upon which generalization to the other polytomous models is trivial. Responses to item i are represented by stochastic variables U_{pij}, $j = 0, \ldots, J$, with realization $u_{pij} = 1$, if the response was in category $j = 0, \ldots, J$ and $u_{pij} = 0$ otherwise. It is assumed that all items have the same number of response categories, but the generalization to a different number of response categories is straightforward. According to the GPCM, the probability of a response in category j is

$$P(U_{pij} = 1; \theta_p, a_i, \mathbf{b}_i) = P_{pij} = \frac{\exp\left(j a_i \theta_p - b_{ij}\right)}{1 + \sum_{h=1}^{J} \exp\left(h a_i \theta_p - b_{ih}\right)} \tag{11.25}$$

for $j = 0, \ldots, J$. Parameter b_{i0} is fixed at zero.

Our current parameterization is somewhat different than the usual parameterization of the GPCM, which uses category-bound parameters b^*_{ij} as item parameters. The alternative sets of parameters are related to each other by the transformation $b_{ij} = \Sigma_{h=1}^{j} b^*_{ij}$. Our motivation for the reparameterization is the estimation equations that are both easier to derive and more comprehensible. The PCM is obtained by setting all parameters a_i equal to one, and the nominal response model is obtained replacing ja_i and ha_i by parameters a_{ij} and a_{ih}, respectively.

For convenience, it is assumed that all P persons respond to all I items, allowing for the same logic as the complete-data case in one of the previous sections. The responses are aggregated in a $I(J+1)$-dimensional response vector $\mathbf{u}_p$ with entries $u_{pij}, i = 1, \ldots, I$ and $J = 0, \ldots, J$. The logarithm of the probability of the response vector is

$$\begin{aligned}
\log \sum_{p=1}^{P} p(\mathbf{u}_p; \theta_p, \mathbf{a}, \mathbf{b}) &= \log \prod_{i=1}^{I} \prod_{j=0}^{J} P_{pij}^{u_{pij}} \\
&= \sum_{p=1}^{P} \left(\sum_{i=1}^{I} \sum_{j=1}^{J} j a_i u_{pij} \right) \theta_p - \sum_{i=1}^{I} \sum_{j=1}^{J} \left(\sum_{p=1}^{P} u_{pij} \right) b_{ij} \\
&\quad - H(\theta, \mathbf{a}, \mathbf{b}) \\
&= \sum_{p=1}^{P} x_p \theta_p - \sum_{i=1}^{I} \sum_{j=1}^{J} s_{ij} b_{ij} - H(\theta, \mathbf{a}, \mathbf{b}) \qquad (11.26)
\end{aligned}$$

with $H(\theta, \mathbf{a}, \mathbf{b})$ a function of the parameters only. If the parameters a_i would be known, the model would be an exponential-family model, where the weighted total scores x_p and the item-category scores s_{ij} are a sufficient statistics for the parameters θ_p and b_{ij}, respectively.

The estimation equations for b_{ij} equate the observed and expected sufficient statistics to zero, that is, the complete-data estimation equation is

$$s_{ij} = \sum_{p=1}^{P} P_{pij}. \qquad (11.27)$$

Assume that the person parameters θ_p have a normal distribution with a mean equal to zero and variance equal to σ^2. The MML estimation equations are derived using Fisher's identity by taking expectations of both sides of the equation, so

$$s_{ij} = \sum_{p=1}^{P} E\left[P_{pij} \,\middle|\, \mathbf{u}_p; \mathbf{a}, \mathbf{b}, \sigma^2 \right]. \qquad (11.28)$$

Deriving the equations for discrimination parameters a_i needs some rearranging of the first term in the second line in Equation 11.26:

$$\sum_{p=1}^{P} \left(\sum_{i=1}^{I} \sum_{j=0}^{J} j a_i u_{pij} \right) \theta_p = \sum_{i=1}^{I} \left(\sum_{p=1}^{P} \theta_p \sum_{j=0}^{J} j u_{pij} \right) a_i.$$

It is now clear that if the parameters θ_p are viewed as missing observations and a_i as parameters to be estimated, the model is an exponential-family model again , and the MML estimation equations can be immediately written down using the sufficient statistics of the complete-data model and Fisher's identity. Therefore,

$$\sum_{p=1}^{P}\sum_{j=0}^{J} j u_{pij} E\left[\theta_p \middle| \mathbf{u}_p; \mathbf{a}, \mathbf{b}, \sigma^2\right] = \sum_{p=1}^{P}\sum_{j=0}^{J} j E\left[\theta_p P_{pij} \middle| \mathbf{u}_p; \mathbf{a}, \mathbf{b}, \sigma^2\right]. \tag{11.29}$$

Finally, standard errors from the information matrix can be directly obtained using formula Equation 11.3.

11.2.4 Computation of MML Estimates

Systems of estimation equations are solved simultaneously. In practice, this can be done using the Newton–Raphson algorithm, EM algorithm, or a combination of the two (Bock and Aitkin, 1981). The EM algorithm (Dempster et al., 1977) is a general iterative algorithm for ML estimation in incomplete-data problems. It handles missing data by (i) computing the posterior distribution of the missing values given the observed data, (ii) estimating new parameters given this distribution, (iii) and reestimating the posterior distribution of the missing values assuming the new parameter estimates are correct. This process is iterated until convergence is achieved. The multiple integrals that appear above can be evaluated using adaptive Gauss–Hermite quadrature (Schilling and Bock, 2005). A critical determinant of the success of Gauss–Hermite quadrature is the dimensionality of the latent space, that is, the number of latent variables analyzed simultaneously. Wood et al. (2002) indicate that the maximum number of variables is 10 for adaptive quadrature, five for nonadaptive quadrature and 15 for Monte Carlo integration.

An extension of MML estimation is Bayes modal estimation. This approach is motivated by the fact that item parameter estimates in the 3PLM are sometimes hard to obtain because their parameters being poorly identified by the available data. In such instances, item response functions can be appropriately described by a large number of different item parameter values, and, as a result, the estimates of the three item parameters in the 3PLM are often highly correlated. To obtain "reasonable" and finite estimates, Mislevy (1986) considers two Bayesian approaches entailing the introduction of prior distributions on the item parameters. For the a_i-, b_i-, c_i-parameters, he recommends the use of log-normal, normal, and beta prior distributions, respectively. These distributions are either fixed or estimated along with the other parameters; for details refer to Mislevy (1986).

11.2.5 CML Estimation

CML estimation was developed in the framework of the Rasch (1960) model, and then also applied to its many generalizations, such as the linear logistic test model (LLTM, Fischer, 1983; Janssen, Chapter 13 *in Handbook of Item Response Theory, Volume One: Models*), the PCM (Masters, 1982; Chapter 7 in *Handbook of Item Response Theory, Volume One: Models*), and the rating scale model (Andrich, 1978; Chapter 5 in *Handbook of Item Response Theory, Volume One: Models*). Perhaps the most general model is the one proposed by Kelderman (1984, 1995). His model will be used here to present CML estimation in its general form. The important difference between MML and CML estimation is that the latter is free of assumptions concerning the distribution of the person parameters. This can be seen as follows. The Kelderman model specifies the probability of a response vector as

$$P(\mathbf{u}_p;\theta_p,\mathbf{b}) = \exp(\mathbf{u}_p^t\mathbf{A}\theta_p - \mathbf{u}_p^t\mathbf{B}\mathbf{b})H(\theta_p,\mathbf{b}), \tag{11.30}$$

where $\theta_p = (\theta_{p1},\ldots,\theta_{nD})$ is a vector of the person's ability parameters, $\mathbf{b}$ is a vector of item parameters, and $\mathbf{A}$ and $\mathbf{B}$ are matrices of fixed, integer scoring weights. Further, $H(\theta_p,\mathbf{b})$ is a function of the parameters which does not depend on response vector $\mathbf{u}_p$. This model is an exponential-family model with sufficient statistics $\mathbf{r}_p$ and $\mathbf{s}_p$ defined by $\mathbf{r}_p^t = \mathbf{u}_p^t\mathbf{A}$ and $\mathbf{s}_p^t = \mathbf{u}_p^t\mathbf{B}$, respectively. Let $\mathbb{B}(\mathbf{r})$ stand for the set of all possible response vectors $\mathbf{u}$ resulting in a value $\mathbf{r}$ of the sufficient statistic for θ_p. Conditioning on the sufficient statistic for the ability parameters results in the conditional response probability

$$\begin{aligned} p(\mathbf{u}_p|\mathbf{r}_p,\mathbf{b}) &= \frac{p(\mathbf{u}_p|\theta_p,\mathbf{b})}{\sum_{\mathbb{B}(\mathbf{r})} p(\mathbf{u}_p|\theta_p,\mathbf{b})} \\ &= \frac{\exp(\mathbf{r}_p^t\theta)\exp(-\mathbf{s}_p^t\mathbf{b})H(\theta_p,\mathbf{b})}{\sum_{\mathbb{B}(\mathbf{r})}\exp(\mathbf{r}_p^t\mathbf{A}\theta)\exp(-\mathbf{s}_p^t\mathbf{b})H(\theta_p,\mathbf{b})} \\ &= \frac{\exp(-\mathbf{s}_p^t\mathbf{b})}{\sum_{\mathbb{B}(\mathbf{r})}\exp(-\mathbf{s}_p^t\mathbf{b})}. \end{aligned} \tag{11.31}$$

Note that $p(\mathbf{u}_p|\mathbf{r}_p,\mathbf{b})$ does not depend on ability parameters θ_p, nor on any assumption regarding the distribution of θ_p. The denominator of Equation 11.31 is an elementary symmetric function. Also, the conditional likelihood is an exponential-family model, and therefore estimation of item parameters $\mathbf{b}$ amounts to equating the observed values of the sufficient statistics with their expected values (see, for instance, Andersen, 1980). So, the estimation equations are

$$s_{ij} = E(S_{ij}|\mathbf{r},\mathbf{b}) \tag{11.32}$$

for $i = 1,\ldots,I$ and $j = 1,\ldots,J$, with

$$\begin{aligned} E(S_{ij}|\mathbf{r},\mathbf{b}) &= \sum_p E(S_{pij}|\mathbf{r}_p,\mathbf{b}) \\ &= \sum_p \frac{\sum_{\mathbb{B}(\mathbf{r}_p,ij)}\exp(-\mathbf{s}_p^t\mathbf{b})}{\sum_{\mathbb{B}(\mathbf{r}_p)}\exp(-s_p^t\mathbf{b})}, \end{aligned} \tag{11.33}$$

where $\mathbb{B}(\mathbf{r}_p,ij)$ is the set of all possible response vectors resulting in a score $\mathbf{r}_p$ with $u_{ij} = 1$.

The standard errors can by computed by inverting the information matrix. In the present case, the information matrix is equal to the covariance matrix of the sufficient statistics, that is, the diagonal elements are given by $Var\left[S_{ij}\middle|\mathbf{r},\mathbf{b}\right]$ and the off-diagonal elements are given by $Cov\left[S_{ij}S_{kl}\middle|\mathbf{r},\mathbf{b}\right]$ (see, for instance, Andersen, 1980). Examples for specific models will be given in the next two sections.

11.2.6 CML Estimation for the Rasch Model

For the Rasch model for dichotomously scored items, $\mathbf{u}_p$ is an I-dimensional vector with dichotomous entries u_{pi}, and θ_p is a scalar. The score function $\mathbf{A}$ becomes an I-dimensional vector with all entries equal to one, and the sufficient statistic for θ_p becomes the total score x_p. Further, $\mathbf{b}$ is a vector of I item parameters and $\mathbf{B}$ an $I \times I$ identity matrix. Thus, the CML estimation equations are

$$s_i = \sum_p \frac{\sum_{\mathbb{B}(x,i)} \exp(-\mathbf{u}_p^t \mathbf{b})}{\sum_{\mathbb{B}(x)} \exp(-\mathbf{u}_p^t \mathbf{b})} \tag{11.34}$$

for $i = 1, \ldots, I$, where s_i is the number of correct responses on item i, and $\mathbb{B}(x)$ and $\mathbb{B}(x,i)$ are the sets of all possible response vectors and all possible vectors with item i correct, respectively. The item parameters need a restriction to make the model identifiable. A simple and straightforward way of achieving this is fixing one of the item parameters at zero and discarding the associated estimation equation. However, this is not necessarily an optimal policy. An item parameter does not have a finite estimate if it is answered correctly or incorrectly by all persons. Analogously, an item parameter becomes difficult to estimate if it is answered correctly or incorrectly by a large proportion of the persons, resulting in estimates with large standard errors. Using the parameter of such an item as an identification restriction will lead to large standard errors for all the other item parameters as well. Therefore, a better policy is to choose an identification restriction which involves all item parameters, such as the restriction $\Sigma_{i=1}^{I} b_i = 0$, and to solve the I estimation equations under this restriction.

To gain more insight in the structure of the equations, consider a reparameterization $\varepsilon_i = \exp(-b_i)$. The elementary function of degree x, denoted by γ_x, is a polynomial formed by adding all distinct products of x terms of the I distinct variables ε_i. So for $I = 4$, the functions are

$$\begin{aligned}
\gamma_0 &= 1 \\
\gamma_1 &= \varepsilon_1 + \varepsilon_2 + \varepsilon_3 + \varepsilon_4 \\
\gamma_2 &= \varepsilon_1\varepsilon_2 + \varepsilon_1\varepsilon_3 + \varepsilon_1\varepsilon_4 + \varepsilon_2\varepsilon_3 + \varepsilon_2\varepsilon_4 + \varepsilon_3\varepsilon_4 \\
\gamma_3 &= \varepsilon_1\varepsilon_2\varepsilon_3 + \varepsilon_1\varepsilon_2\varepsilon_4 + \varepsilon_1\varepsilon_3\varepsilon_4 + \varepsilon_2\varepsilon_3\varepsilon_4 \\
\gamma_4 &= \varepsilon_1\varepsilon_2\varepsilon_3\varepsilon_4.
\end{aligned}$$

Further, define $\gamma_{x-1}^{(i)}$ as an elementary function of degree $x - 1$ of all item parameters except ε_i. These functions can be computed very fast and precisely using recursive algorithms (see, for instance, Verhelst et al., 1984).

Defining n_x as the number of persons obtaining sum score x, Equation 11.34 can be written as follows:

$$s_i = \sum_{x=1}^{I} n_x \frac{\varepsilon_i \gamma_{x-1}^{(i)}}{\gamma_x}. \tag{11.35}$$

Note that $\varepsilon_i \gamma_{x-1}^{(i)} / \gamma_x$ is the expectation of a correct response on item i given a total score x, say $E(U_{pi}|x, \mathbf{b})$. As already indicated, the information matrix is the covariance matrix of the sufficient statistics, so its diagonal elements (i,i) are

$$\sum_{x=1}^{I} n_x Var(U_{pi}|x, \mathbf{b}) = \sum_{x=1}^{I} n_x \left[\frac{\varepsilon_i \gamma_{x-1}^{(i)}}{\gamma_x} \left(1 - \frac{\varepsilon_i \gamma_{x-1}^{(i)}}{\gamma_x} \right) \right]$$

and its off-diagonal elements (i, j) are

$$\sum_{x=1}^{I} n_x Cov(U_{pi}U_{pj}|x, \mathbf{b}) = \sum_{x=1}^{I} n_x \left[\frac{\varepsilon_i \varepsilon_j \gamma_{x-2}^{(i,j)}}{\gamma_x} - \frac{\varepsilon_i \gamma_{x-1}^{(i)}}{\gamma_x} \frac{\varepsilon_j \gamma_{x-1}^{(j)}}{\gamma_x} \right].$$

If one of the items is used as in an identification restriction, the rows and columns related to this item are removed, and inversion of the resulting $(I - 1) \times (I - 1)$ matrix produces the covariance matrix of the parameter estimates. If the identification restriction sets the sum of the item parameters to zero, a generalized inverse can be used (see, for instance, Campbell and Meyer, 1991).

11.2.7 CML Estimation for the PCM

The CML estimation equations for the PCM are relatively straightforward generalizations of the equations for the Rasch model. For the PCM model for polytomously scored items, $\mathbf{u}_p$ is an IJ-dimensional vector with entries u_{pij}, and the person parameter θ_p is a scalar. (Note that it proves convenient to define $\mathbf{u}_p$ somewhat differently then for the GPCM.) The score function $\mathbf{A}$ becomes a IJ-dimensional vector with entry $(i - 1)J + j$ equal to j, for $i = 1, \ldots, I$ and $J = 1, \ldots, J$, such that the sufficient statistic for θ_p becomes the total score $x_p = \sum_{i,j} j u_{ij}$. Further, $\mathbf{b}$ is a vector of IJ item parameters and $\mathbf{B}$ an $IJ \times IJ$ identity matrix. The CML estimation equations are given by

$$s_{ij} = \sum_p \frac{\sum_{\mathbb{B}(x,ij)} \exp(-\mathbf{u}_p^t \mathbf{b})}{\sum_{\mathbb{B}(x)} \exp(-\mathbf{u}_p^t \mathbf{b})}, \tag{11.36}$$

where s_{ij} is the number of responses in category j of item i, and $\mathbb{B}(x)$ and $\mathbb{B}(x,i)$ are the sets of all possible response vectors and all possible vectors with a response in category j on item i, respectively.

To gain some more insight in the structure of the equations, consider a reparameterization $\varepsilon_{ij} = \exp(-b_{ij})$ and define $\varepsilon_{i0} = 1$ for $i = 1, \ldots, I$. Then, the elementary function of degree x, is defined as

$$\gamma_x = \sum_{\{j_1,\ldots,j_I\}} \prod_{i=1}^{I} \varepsilon_{ij_i},$$

where $\{j_1, \ldots, j_I|x\}$ is the set of all permutations of I integers $j_1, \ldots, j_I$ $(j_i < J)$ such that $x = \sum_i j_i$. As an example, the choice of $I = 3$, $J = 2$ and $x = 2$ results in

$$\begin{aligned}\gamma_2 &= \varepsilon_{12}\varepsilon_{20}\varepsilon_{30} + \varepsilon_{10}\varepsilon_{22}\varepsilon_{30} + \varepsilon_{10}\varepsilon_{20}\varepsilon_{32} + \varepsilon_{11}\varepsilon_{21}\varepsilon_{30} + \varepsilon_{11}\varepsilon_{20}\varepsilon_{31} + \varepsilon_{10}\varepsilon_{21}\varepsilon_{31} \\ &= \varepsilon_{12} + \varepsilon_{22} + \varepsilon_{32} + \varepsilon_{11}\varepsilon_{21} + \varepsilon_{11}\varepsilon_{31} + \varepsilon_{21}\varepsilon_{31}.\end{aligned}$$

Further, define $\gamma_{x-j}^{(i)}$ as an elementary function of degree $x - j$ of all item parameters except ε_{ij}, and n_x as the number of persons obtaining a sum score x. Then, Equation 11.34 can be written as follows:

$$s_{ij} = \sum_{x=1}^{I} n_x \frac{\varepsilon_{ij} \gamma_{x-j}^{(i)}}{\gamma_x}.$$

Note the similarity with the equation for the Rasch model given in Equation 11.35. Also, the information matrix resembles the information matrix for the CML estimates in the Rasch model. Its element $((i,j),(k,l))$ is

$$\sum_{x=1}^{I} n_x \left[\frac{\varepsilon_{ij}\varepsilon_{kl}\gamma_{x-j-k}^{(i,k)}}{\gamma_x} - \frac{\varepsilon_{ij}\gamma_{x-j}^{(i)}}{\gamma_x} \frac{\varepsilon_{kl}\gamma_{x-l}^{(k)}}{\gamma_x} \right], \quad \text{if } i \neq k$$

$$\sum_{x=1}^{I} n_x \left[\frac{\varepsilon_{ij}\gamma_{x-j}^{(i)}}{\gamma_x} - \frac{\varepsilon_{ij}\gamma_{x-j}^{(i)}}{\gamma_x} \frac{\varepsilon_{ij}\gamma_{x-j}^{(i)}}{\gamma_x} \right], \quad \text{if } i = k, j = l$$

$$\sum_{x=1}^{I} n_x \left[-\frac{\varepsilon_{ij}\gamma_{x-j}^{(i)}}{\gamma_x} \frac{\varepsilon_{il}\gamma_{x-l}^{(i)}}{\gamma_x} \right], \quad \text{if } i = k,\ j \neq l.$$

Finally, again similar to the case of the Rasch model, a restriction on the item parameters needs to be imposed, say a linear restriction on the parameters b_{ij} (which translates to a product restriction on the ε_{ij} parameters). The estimation equations and information matrix must account for this restriction, as outlined previously.

11.3 Ability Estimation

Once the item parameters and, when applicable, the population parameters are estimated, person parameters are estimated treating the item and population parameters as fixed constants. The theory in this chapter ends with a description of two classes of approaches to estimating person parameters: Frequentist approaches (maximum likelihood and weighted maximum likelihood) and Bayesian approaches (based on either the expectation or the mode of the posterior distribution).

11.3.1 Frequentist Approaches

If the item parameters are known, the likelihood function only involves θ as a parameter, and finding the equations for the maximization of the likelihood with respect to θ becomes relatively easy. For instance, if we consider the class of models defined by Equation 11.30 and, in addition, allow $\mathbf{A}$ to have real-valued entries, we encompass a broad class of IRT models, including the Rasch model, 2PLM, PCM, and GPCM. For this class, the ML estimation equations boil down to setting statistic $\mathbf{u}_p^t\mathbf{A}$ equal to its expected value $E(\mathbf{u}_p^t\mathbf{A}|\theta_p,\mathbf{b})$. For instance, for the Rasch model with incomplete data, the sufficient statistic is the sum score $x_p = \Sigma_i d_{pi} u_{pi}$, and it directly follows that the ML estimation equations are given by $x_p = \Sigma_i d_{pi} P_{pi}$. Similarly, for the GPCM given in Equation 11.25, we obtain

$$\sum_{i=1}^{I} d_{pi} a_i \sum_{j=1}^{J} j u_{pij} = \sum_{i=1}^{I} d_{pi} a_i \sum_{j=1}^{J} j P_{pij}.$$

The second-order derivatives, needed for the Newton–Raphson algorithm and computation of standard errors are

$$\sum_{i=1}^{I} d_{pi}a_i \left[\sum_{j=1}^{J} j^2 u_{pij} - \left(\sum_{j=1}^{J} j P_{pij} \right)^2 \right].$$

For the 3PLM, this approach does not work because the model does not belong to the class defined by Equation 11.30. Therefore, we proceed analogously to the derivation of Equation 11.19, and start with a general expression for the first-order derivative with respect to θ,

$$L'(\theta) = \sum_{i=1}^{I} \frac{d_{pi}(u_{pi} - P_{pi})P'_{pi}}{P_{pi}(1 - P_{pi})}. \tag{11.37}$$

Substituting the expression for P'_{pi} for the 3PLM results in the estimation equation

$$L'(\theta) = \frac{d_{pi}a_{pi}(u_{pi} - P_{ip})(1 - c_i)\Psi_{pi}(1 - \Psi_{pi})}{P_{pi}(1 - P_{pi})} = 0. \tag{11.38}$$

These equations can be solved by using the Newton–Raphson algorithm. Upon convergence, the second-order derivatives can be used to compute standard errors.

Most ML estimators, including those of θ, are biased. Lord (1983) derived an expression for the bias in the case of the 3PLM and showed that its size is of the order I^{-1}. Based on this finding, Warm (1989) derived a so-called weighted maximum-likelihood (WML) estimate, which is less biased than the traditional ML estimate. It removes the first-order bias term by weighting the likelihood function. Consequently, the estimation equations become

$$L'(\theta) + w'(\theta) = 0,$$

where $w'(\theta)$ is the first-order derivative of a weight-function with respect to θ (for the exact expressions for the weight function, refer to Warm, 1989). Penfield and Bergeron (2005) generalized the WML procedure to the GPCM.

11.3.2 Bayesian Approaches

Formula 11.7 is a general expression for the posterior density of person parameter θ_p given response vector $\mathbf{u}_p$, item parameters β and, if applicable, covariates $\mathbf{x}_p$ and population parameters δ. Usually, it is assumed that the population distribution is normal with parameters μ_g and $\mathbf{\Sigma}_g$, but if covariates are present, the population distribution may be specified as normal with the regression coefficients and the covariance matrix of the residuals as parameters. The important difference between Bayesian and frequentist approaches is that the population distribution functions as a prior distribution of θ_p in the Bayesian approach, while the estimate of θ_p solely depends on the observed response vector in the frequentist approach. This difference should guide the choice between the two approaches. In a high-stakes test, it is reasonable and fair that the outcome be dependent only on the performance that was actually delivered and should not involve information on the population from which the person emanates or on background characteristics. If, on the other hand, the ability level of a person is estimated in a survey study, without consequences for the person, the precision of the estimate can be augmented using all information available. How this information is incorporated in the estimate can illustrated best for maximum a posteriori (MAP) estimation. Again, consider the posterior density given in Equation 11.7. The denominator on its right-hand side does not depend on θ_p, so it plays no role in finding the maximum of the posterior. The logarithm of the posterior is proportional to the sum of two terms, that is,

$$\log p(\theta_p|\mathbf{u}_p, \mathbf{x}_p; \beta, \delta) \propto \log p(\mathbf{u}_p|\theta_p, \beta) + \log g(\theta_p; \delta, \mathbf{x}_p). \tag{11.39}$$

Thus, the maximum will be some sort of weighted average of the first (likelihood) and the second term (prior). To make this more concrete, we substitute the multidimensional 3PLM in Equations 11.16 and 11.18. Let $\mathbf{a}_i$ be a column vector with elements a_{id}, for $d = 1, \ldots, D$. In vector notation, the first-order derivatives are

$$\sum_{i=1}^{I} v_i \mathbf{a}_i - \mathbf{\Sigma}^{-1} (\theta_p - \mu) \tag{11.40}$$

with

$$v_i = \frac{[P_{pi} - c_i][u_{pi} - P_{pi}]}{(1 - c_i) P_{pi}}.$$

Equated to zero, the first term of 11.40, that is, the sum over the items, is the ML equation maximizing the likelihood given the response data, while the second term has the effect of moving θ_p toward μ. Such an estimator is called a "shrinkage" estimator. Though Bayesian terminology is used here, the method can also be viewed as a straightforward form of ML estimation. Therefore, standard errors are usually computed using the matrix of second-order derivatives. This matrix is given by

$$\sum_{i=1}^{I} \mathbf{a}_i \mathbf{a}_i^t w_i - \mathbf{\Sigma}^{-1}$$

with

$$w_i = \frac{[1 - P_{pi}][P_{pi} - c_i]\left[c_i u_i - P_{pi}^2\right]}{P_{pi}^2 (1 - c_i)^2}.$$

An alternative to MAP estimation is expected a posteriori (EAP) estimation. EAP estimates are point estimates defined as the posterior expectation given the response vector and possible covariates, $E(\theta_p|\mathbf{u}_p, \mathbf{x}_p; \beta, \delta)$, while in the Bayesian tradition, the confidence in these estimates is expressed as the posterior variance, $Var(\theta_p|\mathbf{u}_p, \mathbf{x}_p; \beta, \delta)$.

11.4 Example

A small simulated example is presented to give an impression of the outcome of the estimation procedure. The data were simulated using the Rasch model. The item parameters

TABLE 11.1
Example of Parameter Estimates

Item	True Value	CML 1PLM b_i	CML 1PLM $se(b_i)$	MML 1PLM b_i	MML 1PLM $se(b_i)$	MML 2PLM b_i	MML 2PLM $se(b_i)$	a_i	$se(a_i)$
1	−1.50	−1.48	0.058	−1.48	0.072	−1.48	0.083	1.08	0.108
2	−1.25	−1.31	0.077	−1.30	0.097	−1.28	0.103	1.02	0.140
3	−1.00	−1.04	0.054	−1.03	0.068	−1.02	0.073	1.04	0.104
4	−0.75	−0.69	0.070	−0.68	0.092	−0.67	0.095	1.04	0.138
5	−0.50	−0.64	0.052	−0.62	0.065	−0.64	0.071	1.18	0.109
6	−0.25	−0.19	0.068	−0.18	0.091	−0.19	0.099	1.31	0.158
7	0.25	0.26	0.051	0.27	0.065	0.27	0.066	1.11	0.101
8	0.50	0.42	0.068	0.42	0.091	0.42	0.091	1.06	0.135
9	0.75	0.78	0.053	0.78	0.067	0.80	0.072	1.14	0.106
10	1.00	0.97	0.072	0.98	0.093	0.95	0.093	0.97	0.130
11	1.25	1.24	0.057	1.23	0.070	1.19	0.072	0.96	0.095
12	1.50	1.68	0.082	1.68	0.103	1.63	0.115	0.96	0.141
Population				μ_g	$se(\mu_g)$	μ_g	$se(\mu_g)$		
1	0.00			0.02	0.064	0.02	0.060		
2	0.00			−0.02	0.077	−0.02	0.078		
Population				σ_g	$se(\sigma_g)$	σ_g	$se(\sigma_g)$		
1	1.00			1.04	0.038	0.97	0.064		
2	1.00			1.08	0.053	1.00	—		

of the 12 items used are given in Table 11.1, in the column labeled "True Value." The parameters of 2000 persons were drawn from a standard normal distribution. For 1000 persons, responses to all items were generated; for the other 1000 persons, only the responses to the items with odd indices were generated. So the data were generated according to an incomplete design, with half of the persons responding to 12 items and the other half of the persons only to six items.

Three estimation runs were made, the first for the Rasch model in combination with CML estimation, the second for the Rasch model in combination with MML estimation, and the third for the 2PLM in combination with MML. In the two MML estimation runs, it was assumed that the persons responding to 12 and six items were drawn from two different populations. The parameters were estimated under the restriction that the b_i-parameters summed to zero. For the 2PLM, the restriction of the variance of one of the two populations equal to one was added. The results are displayed in Table 11.1.

The following findings deserve attention. All three estimates of the b_i parameters were close to their generating values. The lower and upper bound of an approximate 95% confidence interval around the estimates can be determined by subtracting and adding twice the standard error from the estimate, respectively. Most of the estimates were within these intervals. Note that the standard errors for the odd items were smaller than for the even items, which is explained by the fact that the odd items were responded to by twice as many persons. Note further that the standard errors for the CML estimates were smaller than those for the MML estimates. This difference is explained by the fact that the CML standard errors should be interpreted as standard deviations of sampling distributions conditional on the persons' total scores. It is known that conditioning on total scores creates

TABLE 11.2

Example of Person Parameter Estimates

	Population 1					
	ML		WML		EAP	
Score	θ	se(θ)	θ	se(θ)	θ	se(θ)
0			−3.692	1.560	−2.082	0.666
1	−2.78	1.076	−2.442	0.956	−1.666	0.626
2	−1.92	0.819	−1.768	0.788	−1.292	0.598
3	−1.33	0.721	−1.254	0.710	−0.948	0.578
4	−0.85	0.673	−0.811	0.670	−0.622	0.564
5	−0.42	0.650	−0.401	0.650	−0.308	0.557
6	−0.00	0.644	−0.005	0.644	0.001	0.555
7	0.41	0.651	0.392	0.650	0.310	0.558
8	0.85	0.675	0.804	0.671	0.625	0.565
9	1.33	0.724	1.251	0.713	0.951	0.579
10	1.92	0.823	1.770	0.792	1.298	0.599
11	2.79	1.080	2.453	0.961	1.674	0.629
12			3.714	1.567	2.093	0.668
	Population 2					
Score	θ	se(θ)	θ	se(θ)	θ	se(θ)
0			−3.177	1.681	−1.673	0.764
1	−2.05	1.155	−1.781	1.079	−1.116	0.730
2	−1.00	0.949	−0.915	0.941	−0.598	0.712
3	−0.15	0.909	−0.152	0.909	−0.096	0.707
4	0.70	0.952	0.615	0.943	0.409	0.715
5	1.76	1.158	1.490	1.083	0.934	0.737
6			2.894	1.687	1.503	0.774

linear restrictions on the sample space, which are reflected in the size of the standard errors (see, for instance, Glas, 1988). The last two columns give the estimates of the a_i-parameters and their standard errors obtained using the 2PLM. The Rasch model was used to generate the data, so all discrimination parameters should be close to one. It can be seen that they were all within their 95% confidence interval. The MML standard errors obtained using the 2PLM were larger than the standard errors using the 1PLM, a known phenomenon in statistics. Finally, the bottom lines of the table give the estimates of the parameters of the population distributions. These estimates were also well within their 95% confidence intervals.

Table 11.2 gives the estimates of the person parameters using ML, WML, and EAP estimation. The MML estimates of the item and population parameters of the Rasch model were used to compute the estimates. Since the total score is a sufficient statistic for the person parameters, persons with the same total score obtained the same estimate (i.e., as long as they were administered to same items). The ML estimates of a zero and perfect total score do not exist. Related to this is the phenomenon, the ML estimates had an outward bias. For the EAP estimates, the opposite occurred: Both the population distributions had estimates close to zero, and the estimates of the person parameters were "shrunk" toward this a priori value. Note that the magnitudes of the ML and WML standard errors and the posterior standard deviations for the EAP estimates were quite different. This difference is a direct consequence of their conceptual difference. Finally, note that both the standard errors and posterior standard deviations were much larger for the population of persons responding only to half of the items. Standard errors and posterior standard deviations can be seen as local measures of test reliability, and, obviously, a longer test is more reliable than a shorter one.

References

Andersen, E. B. 1980. *Discrete Statistical Models with Social Sciences Applications.* Amsterdam: North-Holland.

Andrich, D. 1978. A rating formulation for ordered response categories. *Psychometrika, 43,* 561–573.

Bock, R. D. 1972. Estimating item parameters and latent ability when responses are scored in two or more nominal categories. *Psychometrika, 37,* 29–51.

Bock, R. D. and Aitkin, M. 1981. Marginal maximum likelihood estimation of item parameters: An application of an EM-algorithm. *Psychometrika, 46,* 443–459.

Bock, R. D. and Zimowski, M. F. 1997. Multiple group IRT. In W. J. van der Linden and R. K. Hambleton (eds.), *Handbook of Modern Item Response Theory.* (pp. 433–448). New York: Springer.

Campbell, S. L. and Meyer, C. D. 1991. *Generalized Inverses of Linear Transformations.* New York: Dover Publications.

Dempster, A. P., Laird, N. M., and Rubin, D. B. 1977. Maximum likelihood from incomplete data via the EM algorithm (with discussion). *Journal of the Royal Statistical Society, Series B, 39,* 1–38.

Efron, B. 1977. Discussion on maximum likelihood from incomplete data via the EM algorithm (by A. Dempster, N. Laird, and D. Rubin). *Journal of the Royal Statistical Society, Series B, 39,* 1–38.

Fischer, G. H. 1981. On the existence and uniqueness of maximum likelihood estimates in the Rasch model. *Psychometrika, 46,* 59–77.

Fischer, G. H. 1983. Logistic latent trait models with linear constraints. *Psychometrika, 48,* 3–26.

Fischer, G. H. and Scheiblechner, H. H. 1970. Algorithmen und Programme für das probabilistische Testmodell von Rasch. *Psychologische Beiträge, 12,* 23–51.

Glas, C. A. W. 1988. The derivation of some tests for the Rasch model from the multinomial distribution. *Psychometrika, 53,* 525–546.

Glas, C. A. W. 1999. Modification indices for the 2-pl and the nominal response model. *Psychometrika, 64,* 273–294.

Kelderman, H. 1984. Loglinear RM tests. *Psychometrika, 49,* 223–245.

Kelderman, H. 1995. Loglinear multidimensional item response model for polytomously scored items. In G. H. Fischer and I. W. Molenaar (eds.), *Rasch Models: Foundations, Recent Developments and Applications.* (pp. 287–304). New York: Springer.

Lord, F. M. 1983. Unbiased estimators of ability parameters, their variance, and of their parallel-forms reliability. *Psychometrika, 48,* 233–245.

Louis, T. A. 1982. Finding the observed information matrix when using the EM algorithm. *Journal of the Royal Statistical Society, Series B, 44,* 226–233.

Masters, G. N. 1982. A Rasch model for partial credit scoring. *Psychometrika, 47,* 149–174.

Mislevy, R. J. 1984. Estimating latent distributions. *Psychometrika, 49,* 359–381.

Mislevy, R. J. 1985. Estimation of latent group effects. *Journal of the American Statistical Association, 80,* 993–997.

Mislevy, R. J. 1986. Bayes modal estimation in item response models. *Psychometrika, 51,* 177–195.

Muraki, E. 1992. A generalized partial credit model: Application of an EM algorithm. *Applied Psychological Measurement, 16,* 159–176.

Neyman, J. and Scott, E. L. 1948. Consistent estimates, based on partially consistent observations. *Econometrica, 16,* 1–32.

Penfield, R. D. and Bergeron, J. M. 2005. Applying a weighted maximum likelihood latent trait estimator to the generalized partial credit model. *Applied Psychological Measurement, 29,* 218–233.

Rasch, G. 1960. *Probabilistic Models for Some Intelligence and Attainment Tests.* Copenhagen: Danish Institute for Educational Research.

Rigdon S. E. and Tsutakawa, R. K. 1983. Parameter estimation in latent trait models. *Psychometrika, 48,* 567–574.

Rubin, D. B. 1976. Inference and missing data. *Biometrika, 63,* 581–592.

Samejima, F. 1969. Estimation of latent ability using a pattern of graded scores. *Psychometrika Monograph Supplement,* No. 17.

Schilling, S. and Bock, R. D. 2005. High-dimensional maximum marginal likelihood item factor analysis by adaptive quadrature. *Psychometrika, 70,* 533–555.

Takane, Y. and de Leeuw, J. 1987. On the relationship between item response theory and factor analysis of discretized variables. *Psychometrika, 52,* 393–408.

Thissen D. 1982. Marginal maximum likelihood estimation for the one-parameter logistic model. *Psychometrika, 47,* 175–186.

Tutz, G. 1990. Sequential item response models with an ordered response. *British Journal of Mathematical and Statistical Psychology, 43,* 39–55.

Verhelst, N. D., Glas, C. A. W., and van der Sluis, A. 1984. Estimation problems in the Rasch model: The basic symmetric functions. *Computational Statistics Quarterly, 1,* 245–262.

Warm, T. A. 1989. Weighted likelihood estimation of ability in item response theory. *Psychometrika, 54,* 427–450.

Wood, R., Wilson, D. T., Gibbons, R. D., Schilling, S. G., Muraki, E., and Bock, R. D. 2002. TEST-FACT: *Test Scoring, Item Statistics, and Item Factor Analysis.* Chicago, IL: Scientific Software International, Inc.

Wright, B. D. and Panchapakesan, N. 1969. A procedure for sample-free item analysis. *Educational and Psychological Measurement, 29,* 23–48.

12

Expectation Maximization Algorithm and Extensions

Murray Aitkin

CONTENTS

12.1 Introduction

The expectation maximization (EM) algorithm (Dempster et al. 1977) has had an enormous effect on statistical modeling and analysis. In this chapter, we discuss only its direct application to item response models, including multilevel and mixture models. In this model family, it has been fundamental in providing computationally feasible methods for the maximum-likelihood analysis of large-scale psychometric tests with item models more complex than the Rasch model. What is even more striking is that nearly all, if not all, of these extensions have required or used this algorithm.

There are three separate reasons for this widespread successful use of EM. The first is the nature of modern psychometric model: They all involve *latent abilities* for which the test items are *indicators*. The EM algorithm is formulated in terms of *incomplete data*, involving latent or unobserved variables which, if observed, would make the analysis much simpler. EM is therefore tailored for just the kinds of models required for psychometric testing.

The second reason is the remarkably tractable and robust convergence properties of the algorithm: It converges from any starting point, however badly chosen, and increases the maximized-likelihood monotonically, so that no additional measures are required, as for other algorithms, to stop the iteration sequence diverging or decreasing. It can, however, converge very slowly and may converge to a local maximum, so varying the starting values, or other methods, are needed to identify the global maximum.

The third reason is the relative simplicity of the algorithm, requiring in particular only first derivatives of the log-likelihood to obtain the maximum-likelihood estimates (MLEs), and frequently using standard maximum-likelihood procedures with small adjustments which are easily programmed. The information matrix, however, requires additional programming and additional data functions.

We describe the general form of the algorithm in the next section, give a very simple illustration with incomplete responses, and then a series of examples relevant to item response models—the two-component normal mixture model, the two-level normal variance component model, and the Rasch model. In subsequent sections, we describe its application to successively more complex psychometric models. The final section describes Bayesian extensions of the algorithm and their implementation using Markov chain Monte Carlo (MCMC) procedures.

12.2 General Form of the EM Algorithm

We adopt, with minor changes, the notation of Dempster, Laird, and Rubin (DLR). Let $\mathbf{x}$ be the unobserved *complete data* from the model $f(\mathbf{x};\phi)$, and $\mathbf{y}$ the observed *incomplete data*. The relation between the distributions of $\mathbf{x}$ and $\mathbf{y}$, and therefore their likelihoods, can be written as

$$g(\mathbf{y};\phi) = \int_{\mathcal{X}(\mathbf{y})} f(\mathbf{x};\phi)d\mathbf{x}$$

where, given the observed data $\mathbf{y}$, g is the *observed-data likelihood* and f is the *complete-data likelihood*. The integral transformation $\mathcal{X}(\mathbf{y})$ is quite general, representing any kind of constraint on the $\mathbf{x}$ space which reduces the data information in the observable data $\mathbf{y}$. We assume this transformation does not depend on the model parameters ϕ.

Then taking logs and differentiating under the integral sign (assuming this is permissible), we have, for the observed-date score function $\mathbf{s}_y(\phi)$ and Hessian matrix $\mathbf{H}_y(\phi)$,

$$\mathbf{s}_y(\phi) = \frac{\partial \log g}{\partial \phi} = \frac{\frac{\partial}{\partial \phi}\int_{\mathcal{X}(\mathbf{y})} f(\mathbf{x};\phi)d\mathbf{x}}{\int_{\mathcal{X}(\mathbf{y})} f(\mathbf{x};\phi)d\mathbf{x}}$$

$$= \frac{\int_{\mathcal{X}(\mathbf{y})} \mathbf{s}_x(\phi) f(\mathbf{x};\phi)d\mathbf{x}}{\int_{\mathcal{X}(\mathbf{y})} f(\mathbf{x};\phi)d\mathbf{x}}$$

$$= \int_{\mathcal{X}(\mathbf{y})} \mathbf{s}_x(\phi) f(\mathbf{x}; \phi \mid \mathbf{y}) d\mathbf{x}$$

$$= \mathrm{E}[\mathbf{s}_x(\phi) \mid \mathbf{y}]$$

$$\mathbf{H}_y(\phi) = \frac{\partial^2 \log g}{\partial \phi \partial \phi'} = \frac{\partial}{\partial \phi'} \int_{\mathcal{X}(\mathbf{y})} \mathbf{s}_x(\phi) f(\mathbf{x}; \phi \mid \mathbf{y}) d\mathbf{x}$$

$$= \int_{\mathcal{X}(\mathbf{y})} \left[\mathbf{H}_x(\phi) f(\mathbf{x}; \phi \mid \mathbf{y}) + \mathbf{s}_x(\phi) \frac{\partial f(\mathbf{x}; \phi \mid \mathbf{y})}{\partial \phi'} \right] d\mathbf{x}$$

$$= \int_{\mathcal{X}(\mathbf{y})} \left[\mathbf{H}_x(\phi) f(\mathbf{x}; \phi \mid \mathbf{y}) + \mathbf{s}_x(\phi) \{\mathbf{s}_x(\phi)' - \mathrm{E}[\mathbf{s}_x(\phi)' \mid \mathbf{y}\} f(\mathbf{x}; \phi \mid \mathbf{y}) \right] d\mathbf{x}$$

$$= \int_{\mathcal{X}(\mathbf{y})} \left[\mathbf{H}_x(\phi) + \mathbf{s}_x(\phi) \{\mathbf{s}_x(\phi)' - \mathrm{E}[\mathbf{s}_x(\phi)' \mid \mathbf{y}\} f(\mathbf{x}; \phi \mid \mathbf{y}) \right] d\mathbf{x}$$

$$= \mathrm{E}[\mathbf{H}_x(\phi) \mid \mathbf{y}] + \mathbf{C}[\mathbf{s}_x(\phi) \mid \mathbf{y}]$$

So,

1. The observed-data score is equal to the conditional expectation of the complete-data score.
2. The observed-data Hessian is equal to the conditional expectation of the complete-data Hessian plus the conditional covariance of the complete-data score, where the conditioning is with respect to the observed data.

Remarkably, these results hold regardless of the form of incompleteness, as long as the conditions above hold. So algorithms like Gauss–Newton can be implemented directly from the complete-data form of the analysis, provided the conditional distribution of the complete-data score and Hessian given the observed data can be evaluated. These results do not, however, guarantee convergence without step-length or other controls on these algorithms.

Write the complete-date log-likelihood as $\ell(\phi)$, and define the function as

$$Q(\phi^* \mid \phi) = E[\ell(\phi^* \mid \mathbf{y}, \phi)]$$

An iteration of the EM algorithm, $\phi^{[p]} \to \phi^{[p+1]}$, is defined by the alternating steps:

1. E step: Given $\phi^{[p]}$, compute $Q(\phi^* \mid \phi^{[p]})$
2. M step: Set $\phi^{[p+1]}$ equal to the maximizer of $Q(\phi^* \mid \phi^{[p]})$

Thus, the complete-data log-likelihood is alternately maximized and replaced by its conditional expectation given the previous MLE. As stated earlier, it is even more remarkable that this algorithm converges monotonically from any starting point.

12.2.1 Example: Randomly Missing Observations

We draw a sample $\mathbf{x}$ of size n from a population modeled by a normal distribution $N(\mu, \sigma^2)$. However, only the first m observations $\mathbf{y}$ are actually recorded, the last $n - m$ being lost by a random process.

Since the lost observations are missing at random, maximum likelihood can be achieved directly from the observed random subsample of m observations, so

$$\hat{\mu} = \bar{y} = \sum_{i=1}^{m} y_i/m, \quad \hat{\sigma}^2 = \sum_{i=1}^{m} (y_i - \bar{y})^2/m$$

Alternatively, we can use the EM algorithm.

The complete-date log-likelihood (omitting constants) is

$$\ell(\mu, \sigma) = -n \log \sigma - \frac{1}{2\sigma^2} \sum_{i=1}^{n} (x_i - \mu)^2$$

Given the current estimates, $\mu^{[p]}$ and $\sigma^{[p]}$, the E step replaces the log-likelihood for the unobserved terms in $\mathbf{x}$ by its conditional expectation given the current parameter estimates. So for the observed terms for $i = 1, \ldots, m$, $x_i = y_i$, and for the missing terms for $i = m + 1, \ldots, n$, each $(x_i - \mu)^2$ is replaced by $\mathrm{E}[(x - \mu)^2 \mid \mathbf{y}, \mu^{[p]}, \sigma^{[p]}] = \sigma^{[p]2}$. So the conditional expected log-likelihood is

$$\begin{aligned} \mathrm{E}[\ell(\mu, \sigma) \mid \mathbf{y}, \mu^{[p]}, \sigma^{[p]}] &= -n \log \sigma - \frac{1}{2\sigma^2} \left[\sum_{i=1}^{m} (y_i - \mu)^2 + \sum_{i=m+1}^{n} \sigma^{[p]2} \right] \\ &= -n \log \sigma - \frac{1}{2\sigma^2} \left[\sum_{i=1}^{m} (y_i - \mu)^2 + (n - m)\sigma^{[p]2} \right] \end{aligned}$$

and the MLEs at the next iteration are

$$\mu^{[p+1]} = \bar{y}, \ \sigma^{[p+1]2} = \left[\sum_{i=1}^{m} (y_i - \bar{y})^2 + (n - m)\sigma^{[p]2} \right] \Big/ n$$

At convergence, the second equation gives

$$\begin{aligned} \hat{\sigma}^2 &= \left[\sum_{i=1}^{m} (y_i - \bar{y})^2 + (n - m)\hat{\sigma}^2 \right] \Big/ n \\ &= \sum_{i=1}^{m} (y_i - \bar{y})^2 \Big/ m \end{aligned}$$

It is easily, if tediously, verified that the conditional expectation of the complete-data score, and the conditional expectation of the complete-data Hessian plus the conditional covariance of the complete-data score are given by

$$E[\mathbf{s}_x(\mu) \mid \mathbf{y}] = \frac{1}{\sigma^2}\sum_{i=1}^{m}(y_i - \mu)$$

$$E[\mathbf{s}_x(\sigma) \mid \mathbf{y}] = \frac{n}{\sigma} + \frac{1}{\sigma^3}\left[\sum_{i=1}^{m}(y_i - \mu)^2 + (n-m)\sigma^2\right]$$

$$E[\mathbf{H}_x(\mu,\mu) \mid \mathbf{y}] = -\frac{n}{\sigma^2}$$

$$E[\mathbf{H}_x(\mu,\sigma) \mid \mathbf{y}] = -\frac{2}{\sigma^3}\left[\sum_{i=1}^{m}(y_i - \mu)\right]$$

$$E[\mathbf{H}_x(\sigma,\sigma) \mid \mathbf{y}] = \frac{n}{\sigma^2} - \frac{3}{\sigma^4}\left[\sum_{i=1}^{m}(y_i - \mu)^2 + (n-m)\sigma^2\right]$$

$$\mathbf{C}[\mathbf{s}_x(\mu,\mu) \mid \mathbf{y}] = \frac{n-m}{\sigma^2}$$

$$\mathbf{C}[\mathbf{s}_x(\mu,\sigma) \mid \mathbf{y}] = 0$$

$$\mathbf{C}[\mathbf{s}_x(\sigma,\sigma) \mid \mathbf{y}] = \frac{2(n-m)}{\sigma^2}$$

which give the observed-data score and Hessian.

This application may seem trivial, but it is a widespread practice to use multiple imputation for randomly missing response values in models of all kinds, with unnecessary loss of precision since randomly missing values do not contribute to the observed-data likelihood.

12.3 Applications Related to Item Response Models

12.3.1 Two-Component Normal Mixture Model

The observed data $\mathbf{y}$ are a random sample from the model

$$f(y_i;\theta) = pf_1(y_i;\mu_1,\sigma_1) + (1-p)f_2(y_i;\mu_2,\sigma_2)$$

with

$$f_j(y_i;\mu_j,\sigma_j) = \frac{1}{\sqrt{2\pi}\sigma_j}\exp\left[-\frac{1}{2\sigma_j^2}(y_i - \mu_j)^2\right]$$

and $\theta = (\mu_1, \mu_2, \sigma_1, \sigma_2, p)$. The likelihood is an awkward product across the observations of sums across the two components. Maximum likelihood is greatly facilitated by introducing a *latent component identifier* Z_i which, if observed, would convert the model to a two-group normal model with group-specific means and variances. We define

$$Z_i = \begin{cases} 1 & \text{if observation } i \text{ is from component 1} \\ 0 & \text{if observation } i \text{ is from component 2} \end{cases}$$

and give Z_i the Bernoulli distribution with parameter p. The complete data $\mathbf{x}$ are then $\mathbf{y}$ and $\mathbf{Z}$, and the complete-data likelihood, log-likelihood, score, and Hessian are (in an obvious shorthand)

$$L_x(\theta) = \prod_{i=1}^{n} [f_1(y_i; \mu_1, \sigma_1)\, p]^{Z_i}\, [f_2(y_i; \mu_2, \sigma_2)\,(1-p)]^{1-Z_i}$$

$$\ell_x(\theta) = \sum_{i=1}^{n} [Z_i \log f_1(y_i; \mu_1, \sigma_1) + (1 - Z_i) \log f_2(y_i; \mu_2, \sigma_2)$$

$$+ Z_i \log p + (1 - Z_i) \log(1 - p)]$$

$$\mathbf{s}_x(\theta) = \sum_{i=1}^{n} \left[Z_i\, \mathbf{s}_{x_i}(\mu_1, \sigma_1) + (1 - Z_i)\, \mathbf{s}_{x_i}(\mu_2, \sigma_2) + \frac{Z_i - p}{p(1-p)} \right]$$

$$\mathbf{H}_x(\theta) = \sum_{i=1}^{n} \left[Z_i\, \mathbf{H}_{x_i}(\mu_1, \sigma_1) + (1 - Z_i)\, \mathbf{H}_{x_i}(\mu_2, \sigma_2) - Z_i/p^2 - (1 - Z_i)/(1-p)^2 \right]$$

The common appearance of Z_i in these terms greatly simplifies the observed-data score and Hessian computations, which require the conditional expectations of Z_i given the observed data. From Bayes' theorem,

$$\begin{aligned}
\mathrm{E}[Z_i \mid \mathbf{y}, \theta] &= \Pr[Z_i = 1 \mid \mathbf{y}, \theta] \\
&= f_1(\mathbf{y}; \theta \mid Z_i = 1) \Pr[Z_i = 1] / f(\mathbf{y}; \theta) \\
&= \frac{f_1(\mathbf{y}; \theta \mid Z_i = 1) \Pr[Z_i = 1]}{f_1(\mathbf{y}; \theta \mid Z_i = 1) \Pr[Z_i = 1] + f_2(\mathbf{y}; \theta \mid Z_i = 0) \Pr[Z_i = 0]} \\
&= Z_i^* \\
\mathbf{s}_y(\theta) &= \mathrm{E}[\mathbf{s}_x(\theta) \mid \mathbf{y}] \\
&= \sum_{i=1}^{n} [Z_i^* \mathbf{s}_{x_i}(\mu_1, \sigma_1) + (1 - Z_i^* (\mathbf{s}_{x_i}(\mu_2, \sigma_2) + Z_i^*/p - (1 - Z_i^*)/(1-p)]
\end{aligned}$$

Formally, we write the observed-data score as

$$s_y(\mu_1) = \frac{1}{\sigma_1^2} \sum_{i=1}^{n} Z_i^* (y_i - \mu_1)$$

$$s_y(\sigma_1) = -\frac{\sum_{i=1}^{n} Z_i^*}{\sigma_1} + \frac{1}{\sigma_1^3} \sum_{i=1}^{n} Z_i^* (y_i - \mu_1)^2$$

$$s_y(\mu_2) = \frac{1}{\sigma_1^2} \sum_{i=1}^{n} (1 - Z_i^*)(y_i - \mu_2)$$

$$s_y(\sigma_2) = -\frac{\sum_{i=1}^n (1 - Z_i^*)}{\sigma_2} + \frac{1}{\sigma_2^3}\sum_{i=1}^n (1 - Z_i^*)(y_i - \mu_2)^2$$

$$s_y(p) = \sum_{i=1}^n Z_i^*/p - \sum_{i=1}^n (1 - Z_i^*)/(1 - p)$$

The score equations are simple weighted versions of the complete-date score equations, with the component membership probabilities as weights. The EM algorithm alternates between evaluating the parameter estimates given the weights, and evaluating the weights given the parameter estimates. At convergence, (asymptotic) variances of the estimates are given by the observed-date Hessian,

$$\begin{aligned}
\mathbf{H}_y(\theta) &= \mathrm{E}[\mathbf{H}_x(\theta) \mid \mathbf{y}] + \mathbf{C}[\mathbf{s}_x(\theta) \mid \mathbf{y}] \\
&= \sum_{i=1}^n [Z_i^* \mathbf{H}_{x_i}(\mu_1, \sigma_1) + (1 - Z_i^*)\mathbf{H}_{x_i}(\mu_2, \sigma_2) - Z_i^*/p^2 - (1 - Z_i^*)/(1 - p)^2] \\
&\quad + \mathrm{E}\left[\sum_{i=1}^n [Z_i - Z_i^*] \times \sum_{j=1}^n [Z_j - Z_j^*]\right] \\
&\quad \times \left[\mathbf{s}_{x_i}(\mu_1, \sigma_1) - \mathbf{s}_{x_i}(\mu_2, \sigma_2) + \frac{1}{p(1-p)}\right] \\
&\quad \times \left[\mathbf{s}'_{x_j}(\mu_1, \sigma_1) - \mathbf{s}'_{x_j}(\mu_2, \sigma_2) + \frac{1}{p(1-p)}\right] \\
&= \sum_{i=1}^n [Z_i^* \mathbf{H}_{x_i}(\mu_1, \sigma_1) + (1 - Z_i^*)\mathbf{H}_{x_i}(\mu_2, \sigma_2) - Z_i^*/p^2 - (1 - Z_i^*)/(1 - p)^2] \\
&\quad + \sum_{i=1}^n Z_i^*[1 - Z_i^*]\left[\mathbf{s}_{x_i}(\mu_1, \sigma_1) - \mathbf{s}_{x_i}(\mu_2, \sigma_2) + \frac{1}{p(1-p)}\right] \\
&\quad \times \left[\mathbf{s}'_{x_i}(\mu_1, \sigma_1) - \mathbf{s}'_{x_i}(\mu_2, \sigma_2) + \frac{1}{p(1-p)}\right]
\end{aligned}$$

Writing the Hessian formally at length, we have (with the expected Hessian terms on the left, the covariance terms on the right)

$$H_y(\mu_1, \mu_1) = -\frac{\sum_{i=1}^n Z_i^*}{\sigma_1^2} + \frac{\sum_{i=1}^n Z_i^*(1 - Z_i^*)(y_i - \mu_1)^2}{\sigma_1^4}$$

$$H_y(\mu_1, \sigma_1) = -\frac{2}{\sigma_1^3}\sum_{i=1}^n Z_i^*(y_i - \mu_1) + \frac{\sum_{i=1}^n Z_i^*(1 - Z_i^*)(y_i - \mu_1)}{\sigma_1^2}\left[-\frac{1}{\sigma_1} + \frac{(y_i - \mu_1)^2}{\sigma_1^3}\right]$$

$$H_y(\mu_1, \mu_2) = -\frac{\sum_{i=1}^{n} Z_i^*(1 - Z_i^*)(y_i - \mu_1)(y_i - \mu_2)}{\sigma_1^2 \sigma_2^2}$$

$$H_y(\mu_1, \sigma_2) = \frac{\sum_{i=1}^{n} Z_i^*(1 - Z_i^*)(y_i - \mu_1)}{\sigma_1^2} \left[\frac{1}{\sigma_2} - \frac{(y_i - \mu_2)^2}{\sigma_2^3} \right]$$

$$H_y(\mu_1, p) = \frac{\sum_{i=1}^{n} Z_i^*(1 - Z_i^*)(y_i - \mu_1)}{p(1 - p)\sigma_1^2}$$

$$H_y(\sigma_1, \sigma_1) = \frac{\sum_{i=1}^{n} Z_i^*}{\sigma_1^2} \left[1 - \frac{3(y_i - \mu_1)^2}{\sigma_1^2} \right] + \frac{\sum_{i=1}^{n} Z_i^*(1 - Z_i^*)}{\sigma_1^2} \left[-1 + \frac{(y_i - \mu_1)^2}{\sigma_1^2} \right]^2$$

$$H_y(\sigma_1, \mu_2) = -\frac{\sum_{i=1}^{n} Z_i^*(1 - Z_i^*)(y_i - \mu_2)}{\sigma_2^2} \left[-\frac{1}{\sigma_1} + \frac{(y_i - \mu_1)^2}{\sigma_1^3} \right]$$

$$H_y(\sigma_1, p) = \frac{\sum_{i=1}^{n} Z_i^*(1 - Z_i^*)}{p(1 - p)} \left[-\frac{1}{\sigma_1} + \frac{(y_i - \mu_1)^2}{\sigma_1^3} \right]$$

$$H_y(\mu_2, \mu_2) = -\frac{\sum_{i=1}^{n}(1 - Z_i^*)}{\sigma_2^2} + \frac{\sum_{i=1}^{n} Z_i^*(1 - Z_i^*)(y_i - \mu_2)^2}{\sigma_2^4}$$

$$H_y(\mu_2, \sigma_2) = -\frac{2}{\sigma_2^3} \sum_{i=1}^{n}(1 - Z_i^*)(y_i - \mu_2) + \frac{\sum_{i=1}^{n} Z_i^*(1 - Z_i^*)(y_i - \mu_2)}{\sigma_2^2} \left[-\frac{1}{\sigma_2} + \frac{(y_i - \mu_2)^2}{\sigma_2^3} \right]$$

$$H_y(\mu_2, p) = \frac{\sum_{i=1}^{n} Z_i^*(1 - Z_i^*)(y_i - \mu_2)}{p(1 - p)\sigma_2^2}$$

$$H_y(\sigma_2, \sigma_2) = \frac{\sum_{i=1}^{n}(1 - Z_i^*)}{\sigma_2^2} \left[1 - \frac{3(y_i - \mu_2)^2}{\sigma_2^2} \right] + \frac{\sum_{i=1}^{n} Z_i^*(1 - Z_i^*)}{\sigma_2^2} \left[-1 + \frac{(y_i - \mu_2)^2}{\sigma_2^2} \right]^2$$

$$H_y(\sigma_2, p) = \frac{\sum_{i=1}^{n} Z_i^*(1 - Z_i^*)}{p(1 - p)} \left[-\frac{1}{\sigma_2} + \frac{(y_i - \mu_2)^2}{\sigma_2^3} \right]$$

$$H_y(p, p) = -\sum_{i=1}^{n} \left[\frac{Z_i^*}{p^2} + \frac{1 - Z_i^*}{(1 - p)^2} \right] + \frac{\sum_{i=1}^{n} Z_i^*(1 - Z_i^*)}{[p(1 - p)]^2}$$

An important point is that while the expected Hessian terms involve only the linear and quadratic terms in y, the covariance matrix of the score involves third and fourth powers of y. Departures of the response model from the assumed normality in each component will therefore affect the observed-data Hessian and the stated precision of the model parameters.

The effort involved in evaluating the covariance terms is considerable, even in this simple model, and increases rapidly in more complex models. Three important features of the Hessian are visible: First, the off-diagonal expected Hessian terms in the mean and standard deviations are zero at the MLEs. Second, the covariance terms are never zero, except by accident in the (μ_1, p) and (μ_2, p) terms. And third, the covariance terms always have signs opposite to the expected Hessian terms. Thus, information is reduced by the presence of the unobserved Z_i, and therefore, not surprisingly, all estimated parameters are positively correlated.

An important point which is not visible is that the mixture likelihood is far from Gaussian in the model parameters. This explains the success and importance of EM in being able to reach a maximum of the likelihood without step-length correction. In their study of composite Gauss–Newton/EM alternatives to EM in this model, Aitkin and Aitkin (1996) found that not much was gained (a reduction of 30% at best in time) by switching to Gauss–Newton after five EM iterations. In fact, divergence occurred rapidly if it was tried at an early stage of the iterations. They commented (p. 130):

> ... we formed the impression of a traveller following the narrow EM path up a hazardous mountain with chasms on all sides. When in sight of the summit, the GN path leapt to the top, but when followed earlier, it caused repeated falls into the chasms, from which the traveller had to be pulled back onto the EM track.

12.3.2 Two-Level Variance Component Model

This type of model is directly relevant to item response models. We consider in detail the normal variance-component model, and comment on its generalization to the exponential family. We change the notation slightly for the two-level model.

We are given a clustered sample of n_p observations from cluster p, for $p = 1, \ldots, P$, with total sample size $n = \sum_p n_p$. The population is more homogeneous within than among clusters, and we observe a response variable Y_{pi} on observation i in cluster p, with explanatory variables $\mathbf{x}_{pi}$. The explanatory variables may be at the observation level or the cluster level, or both.

The model assumes that the observations within a cluster are conditionally independent given a cluster random effect z_p shared by each observation in the cluster, which we further assume to be normally distributed: $z_p \sim N(0, \sigma_A^2)$. The zero mean is a convention, as a nonzero mean will be confounded with any intercept term in the mean model for Y.

We assume further that Y given z is also normal, with homogeneous variance across the clusters:

$$Y_{pi} \mid z_p \sim N(\beta'\mathbf{x}_{pi} + z_p, \sigma^2)$$

The joint distribution of all the observations is then multivariate normal, with a block-diagonal covariance matrix with variances $\sigma^2 + \sigma_A^2$ and an intracluster correlation of $\rho = \sigma_A^2/(\sigma^2 + \sigma_A^2)$. Maximum likelihood in this model is most efficiently achieved by the recursive scoring algorithm of LaMotte (1972), rediscovered and implemented by Longford (1987). While the likelihood can be written directly in closed form through the marginal multivariate normal distribution, MLE is greatly facilitated, as with the finite mixture model, by treating the cluster random effect as an unobserved latent variable. The complete-date log-likelihood, omitting constants, is

$$\ell = \sum_{p=1}^{P} \left[-\log \sigma_P - \frac{z_p^2}{2\sigma_A^2} - \sum_{i=1}^{n_p} \left\{ \log \sigma + \frac{1}{2\sigma^2}(y_{pi} - \beta'\mathbf{x}_{pi} - z_p)^2 \right\} \right]$$

which is linear in z_p and z_p^2. The E step then consists of replacing these terms by their conditional expectations $\widetilde{z_p}$ and $\widetilde{z_{p2}}$, given the observed data and the current parameter estimates, and the M step consists of maximizing this expected complete-data log-likelihood with respect to the model parameters to give new estimates. From the normal/normal model structure, the conditional distribution of z_p given the response is also normal, with straightforward conditional mean $\widetilde{z_p}$ and variance $\widetilde{z_{p2}} + \widetilde{z_p}^2$.

The E step is thus closed form, as is the M step, which is a slightly adjusted (ridge) multiple regression for $\boldsymbol{\beta}$, and the z_p^2 in the score equation for σ_A replaced by $\widetilde{z_{p2}} + \widetilde{z_p}^2$. We do not give details of the score and information matrix computations.

However, there is a different formulation of the complete-date model which gives a different form of the EM algorithm; see the discussions by Aitkin and Hinde in Meng and Van Dyk (1997, pp. 546–550). We may write instead

$$Y_{pi} \mid z_p \sim N(\boldsymbol{\beta}'\mathbf{x}_{pi} + \sigma_A z_p, \sigma^2), \quad z_j \sim N(0, 1)$$

The observed-data model is unchanged, but now σ_A is a regression coefficient instead of a standard deviation. The complete-date log-likelihood is

$$\ell = \sum_{p=1}^{P} \left[-\frac{z_p^2}{2} - \sum_{i=1}^{n_p} \left\{ \log \sigma + \frac{1}{2\sigma^2}(y_{pi} - \boldsymbol{\beta}'\mathbf{x}_{pi} - z_p\sigma_A)^2 \right\} \right]$$

in which the first term can be omitted, since it does not involve the model parameters. The conditional distribution of z_p is unchanged and we still need the conditional expectation of z_p^2, but all parameters are estimated from the ridge regression. The computational effect of this is that the speed of convergence of the second form of the algorithm is unaffected by the value of σ_A, while that of the first is very fast when σ_A is large but very slow when it is small (relative to σ). Aitkin and Hinde discuss a further generalization in which σ_A appears as both a regression coefficient and a variance, allowing an optimal weighting of the two contributions.

The normal variance-component model has quite restrictive assumptions and can be extended to give great generality. We do not give comprehensive details of the changes to the algorithms, or to software implementations.

12.4 Variance-Component Model Extensions

12.4.1 Cluster-Specific Variances

We can modify the Y model to allow for cluster-specific variances:

$$Y_{pi} \mid z_p \sim N(\boldsymbol{\beta}'\mathbf{x}_{pi} + \sigma_A z_p, \sigma_p^2), \quad z_j \sim N(0, 1)$$

This changes the contributions of the explanatory variables to the SSP matrix, which now have to be inversely weighted by the cluster variances, which are themselves estimated from the within-cluster residuals. The conditional distribution of the z_p is also weighted across the clusters by the inverse cluster variances.

12.4.2 Exponential-Family Response Distributions

For exponential-family response distributions other than the normal, the normal random effect distribution is not conjugate, so the observed-data likelihood is not analytic and requires numerical integration to evaluate for likelihood-ratio tests. However, this does not prevent the formulation and fitting of generalized linear model extensions of the variance-component model.

Suppose we have a two-stage random sample with response variable Y_{pi} with $p = 1, \ldots, P$, $i = 1, \ldots, n_p$, and $\sum_p n_p = n$, from an exponential-family distribution $f(y; \theta)$ with canonical parameter θ and mean μ, and covariates $\mathbf{X} = (\mathbf{x}_{pi})$, related to μ through a link function $\eta_{pi} = g(\mu_{pi})$ with linear predictor $\eta_{pi} = \beta' \mathbf{x}_{pi}$ (for exponential families, see Chapter 4). To incorporate the upper-level random effect z_p, we adopt the second approach above and extend the linear predictor, conditional on z_p, to $\eta_{pi} = \beta' \mathbf{x}_{pi} + \sigma_A z_p$. The likelihood is then

$$L(\beta, \sigma_A) = \prod_{p=1}^{P} \left\{ \int \left[\prod_{i=1}^{n_p} f(y_{pi}; \beta, \sigma_A \mid z_p) \right] \phi(z_p) \, dz_p \right\}$$

Approximating the integration by Gauss–Hermite quadrature, we replace the integral over the normal z_p by the finite sum over K Gaussian quadrature mass-points Z_k with masses π_k:

$$L(\beta, \sigma_A) \doteq \prod_{p=1}^{P} \left\{ \sum_{k=1}^{K} \left[\prod_{i=1}^{n_p} f(y_{pi}; \beta, \sigma_A \mid Z_k) \right] \pi_k \right\}$$

The likelihood is thus (approximately) that of a finite mixture of exponential-family densities with known mixture proportions π_k and known mass points Z_k, with the linear predictor for the (p, i)th observation in the k-th mixture component being

$$\eta_{pik} = \beta' \mathbf{x}_{pi} + \sigma_A Z_k$$

Thus, Z_k becomes another observable variable in the regression, with regression coefficient σ_A.

The EM algorithm in this case is a version of that given above for the mixture of normal model, but is specialized as the mixture proportions are not free parameters but fixed by the discrete normal random effect model. The score terms are simple weighted sums of those for an ordinary (single-level) GLM: The E step updates the weights given the current parameter estimates, and the M step reestimates the parameters given the current weights. A full description is given in Aitkin et al. (2005, Chapters 7 through 9) and Aitkin et al. (2009, Chapters 7 through 9).

A further extension allows for general random effect distributions, which are also estimated by nonparametric maximum likelihood, as discrete distributions on a finite number of mass points; the number, locations, and weights of these mass points are all estimated as a finite mixture maximum-likelihood problem. Aitkin (1999) gave a general treatment of this approach, which is also given in the book references above.

12.4.3 Multilevel Models

Normal variance-component models with more than two levels of normal random effects are now widely available in many packages; for instance, Longford's early package VARCL had a nine-level version using the LaMotte recursive algorithm.

Extending the EM algorithm to more than two levels for general exponential-family models by Gaussian quadrature proved much more difficult, but was achieved by Vermunt (2004), and is implemented up to four levels in the Latent Gold package. The GLLAMM extension in Stata (Skrondal and Rabe-Hesketh, 2004) can fit the same models using numerical derivatives and Gauss–Newton methods, but is much slower.

The additional levels allow for clustered survey designs with several levels of nesting, as well as longitudinal or repeated-measures designs, though these may need different covariance structures from the simple intraclasss covariance matrix. A detailed discussion of the formulation and fitting of models for the complex National Assessment of Educational Progress (NAEP) educational surveys is given in Aitkin and Aitkin (2011).

We now deal with item response models as special cases of the two-level exponential family variance-component model.

12.5 Rasch Item Response Model

The Rasch model (Rasch 1960), also known as the one-parameter logistic (PL) model, is the simplest psychometric model, and one of the most widely used (Volume One, Chapter 3). All the other models are extended versions of the Rasch model, or are closely related to it. The model is a two-level Bernoulli variance-component model, though it does not look like one. We observe binary item responses U_{pi} by person $p = 1, \dots, P$ on item $i = 1, \dots, I$. The probability of a correct response q_{pi} is given by the logistic function

$$\begin{aligned}\pi_{pi} &= \Pr[U_{pi} = 1; \theta_p, b_i] \\ &= \frac{\exp(\theta_p - b_i)}{1 + \exp(\theta_p - b_i)} \\ &= \frac{1}{1 + \exp[-(\theta_p - b_i)]}\end{aligned}$$

where θ_p is the ability of person p and b_i is the difficulty parameter for item i. The item responses, both among persons and within a person, are assumed to be independent, so the likelihood function is

$$L(\{b_i\}, \{\theta_p\}) = \prod_{p=1}^{P} \prod_{i=1}^{I} \pi_{pi}^{u_{pi}} (1 - \pi_{pi})^{1-u_{pi}}$$

In this fixed-effect form of the model, it is well-known that maximum likelihood does not give consistent estimates of the model parameters, because as either the number of items or the number of persons increases, so does the dimension of the corresponding parameter space (see Chapter 9).

However, because of the exponential form of the model, the sufficient statistics for each parameter dimension are just the marginal sums of the U_{pi} across the other dimension. This provides a conditional likelihood analysis of the classic Fisherian kind as in the two-way contingency table with both marginal totals random.

This was exploited by Andersen (1973), who was able to compute systematically the likelihood contributions for the large number of permutations of the binary indicators within the table giving the same marginal totals. This provided the standard analysis of the Rasch model for many years; for details, see Glas (see Chapter 11). However, it was not possible to use the conditioning approach with the one-parameter probit model instead of the logistic model.

An alternative approach, discussed much later, was through the random-effect version of this model, as a one-parameter variance-component model:

$$\Pr[U_{pi} = 1; b_i \mid \theta_p] = \frac{\exp(\theta_p - b_i)}{1 + \exp(\theta_p - b_i)}$$

$$\theta_p \sim N(0, \sigma_A^2)$$

Following this development, a natural question arose of the relation of the variance-component analysis to the conditional likelihood analysis. This was resolved in a major paper by Lindsay et al. (1991) which showed their essential identity.

12.6 Two-Parameter Models

The normal ogive or two-parameter probit (2PP) model was proposed by Lord (1952) in test theory, though the same model was used much earlier in psychophysical scaling:

$$\Pr[U_{pi} = 1; \theta_p, a_i, b_i] = \Phi[a_i(\theta_p - b_i)]$$

where $\Phi(\cdot)$ is the normal cdf and a_i is the discrimination parameter for item i. Lawley (1943) had derived the MLEs for this model in its psychophysics application. The latent ability was frequently taken to be given by the scaled observed total-item test score.

Birnbaum (1968), in Lord and Novick (1968), worked with the 2PL model (Volume One, Chapter 2)

$$\Pr[U_{pi} = 1; \theta_p, a_i, b_i] = \frac{\exp[a_i(\theta_p - b_i)]}{1 + \exp[a_i(\theta_p - b_i)]}$$

$$= \frac{1}{1 + \exp[-a_i(\theta_p - b_i)]}$$

For this model, as for the 2PP, there are no sufficient statistics. Birnbaum gave a maximum-likelihood procedure for alternately estimating the item parameters and the person ability parameters, but as with the Rasch model these estimates are not consistent, and the alternating algorithm is unstable. This inherent difficulty, as with the Rasch model, means that additional information about the person parameters is needed to allow consistent estimation of the item parameters.

This can be achieved by a probability model for the ability parameters, usually taken as $\theta_p \sim N(0,1)$. (A variance parameter is not identifiable in this model unless an item discrimination parameter is fixed.) The responses need then to be only conditionally independent given the ability random effect but are marginally positively correlated, a more reasonable assumption than complete independence. The model is then written conditionally as

$$q_{pi} = \Pr[U_{pi} = 1; a_i, b_i \mid \theta_p] = \frac{\exp[a_i(\theta_p - b_i)]}{1 + \exp[a_i(\theta_p - b_i)]}$$

$$\theta_p \sim N(0,1)$$

The likelihood is then

$$L(\{a_i\}\{b_i\}) = \prod_{p=1}^{P} \int \left[\prod_{i=1}^{I} \pi_{pi}^{u_{pi}} (1 - \pi_{pi})^{1-u_{pi}} \right] \phi(\theta)\, d\theta$$

Full maximum likelihood was achieved by Bock and Lieberman (1970) with the 2PP model, but this was computationally limited to 10 items. The EM algorithm used by Bock and Aitkin (1981) takes advantage of the structure of incomplete data, but here as in the two-level variance-component model, there is more than one way of formulating the complete-data model.

Alternative formulations are as a classical single-factor model with the normally distributed test variables Z_{pi} replaced by binary indicators of them, U_{pi}, or as a set of independent probit regression models in which the standard normal regressee variable θ_p is unobserved. Since the classical normal factor model already required substantial computation, the set of probit regressions was much simpler, and was adopted. The numerical integration over the normal ability variable was expressed as a finite mixture with mixing probabilities given by the discrete normal ordinates. The algorithm was quickly extended to the 2PL model, for which the weights needed were simpler.

12.7 Extension of EM to More Complex Models

12.7.1 3PL Model

Birnbaum (1968) also introduced the 3PL model, including a guessing parameter c_i for each item i representing the probability of correctly answering the item for an individual with very low ability. The model is

$$q_{pi} = \Pr[U_{pi} = 1; a_i, b_i, c_i \mid \theta_p] = c_i + (1 - c_i)\frac{\exp[a_i(\theta_p - b_i)]}{1 + \exp[a_i(\theta_p - b_i)]}$$

$$\theta_p \sim N(0,1)$$

(Volume One, Chapter 2). It is clear that the possibility of guessing reduces the information about the item difficulty and discrimination parameters, so this model is more difficult to estimate than the 2PL. Considerable experience with this model has underlined its estimation difficulties, and it is common practice to impose tight prior distributions on the guessing parameters for model identification. (This clearly begs the question of how to specify the centers of the tight priors, and is essentially equivalent to treating them as known.) This point is discussed further in the next section.

If the c_i are known, fitting the 3PL model reduces to fitting the 2PL model to the test data and rescaling the fitted probabilities:

$$\widehat{q_{pi}} = c_i + (1 - c_i)\frac{\exp[\hat{a_i}(\theta_p - \hat{b_i})]}{1 + \exp[\hat{a_i}(\theta_p - \hat{b_i})]}$$

12.7.2 Regression Extensions

The 2PL and 3PL models do not include any covariates—they are pure psychometric models, in which only the item parameters appear. It was very soon clear that the models needed to be extended to include covariates, initially group indicators, later extended to general regression models. These extensions were assumed to operate at the ability level, in the form $\theta_p \sim N(\boldsymbol{\beta}'\mathbf{x_p}, 1)$. As described earlier, the incorporation of the regression function in the linear predictor does not require computational extensions—pure psychometric models are special cases of the general model with a null regression model. Computational speed and storage limitations restricted for many years the use of large-scale regression models.

One important point which was overlooked in these extensions was the possibility that the regression function operated not at the person ability level, but at the response probability level (see Aitkin and Aitkin, 2011, Section 3.2, for a discussion of this point). This change of view allows alternative versions of the 2PL and 3PL models to be considered. For instance, for the 2PL model, we have

$$\Pr[U_{pi} = 1; a_i, b_i, \boldsymbol{\beta} \mid \theta_p] = \frac{\exp[a_i(\theta_p - b_i) + \boldsymbol{\beta}'\mathbf{x}_p]}{1 + \exp[a_i(\theta_p - b_i) + \boldsymbol{\beta}'\mathbf{x}_p]}$$

$$\theta_p \sim N(0, 1)$$

In the econometric literature, this model is called the 2PL model, and the first version is called the MIMIC (multiple indicators, multiple causes) model (Jöreskog and Goldberger, 1975). For the Rasch model with $a_i = 1$ for all i, or if the regression model is null, the model versions are identical. Otherwise, they are not identical.

The implications of the two versions are quite significant psychologically. The second version has a homogeneous ability distribution, in the sense that ability does not vary by sex, ethnicity, or any other person demographic variables. What does vary with these variables is the probability of answering the items correctly, for persons with the same ability. In the first version, the responses are conditionally independent of the demographic variables, given the person ability.

For the Rasch model, we are not able to distinguish these interpretations. For this model, this lack of distinction can be expressed more generally, by having the regression in both the ability and the response probability:

$$\text{logit}\,[U_{pi}; b_i, \boldsymbol{\beta} \mid \theta_p] = \theta_p - b_i + \lambda \cdot \boldsymbol{\beta}'\mathbf{x}_p$$

$$\theta_p \sim N((1 - \lambda) \cdot \boldsymbol{\beta}'\mathbf{x}_p, 1)$$

Here λ can take any positive or negative value without changing the model. Even more generally, the regressions may be different in the ability and response probability:

$$\text{logit}\,[U_{pi}; b_i, \beta \mid \theta_p] = \theta_p - b_i + \beta_1' \mathbf{x}_p$$
$$\theta_p \sim N(\beta_2' \mathbf{x}_p, 1)$$

it is only the sum $\beta_1 + \beta_2$ that can be estimated.

In the two-parameter model, these possibilities are not equivalent and can be discriminated, given sufficient data. This opens the possibility of discriminating the effects on achievement of school environment from person ability in this model.

12.7.3 Mixtures of Item Response Models

In early work with item response models, the item parameters were estimated from a large sample from the population for which the items were relevant. The item parameters were often then taken to be known, that is, set at their MLEs, and the explanatory variable regression fitted under this assumption.

However, the test items are often used in testing heterogeneous population, for example, over a wide child age range, where the difficulty and discrimination parameters of the items might vary across the different age subpopulation. If the subpopulation structure is known, then separate models can be fitted to each subpopulation, and the need for different item parameters can be assessed by the usual (asymptotic) statistical tests. If the nature of the heterogeneity is unknown, then it has to be represented as a latent variable, and fitted as a mixture of item response models.

This second level of latency further reduces the effective sample size, and may require larger samples for identification of the mixture components. MLE in such models can be achieved by an additional EM loop to estimate the class membership probabilities for each person in each latent class. Nested EM algorithms can be expected to run very slowly; Bayesian alternatives to this approach are discussed below.

We do not give full details of specific mixture models. An exposition of mixed Rasch models was given by Rost (1997). An extensive review of many applications of these models, and other extensions of them, are given in the contributed chapters in von Davier and Carstensen (2006). Aitkin and Aitkin (2011) give a detailed discussion of mixed 2PL and MIMIC models and their use in representing student engagement and guessing in the NAEP.

12.8 More than One Ability Dimension

Many tests (e.g., those in the NAEP) use items chosen from distinct domains, for which different latent abilities are thought to be important. The models described above have only one dimension of ability θ, but there is no difficulty in extending this to D dimensions (Volume One, Chapters 11 and 12). We model the latent ability as a multivariate normal distribution: $\theta \sim N(\mathbf{Bx}, \mathbf{P})$, where $\mathbf{B}$ is the matrix of regression coefficients of the explanatory variables on the multiple abilities and $\mathbf{P}$ is the *correlation* matrix of the D abilities; the variances are all fixed at 1.

If we label the ability dimensions or scales by d, with corresponding scale items having parameters indexed by d, we may write the multidimensional (MIMIC) model for the item responses U_{pid} as

$$\begin{aligned} \text{logit}\,\pi_{pid} &= a_{id}(\theta_{pd} - b_{id}) \\ \theta_{pd} &\sim N(\beta_d' \mathbf{x}, 1) \\ \text{corr}\,(\theta_{pd}, \theta_{pd'}) &= \rho_{dd'} \end{aligned}$$

The likelihood function is then

$$L(\{a_i\}, \{b_i\}, \{\beta_d\}, \{\rho_{dd'}\}) = \prod_{p=1}^{P} \int \cdots \int \left[\prod_{i=1}^{I} \pi_{pid}^{U_{pid}} (1 - \pi_{pid})^{1-U_{pid}} \right] \phi(\theta) d\theta$$

If the correlations between scales $\rho_{dd'}$ were zero, the likelihood would reduce to a set of independent 2PL models across the different dimensions. The nonzero case is complex, the difficulty being the estimation of the correlations. The information for the estimation of $\rho_{dd'}$ comes from only those students who take items from both scale dimensions d and d'. In the NAEP, where students receive a small random sample of items spread across multiple dimensions, this may provide very limited information about the correlations, especially as the discrete version of the multivariate normal will require approximating grids in D dimensions. Aitkin and Aitkin (2011) give further discussion.

12.9 More than Two Response Categories

All the preceding models are for binary responses, yes/no or correct/incorrect. More than two response categories require some version of the multinomial rather than the Bernoulli response model. These may be ordered or unordered. For the models for this case, see Chapters 4 through 10 in *Handbook of Item Response Theory, Volume One: Models*. We give details only of the model setup—all the models developed since 1981 use versions of the Bock–Aitkin EM algorithm.

Unordered categories arise, for example, in multiple choice items where the distractors are not in any natural order, or they may be thought to be ordered, but this is to be assessed from the unordered analysis. The natural model is the multinomial form of the 2PL: For a K-category unordered response Y_{pik} on item i, the probability π_{pik} that a person with ability θ_p gives response category k on item i is given by the multinomial logit model:

$$\begin{aligned} \Pr[Y_{pik} = 1 \mid \theta_p] &= q_{pik} \\ \log\left(\frac{\pi_{pik}}{\pi_{piK}}\right) &= a_{ik}(\theta_p - b_{ik}) \\ \theta_p &\sim N(\beta_k' \mathbf{x}_p, 1) \end{aligned}$$

Here the last category K is the reference category; this choice is arbitrary, and any other category can be the reference, requiring only a slight reparameterization of the model. Bock (1989) gave an EM-based approach to the analysis of this model, a straightforward extension of the Bock–Aitkin algorithm.

12.10 Extensions of the EM Approach

We consider two kinds of extensions:

1. More complex ability distributions
2. Bayesian methods

12.10.1 Complex Ability Distributions

A common assumption in almost all the models described above is the normality of the latent ability. Experience with the U.S. NAEP program has shown that on many of these tests the ability distribution has a long left tail of low-ability students. A natural question then is whether the assumption of normality in the analysis of these test items might lead to biased estimates of student-level regression coefficients, on variables like ethnicity and gender. In any case, a further question is whether the ability model assumption can be weakened without losing the computational convenience of the EM algorithm in these models.

Bock and Aitkin (1981) pointed out that because of the Gaussian quadrature approach to the nonanalytic likelihood from the normal ability distribution, other distributions could be substituted using the appropriate mass points and masses for the other distributions, or the distribution itself could be estimated, on a set of mass points and masses which could be determined by finite mixture MLE.

Aitkin and Aitkin (2011) made an extensive study of both these questions. They found in simulations from small NAEP-like models, that estimation of regression parameters at the student level was remarkably robust to even extreme variations in the true ability distribution generating the item responses.

Further, the nonparametric MLE of the ability distribution not only did not improve estimation of the student parameters, but also increased their standard errors, because of the poor precision in the estimation of the ability distribution. There seemed to be no reason not to use the Gaussian quadrature analysis for the NAEP surveys examined.

12.10.2 Bayesian Methods

Foe statistical models generally, as model complexity increases, with constant observed data, there are fewer effective observations per parameter, and so departures from asymptotic (frequentist) behavior must be expected.

Large-scale national tests of the NAEP form, in which students receive a small sample of the test items, take these difficulties to extremes. While the national samples are large, the average item sample per student is small, and the overall model for all the test items is very large—up to 150 items. With a 3PL model, there are already around 400 item parameters, before fitting any covariates in the ability model. Nested survey designs, common in class testing, lead to multiple levels of numerical integration over the random effects used to model these designs (for details on this extended modeling, see Aitkin and Aitkin, 2011). While packages like Latent Gold and GLLAMM in Stata can handle these MLE computations, the standard error precision from the information matrix may not reflect the data sparsity and the departure of the log-likelihood from the asymptotic quadratic form. Model adequacy and model comparisons become unreliable through the usual frequentist procedures for these sparse data models.

The simplest Bayesian parallel to the EM algorithm is MCMC sampling from the posterior distributions using the data augmentation algorithm, a stochastic version of EM (Tanner and Wong 1987; Tanner 1996; see Chapter 15). For the same incomplete-date structure for which an EM algorithm could be used, this approach can be used, with alternate conditional-distribution steps:

1. Given current draws from the conditional posterior distribution of θ given $\mathbf{x}$, make draws from the conditional posterior distribution of the unobserved data $\mathbf{x}$ given θ.
2. Given current draws from the conditional posterior distribution of the unobserved data $\mathbf{x}$ given θ, make draws from the conditional posterior distribution of θ given $\mathbf{x}$.

These steps are the Bayesian versions of the E and M steps. Under certain conditions, the two distributions converge to the posterior distribution of θ as well as the posterior distribution of the missing data $\mathbf{x}$.

Bayesian procedures have much to offer for item response modeling, in several areas. For instance, the extensive item parameterization can be made parsimonious by hierarchical models for the item parameters (variance-component models). Also, MCMC procedures can bypass the difficulty of multiple levels of numerical integration, and also provide a full posterior distribution for the model parameters which does not depend on asymptotics for interpretation. Finally, comparisons between competing item response models, and assessing goodness of fit, can be carried out using the same posterior distribution for the parameters to provide likelihood-based procedures. Aitkin (2010) gives a full discussion of this approach.

As in other areas of applied statistical modeling, Bayesian procedures will increase further in importance and power over time.

References

Agresti, A. 1984. *Analysis of Ordinal Categorical Data.* New York: Wiley.

Aitkin, M. 1999. A general maximum likelihood analysis of variance components in generalized linear models. *Biometrics, 55,* 117–128.

Aitkin, M. 2010. *Statistical Inference: An Integrated Bayesian/Likelihood Approach.* Boca Raton: Chapman and Hall/CRC Press.

Aitkin, M. and Aitkin, I. 1996. A hybrid EM/Gauss-Newton algorithm for maximum likelihood in mixture distributions. *Statistics and Computing, 6,* 127–130.

Aitkin, M. and Aitkin, I. 2011. *Statistical Modeling of the National Assessment of Educational Progress.* New York: Springer.

Aitkin, M., Francis, B. J., and Hinde, J. P. 2005. *Statistical Modelling in GLIM4.* Oxford: Oxford University Press.

Aitkin, M., Francis, B. J., Hinde, J. P., and Darnell, R. E. 2009. *Statistical Modelling in R.* Oxford: Oxford University Press.

Andersen, E. B. 1973. Conditional inference for multiple-choice questionnaires. *British Journal of Mathematical and Statistical Psychology, 26,* 31–44.

Birnbaum, A. 1968. Some latent trait models and their use in inferring an examinee's ability. In F. M. Lord and M. R. Novick (Eds.), *Statistical Theories of Mental Test Scores* (pp. 397–479). Reading, MA: Addison-Wesley.

Bock, R. D. 1989. Measurement of human variation: A two-stage model. In R. D. Bock (Ed.), *Multilevel Analysis of Educational Data* (pp. 319–342). New York: Academic Press.

Bock, R. D. and Aitkin, M. 1981. Marginal maximum likelihood estimation of item parameters: Application of an EM algorithm. *Psychometrika, 46,* 443–459.

Bock, R. D. and Lieberman, M. 1970. Fitting a response model for n dichotomously scored items. *Psychometrika, 35,* 179–197.

Dempster, A. P., Laird, N. M., and Rubin, D. B. 1977. Maximum likelihood from incomplete data via the EM algorithm (with Discussion). *Journal of the Royal Statistical Society, Series B, 39,* 1–38.

Jöreskog, K. G. and Goldberger, A. S. 1975. Estimation of a model with multiple indicators and multiple causes of a single latent variable. *Journal of the American Statistical Association, 70,* 631–639.

LaMotte, L. R. 1972. Notes on the covariance matrix of a random, nested ANOVA model. *Annals of Mathematical Statistics, 43,* 659–662.

Lindsay, B., Clogg, C. C., and Grego, J. 1991. Semiparametric estimation in the Rasch model and related exponential response models, including a simple latent class model for item analysis. *Journal of the American Statistical Association, 86,* 96–107.

Longford, N. T. 1987. A fast scoring algorithm for maximum likelihood estimation in unbalanced mixed models with nested random effects. *Biometrika, 74,* 817–827.

Lord, F. M. 1952. A theory of test scores. *Psychometric Monographs,* No. 7.

Lord, F. M. and Novick, M. R. 1968. *Statistical Theories of Mental Test Scores.* Reading, MA: Addison-Wesley.

Meng, X. L. and van Dyk, D. A. 1997. The EM algorithm – an old folk song sung to a fast new tune (with Discussion). *Journal of the Royal Statistical Society, Series B, 59,* 511–567.

Rasch, G. 1960. *Probabilistic Models for Some Intelligence and Attainment Tests.* Copenhagen, Denmark: Danish Institute for Educational Reserch.

Rost, J. 1997. Logistic mixture models. In W. J. van der Linden and R. K. Hambleton. (Eds.), *Handbook of Modern Item Response Theory* (pp. 449–463). New York: Springer.

Skrondal, A. and Rabe-Hesketh, S. 2004. *Generalized Latent Variable Modeling: Multilevel, Longitudinal and Structural Equation Models.* Boca Raton, FL: Chapman and Hall/CRC.

Tanner, M. A. 1996. *Tools for Statistical Inference* (3rd edn.). New York: Springer.

Tanner, M. and Wong, W. 1987. The calculation of posterior distributions by data augmentation. *Journal of the American Statistical Association, 82,* 528–550.

Vermunt, J. K. 2004. An EM algorithm for the estimation of parametric and nonparametric hierarchical nonlinear models. *Statistica Neerlandica 58,* 220–233.

13

Bayesian Estimation

Matthew S. Johnson and Sandip Sinharay

CONTENTS

The key to statistical inference is to understand the random mechanism by which a given set of data (denoted by y) is generated. Better understanding of the random mechanism, which generates the data, leads to better explanation of the differences observed and potentially better prediction of what one might observe in the future. For example, in item response theory (IRT) applications, the data, y, might be a matrix of item responses of a sample of P examinees to a set of I items. From this set of item responses, we might want to rank order the examinees or determine the extent to which each examinee has acquired a certain skill; we may also be interested in predicting how likely a given student will be to correctly answer the next question, for example, in a computerized adaptive testing environment. Although classical or frequentist statistical approaches (e.g., maximum likelihood, confidence intervals, p values) are commonly used in operational testing and the most widely available methods available, Bayesian statistical methods are recently becoming more readily available and increasingly utilized in psychometrics and IRT.

Bayesian inferential methods, such as frequentist and maximum-likelihood methods, require a sampling model that describes the random phenomenon, which generated the observed data. The sampling model is typically defined by a small number of parameters (small relative to the sample size) and may include information from explanatory variables, such as predictor variables in a regression analysis. For example, in IRT, we may have a matrix of binary responses as the observed data. In this case, we might assume the two-parameter logistic model (2PL) is the sampling model. The 2PL model is defined by $2J$ parameters (J difficulties and J discrimination parameters).

Given the sampling model, we define the likelihood function, which is a function of the parameter vector ψ and the observed data y, by

$$L(\psi; y) = f(y|\psi) = \prod_{i=1}^{N} f(y_i|\psi) \tag{13.1}$$

where $f(y_i|\psi)$ is the probability density or mass function of a single observation in the data set. Although the likelihood defined previously assumes that observations are independent and identically distributed, this is not a requirement of Bayesian statistical models. As long as the joint density (mass) function $f(y|\psi)$ is defined by a set of parameters ψ, Bayesian inference is possible.

Whereas likelihood methods treat the vector of item parameters as a fixed but unknown vector of constants, Bayesian statistical methods treat the parameters as a random vector and must assume a prior distribution. The prior distribution describes the uncertainty that we have about the parameter vector. Let us denote the probability density function corresponding to the prior distribution as $f(\psi)$. The prior distribution can also be utilized to reflect any prior information that we may have about the parameters. For example, in a three-parameter logistic model (3PL), we may want to use the number of choices in a multiple-choice question to define the prior distribution in such a way to reflect the fact that we expect the guessing parameter to be close to the inverse of the number of choices.

The joint probability density of the data y and the parameter vector ψ is obtained by multiplying the likelihood function (defined by the sampling model) and the prior density as follows:

$$f(y, \psi) = f(y|\psi) f(\psi)$$

Integration over the parameter space produces the density of the marginal distribution of the observed data,

$$f(y) = \int f(y, \psi)\, d\psi$$

The conditional density of the parameter vector ψ given the observed data y is then obtained as

$$f(\psi|y) = \frac{f(y, \psi)}{f(y)} = \frac{f(y|\psi) f(\psi)}{\int f(y|\psi) f(\psi)\, d\psi} \tag{13.2}$$

This equation is called Bayes theorem or Bayes rule for finding the conditional distribution (see, e.g., Schervish, 1995, p. 4). This conditional distribution of the parameters given the observed data is called the posterior distribution of the parameter vector ψ. The density $f(\psi|y)$ is referred to as the posterior density of the parameter vector ψ given the observed data y. In Bayesian statistical analyses, all inferences are based on the posterior distribution.

The denominator $f(y)$ in the expression for the posterior density $f(\psi|y)$ does not depend of the parameter vector and is therefore a constant for a given data set for a given sampling model and prior distribution. The term ensures that the posterior density integrates to one. Given that the denominator is constant with respect to the parameter vector ψ, the posterior distribution is often written as proportional to the product of the likelihood and the prior distribution as

$$f(\psi|y) \propto f(y|\psi) f(\psi) = f(y, \psi)$$

By examining the joint density $f(y, \psi)$, the numerator of the posterior density in Equation 13.2, one can often find the denominator without needing to explicitly perform the integration in the denominator of the posterior distribution. For example, if the numerator of the posterior density of a parameter π is proportional to $\pi^3(1-\pi)^2$, $0 \le \pi \le 1$, then it is obvious that the posterior distribution is a Beta(4,3) distribution and the denominator is equal to $6!/(3!2!)$.

The following example demonstrates how the posterior is determined from the likelihood and the prior distribution.

Example 13.1

Consider a simple mathematics assessment with 10 items on single-digit addition. Assume the 10 items are equally difficult and we are interested in estimating the true proportion of single-digit addition items a specific student can correctly answer in 30 seconds. If the student attends the assessment and correctly answers 7 out of the 10 items, as we assume that the items are equally difficult, a reasonable choice for the sampling model is the Binomial$(10, \pi)$ distribution, where π is the student's probability of getting an item correct. Then, the functional form of the sampling model is given as follows:

$$f(y = 7|\pi) = \binom{10}{7} \pi^7 (1-\pi)^3 \qquad (13.3)$$

In the absence of prior information, we might choose to utilize a uniform prior on (0,1) for the parameter π. That is,

$$f(\pi) = \mathbf{1}(0 \le \pi \le 1) \qquad (13.4)$$

where $\mathbf{1}\{\cdot\}$ is the indicator function. Multiplying the likelihood and the prior together, we have

$$f(\pi|y = 7) \propto \pi^7 (1-\pi)^3 \times \mathbf{1}\{0 \le \pi \le 1\} \qquad (13.5)$$

This is the probability density of a Beta(8,4) distribution, which has proportionality constant $11!/(7!3!)$. So the posterior density of π is the Beta(8,4) density:

$$f(\pi|y = 7) = \frac{11!}{3!7!} \pi^7 (1-\pi)^3 \mathbf{1}\{0 \le \pi \le 1\} \qquad (13.6)$$

This density and the Beta(6,6) density, which would be the posterior distribution for an individual who correctly answered five of the 10 items appear in Figure 13.1.

Example 13.1 demonstrates the construction of the posterior distribution in a case where the computation of the normalizing constant is analytically tractable. However, for many statistical models, including the models within IRT (e.g., Rasch, 2PL, 3PL), the integration required to calculate the denominator in the definition of the posterior density is impossible to determine in closed form. The following example demonstrates this issue.

Example 13.2

Consider a reparameterization of the problem described in Example 13.1, where we reparameterize the probability of a correct response in terms of the log-odds of success. That

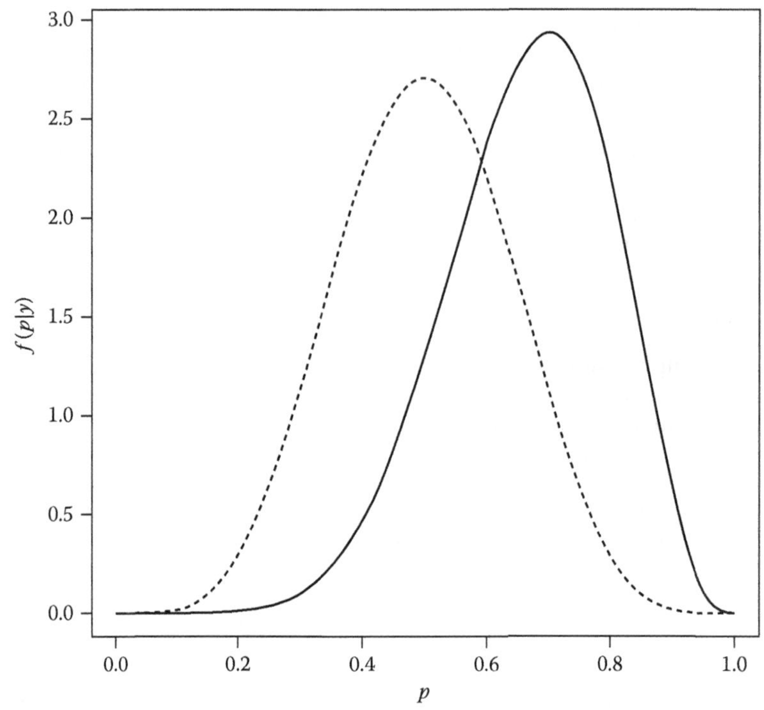

FIGURE 13.1
The posterior distributions for Example 13.1. The solid curve is the Beta(8,4) density and the dashed line is the Beta(6,6) density.

is, let $\theta = \log(p/1-p)$, so $p = 1/(1+e^{-\theta})$. Now, assume we wish to utilize a standard normal prior distribution on the log-odds θ, so we have

$$f(\theta|y=7) \propto \frac{\exp\{7\theta\}}{\{1+\exp\{\theta\}\}^{10}} \times \exp\left\{-\frac{\theta^2}{2}\right\}$$

This function is not proportional to the density of any common distribution (e.g., normal, beta, etc.) and the integral cannot be determined analytically.

Although the normalizing constant for the posterior density in the problem described in Example 13.2 cannot be derived analytically, it is possible to employ numerical integration to compute the normalizing constant. In addition, modern numerical methods allow us to approximate the posterior distribution with Monte Carlo simulation. For the simple one-parameter problem described in Example 13.2, one can generate a sample from the posterior distribution using importance sampling (Geweke, 1989), rejection sampling methods (von Neumann, 1951), or slice sampling (Neal, 2003). When the model is more complex, these methods become too inefficient numerically to utilize.

For more complex models, Markov chain Monte Carlo (MCMC) algorithms are required. Most of these algorithms are based, at least in part, on the Gibbs sampling algorithm (Geman and Geman, 1984), which samples subsets (or blocks) of the parameter vector from the conditional posterior distribution of that block of parameters given the data and all other parameter values. The algorithm then iterates through the blocks of the parameter vector over several thousand steps, generating a Markov chain. The limiting distribution of the Markov chain is the joint posterior distribution of the entire parameter vector.

Gibbs sampling requires the ability to sample from the conditional posterior distributions. However, in many applications, these conditional posterior distributions are intractable and numerical methods are required to sample from the distributions. These methods include the rejection sampling, the slice sampling, and other Markov chain approaches such as the Metropolis–Hastings algorithm (Metropolis et al., 1953; Hastings, 1970) and the adaptive rejection sampling method (Gilks and Wild, 1992).

MCMC methods have been utilized in a number of applications in IRT. Albert and Chib (1993) demonstrated how to utilize a Gibbs sampling algorithm to generate a sample from the posterior distribution of the parameters of probit response models. Patz and Junker (1999a,b) developed a Metropolis–Hastings algorithm within the Gibbs sampling framework to sample parameter values of logistic response models (e.g., 2PL, 3PL, PCM) from a Markov chain that converges to the posterior distribution of the parameter vector. Since these groundbreaking papers demonstrated the use of MCMC for IRT models, a number of extensions have been developed in the literature (see Chapter 15 and Volume Three, Chapter 21).

The following example demonstrates how the posterior distribution of the item difficulties and discrimination parameter from a one-parameter probit (1PP) model are approximated using an MCMC algorithm similar to the one introduced by Albert (1992) and Albert and Chib (1993).

Example 13.3: 1PP Analysis of the LSAT Data

Consider the problem of estimating the parameters of the one-parameter probit model given the LSAT data originally discussed by Bock and Lieberman (1970). The data analyzed here were accessed from the ltm package (Rizopoulos, 2006) in R (R Development Core Team, 2010) and consists of the item responses of $P = 1000$ examinees to $I = 5$ items.

In our Bayesian analysis of the LSAT data, we assume that the 1000 examinee abilities are independent and identically distributed normal random variables with a mean μ and variance σ^2,

$$\theta_p \sim N(\mu, \sigma^2) \quad \text{for } p = 1, 2, \ldots, P \tag{13.7}$$

Under the 1PP model, the probability of a correct response of individual p to item i is given by

$$\Pr(U_{pi} = 1 | \theta_p, \beta_i) = \Phi(\theta_p - \beta_i) \tag{13.8}$$

where Φ denotes the distribution function of the standard normal distribution. The Bayesian analysis presented below assumes an improper uniform prior distribution on the vector of item difficulties $f(\boldsymbol{\beta}) \propto 1$, and an inverse-gamma(1,1) prior distribution for the variance σ^2.

When performing Bayesian analyses of probit models, it is customary to augment the observed item responses with a latent continuous response Z_{pi} associated with subject and item, where we assume

$$Z_{pi} | \theta_p, \beta_i \sim N(\theta_p - \beta_i, 1) \tag{13.9}$$

independently for all $p = 1, \ldots, P$ and $i = 1, \ldots, I$. One then assumes that the observed dichotomous item responses are simply dichotomizations of this continuous outcome, where $U_{pi} = \mathbf{1}\{Z_{pi} > 0\}$. By augmenting the data with the continuous responses, the MCMC algorithm can be purely a Gibbs sampling algorithm, without needing to use Metropolis–Hastings or slice sampling to generate samples from the conditional posterior distributions.

Our MCMC algorithm for the 1PP model generates a sequence of parameter values by repeating the following three steps for a large number of iterations:

1. Sample the latent continuous responses from the complete conditional distributions of Z_{pi} given observed responses u_{pi}, the current estimates of ability θ_p and the current values of the item difficulties, β_i, by sampling each from a truncated normal distribution

$$Z_{pi} \sim N(\theta_p - \beta_i, 1) \times \mathbf{1}\{(2u_{pi} - 1)Z_{pi} > 0\} \tag{13.10}$$

   The truncation forces Z_{pi} to be less than zero whenever an individual gets an item incorrect ($u_{pi} = 0$) and forces Z_{pi} to be greater than zero whenever the individual gets an item correct ($u_{pi} = 1$).
2. Sample the examinee ability vector $\boldsymbol{\theta}$ and the item difficulty vector $\boldsymbol{\beta}$ from their joint conditional posterior given $\mathbf{Z}$ and the ability variance σ^2 according the following substeps.
   a. Sample $\boldsymbol{\beta}$ conditional on $\mathbf{Z}$ by noting the marginal distribution (after integrating over the distribution of the examinee abilities) of *the* vector $\mathbf{Z}_p \sim N(-\boldsymbol{\beta}, \boldsymbol{I} + \sigma^2\boldsymbol{J})$, where $\boldsymbol{I}$ is the $I \times I$ identity matrix and $\boldsymbol{J}$ is the $I \times I$ matrix of ones. This marginal distribution, in conjunction with the improper uniform prior distribution on the item difficulties results in the following conditional posterior distribution:

$$\boldsymbol{\beta}|\mathbf{Z}, \sigma^2 \sim N\left(-\overline{\mathbf{Z}}, \frac{1}{P}(\boldsymbol{I} + \sigma^2\boldsymbol{J})\right) \tag{13.11}$$

   b. Sample $\boldsymbol{\theta}$ conditional on $\mathbf{Z}$, $\boldsymbol{\beta}$, and σ^2 by sampling each examinee's ability independently from a normal distribution

$$\theta_p|\mathbf{Z}_p, \boldsymbol{\beta}, \sigma^2 \sim N\left(\frac{\sigma^2}{I\sigma^2 + 1}\sum_{i=1}^{I}(Z_{pi} - \beta_i)\frac{\sigma^2}{I\sigma^2 + 1}\right) \tag{13.12}$$

   c. Sample the variance of the ability distribution, σ^2, conditional on the vector of ability parameters $\boldsymbol{\theta}$ from an inverse-gamma distribution,

$$\sigma^2|\boldsymbol{\theta} \sim \text{inverse-gamma}\left(1 + \frac{n}{2}, 1 + \frac{1}{2}\sum_p \theta_p^2\right) \tag{13.13}$$

To fit the 1PP model to the LSAT data, we iterated the steps described previously for 10,000 iterations. After discarding the first 1000 iterations as burn-in (the time to converge to the limiting distribution), we approximated the posterior densities of the item difficulties and the variance of the ability distribution using the values from the last 9000 iterations. The histograms and spline-smoothed density estimates of the parameters are displayed in Figure 13.2. The algorithm, which is implemented in *R*, took 137 seconds to run on a PC with an Intel® Core 2™ Q9300 2.5 GHz processor.

In the following sections, we discuss how one utilizes the posterior distribution to make inferences about the parameter vector. We then briefly discuss how to evaluate the fit of the

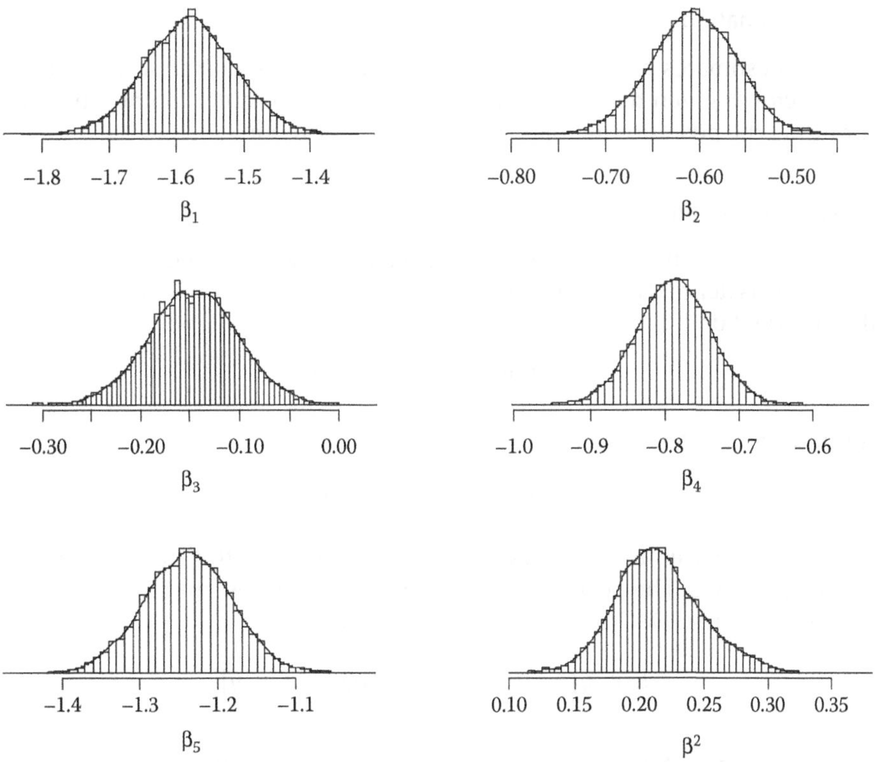

FIGURE 13.2
Approximate posterior densities of the 1PP parameters for the LSAT data.

model and perform model selection within the Bayesian framework. After discussing these inferential techniques, we return to a discussion of how to select the prior distribution.

13.1 Bayesian Point Estimation, Interval Estimation, and Hypothesis Testing

The posterior distribution of the parameters given the observed data reflects the knowledge and uncertainty we have about the parameters after observing the data. For example, the posterior density depicted by the solid curve in Figure 13.1 suggests that there is very little chance that the student's true probability of a correct response falls below 0.2, and the most likely value of the student's probability of a correct response appears to be between 0.60 and 0.85.

While all information we have about the parameter values is reflected in the posterior distribution, the posterior distribution can be difficult to interpret in many cases, especially when dealing with a multidimensional parameter vector. In the following sections, we discuss summaries of the posterior distribution that are usually reported within Bayesian analysis, including point estimation, interval estimation, and hypothesis testing.

13.1.1 Point Estimation

Bayesian point estimates are simply point summaries of the posterior distribution of the parameter vector ψ given the observed data y. The most common point estimates are the posterior mean, the posterior median, and the posterior mode.

13.1.1.1 Posterior Mean

The posterior mean of the parameter vector ψ, sometimes called the expected a posteriori (EAP) estimator, is defined as the mean of the posterior distribution of the parameter vector given the observed data y:

$$E[\psi|y] = \int \psi f(\psi|y)\, d\psi \tag{13.14}$$

for continuous parameter vectors and

$$E[\psi|y] = \sum \psi f(\psi|y) \tag{13.15}$$

for discrete-valued parameter vectors, where the integral or the summation is over the entire parameter space. The posterior mean, if it exists, is the estimator that minimizes the squared expected error (DeGroot, 1986, p. 332). That is,

$$E[\psi|y] = \text{argmin}_{\mathbf{a}} \sum_j E[(\psi_j - a_j)^2|y] \tag{13.16}$$

assuming that the expectations on the right-hand side of the equation exist. The resulting minimum of the sum of posterior squared errors is exactly equal to the trace of the posterior covariance matrix of ψ, or equivalently the sum of the posterior variances of the individual elements of the parameter vector. For this reason, the posterior variances (or standard deviations) are often used to reflect the level of uncertainty we have about the parameter ψ.

The posterior mean of the proportion in Example 13.1 is the posterior mean of the Beta(8,4) distribution, which is $\frac{2}{3}$; the posterior variance is approximately 0.017 and the standard deviation is approximately 0.13.

In situations where MCMC is required, the posterior mean is approximated with the sample mean of the draws obtained from the MCMC algorithm (possibly after a burn-in period). That is, the posterior mean is approximated by

$$E[\psi|y] \approx \overline{\psi} = \frac{1}{M} \sum_m \psi^{(m)} \tag{13.17}$$

where M is the total number of sampled values and $\psi^{(m)}$ is the mth sampled value from the posterior distribution. The posterior variance (or covariance) can be approximated in a similar fashion:

$$\text{cov}(\psi|y) \approx \frac{1}{M} \sum_m \left(\psi^{(m)} - \overline{\psi}\right) \left(\psi^{(m)} - \overline{\psi}\right)' \tag{13.18}$$

Because Equation 13.17 provides a Monte Carlo approximation of the true posterior mean, it is important to report the Monte Carlo error associated with the approximation. See, for example, Geyer (1992) or Patz and Junker (1999a) for further discussion on Monte Carlo error, including recommendations on the estimation of the error.

13.1.1.2 Posterior Median

One of the properties of the mean is that it can be "pulled" toward the tail of a skewed distribution. For this reason, the posterior mean is often not an appropriate point estimate or summary of a skewed posterior distribution. For skewed distributions, the vector of posterior medians of the parameters is often preferred as a point estimate. The posterior median of the parameter ψ_j, denoted m_j, is any number where

$$\Pr\{\psi_j \leq m_j\} \geq \frac{1}{2} \quad \text{and} \quad \Pr\{\psi_j \geq m_j\} \geq \frac{1}{2} \tag{13.19}$$

When more than one value m_j satisfies the equation above, we typically take the midpoint of the set of values to produce a single median value. The posterior median for the probability in Example 13.1 is approximately 0.676.

Whereas the posterior mean minimizes the posterior squared error loss, the posterior median minimizes the posterior mean absolute error. That is,

$$m_j = \text{argmin}_a E\left[|\psi_j - a|\,|y\right] \tag{13.20}$$

If the posterior distribution is symmetric, the posterior mean and posterior median, assuming that they exist, are equal.

When the posterior distribution is approximated using MCMC procedures the median is approximated by the sample median of the Monte Carlo sample, for example,

$$m_j \approx \psi_j^{((M+1)/2)} \tag{13.21}$$

for odd sample sizes M and

$$m_j \approx \frac{\psi_j^{(M/2)} + \psi_j^{(1+(M/2))}}{2} \tag{13.22}$$

for even M, where $\psi_j^{(m)}$ is the mth largest value in the Monte Carlo sample.

13.1.1.3 Posterior Mode

Another common point summary reported in Bayesian analyses is the posterior mode,

$$\tilde{\psi} = \text{argmax}_{\psi} f(\psi|y) \tag{13.23}$$

One of the advantages of the posterior mode is that it can often be approximated using numerical methods such as the Newton–Raphson algorithm or the expectation-maximization (EM) algorithm without calculating the normalizing constant of the posterior distribution. Mislevy (1986) describes such an EM algorithm for approximating the posterior mode of the item parameters in the 1-, 2-, and 3PL models.

In cases where Bayesian analysis is carried out with MCMC methods, approximating the posterior mode is made difficult by the fact that we only have a sample of values from an approximation of the posterior distribution. If the parameter is discrete valued, the posterior mode is the most common value in the sample. However, if the parameter is a continuous random variable, we must first use the sampled values to approximate the joint posterior density and then find the mode of that approximation. While we may be able to approximate the univariate marginal posterior distributions $f(\psi_j|y)$ adequately and thus able to find the model of the marginal posterior distributions, the mode of the joint posterior is in general not equal to the vector of the marginal posterior modes.

13.1.2 Interval Estimation

It is important to provide some sort of summary about how confident we are about the point estimate, whether it is the mean, median, or mode. While a single measure of the uncertainty (e.g., the posterior variance or mean absolute deviation from the median) is useful, an additional measure of the uncertainty is an interval or a set of plausible values for the parameters. Just as frequentist statisticians often report confidence intervals in addition to point estimates and standard errors, Bayesian statisticians often report credible sets in addition to posterior point summaries and posterior measures of uncertainty about the parameter (e.g., Robert, 2007; Carlin and Louis, 2009).

A $(1-\alpha)\times 100\%$ credible set is any subset S of the parameter space such that

$$\Pr\{\psi \in S|y\} = 1-\alpha \qquad (13.24)$$

For continuous parameter vectors, there are uncountable infinite numbers of credible intervals of the level. For example, consider the case of a single parameter ψ_j whose cumulative posterior distribution is given by $F(\psi_j|y)$. We could construct one-sided credible intervals for ψ_j, where $S = \{\psi_j : -\infty < \psi_j < F^{-1}(1-\alpha|y)\}$ or $S = \{\psi_j : F^{-1}(\alpha|y) < \psi_j < \infty\}$, or any two-sided interval of the form

$$S = \left\{\psi_j : F^{-1}(\epsilon) < \psi_j < F^{-1}(1-\alpha+\epsilon)\right\}, \quad 0 \le \epsilon \le 1-\alpha \qquad (13.25)$$

The special case where $\epsilon = \alpha/2$ produces what is typically called the $(1-\alpha)\times 100\%$ equal-tailed credible interval.

Equal-tailed intervals are only defined for single parameters based on their marginal posterior distribution. There is no clear way to generalize the concept of an equal-tailed credible interval to multidimensional parameters. Furthermore, equal-tailed credible intervals are not necessarily the optimal credible intervals, in the sense that they often contain parameter values that have relatively low posterior density when compared to some values left out of the credible set.

The limitations of the equal-tailed credible interval discussed in the previous paragraph are addressed by the highest posterior density (HPD) credible interval (e.g., Robert, 2007; Carlin and Louis, 2009). The $(1-\alpha)\times 100\%$ HPD credible interval seeks a value of the density s such that the probability that the posterior density exceeds that value is equal to $(1-\alpha)$; the HPD credible set is defined as the set of parameter values that has posterior density above the threshold s, that is

$$S = \{\psi : f(\psi|y) \ge s\} \quad \text{such that} \quad \Pr\{\psi \in S|y\} = 1-\alpha \qquad (13.26)$$

The HPD credible set is the smallest possible $(1-\alpha)\times 100\%$ credible interval. However, it can sometimes have strange shapes. For example, for unfolding response models, Johnson (2001, pp. 110–111) demonstrates that the HPD credible set can be the union of nonintersecting sets.

Example 13.4: Credible Intervals

Figure 13.3 shows how the 95% HPD credible interval compares to the 95% equal-tailed credible interval for the probability in Example 13.4. The horizontal line just above 0.5 corresponds to the level s in the definition of the 95% HPD. The area between the two shaded tails is the 95% HPD credible interval (0.4121, 0.9066) for the probability of success and the two vertical line segments are the 2.5th and 97.5th percentiles of the posterior distribution and thus are the endpoints of the 95% equal-tailed credible interval (0.3900, 0.8907). The frequentist 95% confidence interval computed exactly from the binomial distribution (rather than by normal approximation) is (0.3475, 0.9332). The 95% HPD credible interval is the narrowest and the frequentist confidence interval is the widest of the three. Traditional frequentist confidence intervals for models with discrete responses are known to be problematic and some authors have suggested the use of randomization or fuzzy sets for constructing the intervals (see Geyer and Meeden, 2005, and the references therein for a discussion).

Although HPD credible intervals are often superior to equal-tailed credible intervals, they are not often reported. One reason is that they are a bit more complicated to calculate; equal-tailed intervals only require the quantiles of the distribution. Furthermore, computation of the HPD intervals from MCMC simulation is very difficult, because it requires using the MCMC samples to first approximate the posterior density and then to use numerical methods to solve for the level of the density that produces the HPD interval. See Chen and Shao (1999) for more details on the method of approximating HPD credible intervals from MCMC simulations.

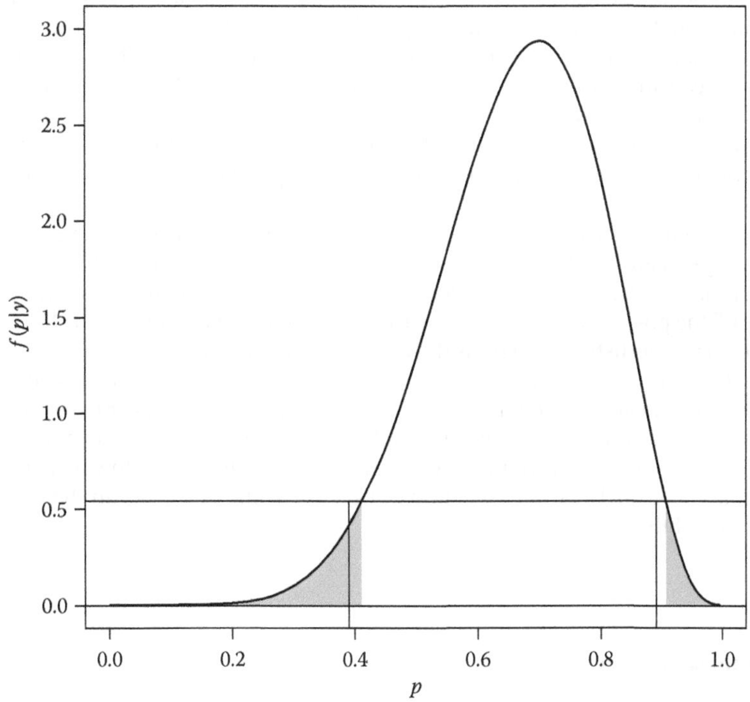

FIGURE 13.3
The posterior density of the probability from Example 13.4. The interval between the two shaded tails of the distribution corresponds to the 95% HPD credible interval. The vertical line segments show the 95% equal-tailed credible interval.

TABLE 13.1

Posterior Estimates of the Item Difficulties and Ability Variance for the 1PP Fit to the LSAT Data

	MML	Posterior Summaries			
	MLE (SE)	Mean	Median	SD	95% Equal-Tailed Credible Interval
Item 1	−1.561 (0.064)	−1.579	−1.579	0.067	(−1.710, −1.448)
Item 2	−0.602 (0.044)	−0.606	−0.606	0.046	(−0.697, −0.519)
Item 3	−0.145 (0.042)	−0.147	−0.146	0.044	(−0.235, −0.060)
Item 4	−0.782 (0.046)	−0.788	−0.787	0.048	(−0.884, −0.694)
Item 5	−1.229 (0.054)	−1.239	−1.238	0.056	(−1.350, −1.132)
σ^2	0.166	0.215	0.213	0.033	(0.155, 0.287)

Example 13.5: Bayesian Inference for the 1PP Fit of the LSAT Data

Table 13.1 contains approximate posterior means and medians for the five item difficulties and the ability variance for the 1PP fit to the LSAT data set. These values were obtained using an MCMC algorithm. For comparison purposes, the marginal maximum-likelihood estimates and standard errors estimated with the lme4 (Bates and Maechler, 2010) package in *R* are also provided. Twenty-one quadrature points were used to obtain these estimates.

Because the posterior distributions of the item difficulties are fairly symmetric, the posterior mean and posterior median are close to one another. In addition, the 95% equal-tailed credible intervals are very close to what would be obtained based on a normal approximation, that is, by using a margin of error $\pm 2SD$. The posterior for the ability variance is not as symmetric, and therefore, we observe a small difference between the posterior mean and median, and a normal approximated credible interval would not be as good an approximation in this case. However, because the posterior distribution of the log-variance is more symmetric, the normal approximated confidence interval for $\log(\sigma^2)$ often produces an interval similar to the equal-tailed credible interval for the log-variance.

Comparing the MLEs to the posterior point estimates, we notice the biggest difference is with the estimates of the ability distribution variance, where the MLE is 0.166 and the posterior mean is 0.215. The difference demonstrates the potential for the prior distribution to pull the posterior estimates away from the value that maximizes the likelihood. In this case, the prior distribution pulls the variance estimate away from zero because of the shape of the prior density (see Figure 13.4), whose mode is 0.5 and mean is not defined.

The posterior means and medians of the item difficulties also appear to be biased away from zero. This is counter to what is typically expected in Bayesian analyses, where estimates are often shrunk toward the prior mean, which is often zero. However, in this case, the estimates are further from zero, in part because of the difference in the estimates of the variance of the ability distribution.

13.1.3 Hypothesis Testing

Consider the problem of comparing the following hypotheses:

$$H_0 : \psi \in \Psi$$
$$H_1 : \psi \notin \Psi$$

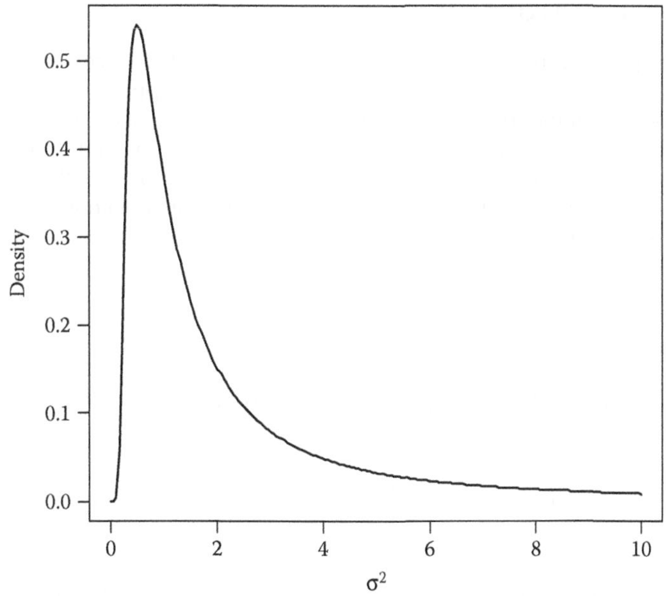

FIGURE 13.4
The inverse-gamma(1,1) prior utilized in the Bayesian estimation of the 1PP for the LSAT data in Example 13.5.

While some Bayesians think that there should be no testing (e.g., Gelfand et al., 1992), and that researchers should simply report the posterior probabilities of the alternative models, Robert (2007) stated that the Bayesian toolbox must also include testing devices, if only because users of statistics have been accustomed to testing as a formulation of their problems. Schervish (1995) and Robert (2007) relate Bayesian testing procedures to decision theory, where losses c_1 and c_2 are assigned to Type I and Type II errors. For example, for true hypothesis testing, we might assign a large loss to a Type I error relative to a Type II error, for example, $c_1 = 20$ and $c_2 = 1$. For model selection, on the other hand, one might assign equal losses to Type I and Type II errors.

Given the costs of making Type I and Type II errors, the Bayesian approach to hypothesis testing is to minimize the "posterior expected loss" to obtain the Bayes rule or Bayes estimator that is given by

$$\text{Reject } H_0 \quad \text{if } \Pr\{\psi \in \Psi | y\} < \frac{c_2}{c_1 + c_2} \tag{13.27}$$

or equivalently

$$\text{Reject } H_0 \quad \text{if } \frac{\Pr\{\psi \in \Psi | y\}}{\Pr\{\psi \notin \Psi | y\}} < \frac{c_2}{c_1} \tag{13.28}$$

For a more detailed description of the decision-theoretic approach to testing, see, for example, Robert (2007).

An alternative to the decision-theoretic approach described previously is to report the amount of evidence for and against a hypothesis. Jeffreys (1961) and Kass and Raftery (1995) offer criteria for evaluating the amount of evidence for and against a hypothesis.

Example 13.6: Hypothesis Testing

Consider the problem of estimating the student's probability of correctly answering a single-digit addition item as described in Example 13.6. Suppose we are interested to know whether or not the student has a better than 50% chance of correctly answering the question. One approach would be to examine whether the 0.5 is in the credible interval, which it is. We could also, as the above discussion suggests, calculate the posterior probability of the region we are interested in. So if we are interested to know if the probability of success is greater than 0.5, our hypotheses might be

$$H_0 : p \leq 0.5$$
$$H_1 : p > 0.5$$

The posterior probability of the null hypothesis, H_0, is

$$\Pr\{p \leq 0.5|\mathbf{y}\} = \int_0^{0.5} \frac{11!}{7!\,3!} p^7 (1-p)^3 \, dp \approx 0.1133$$

If we were to use costs $c_1 = 19$ and $c_2 = 1$, $c_2/(c_1 + c_2) = 0.05$ and we would not conclude that we have sufficient evidence to conclude that the student's probability of success is above 0.5. Note however that if we had assumed $c_1 = c_2 = 1$, then $c_2/(c_1 + c_2)$ would have been 0.5 and we would have concluded that there is sufficient evidence that the student's probability of success is above 0.5.

Now suppose that we have two students who have completed the 10-item assessment, where one student correctly answered 7 of the 10 items and the second student correctly answered 5 of the 10 items. The two posterior distributions for the students' probabilities of success are displayed in Figure 13.1. Now, we are interested in determining if there is evidence to conclude that the student that correctly answered 7 items has a higher probability of success than the student who correctly answered 5 items. That is, suppose we wish to test the hypotheses.

$$H_0 : p_1 \leq p_2$$
$$H_1 : p_1 > p_2$$

The posterior probability of the null hypothesis is

$$\Pr\{p_1 \leq p_2|\mathbf{y}\} = \int_0^1 \frac{11!}{7!\,3!} p_1^7 (1-p_1)^3 \int_p^1 \frac{11!}{5!\,5!} p_2^5 (1-p_2)^5 \, dp_2 \, dp_1$$
$$\approx 0.1935 \tag{13.29}$$

So, if we assume costs of 19 and 1 for Type I and Type II errors, we do not find sufficient evidence to conclude that the first student has a higher ability to correctly answer the single-digit addition items.

The integrals required to calculate the posterior probabilities of the null hypotheses above can be calculated analytically. However, in the majority of hypothesis testing problems in item response theory, the integrals cannot be determined analytically. As discussed earlier, the necessary integrals can be approximated using Monte Carlo procedures. By understanding that a probability is simply the expected value of an indicator, we can use the same approach as described earlier for producing Monte Carlo approximations for means. That is, to approximate the posterior probability that the parameter $\psi \in \Psi$, we approximate the posterior mean of the indicator function $1\{\psi \in \Psi\}$,

$$\Pr\{\psi \in \Psi|y\} = E[1\{\psi \in \Psi\}|y] \approx \frac{1}{M}\sum_{m=1}^{M} 1\left\{\psi^{(m)} \in \Psi\right\} \tag{13.30}$$

The following hypothesis testing example for the 1PP fit of the LSAT data demonstrates the use of this procedure.

Example 13.7: Hypothesis Testing

Examining the approximations of the posterior densities in Figure 13.2 and the Bayesian point estimates in Table 13.1, the LSAT items ordered from least to most difficult are items 1, 5, 4, 2, and 3. With the exception of items 2 and 4, none of the 95% equal-tailed credible intervals intersect one another. We might want to test whether, in fact, we can conclude that the order of difficulty of the items is the same as that described above. That is we may wish to evaluate the hypothesis

$$H_0 : \beta_1 \le \beta_5 \le \beta_4 \le \beta_2 \le \beta_3$$

This sort of hypothesis is very difficult to evaluate under the frequentist paradigm, because the null distribution is difficult to determine; see Silvapulle and Sen (2004) for the discussion of frequentist methods for such hypotheses.

To approximate the posterior probability of the null hypothesis that the items are ordered in difficulty we create an indicator variable for each sampled vector of the item difficulties, for example, we can create the indicator

$$Q_m = 1\left\{\beta_1^{(m)} \le \beta_5^{(m)} \le \beta_4^{(m)} \le \beta_2^{(m)} \le \beta_3^{(3)}\right\} \tag{13.31}$$

for each sampled vector $\beta^{(m)}$ and then calculate the sample average of these indicators over the 9000 samples from the MCMC algorithm. The resulting approximate posterior probability is

$$\Pr\{\beta_1 \le \beta_5 \le \beta_4 \le \beta_2 \le \beta_3|u\} \approx \frac{1}{M}\sum_{m=1}^{9000} Q_m = 0.998 \tag{13.32}$$

Thus, if $c_1 = 19$ and $c_2 = 1$, $c_2/(c_1 + c_2) = 0.05$, then one would conclude that there is sufficient evidence that the difficulties of the items are indeed ordered as described above. This is expected given that the item difficulties are well separated (see, e.g., Table 13.1).

The hypotheses discussed to this point all are composite hypotheses. However, in most applications, researchers are interested in testing point null hypotheses. For example, in a two-parameter logistic model, we might want to test the hypothesis that all discriminations are equal. For example, if the discrimination parameters are denoted by $\alpha_1, \ldots, \alpha_I$, then we might want to test the hypothesis

$$H_0 : \alpha_1 = \alpha_2 = \cdots = \alpha_I$$

However, under the Bayesian paradigm, if the prior distribution on the discrimination parameters assumes a continuous prior distribution, the prior and posterior probability of the null hypothesis is exactly zero. One solution to this problem, suggested by Jeffreys (1961), is to use the Bayes Factor, which can be thought of as giving a prior probability of $\frac{1}{2}$ to the point null hypothesis and then calculating the posterior probability of the null hypothesis (and alternative hypothesis). See, for example, Kass and Raftery (1995) for a general overview of Bayes factors and Chapter 19 for the discussion on the use of Bayes factors for model selection in the context of IRT models.

13.2 Selecting a Prior Distribution

We have assumed till now that a prior distribution already has been selected, and we have provided very little guidance about how the prior distribution is chosen. In some regards, the prior distribution is the most important aspect of the Bayesian analysis, primarily because it is the most controversial aspect of the Bayesian method. Frequentist statisticians observe the introduction of the prior as a way of affecting the final inferences made from an analysis without the inclusion of additional data. It is true that the introduction of prior distributions often biases final estimates and that different priors can lead to different conclusions. However, appropriately selected priors can result in more precise estimates. Given the effect priors may have on final inferences, the choice of them should not be taken likely, especially for small sample sizes when the prior distribution substantially affects the posterior distribution.

Researchers (Schervish, 1995, pp. 21–24; Carlin and Louis, 2009, pp. 27–40) have recommended three broad types of prior distributions or methods of selecting prior distributions: (1) elicited or subjective priors; (2) parametric or convenient priors; and (3) noninformative or flat priors.

The Bayesian analyst that chooses to use subjective priors brings prior knowledge to the statistical model. For example, we might want to use the knowledge that a multiple choice item has five choices to construct the prior distribution of the guessing parameter in the 3PL model (Volume One, Chapter 2). If a student is truly guessing, then we would expect the student to have a one in five chance of correctly answering the problem correctly, and so the prior distribution could reflect that knowledge. For example, Sinharay et al. (2003) utilized a normal prior distribution on the logit (see Chapter 1) of the guessing parameter centered at $\log(0.2/(0.2+0.8))$ for multiple choice items with five choices.

The convenience approach to Bayesian analysis utilizes common probability distributions to model the uncertainty about the parameter vector. In the same way that we often model the sampling distribution with simple distributions such as the normal distribution, binomial distribution, or Poisson distribution, the parametric Bayesian uses convenient distributions for the prior distribution of the parameter ψ. One such class of convenience priors is the class of conjugate prior distributions. A prior distribution is conjugate for a specific sampling distribution, if the prior and the posterior distributions are members of the same class of distributions, for example, both the prior and posterior distributions are normal distributions.

Example 13.8: Conjugate Prior

In Example 13.8, the sampling distribution for the number of successes out of 10 trials is the binomial distribution,

$$f(y = 7|p) \propto p^7(1-p)^3 \tag{13.33}$$

Let us assume that the prior distribution is the Beta distribution given by

$$f(p) \propto p^a(1-p)^b \tag{13.34}$$

Combining the sampling model and the prior we find that the posterior distribution is proportional to

$$f(p|y = 7) \propto p^{7+a}(1-p)^{3+b} \tag{13.35}$$

which is the kernel of the Beta$(7 + a, 3 + b)$ distribution. The prior and posterior are both Beta distributions. Thus, the Beta distributions are the conjugate prior distributions for the binomial sampling distribution.

Conjugate prior distributions are no more justifiable than nonconjugate prior distributions. They simply offer mathematical convenience, because the posterior distributions are often of a simple form. However, in multiparameter models, finding multivariate conjugate distributions is often complicated. In such cases, Bayesians often choose prior distributions that are conditionally conjugate, that is, the conditional prior, for example, $f(\psi_1|\psi_2)$ and the conditional posterior, $f(\psi_1|\psi_2, \boldsymbol{y})$, are members of the same class of distributions. This is especially convenient when MCMC methods are utilized for estimation. Examples where conditionally conjugate priors are used in IRT include the case of the mean and variance of the ability distribution. For example, in many applications, researchers assume that abilities are independent and identically normal random variables, $\theta_p \sim N(\mu, \sigma^2)$. Conditionally on μ, the conditional conjugate prior distributions for σ^2 are the inverse-gamma distributions. Conditional on σ^2, the conjugate priors for μ are normal distributions.

The use of conjugate priors and all parametric priors in general has the potential to produce biased results. For this reason, many authors prefer to use noninformative or flat priors. Whereas elicited and conjugate priors are sometimes called subjective priors, flat priors are often called objective priors, because the final results depend only minimally on the prior (that causes the Bayesian estimates to be close to the frequentist estimates) when flat priors are utilized. There are three general classes of flat priors utilized in the Bayesian literature:

1. Parametric distributions with large variances or long (heavy) tails that have less of an effect on the posterior than parametric prior distributions with lower variances.
2. Improper flat priors assume that the prior distribution is constant for all values of the parameter vector $\boldsymbol{\psi}$. That is, the improper prior assumes $f(\boldsymbol{\psi}) \propto 1$. However, an improper prior distribution does not integrate to one, and therefore is not a proper distribution. However, it often does lead to a proper posterior density.

Jeffreys (1961) noted that the posterior inferences are not invariant to transformations of the parameter vector. For example, assuming an improper uniform prior on the variance σ^2 will not produce the same posterior inferences one would find if instead we placed an improper uniform prior on the precision $\tau = 1/\sigma^2$. Jeffreys went on to demonstrate, that if we choose a prior distribution that is proportional to the square root of the determinant of the Fisher information matrix, $J(\psi)$,

$$f(\psi) \propto |J(\psi)| \tag{13.36}$$

then the posterior distribution would in fact be invariant to transformations of the parameter vector ψ. This choice of prior is often called Jeffreys' prior.

Whatever prior distribution is selected, it is important to evaluate how robust the Bayesian inferences are to slight modifications in the prior distribution. This is often referred to as sensitivity analysis (e.g., Gelman et al., 2003). If inferences are greatly affected by the small differences in the prior distribution, one must take extra care in reporting the results and discussing the rationale for selecting a particular prior distribution.

13.3 Large-Sample Behavior of Bayesian Estimates

Large-sample results are not necessary for performing Bayesian analysis but are often useful as approximations and for understanding the problem. Under some regularity conditions (especially that the likelihood is a continuous function of the model parameters and the true parameter value is not on the boundary of the parameter space), as the sample size becomes large, the following two results hold (see, e.g., Gelman et al., 2003):

- The posterior mode is consistent, that is, it converges in probability to ψ_0.
- The posterior distribution of $\sqrt{n}(\psi - \psi_0)$, where ψ is the parameter vector and n is the sample size, approaches a normal distribution with mean equal to the zero vector and variance equal to the inverse of the Fisher information matrix at ψ_0,

where ψ_0 minimizes the Kullback–Leibler information (see Chapter 7) of the likelihood distribution relative to the true data distribution. Note that, under the same regularity conditions, the maximum-likelihood estimate has very similar properties. These results state that for large samples, the Bayesian and frequentist estimates become indistinguishable and the prior distribution, which is the source of the most controversy regarding Bayesian methods, has almost no effect on the parameter estimates. See, for example, Table 2 of Patz and Junker (1999a), for a demonstration of the closeness of the Bayesian and frequentist estimates for a 2PL model fitted to a moderately large data set with 3000 examinees.

13.4 Concluding Remarks

We discussed the basic ideas in Bayesian inference above. Several related concepts were not covered in this chapter. They include Bayesian decision theory, model fit and model selection methods, computations with Bayesian inference, the Markov chain Monte Carlo algorithm, robust inference, and estimation for missing data sets. For further details these topics, see the other chapters in this volume or a source such as Gelman et al. (2003), Carlin and Louis (2009), or Robert (2007).

We mentioned some important examples of Bayesian estimation earlier. In other notable examples, Bayesian estimation was applied by Swaminathan and Gifford to the Rasch model (1982), the 2PL (1985), and the 3PL (1986), by Beguin and Glas (2001) to multivariate IRT models, by Bolt et al. (2001) to a mixture IRT model, by Fox and Glas (2001, 2003) to multilevel IRT models, by Glas and van der Linden (2003) and Sinharay et al. (2003) to a hierarchical IRT model, by Bradlow et al. (1999) to the testlet model, by Patz et al. (2002) to a hierarchical rater model, by Johnson and Junker (2003) to an unfolding IRT model, by Karabatsos and Sheu (2004) to a nonparametric IRT model, by Henson et al. (2009) to a family of cognitive diagnostic models, Klein Entink et al. (2009) to an IRT model that incorporates response times, and by Mariano et al. (2010) to value-added models.

With faster computers resulting in increased computational speed, we expect to see an increased popularity among researchers of IRT models of Bayesian estimation mostly involving the MCMC algorithm. However, such researchers have to use caution. An application of an MCMC algorithm often does not reveal identifiability problems with the model and unless the investigator is careful, the resulting Bayesian inference may be inappropriate. Therefore, before using a complicated or over-parameterized model, one has to make sure that the model fits the data considerably better than a simpler alternative, both substantively and statistically and that the model is well identified. William Cochran, as quoted in Rubin (1984), was unimpressed with statistical work that produced methods for solving nonexistent problems or produced complicated methods that were at best only imperceptibly superior to simple methods already available. Cochran wanted to see statistical methods developed to help solve existing problems which were without currently acceptable solutions. Bayesian estimation has a great potential to provide methods of the type Cochran wanted to see for IRT models.

References

Albert, J. H. 1992. Bayesian estimation of normal ogive item response curves using Gibbs sampling. *Journal of Educational Statistics, 17*, 251–269.

Albert, J. H. and Chib, S. 1993. Bayesian analysis of binary and polychotomous response data. *Journal of the American Statistical Association, 88*, 669–679.

Bates, D. and Maechler, M. 2010. lme4: Linear mixed-effects models using S4 classes. Available at http://CRAN.R-project.org/package=lme4. R package version 0.999375-37.

Beguin, A. A. and Glas, C. A. W. 2001. MCMC estimation and some model-fit analysis of multidimensional IRT models. *Psychometrika, 66*, 541–561.

Bock, R. and Lieberman, M. 1970. Fitting a response model for n dichotomously scored items. *Psychometrika, 35*, 179–197.

Bolt, D. M., Cohen, A. S., and Wollack, J. A. 2001. A mixture item response model for multiple-choice data. *Journal of Educational and Behavioral Statistics, 26*, 381–409.

Bradlow, E. T., Wainer, H., and Wang, X. 1999. A Bayesian random effects model for testlets. *Psychometrika, 64*, 153–168.

Carlin, B. and Louis, T. 2009. *Bayesian Methods for Data Analysis* (3rd ed.). Boca Raton, FL: CRC Press.

Chen, M. H. and Shao, Q. M. 1999. Monte Carlo estimation of Bayesian credible and HPD intervals. *Journal of Computational and Graphical Statistics, 8*, 69–92.

DeGroot, M. 1986. *Probability and Statistics* (2nd ed.). Reading, MA: Addison-Wesley.

Entink, R. H. K., Fox, J. P., and van der Linden, W. J. 2009. A multivariate multilevel approach to the modeling of accuracy and speed of test takers. *Psychometrika, 74*, 21–48.

Fox, J. P. and Glas, C. A. W. 2001. Bayesian estimation of a multi-level IRT model using Gibbs sampling. *Psychometrika, 66*, 271–288.

Fox, J. P. and Glas, C. A. W. 2003. Bayesian modeling of measurement error in predictor variables using item response theory. *Psychometrika, 68*, 169–191.

Gelfand, A. E., Dey, D. K., and Chang, H. 1992. Model determination using predictive distributions with implementation via sampling-based methods. In Bernardo, J. M., Berger, J. O., Dawid, A. P., and Smith, A. F. M. (Eds.), *Bayesian Statistics*, Oxford University Press, vol. 4, pp. 147–68.

Gelman, A., Carlin, J. B., Stern, H. S., and Rubin, D. B. 2003. *Bayesian Data Analysis*. New York: Chapman & Hall.

Geman, S. and Geman, D. 1984. Stochastic relaxation, Gibbs distributions and the Bayesian restoration of images. *IEEE Transactions on Pattern Analysis and Machine Intelligence, 6*, 721–741.

Geweke, J. 1989. Bayesian inference in econometric models using Monte Carlo integration. *Econometrica, 57*, 1317–1339.

Geyer, C. J. 1992. Practical Markov chain Monte Carlo (with discussion). *Statistical Science, 7*, 473–511.

Geyer, C. J. and Meeden, G. D. 2005. Fuzzy and randomized confidence intervals and p-values. *Statistical Science, 20*, 358–387.

Gilks, W. and Wild, P. 1992. Adaptive rejection sampling for Gibbs sampling. *Applied Statistics, 41*, 337–348.

Glas, C. A. W. and van der Linden, W. J. 2003. Computerized adaptive testing with item clones. *Applied Psychological Measurement, 27*, 247–261.

Hastings, W. 1970. Monte Carlo sampling methods using Markov chains and their applications. *Biometrika, 57*, 97–109.

Henson, R., Templin, J., and Willse, J. 2009. Defining a family of cognitive diagnosis models using log-linear models with latent variables. *Psychometrika, 74*, 191–210.

Jeffreys, H. 1961. *Theory of Probability* (3rd ed.). Oxford: Oxford University Press.

Johnson, M. S. 2001. *Parametric and Non-Parametric Extensions to Unfolding Response Models* (Doctoral thesis). Pittsburgh, PA: Carnegie Mellon University.

Johnson, M. S. and Junker, B. J. 2003. Using data augmentation and Markov chain Monte Carlo for the estimation of unfolding response models. *Journal of Educational and Behavioral Statistics, 88*, 195–230.

Karabatsos, G. and Sheu, C.-F. 2004. Bayesian order constrained inference for dichotomous models of unidimensional non-parametric item response theory. *Applied Psychological Measurement, 28*, 110–125.

Kass, R. E. and Raftery, A. E. 1995. Bayes factors. *Journal of the American Statistical Association, 90*, 773–795.

Mariano, L. T., McCaffrey, D. F., and Lockwood, J. R. 2010. A model for teacher effects from longitudinal data without assuming vertical scaling. *Journal of Educational and Behavioral Statistics, 35*, 253–279.

Metropolis, N., Rosenblush, A., Rosenbluth, M., Teller, A., and Teller, E. 1953. Equations of state calculations by fast computing machines. *Journal of Chemical Physics, 21*, 1087–1091.

Mislevy, R. J. 1986. Bayes modal estimation in item response models. *Psychometrika, 51*, 177–195.

Neal, R. 2003. Slice sampling. *Annals of Statistics, 31*, 705–767.

von Neumann, J. 1951. Various techniques used in connection with random digits. *National Bureau of Standards Applied Mathematics Series, 12*, 36–38.

Patz, R. and Junker, B. 1999a. A straightforward approach to Markov chain Monte Carlo methods for item response models. *Journal of Educational and Behavioral Statistics, 24*, 146–178.

Patz, R. and Junker, B. 1999b. Applications and extension of MCMC in IRT. *Journal of Educational and Behavioral Statistics, 24*, 342–366.

Patz, R., Junker, B., Johnson, M. S., and Mariano, L. 2002. The hierarchical rater model for rated test items and its application to large-scale educational assessment data. *Journal of Educational and Behavioral Statistics, 27*, 341–384.

R Development Core Team. 2010. *R: A Language and Environment for Statistical Computing*. Vienna, Austria: R Foundation for Statistical Computing. Retrieved from http://www.R-project.org/.

Rizopoulos, D. 2006. ltm: An R package for latent variable modeling and item response theory analyses. *Journal of Statistical Software, 17*, 1–25. Retrieved from http://www.jstatsoft.org/v17/i05/.

Robert, C. P. 2007. *The Bayesian Choice* (2nd ed.). New York: Springer.

Rubin, D. R. 1984. William G. Cochranś contributions to the design, analysis, and evaluation of observational studies. In P.S.R.S. Rao and J. Sedransk (Eds.), *William G. Cochran's Impact on Statistics.* New York: Wiley.

Schervish, M. J. 1995. *Theory of Statistics.* New York: Springer.

Silvapulle, M. and Sen, P. 2004. *Constrained Statistical Inference: Order, Inequality, and Shape Constraints.* Hoboken, NJ: John Wiley & Sons.

Sinharay, S., Johnson, M., and Williamson, D. 2003. Calibrating item families and summarizing the results using family expected response functions. *Journal of Educational and Behavioral Statistics, 28,* 295–313.

Swaminathan, H. and Gifford, J. A. 1982. Bayesian estimation in the Rasch model. *Journal of Educational Statistics, 7,* 175–192.

Swaminathan, H. and Gifford, J. A. 1985. Bayesian estimation in the two-parameter logistic model. *Psychometrika, 50,* 349–364.

Swaminathan, H. and Gifford, J. A. 1986. Bayesian estimation in the three-parameter logistic model. *Psychometrika, 51,* 589–601.

14

Variational Approximation Methods

Frank Rijmen, Minjeong Jeon, and Sophia Rabe-Hesketh

CONTENTS

14.1 Introduction

It is generally acknowledged that item response theory (IRT) models are closely related to item factor analysis models (Takane and de Leeuw, 1987; Volume One, Chapter 31). Both types of models can be formulated as generalized linear and nonlinear mixed models (Rijmen et al., 2003; Volume One, Chapters 31 and 33). The adoption of a mixed-model framework has greatly expanded the scope of IRT models because standard IRT models can be readily modified and generalized within the framework. For example, a clustered data structure on the person side can be accounted for by including random effects for clusters of persons (Volume One, Chapters 24 and 30). Other examples of IRT models that were presented in *Handbook of Item Response Theory, Volume One: Models* and fall under the mixed-model framework are the hierarchical rating model with random effects for raters (Volume One, Chapter 27) and item-family models with random item effects (Volume One, Chapter 26).

While it has become relatively straightforward to formulate complex IRT models with many latent variables (at different levels) within a mixed-model framework, estimating the parameters of these models has remained a technical challenge and is still an active area of research. Maximum-likelihood estimation of model parameters in nonlinear mixed models involves integration over the space of all random effects. In general, the integrals have no closed-form solution and have to be approximated. However, numerical integration over the joint space of all latent variables can become computationally very demanding. As an alternative to (full-information) maximum-likelihood estimation, the so-called limited information techniques have been developed in the field of structural equation modeling to deal with binary and ordered categorical observed (indicator) variables (Jöreskog,

1994; Muthén, 1984). Other approximate methods have been suggested in the literature on nonlinear mixed models, such as the Laplace approximation (Pinheiro and Bates, 1995) and penalized quasi-likelihood methods (Breslow and Clayton, 1993). Yet, none of these methods seems to be accurate and applicable in a wide range of situations.

Stochastic approximation methods have been proposed both in a maximum-likelihood and a Bayesian framework (see Chapters 13 and 15). In principle, they can be used to approximate the likelihood (or the posterior in a Bayesian framework) to any required degree of accuracy. In practice, sampling methods can become computationally very demanding for large-scale problems (Bishop, 2006, Chapter 10). In this chapter, we introduce the use of variational approximation methods for estimating the parameters of IRT models. Unlike Markov chain Monte Carlo (MCMC) techniques, variational methods are deterministic in nature, in the sense that they rely on an analytic approximation (i.e., a lower bound) to the likelihood (or the posterior in a Bayesian framework; Bishop, 2006, Chapter 10). Just like MCMC techniques, variational approximation methods have roots in statistical physics (e.g., Parisi, 1988). Variational approximation methods are mostly implemented in a two-step maximization scheme that can be seen as a generalization of the expectation maximization (EM) algorithm. Therefore, we briefly consider the EM algorithm first.

14.2 EM Algorithm

The EM algorithm is a powerful algorithm to obtain parameter estimates for models that contain latent variables (see Chapter 12; Dempster et al., 1977). The algorithm alternates between two steps until convergence: First, in the expectation-step (E-step) of iteration m, the expectation of the complete-data log-likelihood is computed. The complete data consist of both the observed data $\mathbf{u}$, and the latent variables or missing data $\mathbf{z}$. The expectation is computed over the posterior predictive distribution of the missing data $\mathbf{z}$ given the observed data $\mathbf{u}$ and parameter estimates $\psi^{(m-1)}$ from the previous iteration $m-1$,

$$\begin{aligned} Q(\psi;\psi^{(m-1)}) &= E\left\{\log f(\psi;\mathbf{u},\mathbf{z})\,|\mathbf{y}\right\} \\ &= \int_{\mathbf{z}} \log\left[f(\psi;\mathbf{u},\mathbf{z})\right] f(\mathbf{z}|\mathbf{u};\psi^{(m-1)})\,d\mathbf{z} \end{aligned} \tag{14.1}$$

where $f(\psi;\mathbf{u},\mathbf{z})$ is the complete-data likelihood and $f(\mathbf{z}|\mathbf{u};\psi^{(m-1)})$ is the posterior density of $\mathbf{z}$ given $\mathbf{u}$. Second, in the maximization-step (M-step), $Q(\psi;\psi^{(m-1)})$ is maximized with respect to the parameters ψ, leading to new parameter estimates $\psi^{(m)}$.

A primary determinant of the computational complexity of the EM algorithm is the difficulty with which the integral over the latent variables in Equation 14.1 has to be computed. Let us illustrate this for the Rasch model (Volume One, Chapter 3). In the context of this model, the observed data $\mathbf{u}$ are the observed binary responses u_{pi}, where $p = 1,\ldots,P$ denotes the persons and $i = 1,\ldots,I$ refers to the items. The vector of responses for person p is denoted by $\mathbf{u}_p$, and $\mathbf{u}$ is the stacked vector of all $\mathbf{u}_p$ (without loss of generality, we assume a complete data collection design). The unknown abilities θ_p, $p = 1,\ldots,P$, collected in vector θ are treated as missing data. Hence, Equation 14.1 instantiates as

$$Q(\psi;\psi^{(m-1)}) = \int\limits_{\theta} \log\left[f(\psi;\mathbf{u},\theta)\right] f(\theta|\mathbf{u};\psi^{(m-1)})\,d\theta \tag{14.2}$$

Under the assumption of independence between persons, both the complete-data likelihood $f(\psi;\mathbf{u},\theta)$ and the posterior density $f(\theta|\mathbf{u};\psi^{(m-1)})$ factor into a product of person-specific terms. Therefore, $Q(\psi;\psi^{(m-1)})$ can be rewritten as

$$\begin{aligned} Q(\psi;\psi^{(m-1)}) &= \int\limits_{\theta} \log\left[f(\psi;\mathbf{u},\theta)\right] f(\theta||\mathbf{u};\psi^{(m-1)})\,d\theta \\ &= \int\limits_{\theta} \log\left[\prod_p f_p(\psi;\mathbf{u}_p,\theta_p)\right]\left[\prod_p f_p(\theta_p||\mathbf{u}_p;\psi^{(m-1)})\right] d\theta \\ &= \sum_p \int\limits_{\theta_p} \log[f_p(\psi;\mathbf{u}_p,\theta_p)] f_p(\theta_p|\mathbf{u}_p;\psi^{(m-1)})\,d\theta_p \end{aligned} \tag{14.3}$$

It is now clear that, for the Rasch model, the evaluation of $Q(\psi;\psi^{(m-1)})$ involves a numerical integration over unidimensional latent variables.

The factorization of the integrand in Equation 14.3 is rather straightforward; it follows naturally from the assumptions of unidimensionality and independence between persons. For more complex models than the Rasch model, it may be less straightforward or not possible to split the integrand into factors that are low dimensional in the latent variables. It is obvious from Equation 14.3 that such a factorization of the complete-data likelihood is a necessary but insufficient condition. Even when a convenient factorization of the likelihood function is possible, the factorization of the integrand of Equation 14.3 requires a corresponding factorization of the posterior.

It also follows from Equation 14.3 that the number of latent variables is not a good measure for computational complexity. For example, for the bifactor and related models (Gibbons and Hedeker, 1992; Jeon et al., 2013; Rijmen, 2010, 2011; Volume One, Chapter 25), the posterior distributions for each person $f(\theta_p|\mathbf{u}_p;\psi)$ can be split into factors containing a small number of latent variables, even though $\theta_p = (\theta_{p1},\ldots,\theta_{pD})'$ itself can be high dimensional. Another example of a model with a convenient factorization of the posterior distribution is the multilevel IRT model (Rabe-Hesketh et al., 2004; Volume One, Chapter 24).

Thus, a crucial question is whether or not it is possible to rewrite the posterior density as a product of factors that are low dimensional in the latent variables. Because prior independence of latent variables does not necessarily imply posterior independence, the answer is not trivial. Graphical models turn out to be quite useful in answering the question, the reason being a correspondence between conditional independence of (sets of) variables in a statistical model and the property of separation in the graph representing its conditional dependence structure. For an introduction to graphical models, see, for example, Cowell et al. (1999).

14.3 Variational Maximization-Maximization Algorithm

As just noted, several IRT models allow for factorization of their complete-data likelihood into building blocks that are low dimensional in the number of latent variables but miss this feature for their posterior density $f(\mathbf{z}|\mathbf{u};\psi)$. Exact maximum-likelihood inference for this class of models is computationally prohibitive. The remainder of this chapter is about an alternative estimation method for this class of models. For simplicity but without loss of generality, we will discuss the method using a Rasch model with random item effects (e.g., De Boeck, 2008). The model can be written as a generalized linear mixed model for binary data with crossed random effects:

$$\text{logit}\left[P\left\{U_{pi}=1\,|\theta_p,\delta_i\right\}\right]=\theta_p-\delta_i \tag{14.4}$$

where U_{pi} is the random variable for a binary response by person p on item i, θ_p is the ability of person p, and δ_i is the difficulty of item i. All random person effects are assumed to follow independent normal densities $\phi(\theta_p;\beta,\tau_\theta)$ with mean β and standard deviation τ_θ. Similarly, all random item effects are assumed to be independent with normal density $\phi(\delta_i;0,\tau_\delta)$ with mean 0 (to identify the model) and standard deviation τ_δ. We use the symbol δ_i for random item difficulty to contrast it with the fixed item difficulty parameter b_i of the regular Rasch model (Volume One, Chapter 3). The vector of all random item effects is denoted as $\boldsymbol{\delta}$. In terms of the general notation introduced earlier, for the Rasch model with random item effects, $\mathbf{z}$ and ψ are instantiated by (θ,δ') and $(\beta,\tau_\theta,\tau_\delta)'$, respectively.

Conditional on the person and item effects, all responses are assumed to be independent. As a result, the complete-data likelihood factors as

$$f(\psi;\mathbf{u},\theta,\delta)=\left[\prod_i\prod_p P\left\{u_{pi}\,|\theta_p,\delta_i\right\}\right]\left[\prod_p\phi\left(\theta_p;\beta,\tau_\theta\right)\right]\left[\prod_i\phi\left(\delta_p;0,\tau_\delta\right)\right] \tag{14.5}$$

However, the posterior $f(\theta,\delta|\mathbf{u};\psi)$ does not factor as a product over independent person-specific posterior distributions as was the case for the Rasch model (Equation 14.3, second line). Consequently, exact maximum-likelihood inference appears to be computationally prohibitive.

In the variational estimation approach we will present, $f(\mathbf{z}|\mathbf{u};\psi)$ is approximated by an alternative density $g(\mathbf{z})$ that does allow for the required factorization. The quality of the approximation is expressed by the Kullback–Leibler (KL) divergence (see Chapter 7; Kullback and Leibler, 1951) of $f(\mathbf{z}|\mathbf{u};\psi)$ from $g(\mathbf{z})$,

$$\text{KL}(g(\mathbf{z}),f(\mathbf{z}|\mathbf{u};\psi))=\int_{\mathbf{z}}g(\mathbf{z})\log\left[\frac{g(\mathbf{z})}{f(\mathbf{z}|\mathbf{u};\psi)}\right]d\mathbf{z} \tag{14.6}$$

The measure is always nonnegative and equals zero if and only if $g(\mathbf{z})=f(\mathbf{z}|\mathbf{u};\psi)$ almost everywhere. In order to find $g(\mathbf{z})$, it is computationally more convenient to introduce a lower bound $\underline{l}(\psi;\mathbf{u})$ to the marginal log-likelihood $l(\psi;\mathbf{u})$, which can be derived using

Jensen's inequality (Neal and Hinton, 1998),

$$\begin{aligned} l(\psi; \mathbf{u}) &\equiv \log \int_{\mathbf{z}} f(\psi; \mathbf{u}, \mathbf{z})\, d\mathbf{z} \\ &= \log \int_{\mathbf{z}} g(\mathbf{z}) \frac{f(\psi; \mathbf{u}, \mathbf{z})}{g(\mathbf{z})}\, d\mathbf{z} \\ &= \log E_g \left\{ \frac{f(\psi; \mathbf{u}, \mathbf{z})}{g(\mathbf{z})} \right\} \\ &\geq E_g \left\{ \log \left[\frac{f(\psi; \mathbf{u}, \mathbf{z})}{g(\mathbf{z})} \right] \right\} \\ &= \int_{\mathbf{z}} g(\mathbf{z}) \log \left[\frac{f(\psi; \mathbf{u}, \mathbf{z})}{g(\mathbf{z})} \right] d\mathbf{z} \equiv \underline{l}(\psi; \mathbf{u}) \end{aligned} \tag{14.7}$$

where $E_g\{\cdot\}$ denotes an expectation taken with respect to the density $g(\mathbf{z})$ rather than the true posterior density $f(\mathbf{z}|\mathbf{u}; \psi)$. The difference between the marginal log-likelihood $l(\psi; \mathbf{u})$ and its lower bound $\underline{l}(\psi; \mathbf{u})$ amounts to the KL divergence of $f(\mathbf{z}|\mathbf{u}; \psi)$ from $g(\mathbf{z})$,

$$l(\psi; \mathbf{u}) = \underline{l}(\psi; \mathbf{u}) + \mathrm{KL}(g(\mathbf{z}), f(\mathbf{z}||\mathbf{u}; \psi)) \tag{14.8}$$

Hence, for a given set of parameters $\psi^{(m-1)}$, maximizing the lower bound $\underline{l}(\psi^{(m-1)}; \mathbf{u})$ with respect to $g(\mathbf{z})$ is equivalent to minimizing $\mathrm{KL}(g(\mathbf{z}), f(\mathbf{z}|\mathbf{u}; \psi^{(m-1)}))$.

All parameters can now be estimated using an iterative maximization maximization (MM) algorithm. First, given a set of provisional model parameter estimates $\psi^{(m-1)}$, $\underline{l}(\psi^{(m-1)}; \mathbf{u})$ is optimized with respect to $g(\mathbf{z})$. Second, given an updated $g(\mathbf{z})$, $\underline{l}(\psi^{(m-1)}; \mathbf{u})$ is maximized with respect to the model parameters ψ.

The MM algorithm is a generalization of the traditional EM algorithm. The latter is obtained when no restrictions are imposed on the functional form of $g(\mathbf{z})$. In that case, $\underline{l}(\psi^{(m-1)}; \mathbf{u})$ is maximized by choosing $f(\mathbf{z}|\mathbf{u}; \psi^{(m-1)})$ for $g(\mathbf{z})$ in the first M-step, because maximizing the lower bound $\underline{l}(\psi^{(m-1)}; \mathbf{u})$ with respect to $g(\mathbf{z})$ is equivalent to minimizing $\mathrm{KL}(g(\mathbf{z}), f(\mathbf{z}|\mathbf{u}; \psi^{(m-1)}))$. Furthermore, the second M-step of the MM algorithm is equivalent to the M-step of the EM algorithm: Maximizing $\underline{l}(\psi^{(m-1)}; \mathbf{u})$ with respect to the model parameters ψ is equivalent to maximizing $Q(\psi; \psi^{(m-1)})$ (computed with respect to $g(\mathbf{z})$), because $g(\mathbf{z})$ is kept constant and thus does not depend on ψ during the second M-step.

14.4 Mean-Field Approximation

Obviously, the quality of the MM algorithm depends on the choice of the family of distributions for $g(\mathbf{z})$. On the one hand, the family should be sufficiently general so that the first step of the MM algorithm results in a choice for $g(\mathbf{z})$ close to the true posterior $f(\mathbf{z}|\mathbf{u}; \psi^{(m-1)})$. On the other hand, it should be possible to factor each member of the family into a product over factors that are low dimensional in the number of latent variables. Several families of distributions have been proposed (for an overview, see Humphreys and

Titterington, 2003). The mean-field approximation is the simplest one. It approximates the posterior as a product of densities over individual latent variables,

$$g(\mathbf{z}) = \prod_v g_v(z_v) \tag{14.9}$$

where $v = 1, \ldots, V$, with V the number of latent variables in the model. The product density approximation in Equation 14.9 resembles the pseudolikelihood approach of Besag (1975), in which a joint density of observed variables is approximated by the product of the univariate full conditional densities. Archer and Titterington (2002) showed how replacement of the conditioned variables by their mean values results in a mean-field approximation. The product density approximation is equivalent to the assumption of all latent variables being conditionally independent given the observed data.

14.5 Calculus of Variation

Once a family of distributions for $g(\mathbf{z})$ has been specified, it becomes a matter of finding a member of the family that maximizes the lower bound $\underline{l}(\psi; \mathbf{u})$. Traditionally, this has been framed as a functional optimization problem. The lower bound is a functional that maps the function $g(\mathbf{z})$ onto the numerical value $\underline{l}(\psi; \mathbf{u})$. The task at hand is to find the function $g(\mathbf{z})$ that maximizes the functional $\underline{l}(\psi; \mathbf{u})$, subject to the constraints that all individual $g_v(z_v)$ integrate to 1. The constraints stem from the fact that we are optimizing a functional with respect to density functions. The solution to this problem can be found using the calculus of variation and its concept of a functional derivative (for these basic concepts, see Bishop, 2006, Appendix D). We illustrate the idea for the Rasch model with fixed and with random item effects, respectively.

14.5.1 Rasch Model with Fixed Item Effects

In the Rasch model with fixed item effects, the mean-field approximation is

$$g(\theta) = \prod_p g_p(\theta_p) \tag{14.10}$$

The lower bound to the log-likelihood is obtained by substituting Equation 14.10 for $g(\mathbf{z})$ and $\prod_p f_p(\psi; \mathbf{u}_p, \theta_p)$ for $f(\psi; \mathbf{u}, \mathbf{z})$ (per Equation 14.3) in Equation 14.7. The result is

$$\begin{aligned}
\underline{l}(\psi; \mathbf{u}) &= \int_\theta \log\left[\prod_p f_p(\psi; \mathbf{u}_p, \theta_p)\right]\left[\prod_p g_p(\theta_p)\right] d\theta - \int_\theta \log\left[\prod_p g_p(\theta_p)\right]\left[\prod_p g_p(\theta_p)\right] d\theta \\
&= \sum_p \int_{\theta_p} \log\left[f_p(\psi; \mathbf{u}_p, \theta_p)\right] g_p(\theta_p)\, d\theta_p \\
&\quad - \sum_p \int_{\theta_p} \log\left[g_p(\theta_p)\right] g_p(\theta_p)\, d\theta_p
\end{aligned} \tag{14.11}$$

For each $g_p(\theta_p)$, keeping all other $g_q(\theta_q)$ fixed and adding a Lagrange multiplier λ_p for the normalization constraint, we obtain the functional

$$F_p = \int_{\theta_p} \log\left[f_p(\psi; \mathbf{u}_p, \theta_p)\right] g_p(\theta_p)\, d\theta_p - \int_{\theta_p} \log\left[g_p(\theta_p)\right] g_p(\theta_p)\, d\theta_p$$

$$+ \lambda_p \left[\int_{\theta_p} g_p(\theta_p)\, d\theta_p - 1\right] + \text{Constant} \tag{14.12}$$

This expression for F_p has a special form for which the functional derivative vanishes when the derivative of the integrand with respect to $g_p(\theta_p)$ is 0 for all values of θ_p (see Bishop, 2006, Appendix D). Hence, it is stationary for

$$g_p(\theta_p) = f_p(\psi; \mathbf{u}_p, \theta_p) \exp(\lambda_p - 1) \tag{14.13}$$

Integrating over θ_p, we obtain

$$1 = \exp(\lambda_p - 1) \int_{\theta_p} f_p(\psi; \mathbf{u}_p, \theta_p)\, d\theta_p \tag{14.14}$$

and thus

$$\exp(\lambda_p - 1) = \left[\int_{\theta_p} f_p(\psi; \mathbf{u}_p, \theta_p)\, d\theta_p\right]^{-1} \tag{14.15}$$

Substituting Equation 14.15 into Equation 14.13 shows that for the Rasch model with fixed item effects, the solution for each $g_p(\theta_p)$ is the corresponding posterior distribution,

$$g_p(\theta_p) = f(\theta_p | \mathbf{u}_p; \psi) \tag{14.16}$$

For the Rasch model with fixed item effects, the solution is exactly what we expected. The true posterior factors into a product of the individual posteriors because responses of different persons are independent (see Equation 14.3). Therefore, the true posterior distribution belongs to the family of approximating distributions that are considered under the mean-field approximation. As we have seen before, the lower bound $\underline{l}(\psi; \mathbf{y})$ is maximal under the true posterior.

14.5.2 Rasch Model with Random Item Effects

For the case of random item effects, we have seen that there is no elegant factorization of the posterior distribution of the latent variables. Nevertheless, the same principles of the calculus of variation can be applied to obtain a solution for $g(\mathbf{z})$ that maximizes the functional $\underline{l}(\psi; \mathbf{u})$. The mean-field approximation is

$$g(\theta, \delta) = \left[\prod_p g_p(\theta_p)\right]\left[\prod_i g_i(\theta_i)\right] \tag{14.17}$$

The lower bound to the log-likelihood is obtained by substituting Equation 14.17 for $g(\mathbf{z})$ and Equation 14.5 for $f(\psi;\mathbf{u},\mathbf{z})$ in Equation 14.7. Again, for each $g_p(\theta_p)$, keeping all other $g_q(\theta_q)$ and all $g_i(\delta_i)$ fixed, and adding a Lagrange multiplier for the normalization constraint, we obtain the functional

$$F_p = \int_{\theta_p} \log[\phi(\theta_p;\beta,\tau_\theta)] g_p(\theta_p)\,d\theta_p + \int_{\theta_p} \left[g_p(\theta_p) \sum_i \int_{\delta_i} \log[P\{u_{pi}|\theta_p,\delta_i\}] g_i(\delta_i)\,d\delta_i \right] d\theta_p - \int_{\theta_p} \log[g_p(\theta_p)] g_p(\theta_p)\,d\theta_p + \lambda_p \left[\int_{\theta_p} g_p(\theta_p)\,d\theta_p - 1 \right] + \text{Constant} \tag{14.18}$$

The result is again of a special form for which the functional derivative vanishes when the derivative of the integrand with respect to $g_p(\theta_p)$ is 0 for all values of θ_p. The solutions for the mean-field approximations for the individual posteriors for the random person effects are

$$\log g_p(\theta_p) = \log \phi(\theta_p;\beta,\tau_\theta) + \sum_i \int_{\delta_i} \log[P\{u_{pi}|\theta_p,\delta\}] g_i(\delta_i)\,d\delta_i - 1 + \lambda_p \tag{14.19}$$

By taking the exponential and integrating over θ_p, we obtain

$$1 = \exp(\lambda_p - 1) \int_{\theta_p} \phi(\theta_p;\beta,\tau_\theta) \left[\sum_i \int_{\delta_i} \log[P\{u_{pi}|\theta_p,\delta_i\}] g_i(\delta_i)\,d\delta_i \right] d\theta_p \tag{14.20}$$

and thus

$$\lambda_p - 1 = -\log \int_{\theta_p} \phi(\theta_p;\beta,\tau_\theta) \left[\sum_i \int_{\delta_i} \log[P\{u_{pi}|\theta_p,\delta_i\}] g_i(\delta_i)\,d\delta_i \right] d\theta_p \tag{14.21}$$

Substituting Equation 14.21 into Equation 14.19 shows that for the model with random item effects, the optimal choice for $g_p(\theta_p)$ is at all values of θ_p subject to

$$g_p(\theta_p) = \frac{\phi(\theta_p;\beta,\tau_\theta)\exp\left(\sum_i \int_{\delta_i} \log[P\{u_{pi}|\theta_p,\delta_i\}] g_i(\delta_i)\,d\delta_i\right)}{\int_{\theta_p} \phi(\theta_p;\beta,\tau_\theta)\exp\left(\sum_i \int_{\delta_i} \log[P\{u_{pi}|\theta_p,\delta_i\}] g_i(\delta_i) d\delta_i\right) d\theta_p} \tag{14.22}$$

The solutions for $g_i(\delta_i)$ can be derived in a similar way and are given by

$$g_i(\delta_i) = \frac{\phi(\delta_i;0,\tau_\delta)\exp\left(\sum_p \int_{\theta_p} \log[P\{u_{pi}|\theta_p,\delta_i\}] g_p(\theta_p)\,d\theta_p\right)}{\int_{\delta_i} \phi(\delta_i;0,\tau_\delta)\exp\left(\sum_p \int_{\theta_p} \log[P\{u_{pi}|\theta_p,\delta_i\}] g_p(\theta_p) d\theta_p\right) d\delta_i} \tag{14.23}$$

The remaining integrals in Equations 14.22 and 14.23 can be approximated using Gaussian quadrature or adaptive quadrature. In doing so, the masses $g_p(\theta_p = l_q)$ and $g_i(\delta = l_r)$ at each of the locations l_q, $q = 1, \ldots, Q$, and l_r, $r = 1, \ldots, R$, become "variational" parameters. The solution for each of the variational parameters $g_p(\theta_p = l_q)$ depends on the set of variational parameters $g_i(\delta = l_r)$, $r = 1, \ldots, R$. Reversely, the solution for each of the variational parameters $g_i(\delta = l_r)$ depends on the set of variational parameters $g_p(\theta_p = l_q)$, $q = 1, \ldots, Q$. Therefore, after initialization, Equations 14.22 and 14.23 are updated recursively until convergence. For a detailed explanation, see Jeon et al. (2012) and Rijmen and Jeon (2013).

14.6 Application

The variational method was applied to the random-effects version of the linear logistic test model (Janssen et al., 2004), that is, a Rasch model with the item difficulties modeled as a linear function of covariates and a random error term for items (see Volume One, Chapter 13). Janssen et al. (2004) applied the model to the verbal aggression data (Vansteelandt, 2000) used throughout De Boeck and Wilson (2004). These data consist of the responses of 316 first-year psychology students (73 men and 243 women) on 24 items. Each item is a combination of one of four scenarios or situations (e.g., "A bus fails to stop for me") of two types (self-to-blame vs. other-to-blame), a behavior (cursing, scolding, and shouting), and behavior mode (doing vs. wanting). For each combination, students were asked whether they were likely to exhibit the behavior. These questionnaire design factors were the fixed effects in the linear logistic test model with random item effects. Janssen et al. (2004) formulated the model in a Bayesian framework with normal priors for the regression coefficients and inverse-$\chi^2(1)$ distributions for the variance parameters. She fitted the model using an MCMC method. Table 14.1 contains the posterior means reported by Janssen et al. (2004, Table 6.2), along with the parameter estimates that were obtained using the variational method presented in this chapter. The variational method was implemented in MATLAB® and took less than 10 min on a 2.67-GHz processor computer with a 32-bit operating system and 3 GB of usable RAM. We used adaptive quadrature to approximate the integrals in Equations 14.22 and 14.23 with 10 nodes per random effect. Except for τ_δ^2, the parameter estimates obtained with the variational method are nearly identical to the posterior means obtained with MCMC. The variational estimate for τ_δ^2 is slightly smaller than the posterior mean obtained with MCMC (0.12 vs. 0.16). Differences between the mode of the likelihood and the mean of the posterior of this size are not surprising given the small number of items (e.g., Browne and Draper, 2006).

14.7 Discussion

Notwithstanding huge improvements in computational resources and advances in estimation methods, parameter estimation still remains a challenge for complex IRT models, especially for models with random item effects. Due to their crossed structure, with random effects both on the item and person side, maximum-likelihood estimation involves

TABLE 14.1

Estimation of the Item Parameters for the LLTM with Random Item Effects (Verbal Aggression Data)

	Variational	Bayesian
Fixed effects		
Do versus want	0.71	0.71
Other-to-blame	−1.05	−1.05
Blaming[a]	−1.40	−1.39
Expressing[b]	−0.71	−0.70
Intercept	0.33	0.33
Random effects		
τ^2_δ	0.12	0.16
τ^2_θ	1.90	1.89

Note: Bayesian estimates from Janssen et al. (2004, Table 6.2). Variable coding from De Boeck and Wilson (2004, p. 63) was used.

[a] (Curse and scold) versus shout.

[b] (Curse and shout) versus scold.

numerical integration over high-dimensional spaces, quickly resulting in a computational burden if there are more than a few items. Sampling based techniques, such as Monte Carlo EM or MCMC methods, also can become computationally very intensive with increasing dimensionality in this case.

A variational method was presented as an alternative estimation method. In this method, the log-likelihood is approximated by a computationally tractable lower bound. The difference between the log-likelihood function and its lower bound equals the KL divergence of the posterior distribution from its variational approximation. Parameters are estimated using an MM algorithm. In the first maximization step, the lower bound is maximized with respect to the variational approximation while keeping the model parameters constant. In the second step, the lower bound is maximized with respect to the model parameters, with the variational approximation kept constant. The MM algorithm contains the EM algorithm as a special case arising when the variational distribution equals the posterior distribution of the latent variables. We have described the variational approximation method within a maximum-likelihood framework, but note that the method can be applied similarly within a Bayesian framework (see, e.g., Beal, 2003; Minka, 2001).

Many different choices can be made with respect to the variational distribution. Ideally, the distribution is chosen to resemble the true posterior closely, while keeping the evaluation of the resulting lower bound to the log-likelihood function computationally tractable. The parameters that characterize the variational densities are called variational parameters. Their role is very different from the role of model parameters. In particular, there is no risk of over-fitting (Bishop, 2006, Chapter 10). A more flexible variational distribution will provide a tighter lower bound and is worth pursuing as long as it is computationally tractable. In this chapter, we made use of the mean-field approximation, approximating the joint posterior of the latent variables by a product of individual densities, as if the latent variables given the observed responses were independent. An alternative approach is to choose a multivariate normal distribution so that the high-dimensional integrals can be evaluated analytically (Ormerod and Wand, 2012). Humphreys and Titterington (2003) gave an overview of other choices for variational distributions.

A general principle of variational approximation is the approximation of the complicated joint posterior distribution of the latent variables by a product of low-dimensional distributions. Regardless of the specific choice made, a necessary condition is a similar factorization of the complete-data likelihood. If this condition is not met, high-dimensional numerical integration cannot be avoided, as was discussed at some length in the text after Equations 14.2 and 14.3. For example, an exploratory multidimensional IRT model with a large number of dimensions in which every item loads on every dimension will continue to pose a technical challenge, no matter how the posterior distribution of the latent variables is approximated.

The mean-field approximation was used to estimate the parameters of the random-effects version of the linear logistic test model (Janssen et al., 2004). Applied to the verbal aggression data (Vansteelandt, 2000), the parameter estimates were very close to the posterior means obtained from MCMC. The largest difference was observed for the variance of the random item effects. Rijmen and Jeon (2013) applied the mean-field approximation to estimate the parameters of a multigroup IRT model with item parameters that were random over groups (De Jong et al., 2007; De Jong and Steenkamp, 2010). In this application, the variational method was implemented for discrete random effects, and showed fast and accurate performance in a simulation study. Jeon et al. (2012) implemented the method for the linear logistic test model (which includes the Rasch model as a special case) for both discrete and Gaussian random effects. For Gaussian random effects, they implemented Gaussian as well as adaptive quadrature to approximate the remaining integrals. In a simulation study, they found that the variational method performed as well or better than the Laplace approximation.

Further research into the accuracy of variational estimation methods for fitting other complex IRT models is recommended. Moreover, more insights need to be gained into the asymptotic properties of the mean-field and other variational approximations. Nevertheless, the variational approximation method looks promising for the field of psychometrics, either as an estimation method in its own right, or as a fast way to obtain good starting values for more time-consuming methods.

References

Archer, G. E. B. and Titterington, D. M. 2002. Parameter estimation for hidden Markov chains. *Journal of Statistical Planning and Inference*, 108, 365–390.

Beal, M. J. 2003. Variational algorithms for approximate Bayesian inference (unpublished doctoral dissertation). University College, London, UK.

Besag, J. 1975. Statistical analysis of non-lattice data. *Statistician*, 24, 179–195.

Bishop, C. M. 2006. *Pattern Recognition and Machine Learning*. New York: Springer.

Browne, W. J. and Draper, D. 2006. A comparison of Bayesian and likelihood methods for fitting multilevel models. *Bayesian Analysis*, 1, 473–514.

Breslow, N. E. and Clayton, D. G. 1993. Approximate inference in generalized linear mixed models. *Journal of the American Statistical Association*, 88, 9–25.

Cowell, R. G., Dawid, A. P., Lauritzen, S. L., and Spiegelhalter, D. J. 1999. *Probabilistic Networks and Expert Systems*. New York: Springer.

De Boeck, P. 2008. Random item IRT models. *Psychometrika*, 73, 533–559.

De Boeck, P. and Wilson, M. (Eds.). 2004. *Explanatory Item Response Models: A Generalized Linear and Nonlinear Approach*. New York: Springer.

de Jong, M. G., Steenkamp, J.-B. E. M., and Fox, J.-P. 2007. Relaxing measurement invariance in cross-national consumer research using a hierarchical IRT model. *Journal of Consumer Research*, 34, 260–278.
de Jong, M. G. and Steenkamp, J.-B. E. M. 2010. Finite mixture multilevel multidimensional ordinal IRT models for large scale cross-cultural research. *Psychometrika*, 75, 3–32.
Dempster, A. P., Laird, N. M., and Rubin, D. B. 1977. Maximum likelihood from incomplete data via the EM algorithm. *Journal of the Royal Statistical Society, Series B*, 39, 1–38.
Gibbons, R. D. and Hedeker, D. 1992. Full-information item bi-factor analysis. *Psychometrika*, 57, 423–436.
Humphreys, K. and Titterington, D. M. 2003. Variational approximations for categorical causal models with latent variables. *Psychometrika*, 68, 391–412.
Janssen, R., Schepers, J., and Peres, D. 2004. Models with item and item group predictors. In P. De Boeck and M. Wilson (Eds.), *Explanatory Item Response Models: A Generalized Linear and Nonlinear Approach* (pp. 189–212). New York: Springer.
Jeon, M., Rijmen, F., and Rabe-Hesketh, S. 2013. Modeling differential item functioning using a generalization of the multiple-group bifactor model. *Journal of Educational and Behavioral Statistics*, 38, 32–60.
Jeon, M., Rijmen, F., and Rabe-Hesketh, S. 2012, April. Variational maximization-maximization algorithm for generalized linear mixed models with crossed random effects. Paper presented at the *Annual Meeting of the National Council on Measurement in Education*, Vancouver, BC.
Jöreskog, K. G. 1994. On the estimation of polychoric correlations and their asymptotic covariance matrix. *Psychometrika*, 59, 381–389.
Kullback, S. and Leibler, R. A. 1951. On information and sufficiency. *Annals of Mathematical Statistics*, 86, 22–79.
Minka, T. 2001. A family of approximate algorithms for Bayesian inference (unpublished doctoral dissertation). Massachusetts Institute of Technology, Cambridge, MA.
Muthén, B. 1984. A general structural equation model with dichotomous, ordered categorical, and continuous latent variable indicators. *Psychometrika*, 49, 115–132.
Neal, R. M. and Hinton, G. E. 1998. A view of the EM algorithm that justifies incremental, sparse, and other variants. In M. I. Jordan (Ed.), *Learning in Graphical Models* (pp. 355–368). Dordrecht: Kluwer Academic Publishers.
Ormerod, J. T. and Wand, M. P. 2012. Gaussian variational approximate inference for generalized linear mixed models. *Journal of Computational and Graphical Statistics*, 21(1), 2–17.
Parisi, G. 1988. *Statistical Field Theory*. Menlo Park: Addison-Wesley.
Pinheiro, P. C. and Bates, D. M. 1995. Approximations to the log-likelihood function in the nonlinear mixed-effects model. *Journal of Computational and Graphical Statistics*, 4, 12–35.
Rabe-Hesketh, S., Skrondal, A., and Pickles, A. 2004. Generalized multilevel structural equation modeling. *Psychometrika*, 69, 167–190.
Rijmen, F. 2010. Formal relations and an empirical comparison between the bi-factor, the testlet, and a second-order multidimensional IRT model. *Journal of Educational Measurement*, 47, 361–372.
Rijmen, F. 2011. Hierarchical factor item response theory models for PIRLS: Capturing clustering effects at multiple levels. *IERI Monograph Series: Issues and Methodologies in Large-Scale Assessment*, 4, 59–74.
Rijmen, F. and Jeon, M. 2013. Fitting an item response theory model with random item effects across groups by a variational approximation method. *The Annals of Operations Research*, 206, 647–662.
Rijmen, F., Tuerlinckx, F., De Boeck, P., and Kuppens, P. 2003. A nonlinear mixed model framework for item response theory. *Psychological Methods*, 8, 185–205.
Skrondal, A. and Rabe-Hesketh, S. 2004. *Generalized Latent Variable Modeling: Multilevel, Longitudinal and Structural Equation Models*. Boca Raton: Chapman & Hall/CRC.
Takane, Y. and de Leeuw, J. 1987. On the relationship between item response theory and factor analysis of discretized variables. *Psychometrika*, 52, 393–408.
Vansteelandt, K. 2000. Formal models for contextualized personality psychology (unpublished doctoral dissertation). K.U. Leuven, Leuven, Belgium.

15

Markov Chain Monte Carlo for Item Response Models

Brian W. Junker, Richard J. Patz, and Nathan M. VanHoudnos

CONTENTS

15.1 Introduction

Markov chain Monte Carlo (MCMC) has revolutionized modern statistical computing, especially for complex Bayesian and latent variable models. A recent Web of Knowledge search (Thompson ISI, 2012) for "MCMC" yielded 6015 articles, nearly half in statistics, and the rest spread across fields ranging from computational biology to transportation and thermodynamics. Of these, 72 articles appear in *Psychometrika, Journal of Educational and Behavioral Statistics* and *Applied Psychological Measurement*, and another seventeen in *Psychological Methods, Journal of Educational Measurement*, and *Educational and Psychological Measurement*. This is remarkable for an estimation method that first appeared in a psychometrics-related journal around 1990.

For all of its success, it dawned on the broader scientific and statistics community slowly. With roots at the intersection of modern computing and atomic weapons research at Los Alamos, New Mexico during World War II (Los Alamos National Laboratory, 2012; Metropolis, 1987), MCMC began as a method for doing state calculations in physics (Metropolis et al., 1953) and image restoration Geman and Geman (1984), and first came to widespread notice in the field of statistics through the work of Gelfand and Smith (1990), despite earlier pioneering work of Hastings (1970) and Tanner and Wong (1987). A more complete history of MCMC is provided by Robert and Casella (2011), and a very accessible introduction to the method can be found in Chib and Greenberg (1995).

As is evident from its roots (Metropolis, 1987, p. 129; Metropolis et al., 1953), MCMC is not inherently a Bayesian technique. It was used for years as a way of sampling from intractable, often high-dimensional, distributions, and has widespread applications in the integration, estimation, and optimization of functions (e.g., Geyer, 1996). Its success in statistics and psychometrics, however, is driven by the extent to which it makes computation and estimation for complex, novel Bayesian models tractable (e.g., Congdon, 2007; Fox, 2010; Gelman et al., 2003, etc.). In this chapter, we consider the application of MCMC methods to Bayesian item response theory (IRT) and IRT-like models.

15.2 Applied Bayesian Inference

IRT models, and psychometric models generally, deal fundamentally with multiway data. The most common formulation is two-way data consisting of coded responses U_{pi} of persons $p = 1, \ldots, P$ to items (tasks, stimuli, test questions, etc.) $i = 1, \ldots, I$. IRT provides a family of probabilistic models for the two-way array $\mathcal{U} = [U_{pi}]$ of coded item responses,

$$f(\mathcal{U}|\boldsymbol{\Theta}, \boldsymbol{B}, \boldsymbol{\gamma}), \tag{15.1}$$

given a set $\boldsymbol{\Theta}$ of possibly multidimensional person parameters $\boldsymbol{\Theta} = (\boldsymbol{\theta}_1, \ldots, \boldsymbol{\theta}_P)$, a set $\boldsymbol{B}$ of possibly multidimensional item parameters $\boldsymbol{B} = (\boldsymbol{\beta}_1, \ldots, \boldsymbol{\beta}_I)$, and possibly an additional set $\boldsymbol{\gamma}$ of other parameters. (Under mild regularity conditions (e.g., Billingsley, 1995), any parametric model $f(\mathcal{U}; \boldsymbol{\tau})$ for $\mathcal{U}$ expressed in terms of parameters $\boldsymbol{\tau}$ can be identified as a conditional distribution $f(\mathcal{U}|\boldsymbol{\tau})$ for $\mathcal{U}$ given $\boldsymbol{\tau}$. For the Bayesian calculations considered in this chapter, we always assume this identification.)

This basic formulation, and all of the methodology discussed in this chapter, can be modified to account for additional hierarchical structure (e.g., Béguin and Glas, 2001; Fox and

Glas, 2001; Janssen et al., 2000; Kamata, 2001; Maier, 2001), multiple time points (e.g., Fox, 2011; Studer, 2012), ratings of each item (e.g., Mariano and Junker, 2007; Patz et al., 2002), computerized adaptive testing (e.g, Jones and Nediak, 2005; Matteucci and Veldkamp, 2011; Segall, 2002, 2003), missing data (e.g., Glas and Pimentel, 2008; Patz and Junker, 1999b), etc. Each of these changes the structure of $\mathcal{U}$—to be a multiway array, a ragged array, etc.—and may also introduce covariates and other features of designed and realized data collection. To keep the discussion focused on fundamental ideas, however, we will mostly consider complete two-way designs in this chapter.

To make the notation concise, we will collect all of the parameters $\boldsymbol{\Theta}$, $\boldsymbol{B}$, $\boldsymbol{\gamma}$ together into a single J-dimensional vector vector $\boldsymbol{\tau} = (\tau_1, \ldots, \tau_J)^T$, so the basic model in Equation 15.1 becomes $f(\mathcal{U}|\boldsymbol{\tau})$. If we supply a prior distribution $f(\boldsymbol{\tau})$, we can write the joint distribution of the data and parameters as

$$f(\mathcal{U}, \boldsymbol{\tau}) = f(\mathcal{U}|\boldsymbol{\tau}) f(\boldsymbol{\tau}) \tag{15.2}$$

and following Bayes' rule, the posterior distribution of $\boldsymbol{\tau}$ as

$$f(\boldsymbol{\tau}|\mathcal{U}) = \frac{f(\mathcal{U}|\boldsymbol{\tau}) f(\boldsymbol{\tau})}{\int f(\mathcal{U}|\boldsymbol{t}) f(d\boldsymbol{t})} \propto f(\mathcal{U}|\boldsymbol{\tau}) f(\boldsymbol{\tau}) \tag{15.3}$$

as a function of $\boldsymbol{\tau}$.

(Here and throughout, we repeatedly reuse the notation $f()$ to represent various discrete and continuous probability density functions, with the particular role of $f()$ clear from context. In addition, all integrals $\int g(x) f(\mathrm{d}x)$ should be interpreted as Stieltjes integrals, for example, if $f()$ is a continuous density, this is a usual integral, and if $f()$ is a discrete density, this is a sum.)

Applied Bayesian statistics focuses on characterizing the posterior in Equation 15.3 in various ways. For example, we may be interested in the following:

- The posterior mean, or *expected a posteriori* (EAP) estimate of $\boldsymbol{\tau}$,

  $$E[\boldsymbol{\tau}|\mathcal{U}] = \int \boldsymbol{\tau} f(d\boldsymbol{\tau}|\mathcal{U})$$

  or perhaps the posterior mean of a function $E[g(\boldsymbol{\tau})|\mathcal{U}]$, or
- The posterior mode, or *maximum a posteriori* (MAP) estimate of $\boldsymbol{\tau}$,

  $$\arg\max_{\boldsymbol{\tau}} f(\boldsymbol{\tau}|\mathcal{U})$$

  or
- A *credible interval* (CI), that is, a set A of parameter values $\boldsymbol{\tau}$ such that

  $$P(\boldsymbol{\tau} \in A|\mathcal{U}) = 1 - \alpha$$

  for some fixed probability $1 - \alpha$, or
- A *graph* or other characterization of the shape of $f(\boldsymbol{\tau}|\mathcal{U})$ as a function of (some coordinates of) $\boldsymbol{\tau}$, and so forth.

15.3 Markov Chain Monte Carlo—General Ideas

The essential problem is to learn about the posterior distribution $f(\boldsymbol{\tau}|\mathcal{U})$, and this problem is made difficult by the need to compute the integral in the denominator of Equation 15.3. In most applications this integral must be computed numerically rather than analytically, and is usually quite high dimensional—the dimension J of the parameter space is at least as large as $P + I$, the number of persons plus the number of items. Difficult numerical integration can often be sidestepped by Monte Carlo methods, and MCMC offers a straightforward methodology for generating samples from (approximately) the posterior distribution $f(\boldsymbol{\tau}|\mathcal{U})$ in well-behaved Bayesian models, without directly calculating the integral in Equation 15.3.

The essential idea is to define a stationary Markov chain $\mathcal{M}_0, \mathcal{M}_1, \mathcal{M}_2, \ldots$ with states $\mathcal{M}_k = (\boldsymbol{\tau}^{(k)})$, and transition kernel

$$k(\boldsymbol{t}^{(0)}, \boldsymbol{t}^{(1)}) = P[\mathcal{M}_k = (\boldsymbol{t}^{(1)})|\mathcal{M}_{k-1} = (\boldsymbol{t}^{(0)})]\,, \forall k\,,$$

the probability of moving to a new state $\boldsymbol{t}^{(1)}$ given the current state $\boldsymbol{t}^{(0)}$, with *stationary distribution* $\pi(\boldsymbol{t})$, defined by

$$\int k(\boldsymbol{t}^{(0)}, \boldsymbol{t}^{(1)})\pi(d\boldsymbol{t}^{(0)}) = \pi(\boldsymbol{t}^{(1)}). \tag{15.4}$$

A common sufficient condition for Equation 15.4 to hold is *detailed balance* or *reversibility*, that is,

$$\pi(\boldsymbol{t}^{(0)})k(\boldsymbol{t}^{(0)}, \boldsymbol{t}^{(1)}) = \pi(\boldsymbol{t}^{(1)})k(\boldsymbol{t}^{(1)}, \boldsymbol{t}^{(0)})\,. \tag{15.5}$$

The Markov chain is said to be *π-irreducible* if it has positive probability of entering any set A for which $\pi(A) = \int_A \pi(d\boldsymbol{t}) > 0$. It is said to be periodic if there are portions of the state space that it can visit only at regularly spaced intervals; otherwise, it is *aperiodic*. If a Markov chain has stationary distribution $\pi()$ as in Equation 15.4 and it is π-irreducible and aperiodic, then the distribution of $\mathcal{M}_k$ will converge as, $k \to \infty$, to $\pi()$; see Tierney (1994) for details.

If we can define the transition kernel $k(\boldsymbol{t}^{(0)}, \boldsymbol{t}^{(1)})$ so that in Equation 15.4 $\pi(\boldsymbol{t}) = f(\boldsymbol{t}|\mathcal{U})$, then, after throwing away the first K observations—the "burn-in" period before the distribution of $\mathcal{M}_k$ has converged to $\pi(\boldsymbol{t})$—the remaining "good" observations $(\boldsymbol{\tau}^{(1)}) = \mathcal{M}_{K+1}, (\boldsymbol{\tau}^{(2)}) = \mathcal{M}_{k+2}, \ldots, (\boldsymbol{\tau}^{(M)}) = \mathcal{M}_{K+M}$ can be treated like (dependent) draws from $f(\boldsymbol{\tau}|\mathcal{U})$.

For example, an EAP estimate of any integrable function $g(\boldsymbol{\tau})$ can be obtained simply as

$$\int_{\boldsymbol{\tau}} g(\boldsymbol{\tau}) f(d\boldsymbol{\tau}|\mathcal{U}) \approx \frac{1}{M}\sum_{m=1}^{M} g(\boldsymbol{\tau}^{(m)})$$

with convergence guaranteed as $M \to \infty$. Similarly, an approximate graph of $f(\boldsymbol{\tau}|\mathcal{U})$ can be constructed as a (smoothed) histogram of the sample $\boldsymbol{\tau}^{(m)}, m = 1, \ldots, M$, and CIs and MAP estimates can be computed from the graph. More sophisticated methods leading to more efficient and stable estimates are also available, and should be used in practice; see, for example, Section 15.6.

1. Sample $\tau_1^{(k)}$ from $f(\tau_1|\tau_2^{(k-1)},\ldots,\tau_H^{(k-1)},U)$;
2. Sample $\tau_2^{(k)}$ from $f(\tau_2|\tau_1^{(k)},\tau_3^{(k-1)},\ldots,\tau_H^{(k-1)},U)$;
3. Sample $\tau_3^{(k)}$ from $f(\tau_3|\tau_1^{(k)},\tau_2^{(k)},\tau_4^{(k-1)},\ldots,\tau_H^{(k-1)},U)$;

$\vdots$

H. Sample $\tau_H^{(k)}$ from $f(\tau_H|\tau_1^{(k)},\tau_2^{(k)},\tau_3^{(k)},\ldots,\tau_{H-1}^{(k)},U)$.

FIGURE 15.1
A generic-blocked MCMC algorithm based on the partition $(\tau_1, \tau_2, \ldots, \tau_H)$ of τ into H blocks of parameters.

Constructing the transition kernel $\kappa(t^{(0)}, t^{(1)})$, so that the stationary distribution of the Markov chain is the posterior distribution $f(\tau|\mathcal{U})$, is remarkably straightforward. For example, let (τ_1, τ_2) be a disjoint partition of the parameter vector τ into two blocks of parameters. Then, a short calculation verifying Equation 15.4 shows that

$$\kappa(\tau^{(0)}, \tau^{(1)}) = f(\tau_1^{(1)}|\tau_2^{(0)}, \mathcal{U}) f(\tau_2^{(1)}|\tau_1^{(1)}, \mathcal{U}) \tag{15.6}$$

has stationary distribution $f(\tau|\mathcal{U})$.

More broadly, let $(\tau_1, \tau_2, \ldots, \tau_H)$ be any fixed, disjoint partition of the parameter vector τ into $H \leq J$ blocks. In Figure 15.1, we define a sampling scheme to move from $\mathcal{M}_{k-1} = (\tau_1^{(k-1)}, \tau_2^{(k-1)}, \ldots, \tau_H^{(k-1)})$ to $\mathcal{M}_k = (\tau_1^{(k)}, \tau_2^{(k)}, \ldots, \tau_H^{(k)})$ in the Markov chain.

The conditional densities on the right in Figure 15.1 are called *complete conditionals*, because they express the distribution of each partition element τ_h conditional on *all* other parameters and data in the model. To make the notation more concise, we often abbreviate the conditioning variables in a complete conditional as *rest* and write $f(\tau_h|\mathit{rest})$. An extension of the calculation showing that the kernel in Equation 15.6 has $f(\tau|\mathcal{U})$ as its stationary distribution shows that the kernel consisting of the product of the complete conditionals,

$$\begin{aligned}
\kappa(\tau^{(k-1)}, \tau^{(k)}) &= f(\tau_1^{(k)}|\tau_2^{(k-1)}, \ldots, \tau_H^{(k-1)}, \mathcal{U}) \\
&\quad \times f(\tau_2^{(k)}|\tau_1^{(k)}, \tau_3^{(k-1)}, \ldots, \tau_H^{(k-1)}, \mathcal{U}) \\
&\quad \times f(\tau_3^{(k)}|\tau_1^{(k)}, \tau_2^{(k)}, \tau_4^{(k-1)}, \ldots, \tau_H^{(k-1)}, \mathcal{U}) \\
&\quad \times \cdots \times f(\tau_H^{(k)}|\tau_1^{(k)}, \tau_2^{(k)}, \tau_3^{(k)}, \ldots, \tau_{H-1}^{(k)}, \mathcal{U}) \\
&= f(\tau_1^{(k)}|\mathit{rest}) \times f(\tau_2^{(k)}|\mathit{rest}) \times f(\tau_3^{(k)}|\mathit{rest}) \times \cdots \times f(\tau_H^{(k)}|\mathit{rest})
\end{aligned} \tag{15.7}$$

also has $f(\tau|\mathcal{U})$ as its stationary distribution.

Note that each complete conditional density is proportional to the joint density as a function of its block of parameters, for example,

$$\begin{aligned}
f(\tau_1|\tau_2, \ldots, \tau_H, \mathcal{U}) &= \frac{f(\mathcal{U}|\tau_1, \tau_2, \ldots, \tau_H) f(\tau_1, \tau_2, \ldots, \tau_H)}{\int_{\tau_1} f(\mathcal{U}|d\tau_1, \tau_2, \ldots, \tau_H) f(d\tau_1, \tau_2, \ldots, \tau_H)} \\
&\propto f(\mathcal{U}|\tau_1, \tau_2, \ldots, \tau_H) f(\tau_1, \tau_2, \ldots, \tau_H)
\end{aligned} \tag{15.8}$$

as a function of τ_1, holding the other blocks $\tau_2, \ldots, \tau_H$ and the data $\mathcal{U}$ fixed. Thus, when the likelihood $f(\mathcal{U}|\tau_1, \tau_2, \ldots, \tau_H)$ and prior $f(\tau_1, \tau_2, \ldots, \tau_H)$ factor into a product of terms involving separate blocks of the partition $(\tau_1, \tau_2, \ldots, \tau_H)$, it is easy to "pick out" a function proportional to the complete conditional, by simply retaining those terms in the joint density that depend on τ_1. Examples of this idea will be presented in Section 15.7.

It should be noted that there is nothing special about the order in which parameters are sampled in Figure 15.1, or even the fact that each parameter is sampled once per complete step of the algorithm. In designing an MCMC algorithm, we are free to change this scan order in any way that we wish, as long as each parameter is visited a nonvanishing fraction of the time. A basic result discussed by Hastings (1970, p. 102) and Tierney (1994, p. 1710), and referred to as the "product of kernels" principle by Chib and Greenberg (1995, p. 332), guarantees that all such scan orders will have the same stationary distribution. We will return to this idea in Section 15.8.7.

In Sections 15.4 through 15.6 we discuss common, generic approaches to constructing transition kernels for MCMC and making inferences from them, and we illustrate these methods in Sections 15.7 and 15.8. It is not necessary to routinely include the forms of complete conditionals or the methods used to sample from them in published journal articles, except when the complete conditionals or sampling methodology deserve special theoretical or pedagogical attention.

15.4 Gibbs Sampling

An MCMC algorithm is simplest to implement, when the complete conditionals in Figure 15.1 can be written in closed form and can be sampled from directly. In this case, the MCMC algorithm is called a *Gibbs sampler*. This is the kind of sampling scheme first made popular in the statistics literature (Gelfand and Smith, 1990), and it remains the focal method today.

When directly sampling from a complete conditional is not possible, standard alternate Monte Carlo methods can be used. If $f(t)$ is a density that is difficult to sample from, several standard methods can be used, for example:

- If the cumulative distribution function (cdf)

$$F(t) = \int_{-\infty}^{t} f(s)ds$$

  and its inverse $F^{-1}(u)$ can be calculated in closed form, then it is easy to see that $T^* = F^{-1}(U)$, where $U \sim Unif(0,1)$, will have $f(t)$ as its density. This is called *inversion sampling*.
- Suppose $g(t)$ is a density that is easy to sample from, with the property that $f(t) < Cg(t)$ for some constant C and all t. The density $g(t)$ is called the *proposal density*. Let $T^* \sim g(t)$ and $U \sim Unif(0,1)$. If $UCg(T^*) \leq f(T^*)$ then it can be shown, with a little calculus, that T^* has $f(t)$ as its density. Otherwise we reject it and sample another pair (T^*, U), repeating until we accept a T^*. This is called *rejection sampling*.
- Rejection sampling clearly is most efficient (not many rejections per accepted T^*) when $C \approx 1$. However, finding $m(t)$ that is easy to sample from with $C \approx 1$ can

be difficult. If $f(t)$ is a log-concave density, then $\log f(t)$ can be enclosed in a piecewise linear function that, when exponentiated, can serve as $m(t)$. This piecewise linear function can be improved (so that C comes closer to 1) each time T^* is rejected, as outlined in Gilks and Wild (1992); the resulting method is *adaptive rejection sampling*.

Early versions of the BUGS/WinBUGS program relied primarily on these three methods, for one-variable-at-a-time complete conditionals, to automatically create Gibbs samplers for a variety of Bayesian models (Lunn et al., 2009).

An advantage of rejection sampling, especially for sampling from complete conditionals that are known only proportionally as in Equation 15.8, is that $f(t)$ need only be known up to a constant of proportionality. It is easy to see that any required normalizing constant for $f(t)$ can be absorbed into the constant C: $f(t)/B < Cg(t)$ if and only if $f(t) < BCg(t)$, and $UCg(T^*) \leq f(T^*)/B$ if and only if $UBCg(T^*) \leq f(T^*)$. Other direct sampling methods, such as slice sampling (Neal, 2003) and importance sampling (Ripley, 1987) also have this property.

15.5 Metropolis–Hastings

A second popular mechanism for generating an MCMC algorithm is known as the *Metropolis–Hastings* (M–H) algorithm (Chib et al., 1995; Hastings, 1970; Metropolis et al., 1953). The M–H algorithm actually involves a different transition kernel than the one outlined in Equation 15.7 but one with the same stationary distribution. Results reviewed in Tierney (1994) guarantee that using M–H sampling for some complete conditionals and Gibbs sampling for others maintains $f(\boldsymbol{\tau}|\mathcal{U})$ as the stationary distribution.

For the complete conditional $f(\boldsymbol{\tau}_k|rest)$, to implement a M–H step, we first sample $\boldsymbol{\tau}_k^* \sim g_m(\boldsymbol{\tau}_k|\boldsymbol{\tau}_k^{(m-1)})$, where $g_m(\boldsymbol{\tau}_k|\boldsymbol{\tau}_k^{(m-1)})$ is a *proposal density*, which will be discussed further below. Then, we calculate the acceptance probability

$$\alpha^* = \min\left\{\frac{f(\boldsymbol{\tau}_k^*|rest)g_m(\boldsymbol{\tau}_k^{(m-1)}|\boldsymbol{\tau}_k^*)}{f(\boldsymbol{\tau}_k^{(m-1)}|rest)g_m(\boldsymbol{\tau}_k^*|\boldsymbol{\tau}_k^{(m-1)})}, 1\right\} \tag{15.9}$$

and generate $U \sim Unif(0,1)$. If $U \leq \alpha^*$, we set $\boldsymbol{\tau}_k^{(m)} = \boldsymbol{\tau}_k^*$; otherwise, we set $\boldsymbol{\tau}_k^{(m)} = \boldsymbol{\tau}_k^{(m-2)}$.

The proposal density $g_m(\boldsymbol{\tau}_k|\boldsymbol{\tau}_k^{(m-1)})$ can be chosen to be any convenient density. Two common choices are as follows:

- *Independence M–H.* Take the proposal density $g_m(\boldsymbol{\tau}_k|\boldsymbol{\tau}_k^{(m-1)}) = g_m(\boldsymbol{\tau}_k)$, independent of $\boldsymbol{\tau}_k^{(m-1)}$.
- *Normal random walk M–H.* Take the proposal density $g_m(\boldsymbol{\tau}_k|\boldsymbol{\tau}_k^{(m-1)}) = n(\boldsymbol{\tau}_k|\boldsymbol{\mu} = \boldsymbol{\tau}_k^{(m-1)}, \boldsymbol{\Sigma})$, a normal density with mean $\boldsymbol{\tau}_k^{(m-1)}$ and variance matrix $\boldsymbol{\Sigma}$.

Note that if $g_m()$ is constant (or symmetric) in $\boldsymbol{\tau}_k$ and $\boldsymbol{\tau}_k^{(m-1)}$, then the $g_m()$ terms drop out of Equation 15.9, and the algorithm tends to move toward the mode of $f(\boldsymbol{\tau}_k|rest)$.

It is also worth noting that if $g_m(\boldsymbol{\tau}_k|\boldsymbol{\tau}_k^{(m-1)}) = f(\boldsymbol{\tau}_k|rest)$, then the M–H step reduces to a Gibbs step: We are sampling from $f(\boldsymbol{\tau}_k|rest)$ and always accepting $\boldsymbol{\tau}_k^*$ since in this case all terms cancel in Equation 15.9, leaving $\alpha^* = 1$. In this sense, Gibbs sampling is in fact a special case of M–H. A typical MCMC algorithm will intersperse Gibbs steps—for complete conditionals that can be sampled directly—with M–H steps—for complete conditionals that cannot be sampled directly. This class of algorithms is called, somewhat inaccurately, *M–H within Gibbs*.

Results reviewed in Rosenthal (2011) suggest that the MCMC algorithm will achieve good mixing and convergence to the stationary distribution, if $g_m(\tau_k|\tau_k^{(m-1)})$ is chosen so that if τ_k is one dimensional, the rate at which τ_k* is accepted is approximately 0.44, and if τ_k is of dimension d, the acceptance rate should fall to 0.234 for $d \approx 5$ or more. In fact, the mixing and rate of convergence tend to be good as long as the acceptance rate is roughly between 0.1 and 0.6 (Rosenthal, 2011, Figure 4.5).

For example, the normal random walk proposal density for unidimensional τ_k is

$$g_m(\tau_k|\tau_k^{(m-1)}) = \frac{1}{\sqrt{2\pi\sigma_g^2}} e^{-\frac{1}{2}(\tau_k-\tau_k^{(m-1)})^2/\sigma_g^2}.$$

The proposal variance σ_g^2 is a tuning parameter than can be adjusted until the acceptance rate is in the rough neighborhood of 0.44.

Rosenthal (2011) outlines a general theory for adaptive MCMC, in which details of the Markov Chain can be tuned "on the fly" without undermining convergence to the stationary distribution. One such adaptation would be to adjust the variances of normal random walk M–H steps to achieve the above target acceptance rates. Indeed, when M–H steps were introduced into WinBUGS (Lunn et al., 2009), they were introduced in an adaptive form, so that σ_g^2 was adjusted toward a target acceptance rate between 0.2 and 0.4.

15.6 Tricks of the Trade

15.6.1 Initial State, Burn-In, and Convergence

From Equation 15.4, it is clear that the best choice for an *initial state* for the Markov chain would be a draw from the stationary distribution, since then there would be no burn-in segment. We cannot do this (if we could, we would not need MCMC!) but it does suggest starting the chain somewhere near the center of the posterior density $f(\tau|\mathcal{U})$. The default starting values in WinBUGS are draws from the prior distribution; this only makes sense if the prior and the posterior are centered similarly.

In practice, one often proceeds by simulating starting values from normal, t, or similar distributions, centered at provisional parameter estimates (e.g., maximum likelihood or method of moments estimates from a simpler model), overdispersed relative to provisional standard errors. MCMC is fundamentally a local search algorithm, and starting the algorithm from several well-dispersed initial states can help to ensure that the parameter space is fully explored. In addition, starting multiple chains at well-dispersed initial states facilitates correct interpretation of convergence indices such as the $\hat{R}$ statistic (Brooks and Gelman, 1998; Gelman and Rubin, 1992). On the other hand, we found IRT and similar models to be somewhat delicate with respect to starting values, a point we will discuss further below in Sections 15.8.5 and 15.8.6.

As discussed in Section 15.3, the *burn-in* segment of a Markov chain is the initial segment of the chain before it has *converged*—that is, before samples from the chain are like samples from the stationary distribution $\pi(\tau) \equiv f(\tau|\mathcal{U})$. These samples of the chain should be discarded; only samples after burn-in are useful for inference about the posterior distribution $f(\tau|\mathcal{U})$.

Although many methods have been proposed as heuristics for assessing convergence (Cowles and Carlin, 1996), only three or four methods stand out as convenient enough and informative enough to have become standard practice. Apart from perfect sampling (Fill, 1998) and regenerative methods (Mykland et al., 1995), there are no guaranteed methods of determining when an MCMC algorithm has "converged" to its stationary distribution. When developing or applying an MCMC algorithm, one should always look at trace plots, or summaries of trace plots, and autocorrelation function (acf) plots. These are basic graphical tools for assessing whether the MCMC algorithm is functioning well, and they are also useful for assessing burn-in, mixing and convergence to the stationary distribution.

15.6.1.1 Trace Plots

Trace plots or time-series plots are simply plots of $g(\tau^{(m)})$, as a function of m, $m = 1, \ldots, M$. Most often, $g(\tau) = \tau_j$, one of the parameters in the model. But it could be anything. To aid visual detection of some of these problems, it is useful to add a horizontal line at the mean or median of the graphed values, and/or a smoothed running mean or median curve.

Ideal plots look like white (unpatterned) noise centered at the posterior mean or median, as in Figure 15.13 (see later in the chapter). On the other hand, Figure 15.2 illustrates four common problems in trace plots:

a. An initial segment that looks like it "drifted in from somewhere else."
b. A low-frequency cycle or other pattern in the graph (e.g., the white noise is following a large sinusoid rather than being centered on a horizontal line).
c. Excessive "stickiness," that is, the trace plot stays constant for several steps before moving to a new value.
d. An overall trend or drift upwards or downwards across the whole trace plot.

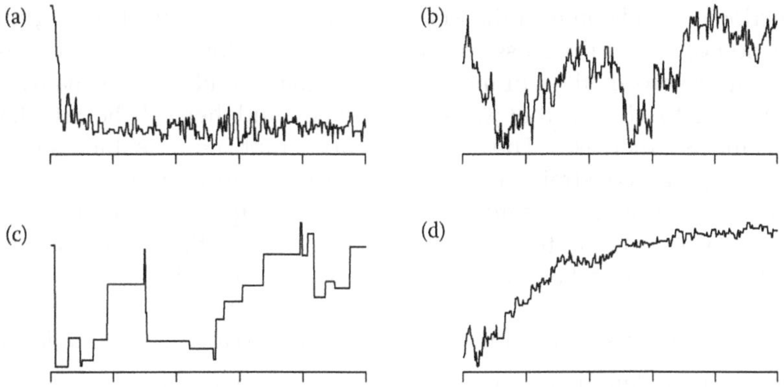

FIGURE 15.2
A collection of trace plots indicating various problems. (a) Drift in the initial (burn-in) segment, (b) Oscillation or other nonconstant pattern, (c) Sticky chain, and (d) Drift in entire chain.

Problem (a) is due to the early part of the chain not being very near the stationary distribution. The usual fix is to throw away the initial (burn-in) part of the chain, and use the stable remaining part for inference. Problem (b) is often due to excessive dependence between steps in the MCMC algorithm. This can happen with parameters that are sampled with random-walk M–H steps if the proposal variance is too small, for example. It can also happen with parameters that are sampled with Gibbs steps. One can live with this problem if MCMC sampling is fast enough that one can take many many more steps M to get precise estimates of posterior quantities. The alternative is to redesign the chain so as to reduce this dependence. Problem (c) is often seen in parameters that are sampled with M–H steps, and it is simply due to too low an acceptance rate, for example, too high a proposal variance in random-walk M–H. Tune or change the proposal density to fix this. Problem (d) may be a special case of problems (a) or (b), which one would discover by running the chain longer to see what happens. If the drift persists even after running the chain a very long time, it may be evidence of problems in the underlying statistical model, Equation 15.2. For example, the posterior density $f(\boldsymbol{\tau}|\mathcal{U})$ may not be sufficiently peaked around a dominant posterior mode (e.g., not enough data), or there may be an identification problem (e.g., the MCMC samples for one parameter drift toward $+\infty$, and the MCMC samples for another drift toward $-\infty$ but their sum stays constant).

15.6.1.2 Autocorrelation Plots

Autocorrelation plots and cross-correlation plots are useful for directly assessing dependence between steps of the chain. Let $\boldsymbol{\tau}^{(m)}$, $m = 1, 2, 3, \ldots$ be the output from the MCMC algorithm, and consider functions $g_1(\boldsymbol{\tau})$ and $g_2(\boldsymbol{\tau})$. The *autocorrelation function* (acf) for $g_1(\boldsymbol{\tau})$ is

$$\rho_k^{g_1} = \frac{\text{Cov}\,(g_1(\boldsymbol{\tau}^{(m)}), g_1(\boldsymbol{\tau}^{(m+k)}))}{\text{Var}\,(g_1(\boldsymbol{\tau}^{(m)}))}$$

and similarly for $g_2(\boldsymbol{\tau})$, and the cross-correlation function is

$$\sigma_k^{g_1 g_2} = \frac{\text{Cov}\,(g_1(\boldsymbol{\tau}^{(m)}), g_2(\boldsymbol{\tau}^{(m+k)}))}{\sqrt{\text{Var}\,(g_1(\boldsymbol{\tau}^{(m)}))\text{Var}\,(g_2(\boldsymbol{\tau}^{(m+k)}))}}$$

(neither function depends on m if the Markov chain is stationary). The acf plot is a plot of $\hat{\rho}_k^{g_1}$ as a function of k, and the cross-correlation plot is a plot of $\hat{\sigma}_k^{g_1 g_2}$. These estimates are usually straightforward method of movements estimates and are reasonably useful if k is not large compared to M. Note that $\hat{\rho}_0^{g_1}$ is forced to be 1 and then $\hat{\rho}_k^{g_1}$ should fall toward zero in absolute value as k increases. The cross-correlation plot should exhibit similar behavior, except that $\sigma_k^{g_1 g_2}$ is not constrained to be equal to 1. To aid in interpreting the plots, it is useful to add a horizontal line at zero correlation to the graph, and also add lines indicating the rejection region for a test of the hypothesis that $\rho_k^{g_1} = 0$, or the hypothesis that $\sigma_k^{g_1, g_2} = 0$.

Problematic behavior in acf and cross-correlation plots include

a. $\rho_k^{g_1}$ (or $\sigma_k^{g_1 g_2}$) remains significantly different from zero for all observable values of k.
b. $\rho_k^{g_1}$ (or $\sigma_k^{g_1 g_2}$) oscillates between values close to zero and values far from zero.

examples are shown in Figure 15.3. Both problems (a) and (b) reflect excessive autocorrelation (or cross-correlation) in the Markov chain. These problems, especially persistent

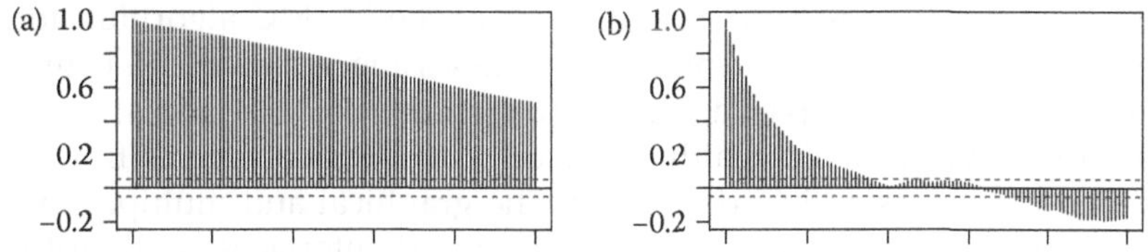

FIGURE 15.3
Two common problems evident in autocorrelation plots. (a) Autocorrelations remain high. (b) Autocorrelations oscillate between values near and far from 0.

positive autocorrelation, are often associated with problems (a), (b), or (c) in Figure 15.2, for example. One can live with this problem if MCMC sampling is fast enough that one can take many many more steps M (and perhaps discard an initial burn-in segment) to get precise estimates of posterior quantities. The alternative is to redesign the chain so as to reduce this dependence. For example, a "good" autocorrelation plot can be seen later in the chapter, in Figure 15.13.

15.6.1.3 Single and Multiple Chain Methods

There exist several heuristic methods for assessing the convergence of an MCMC algorithm to its stationary distribution, based on a single run of the algorithm. A visual inspection of trace plots and autocorrelation plots may be enough: If the trace plots look like white noise centered at the mean or median of the output, and the acf plot drops quickly to zero and stays there, we may have some confidence that the chain has reached its stationary distribution. As a quantitative check, one can break a long chain into two (or more) parts and compare posterior mean and variance estimates, density plots of the posterior, etc., from the different parts of the chain; if the chain has converged and the parts of the chain are long enough, the posterior summaries should be similar. A more formal measure of convergence along the same lines is the time series diagnostic of Geweke (1992).

Another approach to assessing the convergence of an MCMC algorithm is to generate three or more chains from different starting values. If the trace plots overlap one another to a great extent after a suitable burn-in segment is discarded, this is some assurance that the algorithm is mixing well and has reached the stationary distribution. If the chains converge quickly (have short burn-in segments) and not much autocorrelation, then samples from the multiple chains can be pooled together to make inferences about the posterior distribution. Since the chains do not depend on each other in any way, they can be run in parallel on either multiple cores or multiple computers to generate output faster. If convergence to the stationary distribution is slow, it may be more efficient to run one long chain, to avoid spending computation on the burn-in for multiple chains.

An informal quantitative assessment of convergence using multiple chains would be to compare posterior mean and variance estimates, posterior density plots, etc., across the multiple chains. The $\hat{R}$ statistic (Brooks and Gelman, 1998; Gelman and Rubin, 1992) formalizes and quantifies this idea, essentially by examining the ratio of between-chain variation to within-chain variation. When all the chains have achieved stationarity, the between- and within-variation will be comparable; otherwise, the between-variation is likely to be larger than the within-variation. A common heuristic is to declare convergence if $\hat{R} < 1.1$.

The above methods for assessing convergence of the MCMC algorithm for univariate functions $g(\boldsymbol{\tau})$, as well as several methods for assessing multivariate convergence, are implemented in the software package BOA (Bayesian output analysis) of Smith (2007), available at http://www.public-health.uiowa.edu/boa, which derives from the methodology surveyed in Cowles and Carlin (1996). The graphical and multiple chain methods above are also readily available in WinBUGS and R interfaces to WinBUGS such as `R2WinBUGS` (Gelman et al., 2012) and `rube` (Seltman, 2010).

15.6.1.4 Thinning and Saving Values from the Converged Chain

It is fairly common practice to "thin" the Markov chain in some way, for example, to retain only every $m_{\text{thin}}^{\text{th}}$ step of of the converged Markov chain. Thinning is useful for reducing autocorrelation, and for reducing the amount of MCMC data one has to store and calculate with. A common procedure to determine the thinning interval m_{thin} is to examine autocorrelation plots from a preliminary run with no thinning, and then for the final run take m_{thin} to be an interval at which the autocorrelation is near zero. Because of uncertainty of estimation in the acf plot, however, there will still be some autocorrelation in the retained MCMC samples, and this should be addressed using a method like that of Equation 15.11 in Section 15.6.2, to understand the precision of inferences from the MCMC output. Because thinning involves throwing away perfectly good draws from the posterior distribution, some authors advocate against thinning to reduce autocorrelation. For example, in large interesting models, requiring small autocorrelations can result in $m_{\text{thin}} = 50$ or 100 or more; 98% or more of the computation is then wasted. In such cases, it may be possible (and is desirable) to reformulate the model or MCMC algorithm in order to reduce autocorrelation and make the thinning interval and hence computational efficiency more reasonable.

Once one is satisfied that the chain has converged to its stationary distribution, and the burn-in segment has been discarded, the problem remains to decide how many steps of the chain to save, for inference about the posterior distribution. One rule of thumb is to keep roughly 1000 steps from the chain or chains, after burn-in and thinning if any of this is done. For example, if one wishes to calculate 2.5th and 97.5th percentiles in order to report equal-tailed 95% posterior credible intervals, 1000 is a minimal sample size. A more principled approach might be to monitor the Monte Carlo standard error (discussed below in Section 15.6.2) as the length of the chain grows: We want the Monte Carlo standard error to be much smaller than the posterior standard error, and small enough to ensure a couple of digits of accuracy in posterior mean estimates.

It does not seem necessary to publish trace plots, autocorrelation plots, or Monte Carlo standard errors routinely in journal articles (except when one is trying to make a point about the behavior of the MCMC algorithm), but they should be available in on-line supplements for referees or interested readers. Numerical evidence of good mixing and convergence to the stationary distribution (Brooks and Gelman, 1998; Gelman and Rubin, 1992; Geweke, 1992) is useful for published papers.

15.6.2 Monte Carlo Standard Error and Posterior Uncertainty

Two different uncertainty calculations are important to carry out in MCMC estimation of posterior quantities. For specificity, suppose we are interested in estimating a function of $\boldsymbol{\tau}$, $g(\boldsymbol{\tau})$, using MCMC output $\boldsymbol{\tau}^{(m)}$, $m = 1, \ldots, M$. For example, to estimate the kth univariate parameter in $\boldsymbol{\tau} = (\tau_1, \ldots, \tau_K)$, we would take $g(\boldsymbol{\tau}) = \tau_k$.

- *Monte Carlo Uncertainty:* Suppose we are interested in estimating the posterior mean

$$E[g(\boldsymbol{\tau})|\mathcal{U}] \approx \frac{1}{M}\sum_{m=1}^{M} g(\boldsymbol{\tau}^{(m)}) \equiv \bar{g}.$$

How accurate is $\bar{g}$ as an estimate of $E[g(\boldsymbol{\tau})|\mathcal{U}]$? We would like to compute a Monte Carlo standard error, SE_{MCMC} and provide a Monte Carlo confidence interval, for example, $\bar{g} \pm t^*\text{SE}_{\text{MCMC}}$. We can make this Monte Carlo estimate more precise (reduce SE_{MCMC}) simply by increasing the MCMC sample size M, without collecting any more data.

Confidence intervals expressing MCMC estimation error are more readily interpretable to the extent that the central limit theorem (CLT) applies to the chain $\boldsymbol{\tau}^{(m)}, m = 1, 2, 3, \ldots$. If the chain is reversible, that is, satisfies Equation 15.5, then a CLT can be obtained (Geyer, 2011). If not, the CLT depends on *geometric ergodicity* (Flegal and Jones, 2011), which says essentially that the difference (or, technically, the total variation distance) between the distribution of $\boldsymbol{\tau}^{(m)}$ and $f(\boldsymbol{\tau}|\mathcal{U})$ tends to zero like r^m for some number $r \in (0, 1)$, and a moment condition such as $E[g(\boldsymbol{\tau})^{2+\delta}|\mathcal{U}] < \infty$. Many, but not all, common MCMC algorithms satisfy this or stronger sufficient conditions for the CLT to hold.

- *Posterior Uncertainty:* As an inference about $\boldsymbol{\tau}$ from the data, we may be interested in summarizing the posterior distribution of $g(\boldsymbol{\tau})$ using the posterior mean $E[g(\boldsymbol{\tau})|\mathcal{U}]$ and the posterior standard error $\text{SE}_{\text{post}} = \sqrt{\text{Var}\,(g(\boldsymbol{\tau})|\mathcal{U})}$. For example, we may wish to report a posterior credible interval such as $E[g(\boldsymbol{\tau})|\mathcal{U}] \pm t^* \cdot \text{SE}_{\text{post}}$ for some suitable cutoff t^*. We can make our inference about $g(\boldsymbol{\tau})$ more precise (reduce SE_{post}) by collecting more data. Increasing the MCMC sample size M can make our estimates of $E[g(\boldsymbol{\tau})|\mathcal{U}]$ and SE_{post} more precise, but it cannot structurally improve the precision of our inference about $g(\boldsymbol{\tau})$. Intervals based on estimates of $E[g(\boldsymbol{\tau})|\mathcal{U}]$ and SE_{post} are only interpretable, to the extent that a posterior CLT holds as more data is collected (e.g., Chang and Stout, 1993; Walker, 1969).

If there is any concern about the shape of the posterior density, it is better to construct credible intervals directly from the posterior density. For example, the equal-tailed credible interval running from the 0.025 posterior quantile to the 0.975 posterior quantile is guaranteed to produce an interval with 95% posterior probability, regardless of whether the posterior CLT holds or not.

To estimate SE_{MCMC}, we must first estimate $\sigma_g^2 = \text{Var}\,(g_m)$, the variance of the sampled values of the Markov chain (which is due to both the shape of the posterior distribution and Markov sampling). Because the $\boldsymbol{\tau}^{(m)}$ are dependent, the naive estimator

$$\hat{\sigma}^2_{naive} = \frac{1}{M-1}\sum_{m=1}^{M}(g(\boldsymbol{\tau}^{(m)}) - \bar{g})^2 \tag{15.10}$$

is seldom adequate. Several valid methods are currently in use (Flegal and Jones, 2011; Geyer, 2011), but we will focus on only one: The method of *overlapping batch means* (OLBM) (Flegal and Jones, 2011). Define b_M to be the batch length, and define batches $B_1 = (\boldsymbol{\tau}^{(1)}, \boldsymbol{\tau}^{(2)}, \ldots, \boldsymbol{\tau}^{(b_M)})$, $B_2 = (\boldsymbol{\tau}^{(2)}, \boldsymbol{\tau}^{(3)}, \ldots, \boldsymbol{\tau}^{(b_M+1)})$, $B_3 = (\boldsymbol{\tau}^{(3)}, \boldsymbol{\tau}^{(4)}, \ldots, \boldsymbol{\tau}^{(b_M+2)})$, etc. There are $M - b_M + 1$ such batches in an MCMC run of length M. Now let

$$\bar{g}_j = \frac{1}{b_M} \sum_{\tau^{(m)} \in B_j} g(\tau^{(m)}) \quad \text{and}$$

$$\hat{\sigma}^2_{\text{OLBM}} = \frac{M b_M}{(M - b_M)(M - b_M + 1)} \sum_{j=1}^{M-b_M+1} (\bar{g}_j - \bar{g})^2. \tag{15.11}$$

Flegal and Jones (2011) argue that under conditions similar to those needed for the CLT to apply, $\hat{\sigma}^2_{\text{OLBM}}$ will be a consistent estimator of σ^2_g, as long as $B_M \approx M^{1/2}$. In that case, we can take

$$\text{SE}_{\text{MCMC}} \approx \sqrt{\hat{\sigma}^2_{\text{OLBM}}/M}.$$

The MCMC confidence interval is then $\bar{g} \pm t^* \cdot \text{SE}_{\text{MCMC}}$ where t^* is a cutoff from a t-distribution with $M - b_M$ degrees of freedom, although for many practical purposes simply taking $t^* = 2$ gives an adequate impression (since the degrees of freedom $M - b_M$ are likely to be quite large). An alternative method with simpler-to-calculate nonoverlapping batch means is described in Flegal et al. (2008); the nonoverlapping approach described here however is generally thought to be more efficient.

When the Markov chain is ergodic, it is reasonable to estimate $\text{SE}_{\text{post}} \approx \sqrt{\sigma^2_{\text{naive}}}$ in the sense that the right-hand side will converge to the left as M grows. To know how accurate the estimate is, one can exploit the identity $\text{Var}\,(g(\tau)|\mathcal{U}) = E[(g(\tau) - E[g(\tau)|\mathcal{U}])^2|\mathcal{U}] = E[g(\tau)^2|\mathcal{U}] - E[g(\tau)|\mathcal{U}]^2$, separately estimating

$$E[g(\tau)|\mathcal{U}] \approx \frac{1}{M} \sum_{m=1}^{M} g(\tau^{(m)}) \equiv \bar{g} \quad \text{and}$$

$$E[g(\tau)^2|\mathcal{U}] \approx \frac{1}{M} \sum_{m=1}^{M} g(\tau^{(m)})^2 \equiv \bar{g}^2,$$

and monitoring the MCMC standard error of both estimates to make sure that we do not get excessive cancellation in the difference $\bar{g}^2 - \bar{g}^2$. An MCMC standard error for $\widehat{\text{SE}}_{\text{post}} = \sqrt{\sigma^2_{\text{naive}}}$ can then be obtained using the delta method (e.g., Serfling, 1980, p. 124).

When the posterior distribution is not approximately normal, intervals based on (estimates of) $E[g(\tau)|\mathcal{U}]$ and SE_{post} may be misleading; a better measure of posterior uncertainty can be based on quantiles of the posterior distribution, estimated as empirical quantiles of the MCMC sequence $g(\tau^{(m)})$, $m = 1, \ldots, M$. For example, one might report the empirical median as a point estimate for $g(\tau)$ and the empirical 0.025 and 0.975 quantiles as an approximate equal-tailed 95% credible interval. Flegal and Jones (2011) also give methods for estimating the MCMC standard errors of these quantiles.

MCMC standard errors seldom find their way into published empirical papers. Largely speaking, there may be little harm done by this, as long as authors can assure themselves and their readers (or referees) that the magnitude of the posterior uncertainty overwhelms the magnitude of the MCMC uncertainty, as is almost always the case. MCMC standard errors would be important to report in cases where they are comparable in size to posterior standard errors, or in cases where estimating a posterior quantity (such as $E[g(\boldsymbol{\tau})|\mathcal{U}]$, Corr $(\boldsymbol{\tau}_1, \boldsymbol{\tau}_2|\mathcal{U})$, etc.) is of primary interest.

Those who wish to apply MCMC methods in practical testing settings should note carefully that Monte Carlo uncertainty and posterior uncertainty are qualitatively different and do not carry equal weight in evaluating a method's utility. For example, Monte Carlo uncertainty is probably unacceptable in test scoring, unless it can be reduced to a level well below the number of digits used in reporting scores. Stakeholders in testing will not find it acceptable to think that pass/fail decisions or rankings might be sensitive to such factors as the seed of the random number generator, or the length of the chain that was run to obtain the estimate. This is perhaps not entirely unique to MCMC, and we know that maximum-likelihood-based scoring has its own tolerances and sensitivities. MCMC is much more readily usable in item/test calibration, where any imprecision due to estimation method is reflected in the "level playing field" that all examinees encounter when answering test questions. If we consider that such matters as "adequate yearly progress" for schools might be sensitive to Monte Carlo error in calibration and/or linking, there is still reason (as if there were not enough reasons already) to be concerned. Addressing these types of concerns is on the one hand just good psychometric practice (know the precision of any instruments and make decisions accordingly), but on the other hand might be important for adoption of MCMC for IRT models to move from the research literature into the professional practice.

15.6.3 Blocking

As should be clear from our discussion at the end of Section 15.3, it is not necessary for the complete conditional densities to be univariate densities for each single parameter in the model. If some parameters are known (from theory, or from preliminary estimation) to be highly correlated in the posterior $f(\boldsymbol{\tau}|\mathcal{U})$, then it makes sense to group them together into a block and sample them together. Posterior correlation corresponds to a ridge in the posterior distribution, and it is usually more efficient to explore the ridge with MCMC steps parallel to the ridge (as would be the case for blocked MCMC steps) rather than steps parallel to the coordinate axes (as would be the case for one parameter at a time MCMC steps).

If a block of parameters can be sampled in a Gibbs step, the form of the complete conditional for that block will have the appropriate association structure in it. If the block must be sampled using a M–H step, a common strategy is to use a normal random walk proposal, with a proposal variance–covariance matrix that reflects the anticipated posterior correlation in that block. If "good" proposal distributions to update blocks of parameters can be found, block-sampled MCMC can provide a significant improvement over one-parameter-at-a-time MCMC (Doucet et al., 2006).

15.6.4 Data Augmentation

Suppose we have a model of the form Equation 15.2 for a set of parameters $\boldsymbol{\tau}_1$ of interest, $f(\mathcal{U}, \boldsymbol{\tau}_1) = f(\mathcal{U}|\boldsymbol{\tau}_1)f(\boldsymbol{\tau}_1)$, and we wish to calculate the posterior mean $E[g(\boldsymbol{\tau}_1)|\mathcal{U}]$ for the function $g(\boldsymbol{\tau}_1)$. According to the results of Section 15.3, we could do this directly with the output $\tau_1^{(m)}$, $m = 1, \ldots, M$, from an MCMC algorithm for the posterior density $f(\tau_1|\mathcal{U})$, as

$$E[g(\tau_1)|\mathcal{U}] = \int g(\tau_1) f(d\tau_1|\mathcal{U}) \approx \frac{1}{M}\sum_{m=1}^{M} g(\tau_1^{(m)})$$

for any integrable function $g()$.

On the other hand, suppose we expand the model to include additional parameters τ_2, $f(\mathcal{U}, \tau_1, \tau_2) = f(\mathcal{U}|\tau_1, \tau_2) f(\tau_1, \tau_2)$, and we obtain an MCMC sample $(\tau_1^{(m)}, \tau_2^{(m)})$, $m = 1, \ldots, M$, from the expanded posterior $f(\tau_1, \tau_2|\mathcal{U})$. We can still calculate

$$E[g(\tau_1)|\mathcal{U}] = \int g(\tau_1) f(d\tau_1|\mathcal{U}) = \iint g^*(\tau_1, \tau_2) f(d\tau_1, d\tau_2|\mathcal{U})$$

$$\approx \frac{1}{M}\sum_{m=1}^{M} g^*(\tau_1^{(m)}, \tau_2^{(m)}) = \frac{1}{M}\sum_{m=1}^{M} g(\tau_1^{(m)})$$

for any integrable $g()$, where $g^*()$ is simply the function $g^*(\tau_1, \tau_2) = g(\tau_1)$. This shows that the act of "throwing away" some coordinates of an MCMC sample is equivalent to integrating them out of a marginal expected value calculation (Patz and Junker, 1999a).

Moreover, the calculations above suggest a strategy that is often successful: When it is difficult to construct an efficient, well-mixing Markov chain for $f(\tau_1|\mathcal{U})$, it may be much easier to do so for $f(\tau_1, \tau_2|\mathcal{U})$. As long as

$$f(\tau_1|\mathcal{U}) = \int f(\tau_1, d\tau_2|\mathcal{U})$$

then we can just "throw away" $\tau_2^{(m)}$, $m = 1, \ldots, M$ and use $\tau_2^{(m)}$, $m = 1, \ldots, M$ to make inferences on the marginal density $f(\tau_1|\mathcal{U})$.

This strategy is called *data augmentation*, first formally suggested by Tanner and Wong (1987), and implemented for the first time in IRT with the with normal ogive model by Albert (1992). Indeed, we know that the normal ogive item response function for dichotomous responses

$$P[U_{pi} = 1|a_i, b_i, \theta_p] = \frac{1}{\sqrt{2\pi}} \int_{-\infty}^{a_i(\theta_p - b_i)} e^{-t^2/2} dt$$

obtains for U_{pi}, if U_{pi} is defined with the condensation function (Bartholomew and Knott, 1999; Maris, 1995) of $Z_{pi} \sim N(0, 1)$,

$$U_{pi} = \begin{cases} 0, & \text{if } Z_{pi} < -a_i(\theta_p - b_i) \\ 1, & \text{if } Z_{pi} \geq -a_i(\theta_p - b_i). \end{cases}$$

The parameters a_i, b_i, and θ_p would play the role of τ_1, and Z_{pi} would play the role of the data augmentation variables τ_2. This is the basis of the data augmentation approach to analyzing multivariate binary and polytomous probit models (Albert and Chib, 1993), as well as the data augmentation approach to estimating the normal ogive IRT model (Albert, 1992; Fox, 2010; Thissen and Edwards, 2005). Recently, progress has been made on an efficient data augmentation scheme for MCMC estimtation of logistic models as well (Polson et al., 2013).

15.6.5 Rao–Blackwellization

Again consider the model $f(\mathcal{U}, \boldsymbol{\tau}_1, \boldsymbol{\tau}_2) = f(\mathcal{U}|\boldsymbol{\tau}_1, \boldsymbol{\tau}_2) f(\boldsymbol{\tau}_1, \boldsymbol{\tau}_2)$, where $\boldsymbol{\tau}_1$ is of primary interest, and let us examine the problem of estimating $E[g(\boldsymbol{\tau}_1)|\mathcal{U}]$. We have

$$E[g(\boldsymbol{\tau}_1)|\mathcal{U}] = \iint g(\boldsymbol{\tau}_1) f(\mathrm{d}\boldsymbol{\tau}_1, \mathrm{d}\boldsymbol{\tau}_2|\mathcal{U}) = \int \left[\int g(\boldsymbol{\tau}_1) f(d\boldsymbol{\tau}_1|\boldsymbol{\tau}_2, \mathcal{U}) \right] f(\mathrm{d}\boldsymbol{\tau}_2|\mathcal{U})$$
$$\approx \frac{1}{M} \sum_{m=1}^{M} \int g(\boldsymbol{\tau}_1) f(\mathrm{d}\boldsymbol{\tau}_1|\boldsymbol{\tau}_2^{(m)}, \mathcal{U}).$$

This gives an alternative to

$$E[g(\boldsymbol{\tau}_1)|\mathcal{U}] \approx \frac{1}{M} \sum_{m=1}^{M} g(\boldsymbol{\tau}_1^{(m)})$$

for estimating $E[g(\boldsymbol{\tau}_1)|\mathcal{U}]$. By analogy with the Rao–Blackwell theorem of mathematical statistics, in which conditioning an estimator on the sufficient statistic reduces the variance of the estimator, the so-called *Rao–Blackwellized* (Casella and Robert, 1996) estimator

$$E[g(\boldsymbol{\tau}_1)|\mathcal{U}] \approx \frac{1}{M} \sum_{m=1}^{M} \int g(\boldsymbol{\tau}_1) f(\mathrm{d}\boldsymbol{\tau}_1|\boldsymbol{\tau}_2^{(m)}, \mathcal{U}) \tag{15.12}$$

is often better behaved than the simple average. Of course in order to use the Rao–Blackwellized estimator, one has to be able to calculate $\int g(\boldsymbol{\tau}_1) f(\mathrm{d}\boldsymbol{\tau}_1|\boldsymbol{\tau}_2^{(m)}, \mathcal{U})$ many times. This is not usually a problem if the integral can be obtained in closed form or via simple numerical integration; this is often the case if $g()$ is a relatively simple function and conjugate or partially conjugate priors are used (so that the complete conditional in Equation 15.12 can be identified by conjugacy).

A related technique can be used to estimate marginal densities smoothly. Instead of estimating $f(\boldsymbol{\tau}_1|\mathcal{U})$ using a histogram or smooth density estimate using $\boldsymbol{\tau}_1^{(m)}$, $m = 1, \ldots, M$, one can recognize that

$$f(\boldsymbol{\tau}_1|\mathcal{U}) = \int f(\boldsymbol{\tau}_1, \mathrm{d}\boldsymbol{\tau}_2|\mathcal{U}) = \int f(\boldsymbol{\tau}_1|\boldsymbol{\tau}_2, \mathcal{U}) f(\mathrm{d}\boldsymbol{\tau}_2, \mathcal{U}) \approx \frac{1}{M} \sum_{i=1}^{M} f(\boldsymbol{\tau}_1|\boldsymbol{\tau}_2^{(m)}, \mathcal{U}).$$

Again, this latter estimator may be more stable than the simple histogram estimator.

15.7 Item Response Theory Models

The generic model in Equation 15.1 can usually be rewritten in IRT applications as

$$f(\mathcal{U}|\tau) = \prod_{p=1}^{P}\prod_{i=1}^{I} f(U_{pi}|\theta_p, \beta_i),$$

where the product over p is due to the *experimental independence* assumption in IRT (Lord and Novick, 1968) and the product over i is due to IRT's *local independence* assumption. It is also typical to specify independent priors for all parameters, so that the generic joint distribution in Equation 15.2 has the form

$$\begin{aligned} f(\mathcal{U}|\tau)f(\tau) &= \prod_{p=1}^{P}\prod_{i=1}^{I} f(U_{pi}|\theta_p, \beta_i)\prod_{p=1}^{P} f_p(\theta_p|\lambda_\theta)\prod_{i=1}^{I} f_i(\beta_i|\lambda_\beta)f(\lambda_\theta)f(\lambda_\beta) \\ &= \prod_{p=1}^{P}\left\{\prod_{i=1}^{I} f(U_{pi}|\theta_p, \beta_i)f_i(\beta_i|\lambda_\beta)\right\} f_p(\theta_p|\lambda_\theta)f(\lambda_\theta)f(\lambda_\beta), \end{aligned} \quad (15.13)$$

where λ_θ and λ_β are hyperparameters for θ and β, respectively, and $f(\lambda_\theta)$ $f(\lambda_\beta)$ are their hyperprior distributions.

To be concrete, consider the two-parameter logistic (2PL) IRT model, with dichotomous responses $U_{pi} = 1$ if response i from person p is coded as correct or positive; and $U_{pi} = 0$ if the response is coded as incorrect or negative. In this case, $\theta_p \in \Re$ is unidimensional and $\beta_i = (a_i, b_i)$ has two components. The item response function (IRF) is

$$P_i(\theta_p; a_i, b_i) = P[U_{pi} = 1|\theta_p, a_i, b_i] = \frac{\exp(a_i(\theta_p - b_i))}{1 + \exp(a_i(\theta_p - b_i))}$$

and the density for U_{pi} becomes

$$f(u_{pi}|\theta_p, \beta_i) = P_i(\theta_p; a_i, b_i)^{u_{pi}}(1 - P_i(\theta_p; a_i, b_i))^{1-u_{pi}}.$$

In Equation 15.13, we will also use the normal prior distributions for θ_p and b_i and a log-normal prior distribution for a_i,

$$f_p(\theta_p|\lambda_\theta) = n(\theta_p|\mu_\theta, \sigma_\theta^2), \quad (15.14)$$

$$f_i(a_i|\lambda_a) = n(\ln a_i|\mu_a, \sigma_a^2)/a_i, \quad (15.15)$$

$$f_i(b_i|\lambda_b) = n(b_i|\mu_b, \sigma_b^2), \quad (15.16)$$

where $n(x|\mu, \sigma^2)$ is the normal density with mean μ and variance σ^2. (To be clear, $\lambda_\theta = (\mu_\theta, \sigma_\theta^2)$, $\lambda_a = (\mu_a, \sigma_a^2)$, etc.)

Although $f_p(\theta_p|\lambda_\theta)$ plays the mathematical role of a prior distribution in the model, it is usually interpreted in IRT as the population distribution for θ. For this reason, it is not usually taken to be "flat" or uninformative, but rather its shape tells us directly about the distribution of proficiency in the population. Specifying a prior $f(\lambda_\theta)$ allows us to estimate that shape in terms of the parameters λ_θ. When $f_p(\theta_p|\lambda_\theta) = n(\theta_p|\mu_\theta, \sigma_\theta^2)$, it is common to place prior distributions on μ_θ and σ_θ^2. For the purposes of illustrating an MCMC algorithm here, we will assume $\mu_\theta = 0$ and $\sigma_\theta^2 \sim IG(\alpha_\theta, \beta_\theta)$, an inverse-gamma distribution with parameters α_θ and β_θ.

We have now fully specified a Bayesian IRT model, elaborating Equation 15.13 as follows:

$$U_{pi} \overset{\text{indep}}{\sim} \text{Bernoulli}\,(\pi_{pi}) \tag{15.17}$$

$$\ln \frac{\pi_{pi}}{1+\pi_{pi}} = a_i(\theta_p - b_i) \tag{15.18}$$

$$\theta_p \overset{iid}{\sim} \text{Normal}\,(0, \sigma_\theta^2) \tag{15.19}$$

$$a_i \overset{iid}{\sim} \text{Log-normal}\,(\mu_a, \sigma_a^2) \tag{15.20}$$

$$b_i \overset{iid}{\sim} \text{Normal}\,(0, \sigma_b^2) \tag{15.21}$$

$$\sigma_\theta^2 \sim \text{Inverse-Gamma}\,(\alpha_\theta, \beta_\theta) \tag{15.22}$$

$$\mu_a, \sigma_a^2, \sigma_b^2, \alpha_\theta, \beta_\theta \text{ assigned in Section 15.8 below} \tag{15.23}$$

for all $p = 1, \ldots, P$ and all $i = 1, \ldots, I$. Here, $U_{pi} \overset{\text{indep}}{\sim} \text{Bernoulli}\,(\pi_{pi})$ means that the U_{pi}'s are independent with densities $\pi_{pi}^{u_{pi}}(1-\pi_{pi})^{1-u_{pi}}$, $\sigma_\theta^2 \sim \text{Inv-Gamma}\,(\alpha_\theta, \beta_\theta)$ means that σ_θ^2 has inverse-gamma density $IG(\sigma_\theta^2|\alpha_\theta, \beta_\theta)$, and similarly Equations 15.19 through 15.21 correspond to the density specifications in Equations 15.14 through 15.16.

If responses U_{pi} are missing completely at random (MCAR) or missing by design, the corresponding terms may simply be omitted from the products in Equation 15.13 and from Equations 15.17 through 15.23. More complex forms of missingness require additional modeling and may change the structure of the model and complete conditionals (e.g., Glas and Pimentel, 2008).

Complications of this basic framework involve richer latent space structure, and/or dependence induced by constraints on the parameters. MCMC algorithms for Linear logistic test models (De Boeck and Wilson, 2004; Scheiblechner, 1972), latent class models (Lazarsfeld and Henry, 1968), and some conjunctive cognitive diagnosis models (Haertel, 1989; Junker and Sijtsma, 2001; Macready and Dayton, 1977) received early consideration in Junker (1999) and have since been pursued in detail by de la Torre and Douglas (2004), Roussos et al. (2007), Sinharay and Almond (2007), and many others. Random effects models for "testlets" are considered in Bradlow et al. (1999), for example, and for families of automatically generated items are considered by Johnson and Sinharay (2005) and Sinharay et al. (2003).

The basic model of Equation 15.13 can be expanded to additional "levels" of hierarchical structure, by elaborating on the prior densities $f(\lambda_\theta)$ and $f(\lambda_\beta)$, to account for additional covariates, dependence due to administrative clustering such as students within classroom, and so forth. Early explorations of this idea include Kamata (2001) and Maier (2001). In addition, the IRT model itself can be embedded in other hierarchical/Bayesian models, for example MIMIC models. Both of these ideas are nicely developed in Fox (2011).

There is usually not much information available about the item parameters, a_i and b_i in the case of the 2PL model here, and so flat prior distributions are usually chosen for them; we will discuss specific choices below in Section 15.8. Further hierarchical structure is also possible, and is discussed at length in recent monographs such as De Boeck and Wilson (2004) and Fox (2010).

We will discuss particular choices for α_θ and β_θ in Equation 15.23, below in Section 15.8; it is usual to take choices such as $\alpha_\theta = \beta_\theta = 1$, indicating a fairly flat prior, which will allow the data to determine σ_θ^2, and hence, the shape of the normal population distribution $f_p(\theta_p|\lambda_\theta)$. For other prior choices for the parameters of the normal θ distribution, see for example, Casabianca and Junker (see Chapter 3). Non-normal (e.g., Sass et al., 2008; van den Oord, 2005) and nonparametric (e.g., Karabatsos and Sheu, 2004; Miyazaki and Hoshino, 2009; Woods, 2006; Woods and Thissen, 2006) choices for $f_p(\theta_p|\lambda_\theta)$ also appear in the literature.

The factorization displayed in Equation 15.13 into terms involving different blocks of parameters, facilitated by the independence assumptions of IRT, makes the complete conditionals rather uncomplicated. From Equations 15.17 through 15.23, it follows that the complete conditional densities for the individual parameters will be

$$f(\theta_p|rest) \propto \prod_{i=1}^{I} P_i(\theta_p;a_i,b_i)^{u_{pi}}(1-P_i(\theta_p;a_i,b_i))^{1-u_{pi}}n(\theta_p|0,\sigma_\theta^2),$$
$$\forall p = 1,\ldots,P \tag{15.24}$$

$$f(a_i|rest) \propto \prod_{p=1}^{P} P_i(\theta_p;a_i,b_i)^{u_{pi}}(1-P_i(\theta_p;a_i,b_i))^{1-u_{pi}}n(\ln a_i|\mu_a,\sigma_a^2)/a_i,$$
$$\forall i = 1,\ldots,I \tag{15.25}$$

$$f(b_i|rest) \propto \prod_{p=1}^{P} P_i(\theta_p;a_i,b_i)^{u_{pi}}(1-P_i(\theta_p;a_i,b_i))^{1-u_{pi}}n(b_i|0,\sigma_b^2),$$
$$\forall i = 1,\ldots,I \tag{15.26}$$

$$f(\sigma_\theta^2|rest) \propto \prod_{p=1}^{P} n(\theta_p|0,\sigma_\theta^2)IG(\sigma_\theta^2|\alpha_\theta,\beta_\theta)$$
$$= IG\left(\sigma_\theta^2 \,\middle|\, \alpha_\theta + \frac{P}{2}, \beta_\theta + \frac{1}{2}\sum_{p=1}^{P}\theta_p^2\right) \tag{15.27}$$

with σ_a^2, σ_b^2, α_θ, and β_θ to be fixed at values we specify.

The complete conditionals in Equations 15.24 through 15.26 almost always lead to M–H sampling (Section 15.5), because the product-Bernoulli form of the likelihood is typically not reducible to a simple exponential family or similar distribution. Data augmentation using the normal ogive/probit model in place of the logistic curve (Albert, 1992; Fox, 2010) is attractive because they do lead to Gibbs sampling; see for example, Thissen and Edwards (2005) for a sense of the trade-offs between these strategies. Our choice of an inverse-Gamma distribution for σ_θ^2, on the other hand, leads to a conjugate inverse-Gamma complete conditional distribution, so that this part of the algorithm can be implemented using Gibbs sampling (Section 15.4).

It is interesting to note that the complete conditional for σ_θ^2 does not depend on the data, but only θ_p. This is typical for parameters at higher levels in a hierarchical model. It is also interesting to note that, since I is usually much smaller than P, there is much more information to estimate α_i, β_i, and σ_θ^2 than to estimate θ_p. However, computational problems (e.g., floating point underflow) are also more common with complete conditionals that involve more factors.

15.8 Implementing an MCMC Algorithm

Conceptually, MCMC is easy: Identify complete conditionals, develop a sampling scheme like that of Figure 15.1, run one or more chains for "long enough," use the "converged" part of the chains to make inferences about the posterior distribution. As suggested in Section 15.7, IRT models usually suggest a natural separation of parameters into single-parameter blocks with fairly natural Gibbs and M–H steps, and this is a good place to start.

In this section, we will sketch a fruitful approach for developing an MCMC algorithm, using the model and complete conditionals developed in Section 15.7. For concreteness, code is presented in R (R Development Core Team, 2012), but the general recipe will apply to any other programming language, choice of sampling algorithm to implement, or choice of model with suitable translation. After illustrating the development of the algorithm, we will also illustrate its application briefly to item response from a mathematics assessment administered to a U.S. national sample of fifth grade students. Further details are available in an online supplement to this chapter (VanHoudnos, 2012).

15.8.1 Simulating Fake Data

In order to check that an estimation algorithm is working properly, it is useful to see if the algorithm can recover the true parameter values in one or more simulated "test" data sets. Figure 15.4 displays R code, based on Equations 15.17 through 15.19, for simulating dichotomous item responses for $P = 2000$ persons and $I = 30$ items, using the 2PL model, with discrimination parameters a_i sampled randomly from a Unif[0.5, 1.5] distribution, and difficulty parameters b_i equally spaced running from -3 to $+3$. Persons' proficiency parameters θ_p were sampled randomly from a normal distribution $n(\theta|\mu_\theta, \sigma_\theta^2)$ with mean $\mu_\theta = 0$ and standard deviation $\sigma_\theta = 1.25$. The R function `plogis(x)` computes $1/(1+\exp(-x))$; later in the chapter in Figure 15.10, we will also use the inverse function `qlogis(p)` which computes $\log p/(1-p)$.

After simulating the data, it is a good idea to check to see if the simulation was successful, by checking that quantities estimated from the data agree with the simulation setup. In the case of a new model for which MCMC will be the first estimation algorithm, it is useful to check that the simulated data are consistent with predictions from the simulation model. For example, we can check that moments (means, variances, covariances, etc.) of the simulated data agree with the same moments of the simulating model. Although the 2PL IRT model is itself a well-known model, we illustrate this approach in Figure 15.5. The

```
# Set the random-number generator seed,
# to make the results reproducible
set.seed(314159)

# set the number of items I and persons P
I.items      <- 30
P.persons    <- 2000

# set the fixed item and population parameters
a.disc       <- 1 + runif(I.items,-0.5,0.5)
b.diff       <- seq(-3,3,length=I.items)

mean.theta   <- 0
sig2.theta   <- (1.25)^2

# generate thetas and the I x P matrix of response probabilities
theta.abl    <- rnorm(P.persons, mean=mean.theta, sd=sqrt(sig2.theta))
term.1       <- outer(theta.abl, a.disc)
term.2       <- matrix(rep(a.disc*b.diff, P.persons), nrow=P.persons,
                  byrow=TRUE)
P.prob       <- plogis(term.1-term.2)  ### 1/(1 + exp(term.2 - term.1))

# generate the 0/1 responses U as a matrix of Bernoulli draws
U            <- ifelse(runif(I.items*P.persons)<P.prob,1,0)
```

FIGURE 15.4
R code to generate fake 2PL data for testing our MCMC algorithm, based on Equations 15.17 through 15.19.

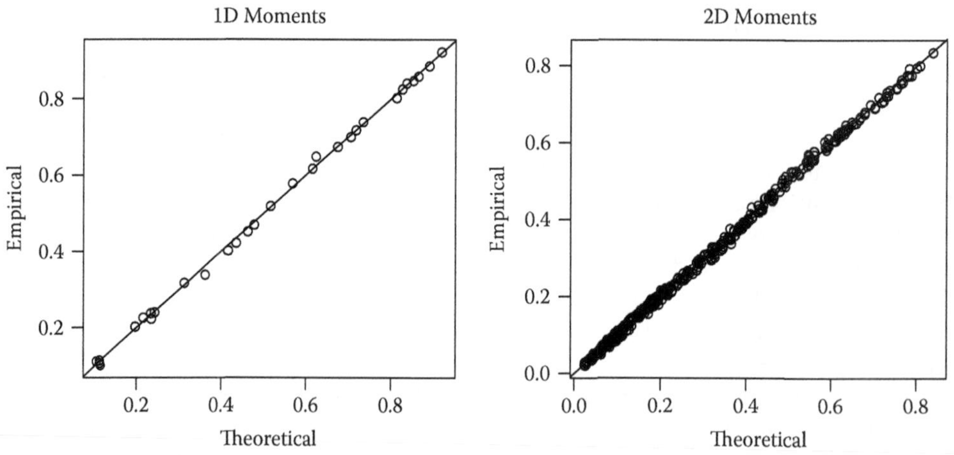

FIGURE 15.5
Checking the fake-data simulation by verifying that empirical moments from the simulated data match theoretical moments from the simulation model. The left panel compares theoretical and empirical estimates of $\pi_i = P[U_i = 1]$, and the right panel compares theoretical and empirical estimates of $\pi_{ij} = P[U_i = 1 \text{ and } U_j = 1]$.

left panel compares the theoretical values

$$\pi_i = \int_{-\infty}^{\infty} \frac{\exp\{a_i(\theta - b_i)\}}{1 + \exp\{a_i(\theta - b_i)\}} \cdot n(\theta|0, \sigma_\theta^2)\, d\theta$$

with their empirical estimates

$$\hat{\pi}_i = \frac{1}{P} \sum_{p=1}^{P} U_{pi}.$$

The right panel compares the theoretical values

$$\pi_{ij} = \int_{-\infty}^{\infty} \frac{\exp\{a_i(\theta - b_i)\}}{1 + \exp\{a_i(\theta - b_i)\}} \cdot \frac{\exp\{a_j(\theta - b_j)\}}{1 + \exp\{a_j(\theta - b_j)\}} \cdot n(\theta|0, \sigma_\theta^2)\, d\theta$$

with their empirical estimates

$$\hat{\pi}_{ij} = \frac{1}{P} \sum_{p=1}^{P} U_{pi} U_{pj}.$$

See VanHoudnos (2012) for computational details. Of course other such "moment matching" checks are possible and should be chosen to check any features of the simulated data that might be in doubt.

In the case of a familiar model such as the 2PL IRT model for which many other estimation algorithms have been written, we can also check to see that parameters are recovered by a known algorithm, such as the R package `ltm` (Rizopoulos, 2006). This is illustrated in Figure 15.6: The left panel compares theoretical discrimination parameters a_i with their ML estimates from `ltm`; the right panel makes the same comparison for difficulty parameters

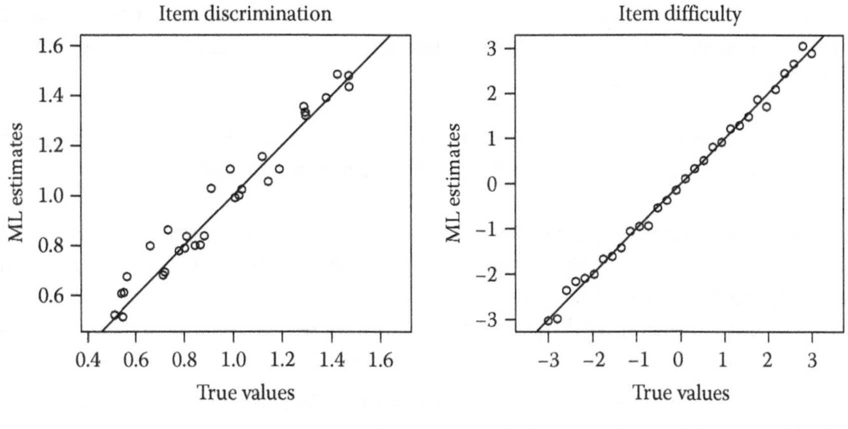

FIGURE 15.6
Checking the fake-data simulation by verifying that the simulation model parameters can be recovered by maximum likelihood (ML). The left panel compares theoretical and ML-estimated values of the discrimination parameters a_i and the right panel makes the same comparison for the difficulty parameters b_i.

b_i. Both panels employ a linear equating using the means and standard deviations of the true and estimated b_i's, as suggested by Cook and Eignor (1991) for example, to account for scale indeterminacy in estimating the IRT model (while this particular transformation is unique to the particular model, we are working with here, it is important to keep similar latent space indeterminacies in mind when working with other IRT and IRT-like models). Figures 15.5 and 15.6 suggest that we simulated fake data from the intended 2PL model; we will use this data to test the MCMC algorithm as we develop it.

15.8.2 The MCMC Algorithm Shell

Before building the individual sampling functions that generate Monte Carlo samples from the complete conditionals in Equations 15.24 through 15.27, it is helpful to think about the context in which they will operate. Figure 15.7 displays the two main components of a working one-variable-at-a-time MCMC sampler.

- `blocked.mcmc.update` collects together four functions, `sample.th`, `sample.a`, `sample.b` and `sample.s2` that will be used to implement one full step of the one-variable-at-a-time MCMC algorithm outlined in Figure 15.1. We will discuss the definitions of these four functions below in Section 15.8.3.

  `U.data` is the matrix $\mathcal{U}$ of item responses. `cur` is a "list" (an R data structure) that contains the current state of the Markov chain (the current set of sampled parameter values), as well as auxiliary information such as hyperparameters, M–H proposal variances and acceptance rate information. The effect of a call to `blocked.sample.mcmc` is to take one step in the Markov chain, that is, to replace the old sampled parameter values with new ones.
- `run.chain.2pl` is a sketch of the main function that "runs" the entire MCMC algorithm. Certain housekeeping details are omitted for clarity, and replaced with "..." in Figure 15.7; they can be seen in detail in Figure 15.8.

  The algorithm runs in two phases. In the *burn-in phase*, `M.burnin` steps of the Markov chain are sampled, and simply thrown away, since the Markov chain is assumed not to have converged yet. In the *converged phase*, `M.keep` × `M.thin` steps are sampled, and every `M.thin`th set of sampled parameter values is kept. The resulting matrix of `M.keep` samples from the posterior distribution is returned in the matrix `chain.keep`.

A complete version of `run.chain.2pl` is shown in Figure 15.8, for completeness. Details present in Figure 15.8, but omitted in Figure 15.7, include the following:

- Additional information passed to `run.chain.2pl`: `U.data`, the data matrix of item responses; `hyperpars`, the vector of hyperparameters for the prior distributions according to Equation 15.23; vectors `th.init`, `a.init`, `b.init` and `s2.init` for all model parameters; M–H proposal variances `MH.th`, `MH.a`, `MH.b`; and a switch `verbose` to indicate whether to report M–H acceptance rates.
- Setting up the "current state" list `cur` and initializing the model parameters in `cur` to the initial values passed to `run.chain.2pl`.
- Setting up the matrix `chain.keep` in which to keep values of the Markov chain once it has converged. Each column of `chain.keep` contains one complete set of model parameters drawn from the posterior distribution.

```
blocked.mcmc.update <-function(U.data,cur){
  cur <- sample.th(U.data, cur) # Equation 1.24
  cur <-  sample.a(U.data, cur) # Equation 1.25
  cur <-  sample.b(U.data, cur) # Equation 1.26
  cur <- sample.s2(U.data, cur) # Equation 1.27
  return(cur)
}
```

```
run.chain.2pl <-function(M.burnin, M.keep, M.thin, ... ){

  ...

  # Burn-in phase: do not keep these results
  if( M.burnin > 0 ) {
    for(ii in 1:M.burnin ) {
      cur <- blocked.mcmc.update(U.data, cur)
  }}

  # Converged phase: Keep these results after thinning
  for (m in 1:M.keep) {
    # Skip the "thinned" pieces of the chain
    if( M.thin > 1 ) {
      for( ii in 1:(M.thin-1) ) {
        cur <- blocked.mcmc.update(U.data, cur)
    }}
    # Generate a "kept" update and save its results
    cur <- blocked.mcmc.update(U.data, cur)
    chain.keep[,m]<- c( cur$th, cur$a, cur$b, cur$s2 )
  }

  ...

  return( chain.keep )
}
```

FIGURE 15.7
The R function `blocked.mcmc.update` collects together four sampling routines that implement the one-variable-at-a-time MCMC algorithm of Figure 15.1, for a 2PL IRT model. `cur` defines the current state of the Markov Chain. Each function call `sample.th`, `sample.a`, etc. implements a set of draws from complete conditionals; the definitions of these functions will be described below in Section 15.8.3. The R function `run.chain.2pl` shows how `blocked.mcmc.update` is used to implement a complete MCMC algorithm. Some housekeeping details have been elided; see Figure 15.8 for these details.

- Code to report `Average acceptance rates` for the M–H portions of the algorithm.

15.8.3 Building the Complete Conditionals

We suggest building and testing each of the parameter-sampling functions `sample.th`, `sample.a`, `sample.b`, and `sample.s2` shown in Figure 15.7 separately. While testing one

```
run.chain.2pl <- function(M.burnin, M.keep, M.thin,
                        U.data, hyperpars,
                        th.init, a.init, b.init, s2.init,
                        MH.th, MH.a, MH.b, verbose=FALSE) {

  # Define and initialize the list of things to keep track of in the
  # "current state "of the chain -- see text for detailed explanation
  cur     <-list(th=th.init, a=a.init, b=b.init, s2=s2.init,
                hyperpars=hyperpars,
                MH = list(th=MH.th,    a=MH.a,    b=MH.b),
                ACC= list(th=0,th.n=0, a=0,a.n=0, b=0,b.n=0))

  # Define matrix to store MCMC results...
  chain.keep <- matrix( NA, ncol = M.keep,
                        nrow = length(th.init) + length(a.init)
                               + length(b.init) + length(s2.init) )
  rownames(chain.keep) <- c( paste('theta.abl', 1:length(th.init)),
                             paste('a.disc',  1:length(a.init)),
                             paste('b.diff',  1:length(b.init)),
                             'sig2.theta')

  # Burn-in phase: do not keep these results
  if(M.burnin > 0 ) {
    for(ii in 1:M.burnin ) {
      cur <- blocked.mcmc.update(U.data, cur)
  }}

  # Converged phase: Keep these results after thinning
  for (m in 1:M.keep) {
    # Skip the "thinned" pieces of the chain
    if( M.thin > 1 ) {
      for( ii in 1:(M.thin-1) ) {
        cur <- blocked.mcmc.update(U.data, cur)
    }}
    # Generate a "kept" update and save its results
    cur <- blocked.mcmc.update(U.data, cur)
    chain.keep[,m] <- c( cur$th, cur$a, cur$b, cur$s2 )

    if( m %% 100 == 0) {
      print(m)
      # Adaptive tuning would go here.
    }

  }

  if (verbose) {
    cat(paste("Average acceptance rates:",
             "\n theta.abl: ", round (cur$ACC$th / cur$ACC$th.n ,3),
             "\n a.disc:    ", round (cur$ACC$a  / cur$ACC$a.n ,3),
             "\n b.diff:    ", round (cur$ACC$b  / cur$ACC$b.n ,3),"\n"))
  }

  return( chain.keep )
}
```

FIGURE 15.8
Function definition for a complete MCMC algorithm for fitting a 2PL model to data. See text for discussion of the various parts of this function.

function, say `sample.th`, the other functions should simply return the values they are given, effectively holding those parameters fixed at their initial values, to isolate the behavior of `sample.th` itself. For example, `sample.a` could be given the provisional definition `sample.a <- function(U.data, cur) { return(cur) }` and identical definitions could be used for `sample.b` and `sample.s2`. These provisional definitions would be replaced with actual MCMC-sampling code as each `sample.*` function is developed (see VanHoudnos, 2012, for details).

For illustration, we show a definition for the function `sample.th`, in Figure 15.9. We sample from the complete conditional shown in Equation 15.24, using a random-walk M–H sampler as described in Section 15.5. All probability calculations have been done on the log scale, to avoid numerical underflow on the computer, which is common when multiplying many probabilities together (for larger problems, logarithmic calculations typically have to

```
log.prob <- function(U.data, th, a,b) {
  term.1     <-outer(th, a)
  term.2     <-matrix(rep(a*b, P.persons), nrow=P.persons,
               byrow=TRUE)
  P.prob <- plogis(term.1 - term.2)

  log.bernoulli <- U.data*log(P.prob) + (1-U.data)*log(1-P.prob)
  return(log.bernoulli)
}
```

```
sample.th <- function(U.data, old) {
 th.old     <- old$th
 MH.th      <- old$MH$th
 P.persons <- length(th.old)
 th.star    <- rnorm(P.persons,th.old,MH.th)

 log.cc.star <- apply(log.prob(U.data, th.star, old$a ,old$b),1,sum)+
                     log(dnorm(th.star,0,sqrt(old$s2)))
 log.cc.old  <- apply(log.prob(U.data, th.old,old$a, old$b),1,sum)+
                     log(dnorm(th.old,0,sqrt(old$s2)))
 log.prop.star <- log(dnorm(th.star,th.old,MH.th))
 log.prop.old  <- log(dnorm(th.old,th.star,MH.th))

 log.alpha.star <- pmin(log.cc.star + log.prop.old - log.cc.old -
                     log.prop.star,0)

 acc.new <- ifelse(log(runif(P.persons))<=log.alpha.star, 1, 0)

 cur           <- old
 cur$th        <- ifelse(acc.new==1, th.star, th.old)
 cur$ACC$th    <- old$ACC$th + mean( acc.new )
 cur$ACC$th.n <-  cur$ACC$th.n + 1

 return(cur)
}
```

FIGURE 15.9
Definition of `sample.th`, for sampling from the complete conditionals for θs, using a normal random-walk M–H strategy. An auxiliary function `log.prob` which is used several times in `sample.th` and other complete-conditional samplers, is also defined here. See text, and VanHoudnos (2012), for details.

be combined with other strategies such as additive offsets, to avoid under flow; see e.g., Monahan, 2011, Chapter 2).

In Figure 15.9, `log.prob` is an auxiliary function that will be used several times in `sample.th`, `sample.a` and `sample.b`. It returns a $P \times I$ matrix of terms

$$\log\left[P_i(\theta_p; a_i, b_i)^{u_{pi}}(1 - P_i(\theta_p; a_i, b_i))^{1-u_{pi}}\right]$$

which are useful in calculating the log-likelihoods following Equations 15.24 through 15.26.

`sample.th` is the main function for sampling θ_ps from the complete conditional densities in Equation 15.24. In `sample.th`,

- `old` contains the "old" current state of the Markov chain, and `cur` contains the "new" current state after `sample.th` is finished. Both variables have the structure of `cur` defined in `run.mcmc.2pl` in Figure 15.8.
- We have exploited the high degree of separation of parameters available in IRT models (due to independence across persons and items in the IRT likelihood) to string together the P calculations needed in Equation 15.24 into a small number of implicit vector calculations in R; this not only shortens the programming task, it also speeds up Rs execution. In other models not enjoying such separation, these calculations would have to be done separately and sequentially.
- The line `th.star <- rnorm(P.persons,th.old,th.MH)` implements the normal random-walk proposal draws

$$\theta_p^* \overset{\text{indep}}{\sim} N(\theta_p^{(m-1)}, \sigma_p^2), \quad p = 1, \ldots, P$$

  where the value of σ_p is specified in `th.MH`.
- The next five assignment statements calculate the vector of M–H acceptance ratios α_p^* for each θ_p (Equation 15.9), using the implicit vector calculations available in R.
- `acc.new` contains a vector of 0s and 1s: 0 in the pth position if the corresponding θ_p^* is not accepted, and 1 if it is.
- Finally, the current state `cur` of the Markov chain is updated with each accepted θ_p^*, the running sum of acceptances, `cur$ACC$th`, is updated with the average acceptance rate across all θ_ps, and the number of updates, `cur$ACC$th.n`, is incremented.

Note that in this algorithm, we are accumulating only average acceptance counts across all P θ_p parameters, rather than individual acceptance rates for each θ_p. In well-behaved IRT models with no missing data in the rectangular array $\mathcal{U}$, this is a workable strategy; in more challenging situations, it may be necessary to keep track of acceptance rates for each θ_p individually and use separate M–H proposal variances for each θ_p (expanding both `old$MH$th` and `old$MH$ACC` to I-dimensional vectors).

After some trial runs to make sure that we are happy with `sample.th`, we would move on to developing `sample.a`, `sample.b`, and `sample.s2`. For `sample.a` and `sample.b`, a strategy similar to that of Figure 15.9 can be used. Since `sample.s2` will be a Gibbs step, simpler direct sampling from the inverse-Gamma distribution in Equation 15.27 can be implemented. More details on all four sampling functions are supplied in VanHoudnos (2012).

15.8.4 Tuning Metropolis–Hastings

The M–H samplers `sample.th`, `sample.a`, and `sample.b` must also be tuned—that is, proposal variances `MH.th`, `MH.a`, and `MH.b` must be chosen to get workable acceptance rates. If the acceptance rate is too high, the chain may not explore very much of the posterior distribution because it is taking too many small, incremental steps. The resulting trace plot may look like plot (b) in Figure 15.2. If the acceptance rate is too low, the chain will also not explore much of the posterior distribution because most proposals are rejected and it stays in place. The resulting trace plot may look like plot (c) in Figure 15.2. In both the too high and too low regions of acceptance rate, the autocorrelation will be high (Figure 15.3). For a single parameter at a time, the optimal acceptance rate is near 0.44; in practice, rates between 0.2 and 0.6 are usually reasonable (Rosenthal, 2011).

In Figure 15.10, we show one possible strategy for tuning the M–H samplers. In the figure, we only keep track of the average acceptance rate for θs, the average rate for as and the average rate for bs. It is possible that some individual parameters do not have reasonable acceptance rate. We could instead tune acceptance rates for each θ_p, a_i, and b_i individually, by recording separate acceptance rates for each of them, and using different component values in `th.MH`. Note that tuning one M–H sampler can affect the acceptance rates of the others.

```
set.seed(314159)

M.burnin <- 250
M.keep <- 1000
M.thin <- 1

# See Section 1.8.5 for details of prior specification
hyperpars <- list( mu.a          =1.185,
                   s2.a          =1.185,
                   s2.b          =100,
                   alpha.th      =1,
                   beta.th       =1 )

# use naive method of moments estimators as initial values for
# theta's and b's, and set the initial values of a's and s2 to 1.
th.tmp       <- qlogis(0.98*apply(U,1,mean) + .01)
a.tmp        <- rep(1,I.items)
b.tmp        <- qlogis(0.98*apply(U,2,mean) + .01)
s2.tmp       <- 1

# specify the MH tuning parameters
MH.th <- .75; MH.a <- .15; MH.b <- .10

tune.run <- run.chain.2pl(M.burnin, M.keep, M.thin, U, hyperpars,
                  th.tmp, a.tmp, b.tmp, s2.tmp,
                  MH.th, MH.a, MH.b, verbose=TRUE)
# Average acceptance rates:
#   theta.abl: 0.416
#   a.disc:    0.47
#   b.diff:    0.436
```

FIGURE 15.10
Example of tuning the completed sampler by iteratively adjusting the values of `MH.th`, `MH.a`, and `MH.b`. The table presents additional sampling runs that were used to find the final tuned values.

TABLE 15.1

Results of Tuning Runs as in Figure 15.10 for Various Values of Tuning Parameters

`MH.th`	Acceptance	`MH.a`	Acceptance	`MH.b`	Acceptance
2.	0.218	1.	0.076	1.	0.072
1.	0.352	0.25	0.306	0.25	0.226
0.75	0.416	0.15	0.470	0.10	0.436

The results of three tuning runs using the code illustrated in Figure 15.10 are shown in Table 15.1. The first run (first row of the table) shows unacceptably low average acceptance rates, especially for the a and b parameters. The second run (second row) shows acceptable but still somewhat low average acceptance rates for all three sets of parameters. The third run (third row) shows nearly optimal average acceptance rates.

In addition to changing a "tuning parameter" for a specific proposal distribution, it can also be advantageous to change the proposal distribution itself, in order to more efficiently explore the posterior. Typically, the proposal distribution would be chosen to mimic the posterior distribution before running the chain, or it can be adaptively adjusted to mimic the posterior distribution as the whole MCMC algorithm runs (see also the discussion of choice of proposal distributions in Section 15.5 above). In the case of one-dimensional complete conditionals, one adaptive approach is to simply adjust the variance of the proposal distribution as the chain runs in order to zero in on optimal acceptance rates. Such adaptations could be built into the function `run.chain.2pl` in Figure 15.7. When the complete conditional is multidimensional, the adaptive procedure would try to mimic the covariance structure of the posterior distribution. Unfortunately, these methods are beyond the scope of the chapter; see Rosenthal (2011) and the references therein for an accessible discussion.

15.8.5 Testing on Simulated Data

We tried the algorithm out on data simulated as in Figure 15.4. We used the prior distributions listed in Equations 15.20 through 15.22 with the hyperparameters listed in Figure 15.10.

The hyperparameters were chosen to create uninformative priors. The b_i prior is a normal with variance $\sigma_b^2 = 100$, which is uninformative because it is very flat for typical values of item difficulty. We similarly dispersed the lognormal prior of the a_i values by matching the 95% quantile of a more familiar distribution: A positive truncated normal distribution with a variance of 100. A quick calculation shows that 95% of the mass of a truncated normal with a variance of 100 is below 19.6 units and that a lognormal with a mode of 1 must have $\mu_a = \sigma_a^2 = 1.185$ to also have 95% of its mass below 19.6 units. Values larger than 1.185 will further flatten the prior, but since the discrimination parameters are typically less variable than difficulty parameters it is not useful to further broaden the discrimination prior. Unlike the previous two priors, which were more uniformative for larger values of the hyperparameters, the inverse-gamma prior for σ_θ^2 becomes less informative as α and β become smaller as can be seen from complete conditional in Equation 15.27. We chose the minimum values of $\alpha = \beta = 1$. A more detailed discussion of these prior specifications can be found in VanHoudnos (2012).

We simulated three chains from starting values that were overdispersed from the method of moments starting values shown in Figure 15.10. On the one hand, in order to get trustworthy evidence of convergence from the Gelman–Rubin convergence statistic R, and moreover, any iterative search algorithm should be started from several different initial values, to gain evidence that algorithm does not get "stuck" in nonoptimal solutions. On the other hand, we have found that MCMC for IRT can be sensitive to excessively poor starting values: typically much of the IRT likelihood and posterior is extremely flat, and starting the MCMC algorithm in a flat part of the posterior distribution will cause the chain to wander a very long time, and/or become numerically unstable, before converging on a local or global mode.

We note in passing that, although simulating multiple chains is extremely valuable—both for diagnosing problems with the MCMC algorithm and for confirming that the chain has converged to the stationary distribution—it is extremely slow computationally, especially if the chains must be generated sequentially. VanHoudnos (2012) suggests some methods for generating the chains in parallel, by exploiting the fact that most modern personal computers are equipped with multiple CPU cores.

To generate starting values for the chains, we began by mildly jittering the method of moments estimates in Figure 15.10. We further spread the starting values out by exploiting the location and scale indeterminacy of logistic IRT models, transforming these jittered values by randomly chosen location and scale values which did not change the probabilities of correct responses.

However, even this mild dispersion caused the chains to converge very slowly, especially for parameters with little support in the data such as σ_θ^2 and the largest item discrimination parameter (max a_i). To speed convergence, we modified the scan order of the M–H algorithm to reduce autocorrelation and cross-correlation in the chain, without incurring a large computational cost, by repeatedly sampling item parameters, holding person parameters fixed, and vice versa (Section 15.8.7 below for details). This improved stability of the Markov chain and sped convergence to the stationary distribution. Figure 15.11 shows typical behavior of the chain, for three sets of overdispersed starting values, using the modified scan order.

Care in starting values and scan order here is partly due to numerical instability in the expression `log.bernoulli <- U.data*log(P.prob) + (1-U.data)*log(1-P.prob)` in the top panel of Figure 15.9, especially when `P.prob` is close to 0 or 1. In R, a more numerically stable expression would be `log.bernoulli <- log(P.prob^U.data) + log((1-P.prob)^(1-U.data));` using this expression instead, we are able to use much more overdispersed starting values, for example. For additional details, see VanHoudnos (2012).

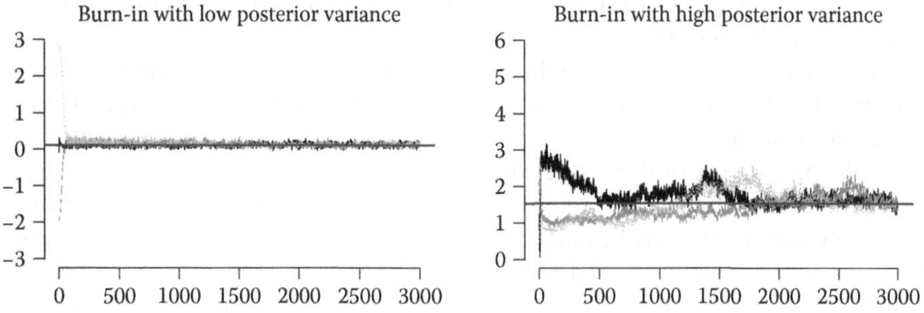

FIGURE 15.11
Trace plots demonstrating the differences in burn-in times between a parameter with small posterior variance, which burns-in within 100 iterations, and a parameter with large posterior variance, which burns-in after 2000 iterations. The equated true values of each parameter are plotted as a horizontal line. See text, and VanHoudnos (2012) for details on the code used to setup, run, and visualize the overdispersed chains.

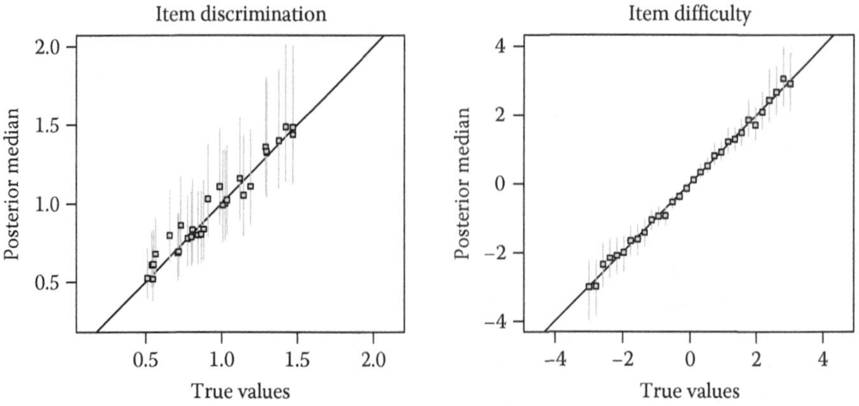

FIGURE 15.12
Checking the MCMC algorithm by verifying that the simulation model parameters can be recovered. The left panel compares equated theoretical and MCMC-estimated values of the discrimination parameters a_i and the right panel makes the same comparison for the difficulty parameters b_i. The error bars are equated 95% posterior quantiles.

From trace plots like Figure 15.11, as well as acf plots, we chose `M.burnin` $= 3000$, `M.thin` $= 1$, and `M.keep` $= 3000$ for the final run of the improved scan order chain. It converged nicely: The Gelman–Rubin convergence statistic $\hat{R} \leq 1.055$ for all parameters and every MCMC standard error was at least an order of magnitude smaller than the associated posterior standard error. A comparison of posterior quantiles with the equated true values from the simulated fake data is shown in Figure 15.12, confirming that our algorithm works well. Additional details, such as the code to generate the graphs, can be found in VanHoudnos (2012).

15.8.6 Applying the Algorithm to Real Data

We analyzed data collected as a part of the national standardization research for the *Comprehensive Test of Basic Skills, Fifth Edition* (CTBS/5), published by CTB/McGraw-Hill (1996). A mathematics test comprised of 32 multiple-choice items and 11 constructed-response items was administered to a U.S. nationally representative sample of 2171 Grade 5 students. All multiple-choice item responses were dichotomously scored, and all constructed responses were professionally evaluated on rubrics that ranged from zero to a maximum of between 2 and 5 points. For this illustration, we analyzed only the 32 multiple-choice questions. There are no missing data.

In principle, all aspects of the model described in Equations 15.17 through 15.23 should be reconsidered when new data are considered. Especially the form of the prior distributions and hyperparameters in Equations 15.20 through 15.23 should receive special care, since they can affect parameter and latent proficiency estimates even if we keep the form of the IRT model in Equations 15.17 through 15.19 the same. For the purposes of illustration in this chapter, however, we used the same prior distributions and chosen hyperparameters $\mu_a = 1.185$, $\sigma_a^2 = 1.185$, $\sigma_b^2 = 100$, $\alpha_\theta = 1$, $\beta_\theta = 1$. The example Grade 5 data set had properties very similar to the simulated "fake data" above (indeed, we designed the "fake data" to mimic the real data), so it was very straightforward to analyze with our improved scan order MCMC algorithm, prior distributions, and so forth.

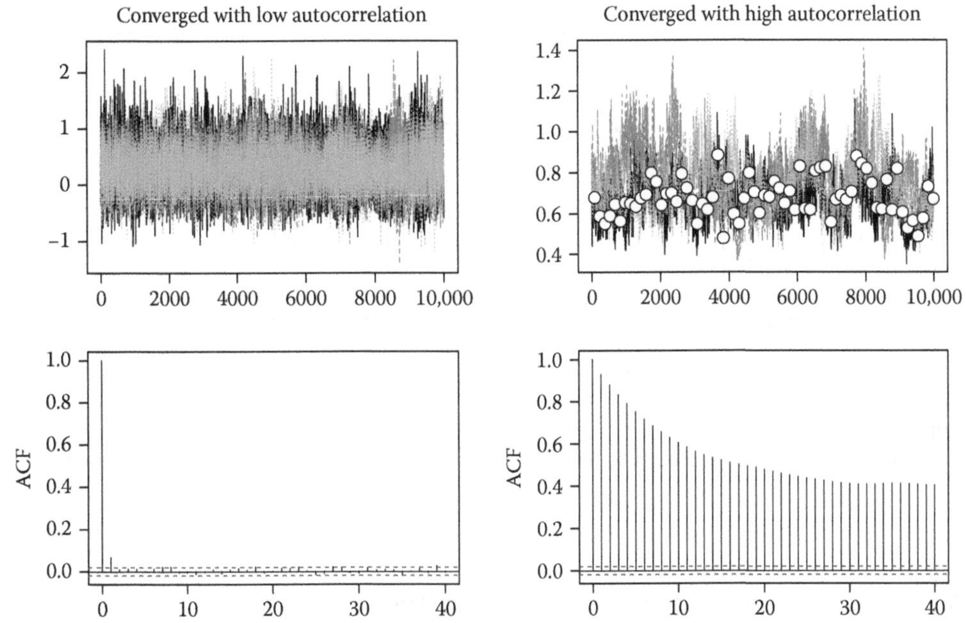

FIGURE 15.13
Converged trace plots on the real data example with the autocorrelation functions calculated from the black chains. The left plots are of a typical person ability parameter (θ_p); the right plots, a typical item discrimination parameter (a_i). The 67 white circles are values that would be kept if the chain were thinned so that its autocorrelation would be as low as that of the left example chain (`M.thin` = 150).

After a 6000 iteration burn-in from overdispersed starting values the sampler had clearly converged both from the inspection of trace plots and the calculation of $\hat{R}$ statistics for each of the parameters, which were all less than 1.05. Even though the chain was properly tuned and it had clearly converged, the autocorrleation of some parameters was still quite high as can be seen in Figure 15.13. To bring the autocorrelation down to nonsignificant levels, the chain would have had to be thinned to keep only every 150th iteration, which would have left only 67 samples per 10,000 iterations. However, this extreme thinning is unnecessary: In all cases, the MCMC error as estimated with overlapping batch means was very small compared to the posterior variance.

Figure 15.14 compares the posterior quantile estimates from pooling all three converged chains to maximum likelihood estimates. Note that in all cases the two sided 95% credible intervals contained the ML estimates. There is also evidence in Figure 15.14 that CI's for as and bs were wider where they were difficult to estimate (larger magnitude as and bs).

15.8.7 Miscellaneous Advice

As we have seen, setting up a functioning MCMC algorithm for an IRT model is not difficult. In practice, the MCMC algorithm may be computationally slow, may converge slowly to the $f(\tau|\mathcal{U})$, and/or may exhibit excessive autocorrelation/cross-correlation. Each of these takes some care to correct, in practice.

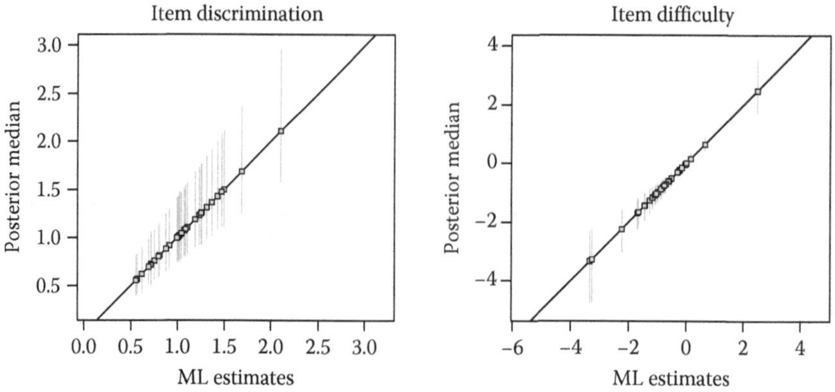

FIGURE 15.14
Posterior medians compared to equated ML estimates. The error bars are equated 95% posterior quantiles.

15.8.7.1 Computational Improvements and Restarting

If the chain is mixing reasonably well but is executing slowly, then some computational adjustments may be needed. Adapting M–H proposal distributions dynamically (Rosenthal, 2011), or replacing M–H steps with adaptive rejection (Gilks and Wild, 1992) or Gibbs (Albert, 1992; Fox, 2010) sampling steps, can speed up computation. Avoiding time-intensive matrix calculations, precomputing quantities that will be used repeatedly, and/or premarginalizing over nuisance parameters, can also help. Parallel computing is an obvious speed-up tool: Simply taking advantage of modern hardware to run multiple instances of existing code simultaneously on a multicore CPU speeds up computation by a simple division of labor (e.g., VanHoudnos, 2012); running a single chain on multiple CPUs is also possible, if one can identify or construct regeneration states (e.g., Brockwell and Kadane, 2005) in the Markov chain. More recently, efficient and highly accurate approximate methods for running a single MCMC algorithm on multiple computers has been developed (Scott et al., 2013).

The order in which the complete conditionals in Figure 15.1 are sampled is called the *scan order* of the MCMC algorithm. *Fixed scan* algorithms sample in the fixed order that the blocks are numbered—1, 2, 3, . . . , H, as in Figures 15.1 and 15.8. Although fixed-scan algorithms do not satisfy the reversibility condition of Equation 15.5, they still do have $f(\boldsymbol{\tau}|\mathcal{U})$ as the stationary distribution, as suggested in the discussion of Equation 15.6. Reversible kernels tend to have better asymptotic properties—it is easier to prove asymptotic normality, and to calculate or estimate convergence rates. *Palindromic scan* algorithms samples the complete conditional in order and then in reverse order, to produce one Markov chain step and is reversible. *Random permutation scan* algorithms sample from the complete conditionals in random order for each step of the Markov chain, and *random scan* algorithms simply sample from a random complete conditional H times, to produce one step. Roberts and Sahu (1997) argue that fixed scan Gibbs samplers generally have the fastest convergence rates, though random scan methods will be faster for some parameterizations of models with linear mean structure.

Sometimes the scan order needs to be tailored to the specific features of the model or data one is working with. For example, in the running IRT example in this chapter, we found that with a simple fixed order

$$\theta_p \rightarrow a_i \rightarrow b_i \rightarrow \sigma_\theta^2, \tag{15.28}$$

samples of the parameter σ_θ^2 were highly affected by variability of samples of θ_ps and in turn samples of the a and b parameters; this slowed the convergence of the chain considerably. We found that the MCMC algorithm stabilized more quickly if we could somewhat isolate samples of θs and σ_θ^2 from samples of as and bs, and vice versa, by sampling k sets of θ_ps and σ_θ^2 for each set of as and bs. Thus, the scan order

$$\left(\theta_p \rightarrow \sigma_\theta^2\right)^k \rightarrow (a_i \rightarrow b_i)^k$$

will still generate an MCMC algorithm that converges to the true posterior distribution, as discussed in Section 15.3, but will converge much faster. The value of k controls a trade-off between reduced autocorrelation and cross-correlations in the Markov chain, and computational time. By trial and error, we found $k = 7$ to be a reasonable choice.

Although it is not a computational improvement per se, it also is a very good practice to save the random number generator seed (or some other sufficient description of the current state of the simulated Markov chain) at regular intervals during an MCMC run. That way, if the simulation is interrupted (by a power failure for the computer running the algorithm, by the need to move to another computer with more memory or a faster processor, etc.), then the simulation can begin from the last saved state rather than from the initial state. This can also be useful in reproducing "bad" behavior to help diagnose computational errors, slow mixing, etc.

15.8.7.2 Blocking, Reparametrization, and Decorrelation Steps

If the chain is mixing slowly and/or exhibiting large autocorrelation or cross-correlation, several remedies are possible. Graves et al. (2011) survey a number of strategies for dealing with slowly mixing or highly autocorrelated or cross-correlated chains, many of which were first gathered together by Gilks and Roberts (1996) and Roberts and Sahu (1997). The most common are blocking, reparametrization, and decorrelation steps.

One suggestion often made is to try a different *blocking* scheme, that is, choose different disjoint subsets of parameters $\tau_1, \ldots, \tau_H$ in Figure 15.1. However, the item parameters for each item are natural blocks that often exhibit strong posterior dependence. These and other groups of parameters that exhibit strong dependence can be revealed in shorter preliminary runs of of a less-than-optimal MCMC algorithm. When blocks of dependent parameters are grouped together and sampled as a block, overall mixing can improve (though of course the dependence between parameters within blocks remains). Exact conditions under which blocking does or does not help, for normal complete conditionals (which are often approximately true for other complete conditionals, by a CLT-like argument), can be found in Roberts and Sahu (1997); see also Doucet et al. (2006).

Another way to deal with dependence between parameters is to *reparametrize* the model so that the new parameters are less dependent. Parameters with a normal or nearly normal posterior distribution can be rotated, via linear transformation, to be uncorrelated for example. This does not remove the dependence among the original parameters, but an MCMC algorithm for the transformed parameters may run better (and the results transformed back to the original parametrization if needed). Hierarchical centering (Gilks and Roberts, 1996; Roberts and Sahu, 1997) involves expressing prior distributions at one level as centered at the hyperparameters at the next level. It is particularly suited to models with linear mean structure such as hierarchical linear models (HLMs), generalized HLMs and HLM-like models, and has the effect of both making parameters at level 2 independent and concentrating more data on estimates at level 3. Data augmentation (Section 15.6.4) simplifies complete conditionals and often reduces both computational time and chain autocorrelation (Fox, 2010), but the gains are not always obvious and probably deserve empirical investigation for each new class of models one encounters (see e.g., Thissen and Edwards, 2005).

Finally, a common source of poor behavior in an MCMC algorithm is that the underlying model is overparametrized, or nearly so. The Markov chain will tend to walk along the ridge in the posterior distribution induced by the overparametrization. In that case Graves et al. (2011) suggest that inserting occasional M–H steps with proposals that are perpendicular to the ridge can reduce autocorrelation in the chain and lead to better exploration of the parameter space. A simpler idea along the same lines (Tierney, 1994) is to insert occasional independence M–H steps as *decorrelation steps*. More general considerations along these lines lead to stochastic relaxation (Geman and Geman, 1984) and simulated annealing (Bertsimas and Tsitsiklis, 1993) algorithms.

15.8.7.3 Model Fit

Applying a sophisticated model to complex data is more than defining the model, as in Section 15.7, and successfully implementing an estimation algorithm, as in Section 15.8. Assessing the fit of the model, informally while working with the estimation algorithm, and formally after estimation is complete, is an essential part of statistical and psychometric practice.

It will sometimes be the case that the model fits the data so poorly that no MCMC algorithm will converge or give sensible answers. MCMC is particularly sensitive to the shape of the posterior distribution: If the posterior is well peaked around a well-defined maximum, MCMC works quite well. If the posterior is relatively flat with ridges and/or poorly defined maximas, as can happen when the model fits the data poorly, then the MCMC sample will tend to wander without converging, often to the edge of the parameter space.

Formal methods can be used to compare the fit of competing models, if more than one model has been fitted to the data. There is a large literature on Bayesian model comparison, focusing mainly on Bayes factor calculations (Han and Carlin, 2001; Kass and Raftery, 1995; Wasserman, 2000) and fit indices such as AIC (Akaike, 1977; Burnham and Anderson, 2002), BIC (Kass and Wasserman, 1995; Schwarz, 1978), and DIC (Ando, 2007; Celeux et al., 2006; Spiegelhalter et al., 2002).

Model criticism is the process of isolating particular features of a model that can be changed to improve the fit of the model; the process is similar in spirit to model modification indices in structural equations modeling, for example. A model criticism method particularly suited to MCMC calculation is posterior predictive checking (Gelman et al., 2003; Gelman et al., 1996; Lynch and Western, 2004). Essentially, one takes the output of the MCMC algorithm and uses it to generate more fake data from the fitted model and compares some summary of the fake data with the same summary of the real data used to fit the model. If the fake data summaries look like the real data summary, this gives confidence that the model is capturing whatever feature of the data the summary was intended to focus on. If not, the difference between the real and fake data summaries usually suggest ways to improve the model. A full discussion is beyond the scope of this chapter;

see Sinharay et al. (2006) or (see Chapter 19) for an introduction to posterior predictive checking of psychometric models.

15.9 Discussion

In this chapter, we have reviewed some theory and useful methodology for implementing MCMC algorithms for inferences with Bayesian IRT models. These ideas were illustrated using a standard 2PL IRT model.

MCMC has been applied successfully in many IRT and IRT-related contexts. To give only a few examples, data-augmentation-based MCMC for normal ogive models (Albert, 1992) has been extended to hierarchical IRT models (Fox, 2010), testlet models (Bradlow et al., 1999), and other settings. Other applications of MCMC include estimation of hierarchical rater models (Volume One, Chapter 27), models for test compromise (Segall, 2002; Shu, 2010), and models involving change over time (Studer, 2012) in IRT and cognitive diagnosis models (CDMs, e.g., Junker and Sijtsma, 2001; Rupp et al., 2010). In a related direction, Weaver (2008) gives an example of MCMC applied to a detailed computational production model for cognitive response data, tied to a particular cognitive architecture (Anderson and Lebiere, 1998).

For modest-sized applications with fairly standard models, it makes sense to work with existing standard software, rather than developing an MCMC algorithm from scratch. WinBUGS (Volume Three, Chapter 22; Lunn et al., 2009) and its cousins OpenBugs (Spiegelhalter et al., 2012) and JAGS (Plummer, 2012a) are readily available and can be used to fit and make inferences with a wide variety of standard models. R packages such as `R2WinBUGS` (Gelman et al., 2012), `rube` (Seltman, 2010), `BRugs` (Ligges, 2012), and `rjags` (Plummer, 2012b) make it convenient to access the power of WinBUGS and its cousins from R. For example, Ayers (2009) sketches a flexible approach to fitting various IRT models and CDMs using WinBUGS from R. A number of other packages for R (Geyer and Johnson, 2012; Hadfield, 2010; Martin et al., 2009) and other computational platforms (Daumé III, 2007; Patil et al., 2010; Stan Development Team, 2012) allow for faster computation for specific models, exploration of specific techniques for improving MCMC, and so forth. For a review, see Volume Three, Chapter 20.

For developing new models, or performing MCMC computations in new or complex settings, it can be preferable to develop MCMC algorithms by hand: One then has greater flexibility concerning sampling methods, scan order, parallelization, diagnostics, etc., and these can be quite important in creating an efficient and trustworthy algorithm. In addition, the methods illustrated here can be applied not only in R but also in C++ or any other computational language, and this can by itself produce significant speed-ups (see e.g., VanHoudnos, 2012). The monographs of Gilks et al. (1996) and Brooks et al. (2011) provide good entry points into the large literature on established MCMC methodology as well as more novel-related approaches.

Acknowledgments

This work was supported in part by a Graduate Training Grant awarded to Carnegie Mellon University by the U.S. Department of Education, Institute of Education Sciences

(#R305B090023). The opinions and views expressed do not necessarily reflect those of IES or the DoEd. The order of authorship is alphabetical.

References

Akaike, H. 1977. On entropy maximization principle, in *Applications of Statistics*, ed. Krishnaiah, P., Amsterdam: North-Holland, pp. 27–41.

Albert, J. H. 1992. Bayesian estimation of normal ogive item response curves using Gibbs sampling, *Journal of Educational Statistics, 17*, 251–269.

Albert, J. H. and Chib, S. 1993. Bayesian analysis of binary and polychotomous response data, *Journal of the American Statistical Association, 88*, 669–679.

Anderson, J. R. and Lebiere, C. 1998. *The Atomic Components of Thought*, Mahwah, NJ: Lawrence Erlbaum Associates, Inc.

Ando, T. 2007. Bayesian predictive information criterion for the evaluation of hierarchical Bayesian and empirical Bayes models, *Biometrika, 94*, 443–458.

Ayers, E. 2009. Using R to Write WinBUGS COde, Tech. rep., American Institutes for Research, Washington DC, http://www.stat.cmu.edu/~eayers/BUGS.html.

Bartholomew, D. J. and Knott, M. 1999. *Latent Variable Models and Factor Analysis Kendalls Library of Statistics*, No. 7, 2nd Edition, New York, NY: Edward Arnold.

Béguin, A. A. and Glas, C. A. 2001. MCMC estimation and some model-fit analysis of multidimensional IRT models, *Psychometrika*, 66, 541–561.

Bertsimas, D. and Tsitsiklis, J. 1993. Simulated annealing, *Statistical Science*, 8, 10–15.

Billingsley, P. 1995. *Probability and Measure*, 3rd Edition, New York, NY: Wiley-Interscience.

Bradlow, E., Wainer, H., and Wang, X. 1999. A Bayesian random effect model for testlets, *Psychometrika*, 64, 153–168.

Brockwell, A. and Kadane, J. B. 2005. Identification of regeneration times in MCMC simulation with application to adaptive schemes, *Journal of Statistical Computation and Graphics*, 14, 436–458.

Brooks, S., Gelman, A., Jones, G. L., and Meng, X.-L. (eds.) 2011. *Handbook of Markov Chain Monte Carlo*, Boca Raton, FL: Chapman and Hall/CRC.

Brooks, S. P. and Gelman, A. 1998. General methods for monitoring convergence of iterative simulations, *Journal of Computational and Graphical Statistics*, 7, 434–455.

Burnham, K. and Anderson, D. 2002. *Model Selection and Multimodel Inference: A Practical Information-Theoretic Approach*, New York: Springer.

Casella, G. and Robert, C. P. 1996. Rao-Blackwellization of sampling schemes, *Biometrika*, 83, 81–94.

Celeux, G., Forbes, F., Robert, C., and Titterington, D. 2006. Deviance information criteria for missing data models (with discussion), *Bayesian Analysis*, 651–674 (disc. 675–706).

Chang, H.-H. and Stout, W. 1993. The asymptotic posterior normality of the latent trait in an IRT model, *Psychometrika*, 58, 37–52.

Chib, S. and Greenberg, E. 1995, Understanding the Metropolis-Hastings algorithm, *The American Statistician*, 49, 327–335.

Chib, S., Greenberg, E., and Chiband, S. 1995. Understanding the Metropolis–Hastings algorithm, *American Statistician*, 49, 327–335.

Congdon, P. 2007. *Bayesian Statistical Modelling*, 2nd Edition, New York: Wiley.

Cook, L. L. and Eignor, D. R. 1991. IRT equating methods, *Educational Measurement: Issues and Practice*, 10, 37–45.

Cowles, M. K. and Carlin, B. P. 1996. Markov chain monte carlo convergence diagnostics: A comparative review, *Journal of the American Statistical Assocation*, 91, 883–904.

CTB/McGraw-Hill. 1996. *Comprehensive Test of Basic Skills*, Fifth Edition, Monterey, CA: Author.

Daumé III, H. 2007. HBC: Hierarchical Bayes Compiler, Tech. rep., http://hal3.name/HBC/.

De Boeck, P. and Wilson, M. E. 2004. *Explanatory Item Response Models: A Generalized Linear and Nonlinear Approach*, New York: Springer-Verlag.
de la Torre, J. and Douglas, J. 2004. Higher-order latent trait models for cognitive diagnosis, *Psychometrika*, 69, 333–353.
Doucet, A., Briers, M., and Sénécal, S. 2006. Efficient block sampling strategies for sequential Monte Carlo methods, *Journal of Computational and Graphical Statistics*, 15, 693–711.
Fill, J. A. 1998. An interruptable algorithm for perfect sampling via Markov chains, *Annals of Applied Probability*, 8, 131–162.
Flegal, J. and Jones, G. 2011. Implementing MCMC: Estimating with confidence, in *Handbook of Markov Chain Monte Carlo*, eds. Brooks, S., Gelman, A., Jones, G. L., and Meng, X.-L., Boca Raton FL: Chapman and Hall/CRC, Chapter 7, pp. 175–197.
Flegal, J. M., Haran, M., and Jones, G. L. 2008. Markov chain Monte Carlo: Can we trust the third significant figure? *Statistical Science*, 23, 250–260.
Fox, J. 2010. *Bayesian Item Response Modeling: Theory and Applications*, New York: Springer.
Fox, J. P. 2011. Joint Modeling of Longitudinal Item Response Data and Survival (Keynote Address), Keynote Address, Escola de Modelos De Regressao. Fortaleza, Brazil.
Fox, J.-P. and Glas, C. A. 2001. Bayesian estimation of a multilevel IRT model using Gibbs sampling, *Psychometrika*, 66, 271–288.
Gelfand, A. E. and Smith, A. F. M. 1990. Sampling-based approaches to calculating marginal densities, *Journal of the American Statistical Association*, 85, 398–409.
Gelman, A., Carlin, J. B., Stern, H. S., and Rubin, D. B. 2003. *Bayesian Data Analysis*, 2nd Edition, New York, NY: Chapman and Hall.
Gelman, A., Meng, X.-L., and Stern, H. S. 1996. Posterior predictive assessment of model fitness via realized discrepancies (with discussion), *Statistica Sinica*, 6, 33–87.
Gelman, A. and Rubin, D. B. 1992. Inference from iterative simulation using multiple sequences, *Statistical Science*, 7, 457–511.
Gelman, A., Sturtz, S., Legges, U., Gorjanc, G., and Kerman, J. 2012. R2WinBUGS: Running WinBUGS and OpenBUGS from R / S-PLUS, http://cran.r-project.org/web/packages/R2WinBUGS/index.html, version 2.1-18.
Geman, S. and Geman, D. 1984. Stochastic relaxation, gibbs distributions, and the bayesian restoration of images, *IEEE Transactions on Pattern Analysis and Machine Intelligence*, 6, 721–741.
Geweke, J. 1992. Evaluating the accuracy of sampling-based approaches to calculating posterior moments, in *Bayesian Statistics 4*, eds. Bernado, J. M., Berger, J. O., Dawid, A. P., and Smith, A. F. M., Oxford, UK: Clarendon Press, pp. 169–194.
Geyer, C. J. 1996. Estimation and optimization of functions, in *Markov Chain Monte Carlo in Practice*, eds. Gilks, W. R., Richardson, S., and Spiegelhalter, D. J., London: Chapman and Hall, pp. 241–258.
Geyer, C. J. 2011. Introduction to Markov Chain Monte Carlo, in *Handbook of Markov Chain Monte Carlo*, eds. Brooks, S., Gelman, A., Jones, G. L., and Meng, X.-L., Boca Raton FL: Chapman and Hall/CRC, Chapter 1, pp. 3–48.
Geyer, G. J. and Johnson, L. T. 2012. MCMC: Markov Chain Monte Carlo (R package), Tech. rep., http://www.stat.umn.edu/geyer/mcmc/.
Gilks, W. R., Richardson, S., and Spiegelhalter, D. J. (eds.). 1996. *Markov Chain Monte Carlo in Practice*, London: Chapman and Hall.
Gilks, W. R. and Roberts, G. O. 1996. Strategies for improving MCMC, in *W. R*, London: Chapman and Hall, pp. 89–114.
Gilks, W. R. and Wild, P. 1992. Adaptive rejection sampling for Gibbs sampling, *Journal of the Royal Statistical Society, Series C (Applied Statistics)*, 41, 337–348.
Glas, C. A. W. and Pimentel, J. 2008. Modeling nonignorable missing data in speeded tests, *Educational and Psychological Measurement*, 68, 907–922.
Graves, T. L., Speckman, P. L., and Sun, D. 2011. Improved mixing in MCMC algorithms for linear models, working paper. http://ibrarian.net/navon/paper/Improved_Mixing_in_MCMC_Algorithms_for_Linear_Mod.pdf?paperid=18098847

Hadfield, J. 2010. MCMC methods for multi-response generalized linear mixed models: The MCMCglmm R package, *Journal of Statistical Software*, 33, 1–22, http://www.jstatsoft.org/v33/i02/.
Haertel, E. H. 1989. Using restricted latent class models to map the skill structure of achievement items, *Journal of Educational Measurement*, 26, 301–321.
Han, C. and Carlin, B. P. 2001. MCMC Methods for computing Bayes factors: A comparative review, *Journal of the American Statistical Association*, 96, 1122–1132.
Hastings, W. 1970. Monte Carlo samping methods using Markov chains and their applications, *Biometrika*, 57, 97–109.
Janssen, R., Tuerlinckx, F., Meulders, M., and de Boeck, P. 2000. A Hierarchical IRT model for criterion-referenced measurement, *Journal of Educational and Behavioral Statistics*, 25, 285–306.
Johnson, M. S. and Sinharay, S. 2005. Calibration of polytomous item families using bayesian hierarchical modeling, *Applied Psychological Measurement*, 29, 369–400.
Jones, D. H. and Nediak, M. 2005. Item Parameter Calibration of LSAT Items Using MCMC Approximation of Bayes Posterior Distributions, Tech. rep., Computerized Testing Report 00-05, Law School Admission Council, Newton, PA, http://www.lsac.org/lsacresources/Research/CT/pdf/CT-00-05.pdf.
Junker, B. 1999. Some statistical models and computational methods that may be useful for cognitively-relevant assessment, Tech. rep., Department of Statistics, Carnegie Mellon University, http://www.stat.cmu.edu/~brian/nrc/cfa/.
Junker, B. W. and Sijtsma, K. 2001. Cognitive assessment models with few assumptions, and connections with nonparametric item response theory, *Applied Psychological Measurement*, 25, 258–272.
Kamata, A. 2001. Item analysis by the hierarchical generalized linear model, *Journal of Educational Measurement*, 38, 79–93.
Karabatsos, G. and Sheu, C. 2004. Order-constrained Bayes inference for dichotomous models of unidimensional nonparametric IRT, *Applied Psychological Measurement*, 28, 110–125.
Kass, R. E. and Raftery, A. E. 1995. Bayes factors, *Journal of the American Statistical Association*, 90, 773–795.
Kass, R. E. and Wasserman, L. 1995. A reference Bayesian test for nested hypotheses and its relation to the Schwarz criterion, *Journal of the American Statistical Association*, 90, 928–934.
Lazarsfeld, P. F. and Henry, N. W. 1968. *Latent Structure Analysis*, Boston: Houghton Mifflin.
Ligges, U. 2012. Package 'BRugs', Tech. rep., Tecnische Universität Dortmund, Germany, http://cran.r-project.org/web/packages/BRugs/index.html.
Lord, F. M. and Novick, M. R. 1968. *Statistical Theories of Mental Test Scores*, Reading, MA: Addison-Wesley, with contributions by Allan Birnbaum.
Los Alamos National Laboratory. 2012. History Center, Tech. rep., http://www.lanl.gov/history/.
Lunn, D., Spiegelhalter, D., Thomas, A., and Best, N. 2009. The BUGS project: Evolution, critique and future directions, *Statistics in Medicine*, 28, 3049–3067.
Lynch, S. M. and Western, B. 2004. Bayesian posterior predictive checks for complex models, *Sociological Methods and Research*, 32, 301–335.
Macready, G. B. and Dayton, C. M. 1977. The use of probabilistic models in the assessment of mastery, *Journal of Educational Statistics*, 2, 99–120.
Maier, K. 2001. A Rasch hierarchical measurement model, *Journal of Educational and Behavioral Statistics*, 26, 307–330.
Mariano, L. T. and Junker, B. 2007. Covariates of the rating process in hierarchical models for multiple ratings of test items, *Journal of Educational and Behavioral Statistics*, 32, 287–314.
Maris, E. 1995. Psychometric latent response models, *Psychometrika*, 60, 523–547.
Martin, A., Quinn, K., and Park, J. 2009. MCMCpack: Markov Chain Monte Carlo (R package), Tech. rep., http://CRAN.R-project.org/package=MCMCpack.
Matteucci, M. and Veldkamp, B. V. 2011. On the use of MCMC CAT with empirical prior information to improve the efficiency of CAT, Tech. rep., AMS Acta, Università di Bologna, http://amsacta.unibo.it/3109/.

Metropolis, N. 1987. The beginning of the Monte Carlo method, *Los Alamos Science (1987 Special Issue Dedicated to Stanisław Ulam)*, 125–130.

Metropolis, N., Rosenbluth, A., Rosenbluth, M., Teller, A., and Teller, E. 1953. Equations of state calculations by fast computing machines, *Journal of Chemical Physics*, 21, 1087–1091.

Miyazaki, K. and Hoshino, T. 2009. A Bayesian semiparametric item response model with dirichlet process priors, *Psychometrika*, 74, 375–393.

Monahan, J. F. 2011. *Numerical Methods of Statistics*, New York, NY: Cambridge University Press.

Mykland, P., Tierney, L., and Yu, B. 1995. Regeneration in markov chain samplers, *Journal of the American Statistical Association*, 90, 233–241.

Neal, R. 2003. Slice sampling, *Annals of Statistics*, 31, 705–767.

Patil, A., Huard, D., and Fonnesbeck, C. J. 2010. PyMC: Bayesian stochastic modelling in python, *Journal of Statistical Software*, 35, http://www.jstatsoft.org/.

Patz, R. J. and Junker, B. W. 1999a. A straightforward approach to markov chain monte carlo methods for item response models, *Journal of Educational and Behavioral Statistics*, 24, 146–178.

Patz, R. J. and Junker, B. W. 1999b. Applications and extensions of MCMC in IRT: Multiple item types, missing data, and rated responses, *Journal of Educational and Behavioral Statistics*, 24, 342–366.

Patz, R. J., Junker, B. W., Johnson, M. S., and Mariano, L. T. 2002. The hierarchical rater model for rated test items and its application to large-scale educational assessment data, *Journal of Educational and Behavioral Statistics*, 27, 341–384.

Plummer, M. 2012a. JAGS Version 3.2.0 user manual, Tech. rep., http://mcmc-jags.sourceforge.net.

Plummer, M. 2012b. Package 'rjags', Tech. rep., http://cran.r-project.org/web/packages/rjags/index.html.

Polson, N. G., Scott, J. G., and Windle, J. 2013. Bayesian inference for logistic models using Pólya-Gamma latent variables, *Journal of the American Statistical Association*, 108, 1339–1349, available at http://arxiv.org/pdf/1205.0310.pdf.

R Development Core Team. 2012. R: A Language and Environment for Statistical Computing, Tech. rep., R Foundation for Statistical Computing, Vienna, Austria, http://www.r-project.org/.

Ripley, B. D. 1987. *Stochastic Simulation*, New York: Wiley and Sons.

Rizopoulos, D. 2006. ltm: An R package for latent variable modelling and item response theory analyses, *Journal of Statistical Software*, 17, 1–25, http://www.jstatsoft.org/v17/i05/.

Robert, C. and Casella, G. 2011. A short history of Markov Chain Monte Carlo: subjective recollections from incomplete data, *Statistical Science*, 26, 102–115.

Roberts, G. and Sahu, S. K. 1997. Updating schemes, correlation structure, blocking and parameterization for the Gibbs sampler, *Journal of the Royal Statistical Society. Series B (Methodological)*, 59, 291–317.

Rosenthal, J. S. 2011, Optimal proposal distributions and adaptive MCMC, in *Handbook of Markov Chain Monte Carlo*, eds. Brooks, S., Gelman, A., Jones, G. L., and Meng, X.-L., Boca Raton FL: Chapman and Hall/CRC, Chapter 4, pp. 93–111.

Roussos, L. A., DiBello, L. V., Stout, W., Hartz, S. M., Henson, R. A., and Templin, J. H. 2007. The fusion model skills diagnosis system, in *Cognitive Diagnostic Assessment for Education: Theory and Applications*, eds. Leighton, J. and Gierl, M., Cambridge University Press.

Rupp, A., Templin, J., and Henson, R. 2010. *Diagnostic Assessment: Theory, Methods, and Applications*, New York: Guilford.

Sass, D. A., Schmitt, T. A., and Walker, C. M. 2008. Estimating non-normal latent trait distributions within item response theory using true and estimated item parameters, *Applied Measurement in Education*, 21, 65–88.

Scheiblechner, H. 1972. Das Lernen und Lösen komplexer Denkaufgaben [The learning and solving of complex reasoning items], *Zeitschrift für Experimentelle und Angewandte Psychologie*, 3, 456–506.

Schwarz, G. E. 1978. Estimating the dimension of a model, *Annals of Statistics*, 6, 461–464.

Scott, S. L., Blocker, A. W., Bonassi, F. V., Chipman, H., George, E., and McCulloch, R. 2013. Bayes and big data: The consensus Monte Carlo algorithm, in *EFaBBayes 250 conference (Vol 16)*, available at http://static.googleusercontent.com/media/research.google.com/en/us/pubs/archive/41849.pdf.

Segall, D. 2002. An item response model for characterizing test compromise, *Journal of Educational and Behavioral Statistics*, 27, 163–179.

Segall, D. O. 2003. Calibrating CAT pools and online pretest items using MCMC methods, in *Annual Meeting of the National Council on Measurement in Education*, Chicago IL.

Seltman, H. 2010. R Package rube (Really Useful WinBUGS Enhancer), http://www.stat.cmu.edu/~hseltman/rube/, version 0.2-13.

Serfling, R. 1980. *Approximation Theorems of Mathematical Statistics*, New York: Wiley.

Shu, Z. 2010. Detecting Test Cheating Using a Deterministic, Gated Item Response Theory Model, Ph.D. thesis, The University of North Carolina at Greensboro, Greensboro, NC.

Sinharay, S. and Almond, R. G. 2007. Assessing fit of cognitive diagnostic models A case study, *Educational and Psychological Measurement*, 67, 239–257.

Sinharay, S., Johnson, M., and Stern, H. 2006. Posterior predictive assessment of item response theory models, *Applied Psychological Measurement*, 30, 298–321.

Sinharay, S., Johnson, M. S., and Williamson, D. M. 2003. Calibrating item families and summarizing the results using family expected response functions, *Journal of Educational and Behavioral Statistics*, 28, 295–313.

Smith, B. J. 2007. BOA: An R package for MCMC output convergence assessment and posterior inference, *Journal of Statistical Software*, 21, 1–37.

Spiegelhalter, D., Thomas, A., Best, N., and Lunn, D. 2012, OpenBUGS User Manual, Tech. rep., http://www.openbugs.info.

Spiegelhalter, D. J., Best, N. G., Carlin, B. P., and van der Linde, A. 2002. Bayesian measures of model complexity and fit (with discussion), *Journal of the Royal Statistical Society, Series B (Statistical Methodology)*, 64, 583–639.

Stan Development Team. 2012. Stan Modeling Language: User's Guide and Reference Manual. Version 1.0. Tech. rep., http://mc-stan.org/.

Studer, C. E. 2012. Incorporating Learning Over Time into the Cognitive Assessment Framework, Ph.D. thesis, Carnegie Mellon University, Pittsburgh, PA.

Tanner, M. A. and Wong, W. H. 1987. The calculation of posterior distributions by data augmentation (with discussion), *Journal of the American Statistical Association*, 82, 528–550.

Thissen, D. and Edwards, M. C. 2005. Diagnostic scores augmented using multidimensional item response theory: Preliminary investigation of MCMC strategies, in *Annual Meeting of the National Council on Measurement in Education*, Montreal, Canada.

Thompson ISI. 2012. Web of Knowlege, Tech. rep., http://wokinfo.com/.

Tierney, L. 1994. Markov chains for exploring posterior distributions, *The Annals of Statistics*, 22, 1701–1728.

van den Oord, E. J. C. G. 2005. Estimating Johnson curve population distributions in MULTILOG, *Applied Psychological Measurement*, 29, 45–64.

VanHoudnos, N. 2012. Online Supplement to Markov Chain Monte Carlo for Item Response Models, Tech. rep., Department of Statistics, Carnegie Mellon University, Pittsburgh, Pennsylvania, USA, http://mcmcinirt.stat.cmu.edu/.

Walker, A. M. 1969. On the asymptotic behaviour of posterior distributions, *Journal of Royal Statistical Society*, 31, 80–88.

Wasserman, L. 2000. Bayesian model selection and model averaging, *Journal of Mathematical Psychology*, 44, 92–107.

Weaver, R. 2008. Parameters, predictions, and evidence in computational modeling: A statistical view informed by ACT-R, *Cognitive Science*, 1349–1375.

Woods, C. M. 2006. Ramsay-curve item response theory (RC-IRT) to detect and correct for nonnormal latent variables, *Psychological Methods*, 11, 253–270.

Woods, C. M. and Thissen, D. 2006. Item response theory with estimation of the latent population distribution using spline-based densities, *Psychometrika*, 71, 281–301.

16

Statistical Optimal Design Theory

Heinz Holling and Rainer Schwabe

CONTENTS

16.1 Introduction

Optimal design allows for estimating parameters of statistical models according to certain optimality criteria, for example, minimal standard errors of estimators. Thus, optimal designs may considerably reduce the number of experimental units and costs of empirical studies.

The first paper explicitly devoted to optimal design goes back to Smith (1918). Much of the pioneering work is due to Kiefer (1959) as well as Kiefer and Wolfowitz (1960). Fedorov (1972) wrote the first book about optimal design. Ever since, several books and a large amount of articles and monographs have been published.

Optimal design is applied in nearly all natural, social, and educational sciences. In the field of item response theory (IRT), it was introduced by Berger (1991) and van der Linden (1994). In IRT, two different types of optimal design problems can be distinguished. The first is known as optimal test design and concerns the selection of items into a test to efficiently estimate person parameters. Optimal sampling or calibration designs, on the other hand, refers to the search of the best sample of test-takers for estimating item parameters.

While Berger (Volume Three, Chapter 1) covers optimal calibration designs and van der Linden (Volume Three, Chapter 9) optimal test designs, this chapter provides a more general review of optimal design theory with emphasis on relevant issues of IRT. First, the basic principles of optimal design are introduced using the general linear model. Both criteria of optimality and the general equivalence theorem as a tool to certify the optimality of designs are outlined. As has been pointed out earlier, for example, by De Boeck and Wilson (2004) or De Boeck et al. (2011), many item response models can be embedded into the family of generalized linear models. Therefore, we will extensively deal with optimal design in generalized linear models. Several item response models that are members of generalized linear models, such as the 1PL and 2PL models, linear logistic test model (LLTM) and Rasch Poisson counts model (RPCM), will be used as examples throughout this chapter. In general, optimal designs for nonlinear designs can only be derived for known parameters. Adaptive, Bayesian and minimax models, which alleviate this restriction, will be introduced. Finally, issues concerning optimal designs within random effects models will be discussed.

16.2 General Linear Model

Optimal design may be best introduced for the general linear model. In this modeling framework, the response Y is related to a systematic part $\mu(\mathbf{x}, \boldsymbol{\beta})$ and an error ϵ by

$$Y(\mathbf{x}) = \mu(\mathbf{x}, \boldsymbol{\beta}) + \epsilon \tag{16.1}$$

and the systematic (true) response is modeled as a linear combination

$$\mu(\mathbf{x}, \boldsymbol{\beta}) = \sum_{j=1}^{p} \beta_j f_j(x_1, \ldots, x_k) = \mathbf{f}^\top(\mathbf{x})\boldsymbol{\beta} \tag{16.2}$$

of the known regression or dummy functions $\mathbf{f} = (f_1, \ldots, f_p)^\top$ of the explanatory variables $\mathbf{x} = (x_1, \ldots, x_k)^\top$, where $\boldsymbol{\beta} = (\beta_1, \ldots, \beta_p)^\top$ contains the unknown parameters. As an example, for the two-factor model $Y(\mathbf{x}) = \beta_1 + \beta_2 x_1 + \beta_3 x_2 + \beta_4 x_1 x_2 + \epsilon$ with interactions, it holds that $k = 2$, $\mathbf{x} = (x_1, x_2)^\top$, $p = 4$, $f_1(\mathbf{x}) = 1$, $f_2(\mathbf{x}) = x_1$, $f_3(\mathbf{x}) = x_2$, $f_4(\mathbf{x}) = x_1 x_2$ and $\boldsymbol{\beta} = (\beta_1, \beta_2, \beta_3, \beta_4)^\top$.

For a study of size n, the ith observation Y_i is then related to the corresponding value $\mathbf{x}_i$ of the explanatory variable through $Y_i = \mathbf{f}^\top(\mathbf{x}_i)\boldsymbol{\beta} + \epsilon_i$. The errors ϵ_i are assumed to be uncorrelated with $E(\epsilon_i) = 0$, $\text{Var}(\epsilon_i) = \sigma^2$ and $\text{Cov}(\epsilon_i, \epsilon_j) = 0$, $i \neq j$. Typically, more restrictively, the errors are assumed to be iid normally distributed: $\epsilon_i \sim N(0, \sigma^2)$.

In vector notation, the general linear model can be expressed as follows:

$$E(\mathbf{Y}) = \mathbf{F}\boldsymbol{\beta}, \tag{16.3}$$

where $\mathbf{Y} = (Y_1, \ldots, Y_n)^\top$ is the n-dimensional vector of responses, $\mathbf{F} = (\mathbf{f}(\mathbf{x}_1), \ldots, \mathbf{f}(\mathbf{x}_n))^\top$ is an $n \times p$ design matrix with the regression function $\mathbf{f}^\top(\mathbf{x}_i)$ evaluated at the value $\mathbf{x}_i$ of the explanatory variables as its ith row, and $\boldsymbol{\beta}$ is the p-dimensional vector of unknown parameters. The variance–covariance matrix $\text{Cov}(\hat{\boldsymbol{\beta}})$ depends on the experimental design, that is, the values of the explanatory variables to be selected for the experiment. Thus, an optimal design will include those values of the explanatory variables that minimize the variance–covariance matrix $\text{Cov}(\hat{\boldsymbol{\beta}})$ with respect to a certain criterion specified below.

The set of all possible design points $\mathbf{x}$ is called the design region and denoted by $\mathcal{X}$. If the design contains $i = 1, \ldots, m$ distinct design points $\mathbf{x}_i$, which occur n_i times in the experiment each, with a total number of $n = \sum_i n_i$ trials, its design is defined exactly as

$$\xi_n = \begin{Bmatrix} \mathbf{x}_1 & \mathbf{x}_2 & \dots & \mathbf{x}_m \\ n_1/n & n_2/n & \dots & n_m/n \end{Bmatrix}. \tag{16.4}$$

According to the Gauss–Markov theorem, ordinary least squares estimation of the parameter vector $\boldsymbol{\beta}$ results in the best linear unbiased (BLUE) estimator

$$\hat{\boldsymbol{\beta}} = (\mathbf{F}^\top \mathbf{F})^{-1} \mathbf{F}^\top \mathbf{Y}, \tag{16.5}$$

provided $\mathbf{F}^\top \mathbf{F}$ is regular. The covariance matrix

$$\mathrm{Cov}(\hat{\boldsymbol{\beta}}) = \sigma^2 (\mathbf{F}^\top \mathbf{F})^{-1} \tag{16.6}$$

of $\hat{\boldsymbol{\beta}}$ is an obvious measure of the performance of the estimator and plays an important role in optimal design.

For a given value $\mathbf{x}$ of the explanatory variables, the best linear unbiased estimator $\widehat{E(Y(\mathbf{x}))}$ of the expectation of a response $Y(\mathbf{x})$ is $\mathbf{f}^\top(\mathbf{x})\hat{\boldsymbol{\beta}}$. This estimated expectation is typically used as predicted response and denoted by $\hat{Y}(\mathbf{x})$. The variance of the predicted response, also an essential element of optimal design, is calculated as follows:

$$\mathrm{Var}(\hat{Y}(\mathbf{x})) = \sigma^2 \mathbf{f}^\top(\mathbf{x}) (\mathbf{F}^\top \mathbf{F})^{-1} \mathbf{f}(\mathbf{x}). \tag{16.7}$$

It follows from the Lehmann–Scheffé theorem that under the assumption of normality the least-squares estimator is also the best unbiased estimator (BUE) and coincides with the maximum-likelihood (ML) estimator for $\boldsymbol{\beta}$.

Under the assumptions of normally distributed errors, the log-likelihood $l(\boldsymbol{\beta}, \sigma)$ for the general linear model is

$$l(\boldsymbol{\beta}, \sigma) = -\frac{n}{2} \log(2\pi) - \frac{n}{2} \log(\sigma^2) - \frac{1}{2\sigma^2} (\mathbf{y} - \mathbf{F}\boldsymbol{\beta})^\top (\mathbf{y} - \mathbf{F}\boldsymbol{\beta}). \tag{16.8}$$

The matrix of the second partial derivatives of l with respect to $\boldsymbol{\beta}$ is given by

$$\frac{\partial^2 l(\boldsymbol{\beta}, \sigma)}{\partial \boldsymbol{\beta} \partial \boldsymbol{\beta}^\top} = -\frac{1}{\sigma^2} (\mathbf{F}^\top \mathbf{F}), \tag{16.9}$$

which is the negative Fisher information matrix for $\boldsymbol{\beta}$, equal to minus the inverse of the variance–covariance matrix $\mathrm{Cov}(\hat{\boldsymbol{\beta}})$. In optimal design, typically the standardized version $\mathbf{M} = \mathbf{F}^\top \mathbf{F}$ of the information matrix, that is, upon disregard of the coefficient $1/\sigma^2$, is considered.

For a single observation at the value $\mathbf{x}$ of the explanatory variables, this information matrix can be written as the outer product $\mathbf{f}(\mathbf{x})\mathbf{f}^\top(\mathbf{x})$ of the regression function $\mathbf{f}$ evaluated at $\mathbf{x}$ with itself. Using the same notation, the information matrix for the whole experiment can be written as

$$\mathbf{M} = \sum_{i=1}^{n} \mathbf{f}(\mathbf{x}_i) \mathbf{f}^\top(\mathbf{x}_i), \tag{16.10}$$

where the $\mathbf{x}_i$ are the values of the explanatory variables for the n observations.

The larger the sample, the more information contained in the information matrix **M**. Thus, it seems reasonable to normalize the information is matrix to an information-per-observation matrix, that is, to divide the total information by the number of runs n. The result is

$$\mathbf{M}(\xi_n) = \frac{1}{n}\mathbf{M} = \sum_{i=1}^{m} \frac{n_i}{n} \mathbf{f}(\mathbf{x}_i)\, \mathbf{f}^\top(\mathbf{x}_i), \tag{16.11}$$

where now the $\mathbf{x}_i$ denote the m distinct design points in ξ_n and n_i the corresponding numbers of occurrence.

The statistical problem of generating optimal designs is simplified when the weights n_i/n are not restricted to be multiples of $1/n$. Thus, following Kiefer (1959), we define an approximate design ξ as a finitely supported discrete measure on $\mathcal{X}$

$$\xi = \begin{Bmatrix} \mathbf{x}_1 & \mathbf{x}_2 & \ldots & \mathbf{x}_m \\ w_1 & w_2 & \ldots & w_m \end{Bmatrix} \tag{16.12}$$

with w_i the weight of design point $\mathbf{x}_i$, where $w_i \geq 0$ and $\sum_i w_i = 1$. These designs are continuous in their weights and include exact designs as special cases. In practice, designs have to be exact of course, but, for a sufficiently large n, efficient integer approximations to continuous designs can be found.

In accordance with Equation 16.11, we may define the normalized information matrix $\mathbf{M}(\xi)$ of an approximate design ξ as

$$\mathbf{M}(\xi) = \sum_{i=1}^{m} w_i\, \mathbf{f}(\mathbf{x}_i)\, \mathbf{f}^\top(\mathbf{x}_i). \tag{16.13}$$

Similarly, the variance of a predicted response $\mathrm{Var}(\hat{Y}(\mathbf{x}))$ can also be normalized, which then is defined as

$$d(\mathbf{x}, \xi) = \mathbf{f}^\top(\mathbf{x})\, \mathbf{M}^{-1}(\xi)\, \mathbf{f}(\mathbf{x}). \tag{16.14}$$

Note that the result is independent of the number of runs and does not incorporate the error variance, which usually is unknown. For exact design, the normalized variance can be computed as

$$d(\mathbf{x}, \xi_n) = \mathbf{f}^\top(\mathbf{x})\, \mathbf{M}^{-1}(\xi_n)\, \mathbf{f}(\mathbf{x}) = \frac{n}{\sigma^2} \mathrm{Var}(\hat{Y}(\mathbf{x})). \tag{16.15}$$

16.3 Design Objectives and Criteria of Optimality

A "good" design should lead to a precise estimation of the unknown parameters and a small variance of the predicted responses. Both objectives depend on the choice of ξ. The set of all approximate designs on the design region $\mathcal{X}$ will be denoted by Ξ. The general optimization problem is now to find an optimal design ξ^* in the convex set Ξ according to a certain criterion. Many different criteria have been proposed in the literature. We follow the convention to define a criterion as a function $\Phi(\mathbf{M})$ of the information matrix rather than the design itself.

In the following, we deal with some important criteria (for a comprehensive overview, see, e.g., Atkinson et al., 2007). To keep things simple, we introduce some mild conditions. The design region $\mathcal{X}$ is assumed to be compact and the regression function $\mathbf{f}$ continuous, so that $\mathbf{f}(\mathcal{X}) = \{\mathbf{f}(\mathbf{x}); \mathbf{x} \in \mathcal{X}\}$ and, hence, $\{\mathbf{f}(x)\mathbf{f}(x)^\top; x \in \mathcal{X}\}$ is compact as well, which, in turn, implies compactness of the set of all normalized information matrices $\mathcal{M} = \{\mathbf{M}(\xi);\ \xi \in \Xi\}$. Thus, we can always use min and max instead of inf and sup and find an optimal solution minimizing a convex criterion $\Phi : \mathcal{M} \to \mathbb{R} \cup \{\infty\}$. Most criteria require regularity of $\mathbf{M}$, and $\Phi(\mathbf{M})$ is set to infinity otherwise.

As the covariance matrix cannot be minimized in the non-negative sense, that is, simultaneously with respect to the variance of every (one-dimensional) linear combination of the parameters, some real-valued functionals have to be considered as design criteria. The most important and popular criterion, used in numerous applications, is D-optimality. According to this criterion, a design is optimal if the determinant of the inverse of the information matrix, that is, the "generalized variance", is minimal. In order to find a D-optimal design, it is favorable to consider the convex function $\Phi_D(\mathbf{M}) = -\log\det(\mathbf{M})$ and define a design ξ^* as D-optimal if

$$\xi^* = \arg\min_{\xi}(-\log\det(\mathbf{M}(\xi))). \tag{16.16}$$

Under normality assumptions, the D-criterion measures the volume of the confidence ellipsoid of the parameters to be estimated. Consequently, a D-optimal design minimizes this volume.

D-optimality can be limited to a subset of $s < p$ parameters while treating the others as nuisance parameters. This criterion is called D_s-optimality (Atkinson et al., 2007). More generally, a D_A-optimal design minimizes the determinant of the covariance matrix $\mathbf{A}\mathbf{M}^{-1}(\xi)\mathbf{A}^\top$ and, hence, the volume of the corresponding confidence ellipsoid for a vector $\mathbf{A}\boldsymbol{\beta}$ of linear combinations of the parameters, $\Phi_{D_A}(\mathbf{M}) = \log\det(\mathbf{A}\mathbf{M}^{-1}\mathbf{A}^\top)$.

For A-optimality, the total (or average) variance of the parameter estimates, $\Phi_A(\mathbf{M}) = \text{tr}\,(\mathbf{M}^{-1})$ is minimized. Thus, a design ξ^* is A-optimal if

$$\xi^* = \arg\min_{\xi} \text{tr}\,(\mathbf{M}(\xi)^{-1}). \tag{16.17}$$

This A-criterion is a special case of a linear (L-) criterion, that is, a linear function of the information matrix, which minimizes the trace (sum of the diagonal entries) of the covariance matrix $\mathbf{A}\mathbf{M}^{-1}(\xi)\mathbf{A}^\top$ of the estimator for a vector $\mathbf{A}\boldsymbol{\beta}$ of linear combinations of the parameters, $\Phi_L(\mathbf{M}) = \text{tr}\,(\mathbf{A}\mathbf{M}^{-1}\mathbf{A}^\top) = \text{tr}\,(\mathbf{L}\mathbf{M}^{-1})$ for $\mathbf{L} = \mathbf{A}^\top\mathbf{A}$.

Another linear criterion of particular interest is c-optimality. For c-optimality, the variance will have to be minimized for one specific linear combination $\mathbf{c}^\top\boldsymbol{\beta}$, $\Phi_c(\mathbf{M}) = \mathbf{c}^\top\mathbf{M}^{-1}\mathbf{c}$. Often, a c-optimal design has a singular information matrix. In that case $\mathbf{M}^{-1}$ in Φ_c has to be replaced by a generalized inverse $\mathbf{M}^-$ if $\mathbf{c}$ is estimable, or $\Phi_c(\mathbf{M}) = \infty$ else.

A less often employed criterion is E-optimality, defined as the minimization of the maximum eigenvalue of the covariance matrix, $\Phi_E(\mathbf{M}) = \lambda_{\max}(\mathbf{M}^{-1})$. Consequently, an E-optimal design minimizes the maximum variance over all one-dimensional linear combinations $\mathbf{a}^\top\boldsymbol{\beta}$ subject to the constraint $\mathbf{a}^\top\mathbf{a} = 1$, that is, it minimizes the (normalized) variance of the least well-estimated linear combination. Due to this minimax formulation, the E-criterion is analytically less tractable (no differentiability).

Within this entire class of alternative design criteria directly based on the covariance matrix, D-optimality has the desirable property of not being affected by a reparameterization of the model, that is, it is independent of the particular parameterization chosen and therefore of the scale of the parameters. On the other hand, all other criteria are sensitive to the parameterization and choice of scale for the parameters.

The next two criteria often considered in optimal design are based on predicted responses. For G-optimality, the maximum (standardized) variance $\max_{\mathbf{x}} d(\mathbf{x}, \xi)$ of the predicted response is minimized over the design region $\mathcal{X}$. Or, formally, $\Phi_G(\mathbf{M}) = \max_{\mathbf{x}} \mathbf{f}^\top(\mathbf{x})\mathbf{M}^{-1}\mathbf{f}(\mathbf{x})$. A design ξ^* is thus G-optimal if

$$\xi^* = \arg\min_{\xi} \max_{\mathbf{x}} d(\mathbf{x}, \xi). \tag{16.18}$$

The average variance of the predicted response over the design region $\mathcal{X}$ is minimized for the IMSE-optimality, also known as I-, V-, or Q-optimality. For a continuous $\mathbf{x}$, a design ξ^* is IMSE-optimal if

$$\xi^* = \arg\min_{\xi} \int_{\mathcal{X}} d(\mathbf{x}, \xi)\, \mathrm{d}\mathbf{x}, \tag{16.19}$$

whereas, for a categorical $\mathbf{x}$, the integral has to be replaced by a sum over its possible values.

The IMSE-criterion (Integrated Mean Squared Error) can be identified as a linear criterion because of $\int_{\mathcal{X}} d(\mathbf{x}, \xi)\, \mathrm{d}\mathbf{x} = \mathrm{tr}\,(\mathbf{L}\mathbf{M}^{-1}(\xi))$ with $\mathbf{L} = \int_{\mathcal{X}} \mathbf{f}(\mathbf{x})\mathbf{f}^\top(\mathbf{x})\, \mathrm{d}\mathbf{x}$. An IMSE-optimal design minimizes the expected mean squared distance between the predicted response $\hat{\mu}(\mathbf{x}) = \mathbf{f}^\top(\mathbf{x})\hat{\boldsymbol{\beta}}$ and the true response $\mu(\mathbf{x}, \boldsymbol{\beta}) = \mathbf{f}^\top(\mathbf{x})\boldsymbol{\beta}$.

Because the variance function $d(\mathbf{x}, \xi)$ is invariant under reparameterization, G- and IMSE-optimality are not affected by reparameterization or rescaling, just as for the D-criterion.

16.4 Equivalence Theory

In practice, numerical optimization methods may have to be used to generate designs according to certain optimality criteria. If so, one cannot always be sure whether or not the designs found are optimal. Similar to the role of derivatives in calculus, equivalence theory can be used to ascertain optimality. It also provides the basis for algorithms that generate optimal or at least efficient designs.

Kiefer and Wolfowitz (1960) were the first to show the general equivalence of D-optimality and G-optimality for linear models. Their results were generalized to heteroscedastic models by Fedorov (1972). The generalization can be used to tackle nonlinearity. Silvey (1980) proposed a general theory of optimal design based on convexity for linear as well as nonlinear models, which is adopted here. Consequently, the optimality criteria Φ considered in the following theorems are always supposed to be convex functions.

Carathéodory's theorem (see Silvey, 1980) facilitates the search of an optimal design considerably. According to it, only designs with a bounded number of design points need to be considered. For every design ξ in Ξ, the information matrix $\mathbf{M}(\xi)$ can be expressed as a convex combination $\sum w_i \mathbf{f}(\mathbf{x}_i)\mathbf{f}^\top(\mathbf{x}_i)$ of maximal $\frac{1}{2}p(p+1)+1$ elements of the form $\mathbf{f}(\mathbf{x})\mathbf{f}^\top(\mathbf{x})$. It is therefore equal to the information matrix for a design with these design points $\mathbf{x}_i$ and corresponding weights w_i. If $\mathbf{M}(\xi)$ is on the boundary of $M(\Xi)$, such a design exists with $\frac{1}{2}p(p+1)$ design points at most.

The following two theorems are based on the concept of the directional derivative of Φ at $\mathbf{M}_1$ in the direction of $\mathbf{M}_2$, defined as

$$F_{\Phi}(\mathbf{M}_1, \mathbf{M}_2) = \lim_{\alpha \downarrow 0} \frac{1}{\alpha} \left(\Phi\left((1-\alpha)\mathbf{M}_1 + \alpha\mathbf{M}_2\right) - \Phi(\mathbf{M}_1)\right). \tag{16.20}$$

Of particular interest are the directional derivatives $F_{\Phi}(\mathbf{M}, \mathbf{f}(\mathbf{x})\mathbf{f}^{\top}(\mathbf{x}))$ in the direction of information matrix $\mathbf{M}(\xi_{\mathbf{x}}) = \mathbf{f}(\mathbf{x})\mathbf{f}^{\top}(\mathbf{x})$ for designs $\xi_{\mathbf{x}}$ that assigns weight 1 to one single point $\mathbf{x}$.

The first fundamental theorem, discussed in more detail by Silvey (1980), provides a condition that can be used to easily verify the optimality of a given design in practice.

Theorem 16.1: cf. Silvey, 1980, Theorem 3.7

If Φ is convex on $\mathcal{M}$ and differentiable at $\mathbf{M}(\xi^*)$, then ξ^* is Φ-optimal if and only if

$$F_{\Phi}(\mathbf{M}(\xi^*), \mathbf{f}(\mathbf{x})\mathbf{f}^{\top}(\mathbf{x})) \geq 0 \quad \text{for all } \mathbf{x} \in \mathcal{X}. \tag{16.21}$$

According to this theorem, an optimal design cannot be improved upon by slightly changing in the direction of any possible design point. Hence, more importantly, to establish optimality it is sufficient to check on whether the design is improved if it is changed in the direction of any possible design point.

Theorem 16.2: cf. Silvey, 1980, Theorem 3.9

If Φ is convex on $\mathcal{M}$, differentiable at all points of $\mathcal{M}^+ = \{\mathbf{M} \in \mathcal{M} : \Phi(\mathbf{M}) < \infty\}$, and let a Φ-optimal design exist, then ξ^* is Φ-optimal if and only if

$$\min_{\mathbf{x}} F_{\Phi}(\mathbf{M}(\xi^*), \mathbf{f}(x)\mathbf{f}^{\top}(x)) = \max_{\xi} \min_{\mathbf{x}} F_{\Phi}(\mathbf{M}(\xi), \mathbf{f}(\mathbf{x})\mathbf{f}^{\top}(\mathbf{x})). \tag{16.22}$$

From this second theorem, it follows that $\min_{\mathbf{x}} F_{\Phi}(\mathbf{M}(\xi^*), \mathbf{f}(\mathbf{x})\mathbf{f}^{\top}(\mathbf{x})) = 0$ and that this minimum occurs at the support points of the design.

Theorem 16.2 is more specific than the previous one. In particular, it establishes the equivalence of D- and G-optimality, as shown as follows: The directional derivative for the D-criterion $\Phi_D(\mathbf{M}) = -\log(\det(\mathbf{M}))$ is given by $F_{\Phi_D}(\mathbf{M}, \mathbf{f}(\mathbf{x})\mathbf{f}^{\top}(\mathbf{x})) = p - \mathbf{f}^{\top}(\mathbf{x})\mathbf{M}^{-1}\mathbf{f}(\mathbf{x})$. Thus, for D-optimality the equivalent condition of Theorem 16.2 can be rewritten as

$$\max_{\mathbf{x}} \mathbf{f}^{\top}(\mathbf{x})\mathbf{M}(\xi^*)^{-1}\mathbf{f}(x) = \min_{\xi} \max_{\mathbf{x}} \mathbf{f}^{\top}(\mathbf{x})\mathbf{M}(\xi)^{-1}\mathbf{f}(\mathbf{x}), \tag{16.23}$$

which is G-optimality.

Observe that the directional derivative of the D-optimality criterion consists of two parts: (i) the constant p and (ii) the standardized variance of the predicted response $d(\mathbf{x}, \xi)$. The latter identifies the points providing the most information for this criterion. Therefore, it is called the sensitivity function of the design.

In general, the directional derivative $F_{\Phi}(\mathbf{M}(\xi), \mathbf{f}(\mathbf{x})\mathbf{f}^{\top}(\mathbf{x}))$ can always be split into two parts, one of which is the sensitivity function $\psi(\mathbf{x}, \xi)$ and the other a constant $\Delta(\xi)$ that may depend on ξ (Fedorov and Hackl, 1997):

$$F_{\Phi}(\mathbf{M}(\xi), \mathbf{f}(\mathbf{x})\mathbf{f}^{\top}(\mathbf{x})) = \Delta(\xi) - \psi(\mathbf{x}, \xi), \tag{16.24}$$

where

$$\Delta(\xi) = -\text{tr}\left(\mathbf{M}(\xi)\frac{\partial\,\Phi(\mathbf{M}(\xi))}{\partial\,\mathbf{M}(\xi)}\right) \tag{16.25}$$

and

$$\psi(\mathbf{x}, \xi) = -\mathbf{f}^{\top}(\mathbf{x})\frac{\partial\,\Phi(\mathbf{M}(\xi))}{\partial\,\mathbf{M}(\xi)}\mathbf{f}(\mathbf{x}) \tag{16.26}$$

and the partial derivatives with respect to a matrix are defined to be componentwise.

The optimality condition Equation 16.21 given by the equivalence theorem then becomes

$$\psi(\mathbf{x}, \xi^*) \leq \Delta(\xi^*). \tag{16.27}$$

In particular, for D-optimality the partial derivative $\frac{\partial\,\Phi_D(\mathbf{M}(\xi))}{\partial\,\mathbf{M}(\xi)}$ is $-\mathbf{M}(\xi)^{-1}$. Hence, the sensitivity function ψ_D coincides with the standardized variance d, and $\Delta_D(\xi^*)$ equals p, so that the optimality condition simplifies to

$$\psi_D(\mathbf{x}, \xi^*) = d(\mathbf{x}, \xi^*) \leq p. \tag{16.28}$$

Similarly, the optimality condition for c-optimality becomes

$$\psi_c(\mathbf{x}, \xi^*) = (\mathbf{c}^{\top}\mathbf{M}^{-1}(\xi^*)\mathbf{f}(\mathbf{x}))^2/(\mathbf{c}^{\top}\mathbf{M}^{-1}(\xi^*)\mathbf{c}) \leq 1. \tag{16.29}$$

Moreover, Theorem 16.2 can be used in the situation of one (continuous) explanatory variable to check, at least, graphically, whether a given design is optimal, and it may provide some analytical conditions, which an optimal design has to satisfy. In particular, for polynomial regression, this condition proves that for all criteria considered here the optimal design is saturated, which means that it is supported on the minimal number p of design points equal to the number of parameters.

The sensitivity functions for the D-criterion plotted in Figure 16.1 illustrate these facts: The function in the left panel is for the linear regression $Y(x) = \beta_1 + \beta_2 x + \epsilon$ on the standardized interval $\mathcal{X} = [-1, 1]$ for a D-optimal design ξ^* (solid curve), which assigns equal weights 1/2 to each of the endpoints -1 and 1. As a competitor, an equidistant three-point design is considered, which assigns equal weights 1/3 to the points $-1, 0$, and 1, with the sensitivity function displayed by the dashed line. The right panel shows the sensitivity function (solid line) for a D-optimal design for the case of quadratic regression $Y(x) = \beta_1 + \beta_2 x + \beta_3 x^2 + \epsilon$ on $\mathcal{X} = [-1, 1]$, which is an equidistant three-point design (dashed curve) with equal weights 1/3 assigned to the points -1, 0, and 1. Again, the dashed curve is for the sensitivity function of an alternative design with equal weights 1/5 at design points $-1, -0.5, 0, 0.5$, and 1. The sensitivity functions of these D-optimal designs are seen to satisfy the optimality condition $\psi(x, \xi^*) \leq p$, while the sensitivity functions of the competing designs obviously exceed these bounds.

In addition to the possibility to check the optimality of a given design offered by it, equivalence theory also plays an important role in constructing algorithms for finding optimal designs based on the following basic idea: Given a design ξ_r that is not Φ-optimal, according to Theorem 16.1 there exists a point $\mathbf{x}_{r+1}$ in its region $\mathcal{X}$ such that the derivative $F_{\Phi}(\mathbf{M}(\xi_r), \mathbf{f}(\mathbf{x}_{r+1})\mathbf{f}^{\top}(\mathbf{x}_{r+1}))$ in the direction of that (potential) design point is negative. Hence, the design ξ_r can be improved with respect to the criterion Φ by moving some

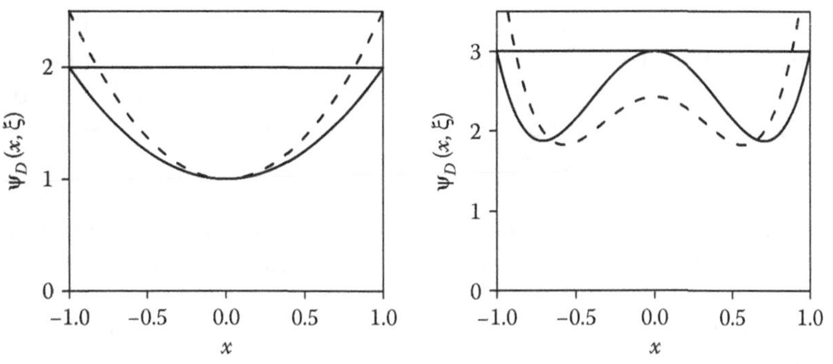

FIGURE 16.1
Sensitivity functions for the D-criterion in linear and quadratic regression.

weight to this point $\mathbf{x}_{r+1}$ from some of the other design points. Formally, this leads to the following algorithm for generating a new (better) design ξ_{r+1} from ξ_r

$$\xi_{r+1} = (1 - \alpha_{r+1})\xi_r + \alpha_{r+1}\xi_{\mathbf{x}_{r+1}}, \tag{16.30}$$

where $\xi_{\mathbf{x}}$ is again the one-point design assigning weight 1 to the single design point $\mathbf{x}$ and α_{r+1} $(0 < \alpha_{r+1} < 1)$ is the suitably chosen additional weight transferred from ξ_r to $\mathbf{x}_{r+1}$. Different methods have been proposed to choose α_{r+1} (e.g., Atkinson et al., 2007). For instance, the choice of harmonic weight $\alpha_{r+1} = 1/(r+1)$ operates such that one additional design point $\mathbf{x}_{r+1}$ is added to an existing exact design ξ_r of size r to form a new exact design ξ_{r+1} of size $r+1$.

As for the example in Figure 16.1, in either of its two cases the sensitivity functions of the competing (nonoptimal) designs exceed the bound p at the endpoints -1 and 1 of the interval. Hence, both can be improved by adding some additional weight to these two points.

16.5 Generalized Linear Model

Obviously, for linear models, the information matrix $\mathbf{M} = \mathbf{F}^\top\mathbf{F}$ does not contain the parameters $\boldsymbol{\beta}$ to be estimated. Therefore, their optimal designs do not depend on these unknown parameters. They are optimal for all feasible values of them.

This property does not hold for nonlinear models. Hence, for this case, only locally optimal designs can be specified. For instance, it is well known that the information for the Rasch model is maximal when the ability parameter meets the difficulty parameter. Thus, even when the difficulty parameter is known, we still need to know the person parameter, or at least have a good initial guess. Several strategies to overcome the restrictions of locally optimal designs will be discussed in later sections. First, however, we deal more generally with locally optimal designs, with results that may serve as a benchmark.

As a preliminary observation, note that locally optimal designs may also arise for linear models, when the interest is in a nonlinear aspect of the parameters. For example, if the interest is in calibration in linear regression, that is, to determine the value x_0 of the explanatory variable that provides a selected mean response $y_0 = \beta_1 + \beta_2 x_0$, then the maximum-likelihood estimate is given by $\hat{x}_0 = (y_0 - \hat{\beta}_1)/\hat{\beta}_2$ which is nonlinear in $\hat{\boldsymbol{\beta}}$. The goal of minimizing the (asymptotic) variance of the ML estimator can be identified as a case of c-optimality with $\mathbf{c} = \mathbf{c}_{\boldsymbol{\beta}} = (1, x_0)^{\top}$ depending on $\boldsymbol{\beta}$. It then follows from the equivalence theorem for c-optimality Equation 16.29 that all designs ξ satisfying $\sum_i w_i x_i = x_0$ are optimal for estimating x_0.

For nonlinear models, locally optimal designs can be generated analogously to linear models by using Taylor expansion; see below. A special class of nonlinear models are generalized linear models. Here, the information matrix can be determined without Taylor expansion, directly from the likelihood as the covariance matrix of the score function. Many IRT models belong to the class of generalized linear models. Thus, the results for optimal design for generalized linear models can be used for IRT models. To discuss this approach, we will consider univariate generalized models as in McCullagh and Nelder (1989).

The structure of generalized linear models is as follows: The first assumption is that the response variable Y has a distribution belonging to an exponential family in canonical representation, which means that a density function f_Y with parameter θ can be written as

$$f_Y(y;\theta) = \exp(\theta y - b(\theta) + c(y)) \tag{16.31}$$

by specific choices for functions b and c (see Chapter 4). We suppress an eventual scale parameter for the sake of simplicity, as it is immaterial for the examples considered here. The second and most specific assumption is the one of the mean response depending on a linear component $\eta(\mathbf{x}, \boldsymbol{\beta}) = \mathbf{f}^{\top}(\mathbf{x})\boldsymbol{\beta}$. Finally, the third assumption is that the linear component and the mean response μ are connected by a monotonically increasing, continuous, two-times differentiable link function $g(\mu) = \eta$ leading to the model equation

$$\mathrm{E}(Y(\mathbf{x})) = \mu(\mathbf{x}, \boldsymbol{\beta}) = g^{-1}(\mathbf{f}^{\top}(\mathbf{x})\boldsymbol{\beta}). \tag{16.32}$$

Or, for short, $\mathrm{E}(Y) = \mu = g^{-1}(\eta)$ (for link functions, see Chapter 1).

The corresponding log-likelihood function l, as a function of θ for given observation y, is computed as

$$l(\theta; y) = \log f_Y(y;\theta) = y\theta - b(\theta) + c(y). \tag{16.33}$$

The parameter θ is related to the mean response by

$$\mathrm{E}(Y) = \mu = b'(\theta) \tag{16.34}$$

and to the response variance using

$$\mathrm{Var}(Y) = b''(\theta). \tag{16.35}$$

When $b' = g^{-1}$, then g is the so-called canonical link function and θ equals η.

The class of generalized linear models includes logit and probit models as well as the Poisson regression model. If an additional scaling factor (σ) is introduced, for normally distributed response variables Y, the classical linear model can be embedded with the identity function as the link function g.

Example 16.1: Logit model

The logit regression model consists of a binary response Y with values $y = 0$ or 1 and success probability $\pi = P(Y = 1) = \exp(\theta)/(1 + \exp(\theta))$. The canonical parameter is $\theta = \log(\pi/(1 - \pi))$. The representation

$$f_Y(y; \theta) = \exp(\theta y - \log(1 + \exp(\theta))) \tag{16.36}$$

of the density by an exponential family is given by $b(\theta) = \log(1 + \exp(\theta))$ and $c(y) = 0$, and the model $E(Y) = \mu = \pi = \exp(\eta)/(1 + \exp(\eta))$ is specified by the logistic link function $g(\mu) = \log(\mu/(1 - \mu))$, which provides the canonical link.

Example 16.2: Poisson regression model

In the Poisson regression setup, the possible values y of the response Y are counts, that is, $y = 0, 1, 2, \ldots$, with probabilities $P(Y = y) = \lambda^y \exp(-\lambda)/y!$, where λ is the Poisson intensity parameter. The canonical parameter is $\theta = \log(\lambda)$. Its density can be specified as an exponential family,

$$f_Y(y; \theta) = \exp(\theta y - \exp(\theta) - \log(y!)) \tag{16.37}$$

with $b(\theta) = \exp(\theta)$ and $c(y) = -\log(y!)$. The logarithmic link $g(\mu) = \log(\mu)$ provided the canonical link and leads to the Poisson regression model $E(Y) = \exp(\eta)$.

The essential components of generalized linear models relevant to IRT are summarized in Table 16.1, where Φ_0 in the probit model denotes the standard normal distribution function and Φ_0^{-1} does not lead to a canonical link.

In generalized linear models, the mean response is given by

$$\mathrm{E}(Y(\mathbf{x})) = g^{-1}(\mathbf{f}^{\top}(\mathbf{x})\boldsymbol{\beta}) \tag{16.38}$$

and the corresponding variance of the response $\sigma^2(\mathbf{x}, \boldsymbol{\beta}) = \mathrm{Var}(Y(\mathbf{x}))$ is

$$\mathrm{Var}(Y(\mathbf{x})) = s^2(\mathbf{f}^{\top}(\mathbf{x})\boldsymbol{\beta}) = b''(b'^{-1}(g^{-1}(\mathbf{f}^{\top}(\mathbf{x})\boldsymbol{\beta}))) \tag{16.39}$$

because of $\theta = b'^{-1}(g^{-1}(\mathbf{f}^{\top}(\mathbf{x})\boldsymbol{\beta}))$. In particular, for a canonical link, the variance function simplifies to $s^2(\eta) = b''(\eta)$. Thus, both the mean and the variance of the response depend on $\boldsymbol{\beta}$ through the linear component $\eta = \mathbf{f}^{\top}(\mathbf{x})\boldsymbol{\beta}$.

Also, the log-likelihood function can be written as a function of η,

$$l(\boldsymbol{\beta}; y) = y\, b'^{-1}(g^{-1}(\mathbf{f}^{\top}(\mathbf{x})\boldsymbol{\beta})) - b(b'^{-1}(g^{-1}(\mathbf{f}^{\top}(\mathbf{x})\boldsymbol{\beta}))) + c(y). \tag{16.40}$$

TABLE 16.1
Selected Generalized Linear Models

Model	$b(\theta)$	$g(\mu)$	$E(Y)$	$\mathrm{Var}(Y)$
Binomial logit	$\log(1 + \exp(\theta))$	$\log(\frac{\mu}{1-\mu})$	$\frac{\exp(\eta)}{1+\exp(\eta)}$	$\frac{\exp(\eta)}{(1+\exp(\eta))^2}$
Binomial probit	$\log(1 + \exp(\theta))$	$\Phi_0^{-1}(\mu)$	$\Phi_0(\eta)$	$\Phi_0(\eta)(1 - \Phi_0(\eta))$
Poisson	$\exp(\theta)$	$\log(\mu)$	$\exp(\eta)$	$\exp(\eta)$

TABLE 16.2
Intensity Function for Selected Generalized Linear Models

Model	$\lambda_0(z)$
Logit	$\exp(z)/(1+\exp(z))^2$
Probit	$\varphi_0(z)^2/(\Phi_0(z)(1-\Phi_0(z)))$
Poisson	$\exp(z)$

The asymptotic covariance matrix of the maximum-likelihood estimator $\hat{\theta}_{\mathrm{ML}}$ is equal the inverse of the Fisher information matrix,

$$E\left(\frac{\partial l}{\partial \boldsymbol{\beta}}\frac{\partial l}{\partial \boldsymbol{\beta}^{\top}}\right) = -\frac{\partial^2 l}{\partial \boldsymbol{\beta}\,\partial \boldsymbol{\beta}^{\top}} = \sum_{i=1}^{n} \lambda(\mathbf{x}_i;\boldsymbol{\beta})\mathbf{f}(\mathbf{x}_i)\mathbf{f}^{\top}(\mathbf{x}_i) \tag{16.41}$$

with intensity function $\lambda(\mathbf{x};\boldsymbol{\beta}) = \lambda_0(\mathbf{f}^{\top}(\mathbf{x})\boldsymbol{\beta})$, where

$$\lambda_0(\eta) = \frac{((g^{-1})'(\eta))^2}{s^2(\eta)} \tag{16.42}$$

is the standardized intensity. Thus, the Fisher information matrix for a design ξ can be defined by

$$\mathbf{M}(\xi;\boldsymbol{\beta}) = \int \lambda(\mathbf{x};\boldsymbol{\beta})\mathbf{f}(\mathbf{x})\mathbf{f}^{\top}(\mathbf{x})\,\xi(d\mathbf{x}). \tag{16.43}$$

For the canonical link $(g^{-1})' = b''^2$, the intensity function simplifies to $\lambda_0 = s^2$ and, hence, $\lambda(\mathbf{x};\boldsymbol{\beta}) = s^2(\mathbf{f}^{\top}(\mathbf{x})\boldsymbol{\beta})$.

The information matrix for the generalized linear model differs from its counterpart for the linear model by the introduction of the additional intensity function, which causes the dependence of the information on the parameter $\boldsymbol{\beta}$.

Table 16.2 lists the standardized intensity functions λ_0 for generalized linear models relevant for IRT. Here, $\varphi_0 = \Phi_0'$ denotes the density function of the standard normal distribution.

The equivalence theory of Section 16.4 can be applied by replacing $\mathbf{f}$ by the weighted regression function $\tilde{\mathbf{f}}_{\boldsymbol{\beta}}(\mathbf{x}) = \sqrt{\lambda(\mathbf{x};\boldsymbol{\beta})}\,\mathbf{f}(\mathbf{x})$, where for each design point the mean response is weighted by the square root of its corresponding intensity. Then, for example, a design ξ^* is locally D-optimal at $\boldsymbol{\beta}$ if $\lambda(\mathbf{x};\boldsymbol{\beta})\mathbf{f}^{\top}(\mathbf{x})\mathbf{M}^{-1}(\xi^*;\boldsymbol{\beta})\mathbf{f}(\mathbf{x}) \leq p$. Now, D- and G-optimality no longer coincide, as the G-criterion minimizes $\mathbf{f}(\mathbf{x})^{\top}\mathbf{M}^{-1}(\xi^*;\boldsymbol{\beta})\mathbf{f}(\mathbf{x})$, which is the left-hand side of the D-optimality condition without the leading intensity term.

Another important approach to derive locally optimal designs for generalized linear models is to transform the design region and parameter structure simultaneously while keeping the regression functions fixed. This canonical transformation (see Ford et al., 1992) is applicable for invariant optimality criteria, in particular, for D-optimality.

The method is best illustrated by the following simple example: Consider the model $E(Y(x)) = g^{-1}(\beta_1 + \beta_2 x)$, with a logistic response on the unrestricted design region $\mathcal{X} = \mathbb{R}$. Reparameterization of this model with $x = \theta$, $\beta_2 = a_i$ and $\beta_1 = -a_i b_i$ leads to the well-known logistic 2PL model from IRT. For this model, the locally D-optimal design ξ_0^* at the standard value $\boldsymbol{\beta}_0 = (0,1)^{\top}$ can be obtained by the optimality condition $\psi_D(x,\xi_0^*) =$

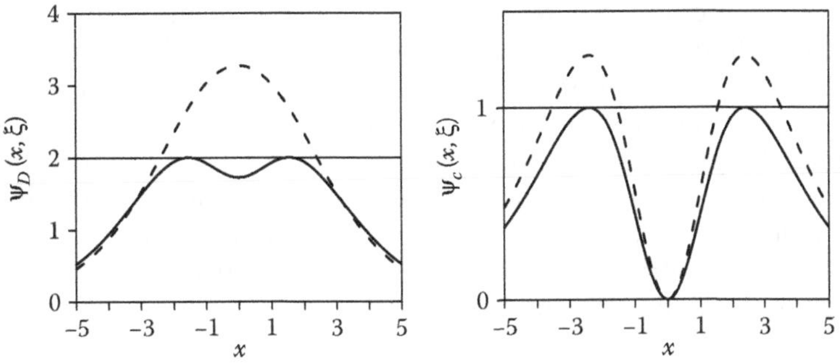

FIGURE 16.2
Sensitivity functions for the 2PL model: D-criterion (left) and c-criterion for the slope (right).

$\lambda_0(x)\mathbf{f}^\top(x)\mathbf{M}(\xi_0^*;\boldsymbol{\beta})\mathbf{f}(x) \le 2$, where $\mathbf{f}(x) = (1,x)^\top$ is the regression function of the linear component. Checking this condition, the locally D-optimal design ξ_0^* at $\boldsymbol{\beta}_0$ is the (saturated) two-point design with $x_1^* = 1.543$ and $x_2^* = -1.543$, which correspond to the 82% and 18% quantiles of the standard logistic distribution function $F(x) = g^{-1}(x)$, with equal weights $w_1 = w_2 = 1/2$ (e.g., Ford et al., 1992).

Similarly, it can be shown that the locally c-optimal design for the slope parameter β_2 is a two-point design on $x_1^* = 2.399$ and $x_2^* = -2.399$, which correspond to the 92% and 8% quantiles, with equal weights $w_1 = w_2 = 1/2$. For illustrative purposes in Figure 16.2, the sensitivity functions of the D-optimal design and of the c-optimal design for the slope are presented: The left panel shows the function ψ_D for the D-optimal design (solid line) and for its competing slope optimal design (dashed line) for the same criterion. The right panel exhibits the sensitivity function ψ_c for the c-optimal design (solid line) and for its competing D-optimal design (dashed line) for the c-criterion for the slope. From Figure 16.2, it is evident that the sensitivity functions $\psi_D(x,\xi^*)$ and $\psi_c(x,\xi^*)$ of the optimal designs do not exceed the corresponding thresholds $\Delta_D = 2$ and $\Delta_c = 1$, respectively, while the competing designs fail to satisfy this optimality condition.

Now, for general $\boldsymbol{\beta} = (\beta_1,\beta_2)^\top$ with $\beta_2 > 0$, the design region $\mathcal{X}$ can formally be transformed by the linear mapping $t(x) = \boldsymbol{\beta}^\top\mathbf{f}(x) = \beta_1 + \beta_2 x$. The mean response can now be written as a logistic regression model $E(Y(t)) = g^{-1}(\mathbf{f}^\top(t)\boldsymbol{\beta}_0)$ with argument t and standard value $\boldsymbol{\beta}_0 = (0,1)^\top$ on the transformed design region $\mathcal{T} = t(\mathcal{X}) = \mathbb{R}$. On the t scale, the D-optimal design τ^* is ξ_0^*, that is, the two-point design on $t_1^* = 1.543$ and $t_2^* = -1.543$ and equal weights $w_1 = w_2 = 1/2$. On the other hand, mapping t can be interpreted to cause a reparameterization, $\mathbf{f}(t(x)) = \mathbf{A}\mathbf{f}(x)$ for a suitable matrix $\mathbf{A}$ independent of x, where

$$\mathbf{A} = \begin{pmatrix} 1 & 0 \\ \beta_1 & \beta_2 \end{pmatrix}.$$

Because of the invariance property of D-optimality, the optimality is not affected by this reparameterization. However, the optimal design τ^* on the t-scale has to be transformed back to the original x-scale, and the D-optimal design ξ^* at $\boldsymbol{\beta}$ is the two-point design on $x_1^* = t^{-1}(t_1^*) = (1.543 - \beta_1)/\beta_2$ and $x_2^* = t^{-1}(t_2^*) = (-1.543 - \beta_1)/\beta_2$ with equal weights $w_1 = w_2 = 1/2$. As the value of the linear component remains the same under the transformation $(\mathbf{f}^\top(x)\boldsymbol{\beta} = \mathbf{f}^\top(t)\boldsymbol{\beta}_0)$, this also holds for the mean response at the optimal design points. Thus, in general, the D-optimal design settings x_1^* and x_2^* represent the 82% and 18% quantiles of the response curve $g^{-1}(\mathbf{f}^\top(x)\boldsymbol{\beta})$.

This construction method can readily be generalized to more than one explanatory variable (e.g., Graßhoff et al., 2007) and to other invariant design criteria. In particular, for the 2PL model, the c-optimal design for its slope parameter will be supported by the 92% and 8% quantiles for every value of the parameter vector $\boldsymbol{\beta}$. Similarly, the locally c-optimal design for the location parameter $x_0 = -\beta_1/\beta_2$, which characterizes the median $(E(Y(x_0)) = 0.5)$, is always the one-point design that assigns weight one to $x = x_0$. However, the latter does not allow for maximum-likelihood estimation because the corresponding information matrix is not regular. Hence, it can only serve as a benchmark, or as a limiting case for sequential designs (see below).

16.6 Nonlinear Models

Optimal designs for generalized linear models can be determined as outlined in the preceding section, but how should one proceed with nonlinear models in general? As is well-known, a Taylor expansion for a nonlinear model can be approximated by a linear model. So it seems as if locally optimal designs can be approximated using the approaches for linear models. However, in order to make the approximation meaningful, asymptotic normality with the appropriate asymptotic variance–covariance matrix is required.

The approximation will be outlined for a nonlinear model $E(Y(x)) = \mu(x,\theta)$ with a single unknown parameter θ. Taylor expansion of the mean response at the value θ_0 yields the linearization

$$E(Y(x)) = \mu(x,\theta) = \mu(x,\theta_0) + (\theta - \theta_0)\frac{\partial\mu(x,\theta)}{\partial\theta}|_{\theta=\theta_0} + \cdots$$
$$\approx \mu(x,\theta_0) + (\theta - \theta_0) f_{\theta_0}(x), \tag{16.44}$$

where $f_{\theta_0}(x) = \frac{\partial\mu(x,\theta)}{\partial\theta}|_{\theta=\theta_0}$ is a regression function depending on θ_0. When θ_0 is fixed, the function can be rearranged into a linear model

$$E(\tilde{Y}(x)) = \beta f_{\theta_0}(x), \tag{16.45}$$

with $\tilde{Y}(x) \approx Y(x) - \mu(x,\theta_0) + \theta_0 f_{\theta_0}(x)$ and $\beta = \theta$. For a one-dimensional parameter β, all optimality criteria coincide and lead to the same optimal design, which maximizes the information $M(\xi;\theta_0) = \int f_{\theta_0}^2(x)\xi(dx)$ under the assumption of homoscedastic errors. Thus, all observations should be made where $|f_{\theta_0}(x)|$ attains its maximum. The information of the linearized model mimics the inverse of the asymptotic variance if θ_0 is chosen as the true parameter value. Hence, only locally optimal designs can be obtained.

In the case of more than one parameter, multivariate Taylor expansion provides a similar linearization, where the gradient results in a vector of regression functions $\mathbf{f}_{\boldsymbol{\theta}_0}(\mathbf{x}) = \frac{\partial\mu(\mathbf{x},\boldsymbol{\theta})}{\partial\boldsymbol{\theta}}|_{\boldsymbol{\theta}=\boldsymbol{\theta}_0}$ depending on $\boldsymbol{\theta}_0$.

For generalized linear models, the information matrix is sometimes computed using the above Taylor expansion. The question then arises whether this approach is in agreement with the direct calculation of the Fisher information matrix as in the preceding section. As will be shown now, both approaches yield the same result provided heteroscedasticity is taken into account.

The generalized linear model $E(Y(\mathbf{x})) = g^{-1}(\mathbf{f}^\top(\mathbf{x})\boldsymbol{\beta})$ is heteroscedastic with $\mathrm{Var}(Y(\mathbf{x})) = s^2(\mathbf{f}^\top(\mathbf{x})\boldsymbol{\beta})$. A first-order Taylor expansion of the mean response at $\boldsymbol{\beta}_0$ leads to

$$g^{-1}(\mathbf{f}^\top(\mathbf{x})\boldsymbol{\beta}) \approx g^{-1}(\mathbf{f}^\top(\mathbf{x})\boldsymbol{\beta}_0) + \mathbf{f}_{\boldsymbol{\beta}_0}^\top(\mathbf{x})\,(\boldsymbol{\beta} - \boldsymbol{\beta}_0), \tag{16.46}$$

where

$$\mathbf{f}_{\boldsymbol{\beta}_0}(\mathbf{x}) = \frac{\partial g^{-1}(\mathbf{f}^\top(\mathbf{x})\boldsymbol{\beta})}{\partial \boldsymbol{\beta}}\,|_{\boldsymbol{\beta}=\boldsymbol{\beta}_0} = (g^{-1})'(\mathbf{f}^\top(\mathbf{x})\boldsymbol{\beta}_0)\,\mathbf{f}(\mathbf{x}) \tag{16.47}$$

is the gradient of $g^{-1}(\mathbf{f}^\top(\mathbf{x})\boldsymbol{\beta})$ with respect to $\boldsymbol{\beta}$. As above, this linearization leads to the linear model

$$E(\tilde{Y}(\mathbf{x})) = \mathbf{f}_{\boldsymbol{\beta}_0}^\top(\mathbf{x})\boldsymbol{\beta} = (g^{-1})'(\mathbf{f}^\top(\mathbf{x})\boldsymbol{\beta}_0)\,\mathbf{f}^\top(\mathbf{x})\boldsymbol{\beta} \tag{16.48}$$

with $\tilde{Y}(\mathbf{x}) = Y(\mathbf{x}) - g^{-1}(\mathbf{f}^\top(\mathbf{x})\boldsymbol{\beta}_0) + \mathbf{f}_{\boldsymbol{\beta}_0}^\top(\mathbf{x})\boldsymbol{\beta}_0$ and response variance $\sigma^2(\mathbf{x}, \boldsymbol{\beta}) = s^2(\mathbf{f}^\top(\mathbf{x})\boldsymbol{\beta})$.

As for the consistency of the estimators, the appropriate choice for the node of the Taylor expansion is the true parameter vector $\boldsymbol{\beta}_0 = \boldsymbol{\beta}$, so that the linearized model becomes

$$E(\tilde{Y}(\mathbf{x})) = (g^{-1})'(\mathbf{f}^\top(\mathbf{x})\boldsymbol{\beta})\,\mathbf{f}^\top(\mathbf{x})\boldsymbol{\beta}. \tag{16.49}$$

Taking the heteroscedasticity into account, we get for the corresponding information matrix

$$\begin{aligned}\tilde{\mathbf{M}}(\xi;\boldsymbol{\beta}) &= \int \frac{1}{\sigma^2(\mathbf{x},\boldsymbol{\beta})}\,\mathbf{f}_{\boldsymbol{\beta}}(\mathbf{x})\,\mathbf{f}_{\boldsymbol{\beta}}^\top(\mathbf{x})\,\xi(d\mathbf{x}) \\ &= \int \frac{[(g^{-1})'(\mathbf{f}^\top(\mathbf{x})\boldsymbol{\beta})]^2}{s^2(\mathbf{f}^\top(\mathbf{x})\boldsymbol{\beta})}\,\mathbf{f}(\mathbf{x})\,\mathbf{f}^\top(\mathbf{x})\,\xi(d\mathbf{x}).\end{aligned} \tag{16.50}$$

This result establishes the following identity:

Theorem 16.3

In generalized linear models, the information matrix obtained by linearization coincides with the Fisher information matrix.

16.7 Locally Optimal Designs for IRT models

We now derive a few exemplary locally optimal designs for some IRT models by embedding them in particular generalized linear models. The examples include optimal test or sampling designs for the logistic Rasch (1PL) model, the LLTM, and the RPCM.

We start with the Rasch model with the well-known item response function

$$P(U_{pi} = 1) = \frac{\exp(\theta_p - b_i)}{1 + \exp(\theta_p - b_i)}, \tag{16.51}$$

where U_{pi} is a binary response variable with $U_{pi} = 1$ if person p solves an item i correctly and $U_{pi} = 0$ otherwise (Volume One, Chapter 3). The two parameters θ_p and b_i describe the person's ability and the difficulty of the given item, respectively. Using log odds, the model can be written as

$$\log \frac{P(U_{pi} = 1)}{P(U_{pi} = 0)} = \theta_p - b_i. \tag{16.52}$$

Instead of the logit link, often a probit link is chosen. Both models, as well as models with any other feasible link functions, are of course generalized linear models.

In a test design problem, the abilities θ_p will have to be estimated while the item difficulties b_i are assumed to be known and can be selected at will. Calibration designs for estimating the difficulties b_i of given items require known abilities θ_p. Both design problems can be solved similarly, just by changing the role of the difficulty and ability parameters. Therefore, only the test design problem is considered. The problem of how to deal with the unknown ability parameters is postponed until later sections.

An exact experimental test design $\xi = (b_1, \ldots, b_I)$ of sample size I consists of I items with difficulties $b_1, \ldots, b_I$. The performance of the design is measured by its normalized Fisher information $M(\xi; \theta)$, which is inversely proportional to the asymptotic variance of the maximum likelihood estimator of θ. In the Rasch model, for a given ability θ, the normalized Fisher information is equal to

$$M(\xi; \theta) = \frac{1}{I} \sum_{i=1}^{I} f_0(\theta - b_i), \tag{16.53}$$

where $f_0(z) = \exp(z)/(1 + \exp(z))^2$ denotes the density of the standard logistic distribution. A locally optimal test design ξ^* at a specified ability $\theta = \theta_0$ maximizes the information $M(\xi; \theta_0)$. For each observation, the information $M(b; \theta) = f_0(\theta - b)$ achieves its maximal value 1/4 for $b = \theta$. Therefore, for any given ability θ, the design ξ_{b^*} is locally optimal at θ when all observations are assigned to items with difficulty $b^* = \theta$.

Formally, this optimality result can also be checked by the equivalence theory developed in Section 16.4. For the optimal design ξ^* with $b = \theta$ for all items, the sensitivity function $\psi(b, \xi^*) = 4 f_0(\theta - b)$ is depicted in Figure 16.3 (solid line). As a competitor, the sensitivity function for the earlier D-optimal design for the 2PL model, with half of the items with difficulty $b = \theta + 1.543$ and the other half $b = \theta - 1.543$ is also plotted (dashed line).

An important extension of the Rasch model, especially for generating rule-based tests is the LLTM

$$P(U_{pi} = 1) = \frac{\exp(\theta_p - \sum_{k=1}^{K} \eta_k x_{ik} - c)}{1 + \exp(\theta_p - \sum_{k=1}^{K} \eta_k x_{ik} - c)} \tag{16.54}$$

(Volume One, Chapter 13). According to this model, the difficulty b_i of an item i is determined by a weighted sum of basic parameters, $\eta_k, k = 1, \ldots, K$, representing the properties of the item,

$$b_i = \sum_{k=1}^{K} \eta_k x_{ik} + c \tag{16.55}$$

with a norming constant c.

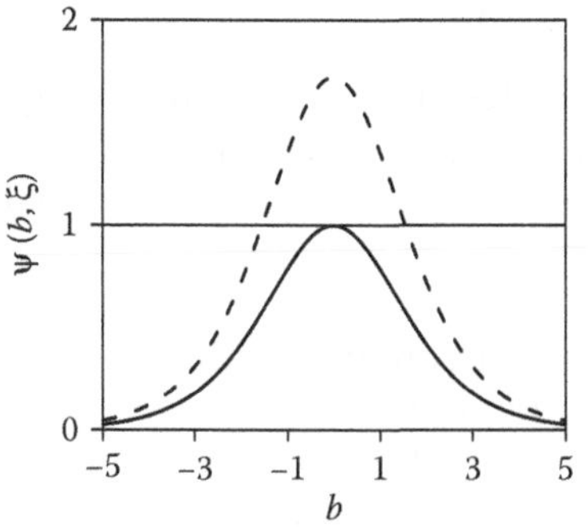

FIGURE 16.3
Sensitivity functions for the Rasch model.

The calibration design problem for an LLTM is how to select the items for estimating the basic parameters of an item while the abilities are assumed to be known. Every item in an LLTM is described by its properties and represented by a row in a design matrix. Often, this matrix consists of binary elements (0: property is not given, 1: property is given). To derive locally optimal calibration test designs, we embed the LLTM in the generalized linear model, as

$$E(Y(\mathbf{x}_1, x_2)) = \mu(\mathbf{x}_1, x_2) = g^{-1}(\mathbf{f}^\top(\mathbf{x}_1)\boldsymbol{\beta} + x_2) \tag{16.56}$$

with $\mathbf{x}_1 \in \mathcal{X}$, $g^{-1}(z) = \exp(z)/(1+\exp(z))$, $\mathbf{f}^\top(\mathbf{x}_{1i})\boldsymbol{\beta} = -\sum_{k=1}^{K} \eta_k x_{ik} - c$ and $x_2 = \theta_p \in \mathbb{R}$. As for $\mathbf{x}_1 = (x_{11}, \ldots, x_{1K})^\top$, we should think of this vector as representing the results from K attributes with the vector of regression functions $\mathbf{f} = -(1, \mathbf{f}_1^\top, \ldots, \mathbf{f}_K^\top)^\top$, where x_{1k} is the value of the kth attribute. The regression functions $\mathbf{f}_k$ represent either dummy variables for qualitative factors or real-valued functions when quantitative factors are involved. The second part of the linear component, x_2, is a continuous factor with, due to the assumptions of the Rasch model, the corresponding effect for a known ability parameter with value 1; hence, x_2 is a free additional variable to choose. In the following, we assume that each item is presented to exactly one person, which means $p = i$.

The vector of unknown parameters is denoted by $\boldsymbol{\beta} = (\beta_0, \boldsymbol{\beta}_1^\top, \ldots, \boldsymbol{\beta}_K^\top)^\top$, where $\beta_0 = c$ is the normalizing constant and $\boldsymbol{\beta}_k$ is related to the effect of the kth attribute. Both the mean response μ and response variance $b^2 = \mu(1-\mu)$ depend on the linear effect $\mathbf{f}^\top(\mathbf{x}_1)\boldsymbol{\beta} + x_2$ only, $\mathrm{Var}(Y(\mathbf{x}_1, x_2)) = b^2(\mathbf{f}^\top(\mathbf{x}_1)\boldsymbol{\beta} + x_2)$. $\lambda_0(z) = \mu'^2/b^2(z)$ is the intensity function evaluated at $z = \mathbf{f}^\top(\mathbf{x}_1)\boldsymbol{\beta} + x_2$. It gives the intensity of the information of an observation with setting $(\mathbf{x}_1, x_2)$.

To measure the quality of an exact design $\xi = ((\mathbf{x}_{11}, x_{21}), \ldots, (\mathbf{x}_{1I}, x_{2I}))$, we use the D-criterion applied to the normalized information matrix

$$\mathbf{M}(\xi; \boldsymbol{\beta}) = \frac{1}{I} \sum_{i=1}^{I} \lambda_0(\mathbf{f}^\top(\mathbf{x}_{1i})\boldsymbol{\beta} + x_{2i})\, \mathbf{f}(\mathbf{x}_{1i})\mathbf{f}^\top(\mathbf{x}_{1i}). \tag{16.57}$$

For a design ξ, we use ξ_1 to denote the marginal design with respect to the first component $\mathbf{x}_1$ and $\mathbf{M}_1(\xi_1) = 1/I \sum_{i=1}^{I} \mathbf{f}(\mathbf{x}_{1i})\mathbf{f}^\top(\mathbf{x}_{1i})$ for the marginal "linear" information matrix, which does not depend on $\boldsymbol{\beta}$. Everything is supposed to hold for the corresponding linear response $E(Y_1(\mathbf{x}_1)) = \mathbf{f}^\top(\mathbf{x}_1)\boldsymbol{\beta}$.

Theorem 16.4: Graßhoff et al., 2010

Let ξ_1^* be an exact D-optimal design on $\mathcal{X}$ for the marginal linear model $E(Y_1(\mathbf{x}_1)) = \mathbf{f}^\top(\mathbf{x}_1)\boldsymbol{\beta}$ given by I design points $(\mathbf{x}_{11}^*, \ldots, \mathbf{x}_{1I}^*)$. For given $\boldsymbol{\beta}$ set $x_{2i}^* = -\mathbf{f}^\top(\mathbf{x}_{1i}^*)\boldsymbol{\beta}$. Then the combined design $\xi^* = ((\mathbf{x}_{11}^*, x_{21}^*), \ldots, (\mathbf{x}_{1I}^*, x_{2I}^*))$ is D-optimal at $\boldsymbol{\beta}$ for the model (16.56) on $\mathcal{X} \times \mathbb{R}$.

The result in this theorem is in accordance with the earlier argument for the 2PL model, when an optimal design was found for a standardized setting and then shifted to a region specified by the parameter $\boldsymbol{\beta}$. Also, Theorem 16.4 justifies the usual practice of using a D-optimal design to determine both the design matrix of the item components and then a person parameter corresponding to the resulting difficulty of the given item. Although the theorem is formulated in terms of exact designs, it is valid for approximate designs as well.

The RPCM was first published in the classical monograph by Rasch (1960) and allows for the analysis of count data, for example, from mental speed tests. According to the RPCM, the number of correct answers Y follows a Poisson distribution

$$P(Y = y) = \lambda^y \exp(-\lambda)/(y!), \quad y = 0, 1, 2, \ldots \tag{16.58}$$

with intensity $\lambda = \theta b$, where, again, θ is the ability of the person, but b now represents the easiness of the item (Volume One, Chapter 15). This parameterization reflects the fact that the number of correct answers is expected to be larger for persons with a higher ability and items with easier tasks.

The Fisher information about θ in an item, $M(\theta)$, equals b/θ. Hence, the easiest item in the pool should be select to maximize the information of any ability θ. Likewise, when calibrating an item, Fisher information $M(b)$ is equal to θ/b, and the person with the highest ability available maximizes the information for any item easiness b.

Graßhoff et al. (2013) incorporated covariates in the RPCM to explain the easiness of items in rule-based testing as $b = \exp(\mathbf{f}^\top(\mathbf{x})\boldsymbol{\beta})$, and, hence, $\lambda(\mathbf{x}; \boldsymbol{\beta}) = \theta \exp(\mathbf{f}^\top(\mathbf{x})\boldsymbol{\beta})$, where $\mathbf{x}$ is the combination of rules, $\mathbf{f}$ the vector of regression functions, and $\boldsymbol{\beta}$ the vector of parameters for these rules to be estimated.

By embedding this extended version of the RPCM into the generalized linear model for Poisson regression, the authors derive locally D-optimal sampling designs using a K-way layout with binary explanatory variables x_k, where $x_k = 1$, if the kth rule was applied and $x_k = 0$ otherwise. Thus, $\mathbf{x} = (x_1, \ldots, x_K) \in \mathcal{X} = \{0, 1\}^K$ and, disregarding possible interactions, the vector of regression functions is given by $\mathbf{f}(\mathbf{x}) = (1, x_1, x_2, \ldots, x_K)^\top$. The parameter vector $\boldsymbol{\beta}$ contains a constant term β_0 and K main effects β_k and thus has dimension $p = K + 1$. The expected response $E(Y(\mathbf{x}))$ is equal to the intensity $\lambda(\mathbf{x}; \boldsymbol{\beta}) = \theta \exp(\beta_0 + \sum_{k=1}^K \beta_k x_k)$.

For approximate designs ξ with design points $\mathbf{x}_1, \ldots, \mathbf{x}_m$ and corresponding weights w_i, the Fisher information is $\mathbf{M}(\xi; \boldsymbol{\beta}) = \sum_{i=1}^m w_i \lambda(\mathbf{x}_i; \boldsymbol{\beta}) \mathbf{f}(\mathbf{x}_i)\mathbf{f}^\top(\mathbf{x}_i)$. The intensity and, consequently, the information is proportional to θ and $\exp(\beta_0)$, $\mathbf{M}(\xi; \boldsymbol{\beta}) = \theta \exp(\beta_0)\mathbf{M}_0(\xi; \boldsymbol{\beta})$, where $\mathbf{M}_0(\xi; \boldsymbol{\beta}) = \sum_{i=1}^m w_i \exp(\sum_{k=1}^K \beta_k x_{ik})\mathbf{f}(\mathbf{x}_i)\mathbf{f}^\top(\mathbf{x}_i)$ is the information matrix for the standardized case $\theta = 1$ and $\beta_0 = 0$. Thus, θ and β_0 are immaterial for the optimization; $\det(\mathbf{M}_0(\xi; \boldsymbol{\beta}))$ can be optimized for a fixed person, and the standardized case can be considered without loss of generality. When more than one examinee is involved, the obtained optimal design should be applied to each of them.

The D-optimal design for one rule ($K = 1$) assigns equal weights $w_i^* = 1/2$ to the only two possible settings $x = 1$ and $x = 0$, indicating whether or not the rule is applied, respectively. The result holds since, according to a basic finding of optimal design theory, D-optimal designs that are saturated ($m = p$) have equal weights $w_i = 1/p$.

In the case of $K = 2$ rules, the four possible settings are $\mathbf{x}_1 = (0,0)$ when no rule is applied, $\mathbf{x}_2 = (1,0)$ and $\mathbf{x}_3 = (0,1)$, when only one of the rules is applied, and $\mathbf{x}_4 = (1,1)$, when both are applied. Thus, any design ξ is completely determined by the corresponding weights $w_1, \ldots, w_4 \geq 0$. Denoting the related intensities by $\lambda_i = \exp(\mathbf{x}_i \boldsymbol{\beta})$, the determinant $\det(\mathbf{M}(\xi; \boldsymbol{\beta}))$ of the information matrix can be calculated as

$$w_1 w_2 w_3 \lambda_1 \lambda_2 \lambda_3 + w_1 w_2 w_4 \lambda_1 \lambda_2 \lambda_4 + w_1 w_3 w_4 \lambda_1 \lambda_3 \lambda_4 + w_2 w_3 w_4 \lambda_2 \lambda_3 \lambda_4. \tag{16.59}$$

Potential optimal designs are either saturated designs on any three of these settings with corresponding weights $w_i = 1/3$ or "true" four-point designs with suitable positive weights for all four settings.

When both rules make the item more difficult, which typically will be the case, the values for β_1 and β_2 will be nonpositive. As larger values for these parameters result in higher intensity and, hence, an increase in the Fisher information, it seems reasonable to prefer items for which at most one rule is applied. Let ξ_0 denote the corresponding saturated design with weights $w_1 = w_2 = w_3 = 1/3$ on $\mathbf{x}_1 = (0,0)$, $\mathbf{x}_2 = (1,0)$ and $\mathbf{x}_3 = (0,1)$. It follows, by using the equivalence theory, that ξ_0 is locally D-optimal if and only if $\lambda_1\lambda_2\lambda_3 - \lambda_1\lambda_2\lambda_4 - \lambda_1\lambda_3\lambda_4 - \lambda_2\lambda_3\lambda_4 \geq 0$, a condition that is satisfied if and only if $\exp(\beta_1) + \exp(\beta_2) + \exp(\beta_1 + \beta_2) \leq 1$. Or, equivalently, $\beta_2 \leq \log((1 - \exp(\beta_1))/(1 + \exp(\beta_1)))$. Otherwise, a four-point design will be optimal with a small weight $w_4 > 0$ for the $\mathbf{x}_4 = (1,1)$, that is, when both rules are used. In Figure 16.4, the parameter region of nonpositive β_1 and β_2, where ξ_0 is locally D-optimal, is below the curved border line, for which equality holds in the above condition. For parameter combinations above this curve, a four-point design is locally D-optimal. In particular, for $\beta_1 = \beta_2 = 0$ the equireplicated design, which assigns weights $w_i = 1/4$ to all four possible settings, is D-optimal.

For ξ_0 to be D-optimal, the effect sizes have to be large and the ratio between the highest and the lowest intensity must be at least $(1 + \sqrt{2})^2 \approx 5.83$. However, for many applications of the RPCM, for example, in rule-based testing of mental speed, such ratios seem unrealistic. Then, for calibration studies that involve test items for which both rules are used, four-point designs will be required. Further results, including explicit optimal weights for particular parameter constellations were derived by Graßhoff et al. (2013).

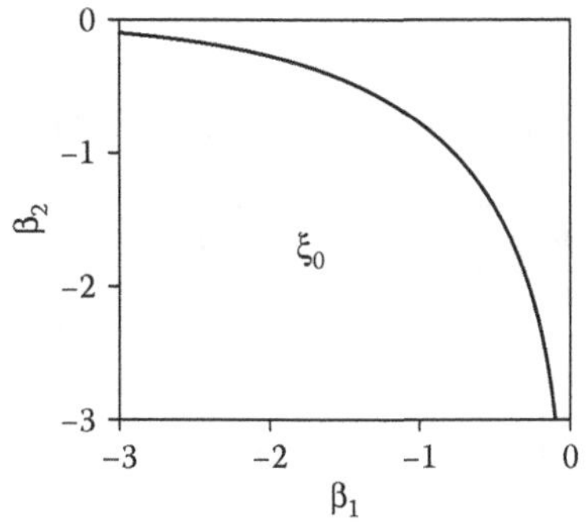

FIGURE 16.4
Locally D-optimal designs for the Rasch Poisson counts model for $\beta_1, \beta_2 \leq 0$.

16.8 Bayesian Optimal Designs

One approach to avoid local optimality, which requires exact prior information, is Bayesian optimal design (for an introduction, see Chaloner and Verdinelli, 1995). Most of the Bayesian optimal design criteria are classical versions of optimality integrated with respect to a prior or weight distribution Π for the unknown parameters. Hence, the optimality criteria have to be rewritten in terms of the design ξ rather than as a functional of the information matrix.

It is common practice to use a prior distribution for the design but not for inference. If so, no posterior distribution is considered, and, as a matter of fact, these design criteria do not describe a fully Bayesian approach.

As derived earlier, for a given value of parameter $\boldsymbol{\beta}$, the information matrix can be specified as

$$\mathbf{M}_{\beta}(\xi) = \mathbf{M}(\xi; \boldsymbol{\beta}) = \int \mathbf{f}_{\beta}(\mathbf{x})\mathbf{f}_{\beta}^{\top}(\mathbf{x})\,\xi(d\mathbf{x}), \tag{16.60}$$

where the local regression function $\mathbf{f}_{\beta}(\mathbf{x})$ may account for both nonlinearity and heteroscedasticity.

In the Bayesian optimal design literature, mostly D-optimality has been dealt with. Unlike local D-optimality, several criteria for Bayesian D-optimality can be derived, dependent on where the weighting with the prior information occurs in the formula. Atkinson et al. (2007) presented the five criteria in Table 16.3. Each of them typically leads to different optimal designs.

The first two criteria in Table 16.3 are most appealing. The first minimizes the averaged asymptotic generalized variance, $\exp(\Phi_1(\xi)) = \int \det(\mathbf{M}(\xi; \boldsymbol{\beta})^{-1})\Pi(d\boldsymbol{\beta})$; the second criterion, which is used more frequently, minimizes the averaged logarithm of the generalized variance, $\Phi_2(\xi) = -\int \log(\det(\mathbf{M}(\xi; \boldsymbol{\beta})))\,\Pi(d\boldsymbol{\beta})$ with respect to the weight distribution Π. The latter is analytically more convenient and does have a clear interpretation because it maximizes the expected increase in the Shannon information.

Firth and Hinde (1997) showed that these two criteria are convex, while the other may fail to meet this requirement. Thus, for these two criteria, Φ_1 and Φ_2, convex optimization techniques can be applied to search for optimal designs, and the general equivalence theory of Section 16.4 can be extended to these criteria. The so-called equivalence theorems allow for verifying global optimality. However, the theorem by Carathéodory, which restricts the

TABLE 16.3
Various Bayesian D-Optimality Criteria ($\mathbf{M}_{\beta} = \mathbf{M}(\xi; \boldsymbol{\beta})$) : $\psi(\mathbf{x}, \xi^*) \leq p$ for Optimal ξ^*

	Criterion $\Phi(\xi)$	Sensitivity Function $\psi(\mathbf{x}, \xi)$
1	$\log(\int \det(\mathbf{M}_{\beta}^{-1})d\Pi)$	$\int \det(\mathbf{M}_{\beta}^{-1})\mathbf{f}_{\beta}^{\top}(\mathbf{x})\mathbf{M}_{\beta}^{-1}\mathbf{f}_{\beta}(\mathbf{x})d\Pi / \int \det(\mathbf{M}_{\beta}^{-1})d\Pi$
2	$\int \log(\det(\mathbf{M}_{\beta}^{-1}))d\Pi$	$\int \mathbf{f}_{\beta}^{\top}(\mathbf{x})\mathbf{M}_{\beta}^{-1}\mathbf{f}_{\beta}(\mathbf{x})d\Pi$
3	$\log(\det(\int \mathbf{M}_{\beta}^{-1}d\Pi))$	$\int \mathbf{f}_{\beta}^{\top}(\mathbf{x})\mathbf{M}_{\beta}^{-1}(\int \mathbf{M}_{\beta}^{-1}d\Pi)\mathbf{M}_{\beta}^{-1}\mathbf{f}_{\beta}(\mathbf{x})d\Pi$
4	$\log(\int \det(\mathbf{M}_{\beta})d\Pi)^{-1}$	$\int \det(\mathbf{M}_{\beta})\mathbf{f}_{\beta}^{\top}(\mathbf{x})\mathbf{M}_{\beta}^{-1}\mathbf{f}_{\beta}(\mathbf{x})d\Pi / \int \det(\mathbf{M}_{\beta})d\Pi$
5	$\log(\det(\int \mathbf{M}_{\beta}d\Pi)^{-1})$	$\int \mathbf{f}_{\beta}^{\top}(\mathbf{x})(\int \mathbf{M}_{\beta}d\Pi)^{-1}\mathbf{f}_{\beta}(\mathbf{x})d\Pi$

number of required design points, does no longer hold, and, when the weight distribution is quite dispersed, the number of design points can become arbitrarily large for a Bayesian optimal design (Dette and Neugebauer, 1996).

Graßhoff et al. (2012) derived Bayesian optimal designs for the Rasch model using different discrete and continuous weight functions and the two criteria Φ_1 and Φ_2. More specifically, they provided conditions for the optimality of a one-point test design ξ_{b^*} at $b^* = \theta_0$, which is locally optimal at θ_0, for weight distributions centered at θ_0. The corresponding sensitivity functions for the two criteria are

$$\psi_1(b, \xi_{b^*}) = \int f_0(\theta - b^*)^{-2} f_0(\theta - b)\Pi(d\theta) \Big/ \int f_0(\theta - b^*)^{-1}\Pi(d\theta) \qquad (16.61)$$

and

$$\psi_2(b, \xi_{b^*}) = \int f_0(\theta - b)/f_0(\theta - b^*)\Pi(d\theta), \qquad (16.62)$$

where f_0 denotes the standard logistic density. For the normal distribution as weight function, Π, the integrals in the sensitivity function have to be solved numerically. The one-point design at $b^* = \theta_0$ will be optimal for the first criterion when the variance τ^2 of the normal weight function $N(\theta_0, \tau^2)$ does not exceed 1.177. For the second criterion, optimality of the one-point design ξ_{b^*} is preserved when $\tau^2 \leq 1.683$.

Figure 16.5 shows the sensitivity functions for different values of the variance τ^2 for the normal weight function. In the left panel, the sensitivity function ψ_1 is shown for the normal weight function $N(0, \tau^2)$ for $\tau = 1.3$ (dashed line), $\tau = 1.177$ (solid line), and $\tau = 0.5$ (dotted line). The right panel shows the sensitivity function ψ_2 for $\tau = 2$ (dashed line), $\tau = 1.687$ (solid line), and $\tau = 1$ (dotted line). According to the corresponding equivalence theorems optimality with respect to both criteria is established when the sensitivity functions are bounded by $p = 1$, since one parameter has to be estimated here. If the variance exceeds the thresholds of 1.177 and 1.687 for the first and second criterion respectively, the sensitivity functions fail to be bounded by 1. Then, the one-point design is no longer optimal for the normal weight function.

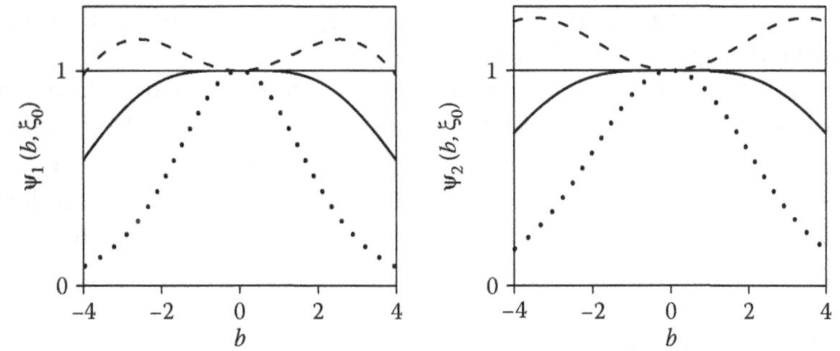

FIGURE 16.5
Sensitivity functions for the Bayesian criteria in the Rasch model.

16.9 Maximin Efficient Designs

A drawback of Bayesian optimal designs is that prior distributions for the unknown parameters have to be specified. Another approach to overcome the dependence of the optimal design on the unknown parameters is to develop maximin efficient designs. These designs only require the specification of a range of plausible parameter values. Within this range, the loss for the worst case is minimized according to a certain criterion. Or, in other words, the minimal efficiency is maximized.

In case there is only one (one-dimensional) parameter, the efficiency of a design ξ is defined as the ratio

$$\mathrm{eff}(\xi; \beta) = M(\xi; \beta)/M(\xi^*_\beta; \beta) \tag{16.63}$$

of its information relative to the information for the locally optimal design ξ^*_β, where β is the true parameter. More generally, D-efficiency of a design ξ is defined as

$$\mathrm{eff}(\xi; \boldsymbol{\beta}) = \mathrm{D_{eff}}(\xi; \boldsymbol{\beta}) = \left(\frac{\det(\mathbf{M}(\xi; \boldsymbol{\beta}))}{\det(\mathbf{M}(\xi^*_{\boldsymbol{\beta}}; \boldsymbol{\beta}))}\right)^{\frac{1}{p}}, \tag{16.64}$$

where $\xi^*_{\boldsymbol{\beta}}$ is the locally D-optimal design at $\boldsymbol{\beta}$ and p is the total number of parameters. For the other criteria, efficiency can be defined similarly.

In any case, the efficiency represents the ratio of observations needed to obtain the same value of the criterion function based upon the true parameter vector $\boldsymbol{\beta}$, when the locally optimal design $\xi^*_{\boldsymbol{\beta}}$ is used instead of the design ξ under consideration. Or, in other words, the deficiency

$$\mathrm{eff}(\xi; \boldsymbol{\beta})^{-1} - 1 \tag{16.65}$$

indicates the proportion of additional observations required when the design ξ is used, in order to get the same information as from the locally optimal design $\xi^*_{\boldsymbol{\beta}}$.

A design ξ^* is called maximin efficient, if it maximizes the minimal efficiency

$$\min_{\beta_1 \le \beta \le \beta_2} \mathrm{eff}(\xi^*; \beta) = \max_{\xi} \min_{\beta_1 \le \beta \le \beta_2} \mathrm{eff}(\xi; \beta) \tag{16.66}$$

over the range of parameters $\boldsymbol{\beta}_1 \le \boldsymbol{\beta} \le \boldsymbol{\beta}_2$. A maximin efficient design thus controls the maximal loss compared to the locally optimal design over a prespecified region, while a Bayesian optimal design deals with the average loss.

Graßhoff et al. (2012) derived maximin efficient one-point test designs for the Rasch model with θ as ability and b as difficulty parameter. For estimating ability parameter θ, the one-point design ξ_{b^*}, concentrated at the midpoint $b^* = \frac{1}{2}(\theta_1 + \theta_2)$ of the range $\theta_1 \le \theta \le \theta_2$, is maximin efficient, provided the range of values for the ability parameter does not get too large. The maximal size of the range is specified by the following theorem.

Theorem 16.5: Graßhoff et al., 2012

In the Rasch model, the one-point test design ξ_{b^*} at $b^* = \frac{1}{2}(\theta_1 + \theta_2)$ is maximin efficient on $[\theta_1, \theta_2]$ if $\theta_2 - \theta_1 \le 2\log(2 + \sqrt{3}) \approx 2.634$.

Obviously, the minimal efficiency of a maximin efficient design depends on the length of the interval $\theta_2 - \theta_1$. It varies from, at least, 0.667 for the least favorable possible situation of $\theta_2 - \theta_1 = 2.634$ to 1 if $\theta_2 - \theta_1$ is getting smaller. Thus, for the least favorable situation, at most 50 percent more observations have to be made when the maximin efficient design ξ_{b^*} is used instead of the locally optimal design ξ^*_θ. If the range becomes narrower, the maximin efficient design gets even more efficient.

16.10 Sequential Optimal Designs

The development of sequential designs is a further possibility to cope with the problem of not knowing the parameters. A sequential design consists of different stages. In the first step, a preliminary point estimate or a prior distribution for the vector of parameters is determined. This estimate, which is usually based on available experience, is then used to build an optimum design for the next few trials according to a certain optimality criterion. After executing these trials, the parameter estimates are updated and further trials are planned based on these updates. If only one batch of further trials is executed, a two-stage design should be given. A multistage design consists of more than two batches. The number of batches to be used depends on a predetermined criterion or a stopping rule that is satisfied when all parameters have been estimated with sufficient accuracy.

As Silvey (1980) has pointed out, when a sequential design is used, the matrix $\mathbf{M}(\xi_{r+1})$ based on $r + 1$ observations is not the Fisher information matrix. If the $(r + 1)$th observation depends on the outcome of the previous r observations, the $(r + 1)$th observation is not independent of the past. For calculating the Fisher information matrix the expectation over the first r observations should be used, which, however, is generally very difficult to calculate. This dependence between observations is an important point when frequentist inferences from data based on a sequential design will be made.

Furthermore, the design points, that is, the item and person parameters in calibration and test designs, respectively, are usually estimates and consequently subject to measurement errors. Thus, from a statistical point of view, the development of sequential designs is a challenging task.

Sequential test designs for adaptive testing go back to Lord (1970), who introduced the method of stochastic approximation by Robbins and Monro (1951) to estimate the ability parameter in IRT models. In each step, the next item is chosen to have difficulty equal to the estimated ability since the information on the ability is maximized when the difficulty meets the ability. The sequence of estimated abilities obtained by a Robbins–Monro procedure does converge to the true ability parameter. For this convergence to hold, no knowledge of the discrimination parameter is required. But this handy procedure may converge fairly slowly if the tuning parameter for the step sizes is not optimally chosen, which in fact depends on the discrimination and guessing parameter. The problem may be overcome by modifying the Robbins–Monro approach using averaging or an adaptive estimate of the discrimination parameter, but the process may still show unsatisfactory convergence if the initial guess of the ability is far from truth.

Thus, Lord (1970) developed the maximum-information approach which selects the $(i + 1)$th item maximizing the Fisher information. For the 1PL and 2PL models, the information on θ is maximal if the item has difficulty $b = \theta$ (see Lord, 1980). Bickel et al. (2001) proved that this property holds for the normal ogive versions of these models as well. Irrespective of a logit or probit link function, the information for θ is maximal if the

$(i+1)$th item has difficulty $b_{i+1} = \hat{\theta}_i$ for the models with one or two item parameters, where $\hat{\theta}_i$ is the present estimate of θ. For the 3PL model, Birnbaum (1968) proved that an item gives maximal information when $b_{i+1} = \hat{\theta}_i + 1/a_i \log((1+\sqrt{1+8c})/2)$, where a is the discrimination and c the guessing parameter. Many articles published since then have been concerned with almost sure convergence and asymptotic normality of the estimators. Important contributions were given especially by Wu (1985) and Ying and Wu (1997). Based upon their results, Chang and Ying (2009) proved that, for maximum-information item selection, the estimator of θ is strongly consistent and asymptotically normal for the logistic 1PL model when an infinitely large item pool is given. Consistency also holds for the 2PL model when the discrimination parameters are in a bounded interval, which is usually a realistic assumption. For the 3PL model, the same results hold when some additional, reasonable regularity conditions are introduced.

Chang and Lu (2010), using locally optimal design for the 2PL model, proved that the difficulty and discrimination estimators are strongly consistent even if the ability parameters include measurement error, which typically is the case in practice when estimates instead of the true parameter values are used. Furthermore, the estimators of the parameters are asymptotically jointly normally distributed. These results can be generalized to other sequential designs when some regularization conditions are fulfilled.

16.11 Concluding Remarks

In this chapter, optimal design theory for models in IRT was reviewed. First, the concept of optimal design was introduced for the framework of linear models. Creating an optimal design is usually much easier for linear models than for such nonlinear models as in IRT, because for the former the optimal design typically does not depend on the parameters that are estimated. Results from linear models are useful, however, as nonlinear models can often be linearized by means of a first-order Taylor expansion.

Several criteria are available to measure the quality of a design. The most popular design criterion is the D-criterion. It minimizes the determinant of the covariance matrix of the estimators, which is equivalent to minimizing the volume of their confidence ellipsoid. This criterion has the desirable property to be independent of parameterization and scaling. However, it is not an all-purpose tool; the selection of an adequate criterion has to depend on the specific goal of the study. For instance, when one aims at small confidence intervals for the difficulty and discrimination parameter in a calibration study for the 2PL model, the D-criterion might lead to unsatisfactory results. When the estimators correlate highly, the volume of the joint confidence ellipsoid for both estimators can be small while the length of their individual confidence intervals may be quite large. If one is interested only in a small variance for a given parameter, c-optimality is the adequate criterion. Furthermore, invariance of the criteria should be desirable if the interest is not in particular parameters.

Mulder and van der Linden (2009) analyzed the suitability of several optimality criteria criteria for multidimensional adaptive testing when all abilities, a subset of them, or a composite of the abilities are of interest. Based on theoretical analyses and simulation studies, the authors recommend A-optimality and D-optimality when all abilities are intentional. Because of its sometimes erratic behavior, E-optimality was not recommended. If some of the abilities are considered a nuisance, A-optimality and D-optimality should be limited to the corresponding subset of parameters. For linear combinations of abilities c-optimality was recommended.

Many IRT models are special cases of generalized linear models. Hence, a large body of research on optimal design for this general class of statistical models is available for use with IRT models. This relationship has been demonstrated especially for the 1PL model as well as for the LLTM, but application of the results on optimal designs for generalized linear models is not limited to the above cases. For example, Yang (2008) developed A-optimal designs for generalized linear models with two parameters, which include the 2PL model as well as its probit counterpart.

We have derived locally optimal designs for the logistic 1PL and 2PL model as well as for the linear logistic test model. By embedding IRT models into generalized linear models locally optimal designs for further IRT models may be conveniently obtained. Three strategies to circumvent the problem of locally optimal designs depending on unknown parameters are Bayesian optimal designs, maximin designs and sequential designs. Only a few efforts have been undertaken to develop Bayesian optimal designs and maximin designs for IRT models so far. Both approaches deserve more attention, whereby maximin designs have to meet less specific prerequisites. Sequential designs, however, have a long practical tradition in IRT. Meanwhile, the consistency and asymptotic normality of the estimators have been proved for important IRT models, which allows for statistical inference (confidence intervals and testing hypotheses) beyond pure point estimation.

Another topic, which has not yet been addressed much in the theory of optimal designs for IRT, is the impact of random effects in the variables as well as in the parameters. Typically, such random effects occur in IRT both in calibration designs, when the ability of the selected examinees is not given exactly, but they are known to come from a population with a prespecified mean ability and dispersion, and in test design, when similar items are generated by the same (combination of) rules, but still may vary slightly in difficulty. Generally, both problems require the introduction of random coefficients to describe the individual variability of examinees in a population or test items from a larger stock of similar items. For linear models, some general results for optimal designs in the presence of random effects are available (e.g., Fedorov and Hackl, 1997, Section 5.2). However, for nonlinear models, including generalized linear models, the situation is quite unclear as a straightforward linearization by Taylor approximation cannot be justified. The main problem here is the fact that for nonlinear mixed models the likelihood and, therefore the Fisher information, cannot be represented in closed form. Various approximations have been proposed in the literature, but Mielke (2012) showed that each of them may fail for certain parameter constellations and none of them is thus universally adequate.

An alternative approach is based on the use of generalized estimation equations, such as quasi likelihood (McCullagh and Nelder, 1989), rather than maximum likelihood, and derives optimal designs with respect to the asymptotic covariance matrix for the corresponding estimators (see Niaparast and Schwabe, 2013, for a Poisson count model). The derivation of a general theory of optimal design for IRT in the presence of random effects is an important subject of future investigation.

Acknowledgment

This research was partly supported by the Deutsche Forschungsgemeinschaft (DFG) under grant HO 1286/6-2.

References

Atkinson, A. C., Donev, A. N., and Tobias, R. D. 2007. *Optimum Experimental Designs, with SAS*. Oxford: Oxford University Press.

Berger, M. P. F. 1991. On the efficiency of IRT models when applied to different sampling designs. *Applied Psychological Measurement, 15*, 293–306.

Berger, M. P. F., King, C. Y. J., and Wong, W. K. 2000. Minimax D-optimal designs for item response theory models. *Psychometrika, 65*, 377–390.

Bickel, P., Buyske, S., Chang, H.-H., and Ying, Z. 2001. On maximizing item information and matching difficulty with ability. *Psychometrika, 66*, 69–77.

Birnbaum, A. 1968. Estimation of an ability. In F. M. Lord and M. R. Novick (Eds.), *Statistical Theories of Mental Test Scores*. Reading: Addison-Wesley, pp. 423–479.

Chaloner, K. and Verdinelli, I. 1995. Bayesian experimental design: A review. *Statistical Science, 10*, 273–304.

Chang, Y.-C.I. and Lu, H.-Y. 2010. Online calibration via variable length computerized adaptive testing. *Psychometrika, 75*, 140–157.

Chang, H.-H. and Ying, Z. 2009. Nonlinear sequential designs for logistic item response theory models with applications to computerized adaptive tests. *The Annals of Statistics, 37*, 1466–1488.

De Boeck, P. and Wilson, M. 2004. *Explanatory Item Response Models: A Generalized Linear and Nonlinear Approach*. New York: Springer.

De Boeck, P., Bakker, M., Zwitser, R., Nivard, M., Hofman, A., Tuerlinckx, F., and Partchev, I. 2011. The estimation of item response models with the lmer function from the lme4 package in R. *Journal of Statistical Software, 39*, 1–28.

Dette, H. and Neugebauer, H.-M. 1996. Bayesian optimal one point designs for one parameter nonlinear models. *Journal of Statistical Planning and Inference, 52*, 17–31.

Fedorov, V. V. 1972. *Theory of Optimal Experiments*. New York: Academic Press.

Fedorov, V. V. and Hackl, P. 1997. *Model-Oriented Design of Experiments*. New York: Springer.

McCullagh, P. and Nelder, J. A. 1989. *Generalized Linear Models* (2nd edn.). London: Chapman and Hall.

Firth, D. and Hinde, J. P. 1997. On Bayesian D-optimum design criteria and the equivalence theorem in nonlinear models. *Journal of the Royal Statistical Society, Series B, 59*, 793–797.

Ford, I., Torsney, B., and Wu, C. F. J. 1992. The use of a canonical form in the construction of locally optimal designs for nonlinear problems. *Journal of the Royal Statistical Society, Series B, 54*, 569–583.

Graßhoff, U., Großmann, H., Holling, H., and Schwabe, R. 2007. Design optimality in multi-factor generalized linear models in the presence of an unrestricted quantitative factor. *Journal of Statistical Planning and Inference, 137*, 3882–3893.

Graßhoff, U., Holling, H., and Schwabe, R. 2010. Optimal designs for linear logistic test models. In A. Giovagnoli, A. C. Atkinson, B. Torsney and C. May (Eds.), *mODa 9—Advances in Model-Oriented Design and Analysis*. Heidelberg: Physica, pp. 97–104.

Graßhoff, U., Holling, H., and Schwabe, R. 2012. Optimal designs for the Rasch model. *Psychometrika, 77*, 710–723.

Graßhoff, U., Holling, H., and Schwabe, R. 2013. Optimal design for count data with binary predictors in item response theory. In D. Uciński, A. C. Atkinson and M. Patan (Eds.), *mODa 10—Advances in Model-Oriented Design and Analysis*. Heidelberg: Physica, pp. 117–124.

Kiefer, J. 1959. Optimum experimental designs. *Journal of the Royal Statistical Society, Series B, 21*, 272–319.

Kiefer, J. and Wolfowitz, J. 1960. The equivalence of two extremum problems. *Canadian Journal of Mathematics, 42*, 363–366.

Lord, F. M. 1970. Some test theory for tailored testing. In W. H. Holzman (Ed.), *Computer Assisted Instruction, Testing, and Guidance*. New York: Harper and Row, pp. 139–183.

Lord, F. M. 1980. *Applications of Item Response Theory to Practical Testing Problems*. Hillsdale, NJ: Lawrence Erlbaum.

Mielke, T. 2012. Approximations of the Fisher information for the construction of efficient experimental designs in nonlinear mixed effects models. PhD thesis, Otto-von-Guericke University Magdeburg, Faculty of Mathematics.

Mulder, J. and van der Linden, W. J. 2009. Multidimensional adaptive testing with optimal design criteria for item selection. *Psychometrika, 74*, 273–296.

Niaparast, M. and Schwabe, R. 2013. Optimal design for quasi-likelihood estimation in Poisson regression with random coefficients. *Journal of Statistical Planning and Inference, 143*, 296–306.

Rasch, G. 1960. *Probabilistic Models for Some Intelligence and Attainment Tests*. Copenhagen: Danish Institute for Educational Research.

Robbins, H. and Monro, S. 1951. A stochastic approximation method. *The Annals of Mathematical Statistics, 29*, 373–405.

Silvey, S. D. 1980. *Optimal Design*. London: Chapman and Hall.

Smith, K. 1918. On the standard deviations of adjusted and interpolated values of an observed polynomial function and its constants and the guidance they give towards a proper choice of the distribution of observations. *Biometrika, 12*, 1–85.

van der Linden, W. J. 1994. Optimum design in item response theory: Applications to test assembly and item calibration. In G. H. Fischer and D. Laming (Eds.), *Contributions to Mathematical Psychology, Psychometrics, and Methodology*. New York: Springer, pp. 308–318.

Wu, C. F. J. 1985. Efficient sequential designs with binary data. *Journal of the American Statistical Association, 80*, 974–984.

Yang, M. 2008. A-optimal designs for generalized linear models with two parameters. *Journal of Statistical Planning and Inference, 138*, 624–641.

Section IV

Model Fit and Comparison

17

Frequentist Model-Fit Tests

Cees A. W. Glas

CONTENTS

17.1 Introduction

There is little doubt that the well-known quote "All models are wrong, but some are useful" also applies to item response theory (IRT) models. For instance, in large-scale educational surveys, such as those conducted in the Programme for International Student Assessment (PISA) project, tests of model fit easily become significant. Since item calibration in the PISA project is typically carried out using samples of more than 17,000 students from some 34 OECD (Organisation for Economic Co-operation and Development) countries, this should come as no surprise. Rather than in testing, the fit of IRT models against unspecified alternatives, our interest should therefore be in assessing specific discrepancies between observations and model predictions, known as residuals, to evaluate whether the intended inferences made from the model are trustworthy. A plethora of residuals can be defined, including residuals that target potential violations of specific well-known model assumptions.

In IRT, the most important assumptions to be evaluated are the ones of subpopulation invariance—a violation often labeled as differential item functioning (DIF) (Volume Three, Chapter 4)—the shape of the item response functions, and local stochastic independence. The first assumption implies that the item responses can be described by the same parameters in all possible subpopulations; or, in other words, subpopulations defined on the basis of background variables such as gender, race, and socioeconomic states, should not be relevant in a specific testing situation. The second assumption addresses the appropriateness of the functions that describes the relationship between a latent variable, say proficiency, and observable responses to an item. Evaluation of the appropriateness is usually done by comparing observed and expected item response frequencies given some measure of the latent trait level. The third assumption of local stochastic independence implies that responses to different items should be independent given the latent trait value, so that the proposed latent variable completely describes the responses and no additional variables are necessary to do so.

This chapter is organized as follows. First, in order to give a flavor of the model fit procedures, tests targeted at the item response function and local independence are presented for one of the most commonly used IRT models, the 3-parameter logistic model (3PLM; Birnbaum, 1968; Volume One, Chapter 2). Next, a general framework for the construction of goodness-of-fit tests based on the Lagrange multiplier (LM) test for a broad class of IRT models is presented. The approach is then illustrated with tests of the shape of the response function, local independence, and DIF, as well as one of an assumption on the distribution of ability parameters. Finally, alternative approaches based on the likelihood-ratio (LR) test, Wald test, a uniformly most powerful (UMP) test, and a limited-information goodness-of-fit test are discussed.

17.1.1 Fit of the 3PLM Response Function

In the 1-, 2-, and 3-parameter logistic models (1PLM, 2PLM, and 3PLM) (Birnbaum, 1968; Volume One, Chapter 2), it is assumed that the proficiency level of test takers $p = 1, \ldots, P$ on items $i = 1, \ldots, I$ can be represented by a one-dimensional parameter θ_p. More specifically, the 3PLM explains the probability of a correct response as

$$\Pr\left(U_{pi} = 1\right) = p_i(\theta_p) = c_i + (1 - c_i)\frac{\exp(a_i(\theta_p - b_i))}{1 + \exp(a_i(\theta_p - b_i))}, \tag{17.1}$$

where U_{pi} is a random variable taking a value equal to one if a correct response is given by p to i, and zero otherwise. The three types of item parameters are interpreted as discrimination, difficulty and guessing parameters, respectively. The 2PLM follows upon setting the guessing parameter equal to zero, and the 1PLM follows upon introducing the additional constraint of all discrimination parameters being equal. Ideally, a test of the fit of the response function would assess the degree to which the observed proportion of correct responses to item i of test takers with a proficiency level θ_p matches $p_i(\theta_p)$. But for the 2PLM and 3PLM, we have to face the problem of estimates of the proficiency parameters for all test takers that are virtually unique, so that for every available θ value at most one observed response on each item may be available. As a result, we cannot meaningfully compute proportions of correct responses given a value of θ.

However, number-correct scores and estimates of θ are usually highly correlated and, therefore, tests of model fit may be based on the assumption that groups of test takers with the same number-correct score x are homogeneous with respect to their proficiency level. The assumption seems supported by the analysis of a test of four items with the results shown in Table 17.1. The columns in the table pertain to the number-correct scores of one, two, and three. The zero and perfect scores are omitted because they do no contribute any information with respect to the items.

The columns labeled "Obs" give the proportion of correct responses in the three score groups, respectively. The columns labeled "Exp" give the associated expectations. Details on how to compute these expectations are addressed below. For every item, the pattern

TABLE 17.1
Test Statistics for Item Response Functions

	Number-Correct Score								
	1		2		3				
Item	Obs	Exp	Obs	Exp	Obs	Exp	Res	Q_1	Prob
1	0.47	0.37	0.56	0.53	0.60	0.69	0.07	30.22	0.00
2	0.16	0.13	0.24	0.25	0.36	0.42	0.03	6.84	0.03
3	0.37	0.31	0.46	0.47	0.60	0.64	0.04	8.55	0.01
4	0.09	0.16	0.23	0.27	0.54	0.44	0.07	29.99	0.00

of the differences between the observed and expected proportions gives an indication of the type of misfit. For instance, for Item 1, the observed proportion correct is higher than expected for the lowest-scoring group, and lower than expected for the highest group. So the response curve predicted by the model is too steep. Item 4 shows the opposite result, that is, its predicted curve is too flat. The column labeled "Res" gives residuals aggregated over the three groups (in this case the average difference between the observed and expected proportions, but, of course, other choices are possible too). Differences between observed and expected proportions of correct responses, say $O_{xi} - E_{xi}$, generally give rise to a Pearson statistic of the form

$$Q_1 = \sum_{x=1}^{I-1} n_x \frac{(O_{xi} - E_{xi})^2}{E_{xi}(1 - E_{xi})}, \tag{17.2}$$

where n_x is the number of test takers with number-correct score x, O_{xi} is the proportion of test takers with a score x and a correct score on item i, and E_{xi} is the analogous probability under the used IRT model.

Examples of this type of statistic were proposed for the 1PLM in a CML framework by van den Wollenberg (1982) and for the 2PLM and 3PLM in an MML framework by Orlando and Thissen (2000). Simulation studies have shown that their null distributions are well approximated by a Chi-square-distribution (van den Wollenberg, 1982; Glas and Suárez-Falćon, 2003; Suárez-Falćon and Glas, 2003).

The values for the Q_1 statistic by Orlando and Thissen (2000) and their associated significance probabilities are given in the last two columns of Table 17.1. This statistic is thus assumed to have an approximate Chi-square-distribution with two degrees of freedom. The expectations E_{xi} were computed using MML estimates (Volume One, Chapter 11). One of the main features of MML estimation is the introduction of a distribution of θ with density $g(\theta)$ typically assumed to be normal. The likelihood is then marginalized with respect to θ. For this framework, Orlando and Thissen (2000) showed that the conditional expectation E_{xi} is given by

$$E_{xi} = P(U_i = 1|x) = \frac{\int p_i(\theta) G^{(i)}(x-1, \theta) g(\theta) d\theta}{\int G(x, \theta) g(\theta) d\theta},$$

where

$$G(x,\theta) = \sum_{\mathbb{B}(x)} \prod_{i=1}^{I} p_i(\theta)^{u_{pi}} (1 - p_i(\theta))^{1-u_{pi}},$$

that is, $G(x,\theta)$ is the sum of the probabilities of all possible response patterns with number-correct score x. Furthermore,

$$G^{(i)}(x-1,\theta) = \sum_{\mathbb{B}(x,i)} \prod_{j \neq i}^{I} p_j(\theta)^{u_{pj}} (1 - p_j(\theta))^{1-u_{pj}},$$

which is the sum of the probabilities of all possible partial response patterns without item i. The functions $G(x,\theta)$ and $G^{(i)}(x-1,\theta)$ can be computed using the recursion formula by Lord and Wingersky (1984).

For substantial numbers of items, more than 20, say, the need to consider each score separately might produce voluminous and, dependent on the sample size, unreliable output. It might then be necessary to combine adjacent number-correct scores to obtain a smaller number of score groups (three to six, say). This operation replaces the original number-correct scores x by sums over score levels $g = 1, \ldots, G$, and leads to a test be based on the differences $O_{gi} - E_{gi}$. Unfortunately, the variances of these differences cannot be properly estimated using an expression analogous to the denominator of Equation 17.2. The problem of how to define a proper test statistic for combined score-level groups will be addressed in the framework of the LM statistic below.

17.1.2 Local Independence in the 3PLM

The next assumption underlying the IRT models presented above is the one of unidimensionality. If the test taker's position on one latent trait is fixed, the assumption of local stochastic independence requires the association between the responses to the items to vanish. Suppose the assumption is violated and more than one dimension is necessary to describe the test takers' position in the latent space. The association between the responses to the items given one proficiency parameter does then not vanish. Therefore, tests of local independence between items serve as test of the unidimensionality assumption as well.

van den Wollenberg (1982) and Yen (1984, 1993) showed that violation of local independence can be tested using a test statistic based on the association between items in a 2×2 table. Applying the same idea to the 3PLM in an MML framework, a statistic based on the differences between observed and expected frequencies is

$$\begin{aligned} d_{ij} &= n_{ij} - E(N_{ij}) \\ &= n_{ij} - \sum_{x=2}^{I-2} n_x p(U_i = 1, U_j = 1|x), \end{aligned} \tag{17.3}$$

where $N_{ij} = n_{ij}$ is the observed number of test takers with item i and item j correct in the group with a number-correct score between two and $I-2$, and $E(N_{ij})$ is its expectation. Only scores between two and $I-2$ are considered because test takers with a lower or higher score cannot make both items correct. So both categories of test takers do not contribute any information to the 2×2 table.

Using Pearson's Chi-square-statistic for association in a 2 × 2 table results in

$$Q_2 = \frac{d_{ij}^2}{E(N_{ij})} + \frac{d_{ij}^2}{E(N_{i\bar{j}})} + \frac{d_{ij}^2}{E(N_{\bar{i}j})} + \frac{d_{ij}^2}{E(N_{\bar{i}\bar{j}})},$$

where $E(N_{i\bar{j}})$ is the expected number of test takers with item i correct and j wrong, and $E(N_{\bar{i}j})$ and $E(N_{\bar{i}\bar{j}})$ are defined analogously. Simulation studies by Glas and Suárez-Falćon (2003) showed that this statistic is well approximated by a Chi-square-distribution with one degree of freedom.

The conditional probability in Equation 17.3 is computed as

$$p(U_i = 1, U_j = 1|x) = \frac{\int p_i(\theta)p_j(\theta)G^{(i,j)}(x-2,\theta)g(\theta)d\theta}{\int G(x,\theta)g(\theta)d\theta},$$

where

$$G^{(i,j)}(x-2,\theta) = \sum_{\mathbb{B}(x,i,j)} \prod_{k \neq i,j}^{I} p_k(\theta)^{u_{pk}}(1-p_k(\theta))^{1-u_{pk}},$$

that is, $G^{(i,j)}(x-2,\theta)$ is the sum of the probabilities of all possible partial response patterns without items i and j resulting in number-correct score $x-2$. As before, $G^{(i,j)}(x-2,\theta)$ can be computed using the recursion formula by Lord and Wingersky (1984). In the next section, we consider an alternative to this test based on a similar logic but for a framework that more generally applies to a broad class of IRT models and does not require any simulations to justify claims regarding its asymptotic distributions.

17.2 A General Framework for Evaluating Model Fit

The basic idea underlying the general approach in this section is the one of an IRT model as null model tested against an alternative model that incorporates a violation of the specific assumption that is tested. For the assumption of local independence. Kelderman (1984, 1989) and Jannarone (1986) proposed an alternative to the 1PLM, where the probability of a correct response on item i given a response on item j is given as

$$\Pr(U_{pi} = 1|U_{pj} = u_{pj}; \theta_p, b_i, \delta_{ij}) = \frac{\exp(\theta_p - b_i + u_{pj}\delta_{ij})}{1+\exp(\theta_p - b_i + u_{pj}\delta_{ij})} \tag{17.4}$$

with U_{pi} and U_{pj} the responses of test taker p to items i and j, respectively. Note that the magnitude of δ_{ij} gauges the dependency between the two responses, and that setting $\delta_{ij} = 0$ results in the 1PLM. The null hypothesis $\delta_{ij} = 0$ can be tested using a LR, Wald, and LM test (for a review of their differences, see Buse, 1982). The first two types of tests require explicit evaluation of the alternative model. However, the LM test requires evaluation of the null model only. Consequently, its test statistics can be computed from one estimation run to test various kinds of model violations for all items.

LM tests are based on an evaluation of the first-order partial derivatives of the log-likelihood function of the alternative model using the maximum-likelihood estimates of the null model. Let the full vector of all parameters of the alternative model, say $\boldsymbol{\eta}$, be partitioned as $\boldsymbol{\eta} = (\boldsymbol{\eta}_1, \boldsymbol{\eta}_2)$, with the null hypothesis represented by $\boldsymbol{\eta}_2 = 0$. Further, let $\mathbf{h}(\boldsymbol{\eta})$ denote the first-order partial derivatives of the log-likelihood of the alternative model, that is, $\mathbf{h}(\boldsymbol{\eta}) = \partial \log L(\boldsymbol{\eta})/\partial\boldsymbol{\eta}$. Partition $\mathbf{h}(\boldsymbol{\eta})$ likewise as $(\mathbf{h}(\boldsymbol{\eta}_1), \mathbf{h}(\boldsymbol{\eta}_2))$. Then, $\mathbf{h}(\boldsymbol{\eta}_1) = 0$, because $\boldsymbol{\eta}_1$ is estimated by maximum likelihood. Consequently, the size of the elements of $\mathbf{h}(\boldsymbol{\eta}_2)$ determine the value of the statistic: The closer they are to zero, the better the fit of the model to the data.

The test statistic is

$$LM = \mathbf{h}(\boldsymbol{\eta}_2)'\boldsymbol{\Sigma}^{-1}\mathbf{h}(\boldsymbol{\eta}_2), \tag{17.5}$$

where

$$\boldsymbol{\Sigma} = \boldsymbol{\Sigma}_{22} - \boldsymbol{\Sigma}_{21}\boldsymbol{\Sigma}_{11}^{-1}\boldsymbol{\Sigma}_{12} \tag{17.6}$$

and

$$\boldsymbol{\Sigma}_{pq} = -\frac{\partial^2 \log L(\boldsymbol{\eta})}{\partial\boldsymbol{\eta}_p \partial\boldsymbol{\eta}_q'}$$

for $p = 1, 2$ and $q = 1, 2$. Under certain mild regularity assumptions, the LM statistic has an asymptotic Chi-square-distribution with degrees of freedom equal to the number of parameters in $\boldsymbol{\eta}_2$ (Rao, 1947; Aitchison and Silvey, 1958).

A test of the null hypothesis $\delta_{ij} = 0$ for the model in (17.4) proceeds as follows: First, note that, conditionally on $U_{pj} = u_{pj}$, (17.4) is an exponential family model (see Chapter 4). The sufficient statistic for δ_{ij} is

$$N_{ij} = \sum_{p=1}^{N} u_{pi}u_{pj},$$

while the ML estimate of δ_{ij} is obtained by equating the observed value of $N_{ij} = n_{ij}$ to its expectation

$$E(N_{ij}) = \sum_{p=1}^{N} u_{pj} \Pr(U_{pi} = 1 | U_{pj} = u_{pj}; \theta_p, b_i, \delta_{ij}). \tag{17.7}$$

Fisher's identity (Efron, 1977; Louis, 1982) can be used as a device for easy derivation of MML estimation equations (Glas, 1999; see Chapter 11). In the current context, the identity equates the first-order derivatives of the marginal log-likelihood to the posterior expectation of the first-order derivatives of the conditional log-likelihood given θ. The advantage of using the identity resides in the fact that, conditional on θ, first-order derivatives are often extremely easy to derive for exponential families. The posterior expectation of Equation 17.7 is

$$\sum_{p=1}^{N} u_{pj} E(P_i(\theta) \mid \mathbf{u}_p), \tag{17.8}$$

where $P_i(\theta) = \Pr(U_{pi} = 1 | U_{pj} = u_{pj}; \theta_p, b_i, \delta_{ij})$ is evaluated at $\delta_{ij} = 0$, and $\mathbf{u}_p$ is the response pattern of test taker p. So, the LM test statistic is based just on an evaluation of the difference between $N_{ij} = n_{ij}$ and its posterior expectation.

TABLE 17.2

Test Statistics for Local Independence

Score on Item B		Incorrect		Correct				
Item A	Item B	Obs	Exp	Obs	Exp	Res	LM	Prob
1	1	0.23	0.22	0.26	0.29	0.01	0.68	0.41
2	2	0.46	0.45	0.52	0.57	0.01	1.68	0.20
3	3	0.47	0.48	0.61	0.60	−0.01	0.20	0.65
4	4	0.43	0.49	0.67	0.65	−0.06	11.44	0.00

An example of the test for the 3PLM is given in Table 17.2. (All examples in this and the following tables were computed using the public domain software MIRT [Glas, 2010a].) The columns labeled "Obs" and "Exp" give the observed proportions correct on item A, given an incorrect or a correct score on item B.

The column labeled "Res" gives the residuals indicating whether the observed associations were too high or too low. The last two columns give the values of the test statistic and the associated significance probability, respectively.

In the following paragraphs, three additional LM tests will be defined. The first is targeted at the item response functions, the second is at DIF, and the last one at an assumption about the distribution of θ. The tests will be presented in the framework of MML estimation of the generalized partial credit model (GPCM) (Muraki and Muraki, 1992; Volume One, Chapter 8). However, at the end of this section, it will be pointed out that the framework also applies to other models, including multidimensional models, as well as to CML estimation.

For a general definition of the approach for the GPCM in an MML framework, define covariates $y_{pc}, c = 1, \ldots, C$. Special cases of these covariates lead to the specific statistics given below. The covariates may be separately observed person characteristics. They may also depend on the observed response pattern, but without the response to the item i that is targeted. The null model is the GPCM in the parameterization used in Glas (see Chapter 11, Equation 11.25). The responses are coded $U_{pij} = 1,\ j = 0, \ldots, J$, if they were in category j and $U_{pij} = 0$ otherwise. As an alternative to the null model, we consider a model, where conditional on the covariates, the probability of scoring in category j of item i is

$$P(U_{pij} = 1 \mid \theta_p) = \frac{\exp(ja_i\theta_p - b_{ij} + j\sum_c y_{pc}\delta_c)}{1 + \sum_{h=1}^{J} \exp(ha_i\theta_p - b_{ih} + h\sum_c y_{pc}\delta_c)}. \tag{17.9}$$

Thus, under the null model, the additional parameters δ_c are equal to zero.

Notice that parameter δ_c is multiplied by j, so it can be interpreted as a uniform shift in θ. In other words, the alternative model implies item parameters and the latent parameters θ_p that together are insufficient to describe response behavior, unless some shift related to the response level is incorporated. If the person parameters θ_p would be known, the response probabilities in Equation 17.9 define an exponential family model. So the first-order derivatives with respect to the parameters δ_c are the differences between the sufficient statistic for a parameter and its expected value, that is

$$-\sum_p y_{pc} \sum_{j=1}^{J} ju_{pij} + \sum_p y_{pc} \sum_{j=1}^{J} jP(U_{pij} = 1 \mid \theta_p). \tag{17.10}$$

Taking posterior expectations results in

$$-\sum_p y_{pc} \sum_{j=1}^{J} j u_{pij} + \sum_p y_{pc} \sum_{j=1}^{J} j E(P_{ij}(\theta) \mid \mathbf{u}_p) \tag{17.11}$$

for $i = 1,\ldots,I$, and $c = 1,\ldots,C$, where $P_{ij}(\theta)$ is the GPCM response function defined in Glas (see Chapter 11, Equation 11.25). The expected value is computed using the GPCM without the additional parameter—the null model. An important requirement is that the alternative model be identified. This can be accomplished by setting the first additional parameter, δ_1 equal to zero. A large difference between the observed and expected values means that the GPCM model did not fit the data, that is, the additional parameters δ_c are necessary to obtain fit of the model, so the null hypothesis $\delta_c = 0$, $c = 1,\ldots,C-1$, is rejected. The LM statistic in Equation 17.5 is based on $C-1$ residuals and has an asymptotic Chi-square-distribution with $C-1$ degrees of freedom.

Although the test outlined here was presented in the framework of the GPCM with MML estimation, it can be generalized into several directions. First of all, there exist two alternatives to the GPCM: The graded response model (GRM) (Samejima, 1969; Volume One, Chapter 9) and the sequential model (SM) (Tutz, 1990; Volume One, Chapter 6). The rationales underlying these models are quite different (Mellenbergh, 1995), though the practical implications of them are often negligible because their item response functions are frequently so close each other that they can hardly be distinguished on the basis of empirical data (Verhelst et al., 1997). The principle of adding covariates y_{pc} to derive an alternative to the GRM or SM to check if an assumption of the original model is violated can also be applied to these models (Glas, 2010b). However, in practice, the resulting LM tests gives comparable results for all three models.

The second generalization pertains to the assumption of unidimensionality made in the GPCM, SM, and GRM. Each of these models contains a term $a_i\theta_p$. Multidimensional versions of them are obtained by replacing the term by $\sum_{q=1}^{Q} a_{iq}\theta_{pq}$, where Q is the number of dimensions, θ_{pq} is the level of person p on dimension q, and a_{iq} gauges the extent to which item i relates to the qth dimension. It is easily verified that the framework outlined previously does not essentially change when $a_i\theta_p$ is replaced by $\sum_{q=1}^{Q} a_{iq}\theta_{pq}$.

Finally, the framework outlined above can also be applied to models estimated by CML, such as for the 1PLM (Rasch, 1960), the PCM (Masters, 1982), the OPLM (Verhelst and Glas, 1995) and the family of models developed by Kelderman (1984, 1989). To test these models, Glas (1988) and Glas and Verhelst (1989, 1995) developed a class of so-called generalized Pearson statistics. However, these are equivalent to the item-oriented LM statistics presented in this chapter (Glas, 2006).

17.2.1 Evaluation of the Fit of the Response Functions

In the introduction of this chapter, we considered a test of fit of the item response functions for the 3PLM. We now introduce a similar test for polytomously scored responses. Again, we partition the test takers on the basis of their total scores. The test is performed for every item. Let i denote the item on which we currently focus, while the other items are labeled $k = 1,\ldots,i-1,i+1,\ldots,I$. Define $S_p^{(i)}$ as the unweighted score for the partial response pattern without item i, that is,

$$S_p^{(i)} = \sum_{k \neq i} \sum_{j=1}^{J} j u_{pkj}.$$

The possible scores $S_p^{(i)}$ are partitioned into C disjoint score levels and y_{pc} is defined as the set of indicator variables that take a value one if the score is in subset c and zero otherwise. From the theory in the previous section, it follows that the test is based on the squared difference between the observations and their posterior expectations weighted by their covariance matrices. These differences are

$$\sum_p y_{pc} \sum_{j=1}^{J} j u_{pij} - \sum_p y_{pc} \sum_{j=1}^{J} j E(P_{ij}(\theta) \mid \mathbf{u}_p) \tag{17.12}$$

for $i = 1, \ldots, I$, and $c = 1, \ldots, C$. Examples of the test will be given in the next paragraph, where its approach is generalized further.

17.2.2 Evaluation of DIF

In educational measurement, DIF is a difference in response probabilities between equally proficient members of two or more groups; for an overview refer to Holland and Wainer (1993), Camilli and Shepard (1994), or Gamerman et al. (Volume Three, Chapter 4). As an example, one could think of a test of foreign language comprehension, where girls are impeded by items referring to the game of a football. A possible poor performance of girls on such items might not must be indicative of poor comprehension of the foreign language but of their lack of knowledge of football.

In the sequel, we will consider the case of two populations, labeled the focal and the reference populations, but the argument is easily extended to more populations. DIF can be uniform or nonuniform (Mellenbergh, 1982). With uniform DIF, the difficulty parameter b_i is shifted uniformly for the focal population. In the case of nonuniform DIF, there is an additional interaction between the item and the level of the populations; in terms of the IRT model, both the parameters a_i and b_i differ across populations. It is important to note that the two populations need not have the same distribution of the θ parameters. However, DIF pertains to differences in response behavior conditional on θ, not on different distributions of θ. So in an MML framework, their distributions are usually modeled as normals with different means and variances.

To assess uniform DIF using an LM test, the covariates y_{pc} are indicators of the different populations, enabling us to test for the differences between the observed and expected item scores in samples from them. An example of the test for nine items is given in Table 17.3. The model was estimated using MML estimation with two different ability distributions. The values of the LM statistic and their significance probabilities are given in the columns labeled "LM" and "Prob." Note that item 1 had the largest value of the LM statistic, so this item is most suspicious. The next columns show the seriousness of the model violations in the form of average item scores and expectations. The maximum score on each item was three, so theoretically the average score is between zero and three. On item 1, the sample from the focus group scored lower than expected; so this group was disadvantaged by the item. Effect sizes are reported in the last column. Whether the effect size of −0.15 for item 1 is important or not depends on the intended use of the test and is a matter for judgement.

TABLE 17.3
Lagrange Tests of Uniform DIF ($df = 1$)

			Focus		Reference		
Item	**LM**	**Prob**	**Obs**	**Exp**	**Obs**	**Exp**	**Res**
1	11.73	0.00	1.67	1.82	2.05	1.97	−0.15
2	1.56	0.21	0.96	1.04	1.22	1.18	−0.08
3	1.77	0.18	1.69	1.70	1.78	1.78	−0.01
4	0.31	0.57	0.33	0.34	0.47	0.46	−0.01
5	5.09	0.02	1.75	1.61	1.70	1.78	0.14
6	1.12	0.28	2.16	2.09	2.15	2.18	0.07
7	2.91	0.08	1.84	1.82	1.92	1.93	0.02
8	1.57	0.21	0.87	0.79	0.89	0.93	0.08
9	0.08	0.77	1.56	1.57	1.72	1.72	−0.01

TABLE 17.4
Lagrange Tests of Nonuniform DIF ($df = 2$)

			Level 1		Level 2		Level 3		
Item	**LM**	**Prob**	**Obs**	**Exp**	**Obs**	**Exp**	**Obs**	**Exp**	**Res**
1	17.02	0.00	1.53	1.36	1.56	1.76	1.96	2.05	0.16
2	3.30	0.19	0.72	0.79	0.86	1.00	1.27	1.29	0.08
3	3.23	0.20	1.48	1.58	1.61	1.70	2.00	1.84	0.12
4	2.02	0.36	0.30	0.21	0.25	0.30	0.43	0.51	0.07
5	2.12	0.35	1.43	1.29	1.56	1.55	2.13	1.86	0.14
6	0.71	0.70	2.00	1.93	2.03	2.08	2.42	2.25	0.10
7	4.20	0.12	1.34	1.58	1.78	1.79	2.00	2.23	0.16
8	3.62	0.16	0.76	0.56	0.67	0.72	1.14	1.00	0.13
9	0.34	0.84	1.28	1.32	1.61	1.56	1.79	1.83	0.04

The second suspicious item was item 5. Note that here the focal group is the advantaged group.

To test for nonuniform DIF, covariates y_{pc}, $c = 1,\ldots,C$, are defined combining of the earlier covariates for the tests of the response functions and group membership, that is, y_{pc} takes the value of one when the test taker is the group for which the test is reported and obtains a given total score (disregarding the studied item i) and is equal to zero otherwise. Table 17.4 gives the results for focal group for the same data set as in Table 17.3. A similar table can be produced for the reference group. Note again that item 1 shows the largest DIF. Its estimated response function is too steep for the focal group; the observed scores on it are too high for the low scoring group and too low for the high scoring group. The last column labeled "Res" gives the residuals, in the present case the average differences between the observed and expected items scores across the three score levels.

17.2.3 Evaluation of the Assumption of an Ability Distribution

In MML estimation, the assumptions on the distribution of θ are an integral part of the statistical model and therefore need to be tested. One of the ways in which a misspecification of the distribution can show up is through poor representation of the total score distribution. This can be evaluated using an LM statistic. Because the test is not item oriented, the framework introduced above needs some modification. We no longer consider the conditional response probability for a specific item, but rather the probabilities of the entire response pattern conditional on θ. Define indicator variables y_{pc} equal to one when the response pattern has a total score of level c and zero otherwise. Usually, each possible score defines one level, but different scores may be combined when their individual frequencies are low,which most likely may happen for the highest or lowest scores. The conditional probability of a response pattern can be written as

$$p(\mathbf{u}_p|\theta) \propto \exp\left[\left(\sum_{ij} u_{pij}(a_i j\theta - b_{ij})\right) + \sum_{c} y_{pc}\delta_c\right]. \tag{17.13}$$

For $\boldsymbol{\delta} = \mathbf{0}$ this simplifies to

$$p(\mathbf{u}_p|\theta) \propto \exp\left(\sum_{ij} u_{pij}(a_i j\theta - b_{ij})\right). \tag{17.14}$$

Further, if we consider $\boldsymbol{\delta}$ as the only parameter, the model defined by Equation 17.13 is an exponential family model, and its first-order derivatives with respect to $\boldsymbol{\delta}$ are the differences between a sufficient statistic and its expectation. Define

$$f_c(\theta) = \sum_{\mathbb{B}(c)} p(\mathbf{u}|\theta),$$

where the sum is over all response patterns with a total score on level c. As earlier, the sum can be computed using the recursion formula by Lord and Wingersky (1984). The sufficient statistic for parameter δ_c is the vector of the number of observations N_c of score levels $c = 1, \ldots, C$. Application of Fishers' identity results in first-order derivatives equal to the difference between N_c and its posterior expectation,

$$\sum_{p=1}^{N} y_{pc}\, E\left(f_c(\theta)|\, \mathbf{u}_p\right).$$

An example of the test is given in Table 17.5. Note that the scores of zero, one, and two as well as those of 13, 14, and 15 are combined in score levels but all other scores enter the statistic separately. The value of the test statistic is given at the bottom of the table, together with its degrees of freedom ($C - 1 = 11$) and significance probability. In the present example, the test is not significant and the model is not rejected. If the test would indicate serious misfit, several options are open. When relevant background variables are available, one of the possibilities is to distinguish between more than one distribution of θ. Another option is to consider a wider family of distributions (Thissen, 1991; van den Oord, 2005).

TABLE 17.5

Lagrange Test of θ Distribution

Score	Range	Frequency	Expected
0	2	30	30.90
3	3	37	38.14
4	4	66	59.92
5	5	80	83.16
6	6	96	104.49
7	7	126	120.31
8	8	136	127.47
9	9	126	124.06
10	10	105	110.04
11	11	78	87.55
12	12	68	60.73
13	15	52	53.22
LM	df	Prob	
4.79	11	0.94	

17.3 Other Approaches to Evaluating the Fit of IRT Models

17.3.1 Likelihood-Ratio Tests

The tests presented above were of a specific model against a more general alternative. The null hypotheses of these specific models holding can be tested using two alternative, asymptotically equivalent tests: The LR and Wald test (Buse, 1982). As already noted, the LM test was the test of choice above because it only needs parameter estimates under the specific model and therefore can be used to test the assumptions on many items using the results from one parameter estimation run. However, the LR and Wald test are also useful for IRT models, and some of their applications are reviewed below.

The well-known principle of a LR test can be summarized as follows: Let $L_0(\phi_0)$ be the likelihood function of a model that is a special case of some more general model with likelihood function $L_1(\phi_1)$. It is assumed that $L_0(\phi_0)$ and $L_1(\phi_1)$ are functions of s_0 and s_1 parameters, respectively, with $s_0 < s_1$. The validity of the special, more restricted model can be tested against the more general alternative using the statistic $LR = -2\log(L_0(\hat{\phi}_0/L_1(\hat{\phi}_1))$, with the likelihood functions of both models evaluated using ML estimation. Under certain mild regularity assumptions, the LR statistic has an asymptotic Chi-square distribution with $s_1 - s_0$ degrees of freedom (see, for instance, Lehmann, 1986).

A straightforward application in an MML framework is evaluation of the fit of the 1PLM against the 2PLM in the form of a test of the hypothesis $a_i = 1\ (i = 1, \ldots, I)$. Two versions of the test for the 1PLM in a CML framework will be discussed: Andersen's and Martin-Löf's LR test.

The LR test by Andersen (1973) exploits the feature of CML that, under very mild regularity assumptions (see Pfanzagl, 1994), the item parameters can be estimated consistently from any (not necessarily random) sample from a population for which the 1PLM holds. Therefore, a test can be based on evaluation of the differences between the CML estimates of the item parameters in different subgroups formed, for instance, on the basis of score levels or on some external criterion. If the model does fit the items, the parameter estimates are equal up to random fluctuation, whereas they will generally not be equal for the estimates of items for which the model does no fit. In the latter case, the pattern of the parameter estimates may reveal the cause of misfit. For example, if they are too low for a low-scoring group, the group may have produced more correct responses than expected due to guessing on the item.

The test statistic is defined as follows. Assume that $L_c(\mathbf{b};\mathbf{U})$ is the conditional likelihood function evaluated using the CML estimates of the item parameters obtained from all data, say $\mathbf{U}$. Further, let $\mathbf{U}_g$ be the data of subgroup g and $L_c(\mathbf{b}_g;\mathbf{U}_g)$ the conditional likelihood function evaluated at the CML estimates of the item parameters obtained from g. If the item parameters cannot be estimated from every subgroup due to lack of data, CML estimates may be obtained by restricting the parameters to be equal for some subgroups in the test. Using these definitions, the LR statistic defined by

$$\mathrm{LR} = 2\left(\sum_{g=1}^{G} \log L_c(\mathbf{b}_g;\mathbf{U}_g) - \log L_c(\mathbf{b};\mathbf{U})\right) \tag{17.15}$$

has an asymptotic Chi-square-distribution with degrees of freedom equal to the number of parameters estimated from the subgroups minus the number of parameters estimated from the total data set. Comparing Andersen's LR test with the alternative LM test, it may come as no surprise that both tests have power against the same model violations (Glas, 1988). For instance, if the subgroups are score regions, all three tests are sensitive to differences in discrimination between the items, and their results are usually not much different.

The second test is the one proposed by Martin-Löf (1973). Since its statistic is devised to test whether two sets of items load on the same latent variable θ, it can be viewed as a test of the dimensionality assumption. Let the items be partitioned into two subsets of I_1 and I_2 items. Let $\mathbf{r} = (r_1, r_2)'$, $r_1 = 0, \ldots, I_1$, and $r_2 = 0, \ldots, I_2$, be score patterns on the two subtests and let $n_{\mathbf{r}}$ be the number of persons who had score pattern $\mathbf{r}$. It is assumed that the items are scored dichotomously, although the assumption is not essential. As above, let r denote the test takers' total sum score (so $r = r_1 + r_2$) and n_r the number of persons attaining this score. The statistic is defined as

$$\mathrm{LR} = 2\left(\sum_{\mathbf{r}} n_{\mathbf{r}} \log(n_{\mathbf{r}}/N) - \sum_{r} n_r \log(n_r/N) - \log L_c + \log L_c^{(1)} + \log L_c^{(2)}\right), \tag{17.16}$$

where L_c, $L_c^{(1)}$, and $L_c^{(2)}$ are the likelihood functions evaluated at the CML estimates obtained from the complete test, the first subtest and the second subtest, respectively. If the items are on a unidimensional scale, Equation 17.16 has an asymptotic Chi-square-distribution with $I_1 I_2 - 1$ degrees of freedom. Superficially, the last three terms of Equation 17.16 alone may seem to represent a proper LR statistic. However, adding the first two sums is necessary because L_c is defined conditionally on the frequency distribution of sum scores on the entire test of I items, while $L_c^{(1)}$ and $L_c^{(2)}$ are defined conditionally on the frequency distributions of the subtests only.

Both previous tests are only two possible examples of application of the LR principle for the case of the 1PLM. In the framework of CML estimation, LR tests can also used to evaluate the validity of a 1PLM with linear restrictions on the item parameters, the so-called linear logistic test model (LLTM) (Fischer, 1983; Volume One, Chapter 13), against the unrestricted 1PLM. Also, nested LLTMs can be tested against each other. Kelderman (1984) took the approach even further, by added extra parameters to the 1PLM to construct a class of log-linear 1PLMs. The validity of models in this class can also be evaluated using LR statistics with CML estimation.

17.3.2 Wald Tests

Wald tests have typically been used for the evaluation of DIF. In this section, only the case of two subgroups will be considered; the generalization to more subgroups is straightforward. Let the model parameters for the gth subgroup be denoted as $\boldsymbol{\phi}_g = (\phi_{g1}, \ldots, \phi_{gm})'$, $g = 1, 2$, and m be the number of free parameters. For simplicity, we consider the 1PLM; generalization to the 2PLM is straightforward. Assume that the parameter of the last item is fixed to zero to identify the model and that $\phi_{g1}, \ldots, \phi_{g,I-1}$ are the parameters of items 1 to $I-1$, respectively. Then, $m = I - 1$ for CML estimation, but $m = I + 1$ for MML estimation with a normal ability distribution.

Estimating the parameters in the two subgroups separately amounts to estimating $2m$ parameters. In a CML framework, the hypothesis of two different sets of item parameters for the two subgroups can be translated into a design with the two groups taking two different sets of items. However, the two sets of items are not linked and only $2(I-1)$ parameters are free to vary. In an MML framework, the same design is considered and the two means and variances are treated as free parameters. Finally, let $\boldsymbol{\phi}' = (\boldsymbol{\phi}_1', \boldsymbol{\phi}_2')$. The null hypothesis is the one of functions $h_j(\boldsymbol{\phi})$, $j = 1, \ldots, q$, equal to zero. If only equality of the item parameters in the two groups is tested, the $q = I - 1$ restrictions are

$$h_j(\boldsymbol{\phi}) = \phi_{1j} - \phi_{2j} = 0, \quad j = 1, \ldots, q. \tag{17.17}$$

Let $\boldsymbol{h}(\boldsymbol{\phi})' = (h_1(\boldsymbol{\phi}), \ldots, h_q(\boldsymbol{\phi}))$, and $\boldsymbol{\Sigma}_g$, $g = 1, 2$ denote the variance–covariance matrix of the ML estimator of $\boldsymbol{\phi}_g$. Since the responses of the two subgroups are independent, it follows for the variance–covariance matrix of the CML of MML estimator of $\boldsymbol{\phi}$ that

$$\boldsymbol{\Sigma} = \begin{pmatrix} \boldsymbol{\Sigma}_1 & 0 \\ 0 & \boldsymbol{\Sigma}_2 \end{pmatrix}. \tag{17.18}$$

The Wald test statistic is

$$W = \boldsymbol{h}'(\hat{\boldsymbol{\phi}})[\boldsymbol{T}'(\hat{\boldsymbol{\phi}})\boldsymbol{\Sigma}\boldsymbol{T}(\hat{\boldsymbol{\phi}})]^{-1}\boldsymbol{h}(\hat{\boldsymbol{\phi}}), \tag{17.19}$$

where $T(\boldsymbol{\phi})$ is a $2m \times q$ matrix $[\, t_{gj}]$ defined as

$$t_{gj} = \frac{\partial h_j(\boldsymbol{\phi})}{\partial \phi_g}. \tag{17.20}$$

For the null hypothesis in Equation 17.17, it is easily verified that Equation 17.19 simplifies to

$$W = (\hat{\boldsymbol{\phi}}_1^* - \hat{\boldsymbol{\phi}}_2^*)'[\boldsymbol{\Sigma}_1^* + \boldsymbol{\Sigma}_2^*]^{-1}(\hat{\boldsymbol{\phi}}_1^* - \hat{\boldsymbol{\phi}}_2^*)', \tag{17.21}$$

where the asterisk signifies (pairs of) parameters not involved in any restriction being deleted from all vectors and matrices. Under mild regularity assumptions, W is asymptotically Chi-square distributed with q degrees of freedom (Wald, 1943). In general, the degrees of freedom are equal to the number of restrictions. In the particular case discussed above, the degrees of freedom are $I - 1$.

As pointed out by Glas and Verhelst (1995), identification of the models for the two groups using one item only has several problems. First, the item parameter may be poorly identified from the data, that is, hard to estimate, with a standard error that is very large. Second, the item itself may be biased, which can, for instance, lead to the conclusion that all items are biased, except the one used to identify the latent scale. The first problem is solved when adding the restriction of the sum of the item parameters equal to zero. Further, the authors suggested to use an additional restriction $h(\cdot)$ independent of the chosen identification restriction, for instance, one defined on the differences between a particular item parameter estimate and the estimates of the other item parameters, such as

$$h_{1i} = \sum_{j \neq i} (\hat{b}_{1i} - \hat{b}_{1j}) - \sum_{j \neq i} (\hat{b}_{2i} - \hat{b}_{2j}) = 0, \tag{17.22}$$

$i = 1, \ldots, I$ where the first subscript of the b refers to the sample and subscript j takes the values 1 to I, except i. However, both sums may cancel. To avoid this, their terms can be squared, yielding

$$h_{2i} = \sum_{j \neq i} (\hat{b}_{1i} - \hat{b}_{1j})^2 - \sum_{j \neq i} (\hat{b}_{2i} - \hat{b}_{2j})^2 = 0, \tag{17.23}$$

$i = 1, \ldots, I$, as an alternative set of restrictions If the scale is chosen such that the sum of the parameters is zero, Equations 17.22 and 17.23 reduce to

$$h_{1i} = 2(\hat{b}_{1i} - \hat{b}_{2i}) = 0 \tag{17.24}$$

and

$$h_{2i} = \left(I\,\hat{b}_{1i}^2 + \sum_j \hat{b}_{1j}^2 \right) - \left(I\,\hat{b}_{2i}^2 + \sum_j \hat{b}_{2j}^2 \right) = 0, \tag{17.25}$$

$i = 1, \ldots, I$, respectively. As is clear from the latter, the test will be sensitive to differences in the variance of the item parameters as well as the square of a particular item parameter estimate.

We first substitute Equation 17.24 into the general expression for the Wald statistic. If $\boldsymbol{\Sigma}$ denotes the variance–covariance matrix for a solution normalized to have zero sums in both groups (17.19) reduces to

$$W_{1i} = \frac{(\hat{b}_{1i} - \hat{b}_{2i})^2}{\sigma_{1i}^2 + \sigma_{2i}^2}, \tag{17.26}$$

where σ_{1i} and σ_{2i} denote the ith diagonal element of $\boldsymbol{\Sigma}_1$ and $\boldsymbol{\Sigma}_2$, respectively. The result is a statistical test of DIF for item I_i, with W_{1i} asymptotically Chi-square distributed with one degree of freedom.

Substituting Equation 17.25 in the general expression for the Wald statistic gives

$$W_{2i} = \frac{h_{2i}^2}{\mathbf{t}'_1 \boldsymbol{\Sigma}_1 \mathbf{t}_1 + \mathbf{t}'_2 \boldsymbol{\Sigma}_2 \mathbf{t}_2}, \tag{17.27}$$

where $\mathbf{t}_g$, $g = 1, 2$, is a I-dimensional vector defined as

$$t_{gj} = 2(1 + \delta_{ij} I)\beta_{gj} \tag{17.28}$$

and δ_{ij} is defined as Kronecker delta.

17.3.3 UMP Tests for the 1PLM

Ponocny (2001) proposed a test for the 1PLM against an alternative model representing the case of violation of the assumption of local independence between the responses on a pair of items i and j. More specifically, the alternative model is the 1PLM version of the model in Equation 17.4 with the probability of a correct response on item i given a response on item j equal to

$$\Pr(U_{pi} = 1 | U_{pj} = u_{pj}; \theta_p, b_i, \delta_{ij}) = \frac{\exp(\theta_p - b_i + u_{pj}\delta_{ij})}{1 + \exp(\theta_p - b_i + u_{pj}\delta_{ij})}.$$

Since the 1PLM is an exponential family model, a UMP test of the hypothesis $\delta_{ij} = 0$ can be constructed (Lehman, 1986, Chapter 3). The test is based on the sufficient statistic for δ_{ij}, which is the number of persons responding correctly to both items, that is, $T_{ij} = \sum_p U_{pi} U_{pj}$. A realization of T_{ij} will be labeled as $t_{ij}^{(obs)}$. The distribution of T_{ij} given the realizations of the minimal sufficient statistics for θ_p $(p = 1, \ldots, N)$ and b_i $(i = 1, \ldots, I)$, which as the item and subject marginals, is considered. As the distribution of T_{ij} does not depend on any parameters, the test is nonparametric.

Ponocny (2001) presented a Monte Carlo algorithm for generating all matrices with the same marginals. For every generated matrix, a value of the sufficient statistic T_{ij}, say $t_{ij}^{(rep)}$, is computed. The significance probability of the test is the percentile of the observed value $t_{ij}^{(obs)}$ in the distribution of $t_{ij}^{(rep)}$ generated under the null-model of the 1PLM. Finally, a global test statistic of item fit, is created by determining the percentage of significant T_{ij} tests for a number of item pairs (i, j).

17.3.4 Limited-Information Test Statistics

Item response models can be viewed as multinomial models with the set of all possible response patterns as observed categories (Cressie and Holland, 1983), that is, a multinomial model with C mutually exclusive outcomes occurring with probabilities $\pi_1(\boldsymbol{\phi}), \ldots, \pi_C(\boldsymbol{\phi})$, with $\boldsymbol{\phi}$ denoting a q-vector of model parameters. The sample space is the set of all possible response patterns $\boldsymbol{u}$. If a test consists of I dichotomously scored items, the number of possible response patterns C equals 2^I. Let N observations be made, where the C possible outcomes occur with sample proportions $p_1, \ldots, p_C$, and let $\hat{\pi}_c$ denote $\pi_c(\boldsymbol{\phi})$ evaluated at a best asymptotically normal (BAN) estimate, for instance a maximum-likelihood estimate or a minimum Chi-square estimate of $\boldsymbol{\phi}$. It is a well-known result of asymptotic theory that under very mild regularity conditions and for $N \to \infty$,

$$X^2 = N \sum_{c=1}^{C} \frac{(p_c - \hat{\pi}_c)^2}{\hat{\pi}_c} \tag{17.29}$$

is asymptotically Chi-square distributed with $C - q - 1$ degrees of freedom (Birch, 1964). The regularity conditions are assumed, and in applications typically will be satisfied; for a discussion of them, refer to Bishop et al. (1975, pp. 509–511).

Notice that in the case of a test of an IRT model, the sum runs over all $C = 2^I$ possible response patterns $\boldsymbol{u}$. Even for the tests of moderate length, the number of possible response patterns is prohibitively large, which leads to two drawbacks. Firstly, the expected frequencies will be very small and thus the claims concerning the asymptotic distribution of the statistic are jeopardized. Secondly, the number of frequencies on which the statistic is based, is too large to be informative with respect to model violations. Therefore, as an alternative, Glas and Verhelst (1989) introduced a class of asymptotically Chi-square distributed test statistics that evade these problems. Essentially, these statistics were special cases of the LM statistics presented above.

The same problem was also tackled from a different perspective. Maydue-Olivares and Joe (2005, 2006) and also see, Maydue-Olivares and Montag (2013) introduced a family of goodness-of-fit statistics for testing hypotheses in multidimensional contingency tables that can also be applied to IRT models. These statistics are quadratic forms in marginal residuals up to order r. They are asymptotically Chi-square distributed provided the model parameters are estimated using a consistent, asymptotically normal estimator. To test the null hypothesis $\pi = \pi(\phi)$ against the alternative $\pi \neq \pi(\phi)$ Maydeu-Olivares and Joe (2005) also proposed a test based on a quadratic form in the residual moments of the multivariate Bernoulli distribution (Teugels, 1990) up to the smallest order at which the model is identified. For $r = 1, \ldots, I$, the test statistic is defined as

$$M_r = (\mathbf{p}_r - \boldsymbol{\pi}_r(\boldsymbol{\phi}))' C_r (\mathbf{p}_r - \boldsymbol{\pi}_r(\boldsymbol{\phi})), \tag{17.30}$$

where $\mathbf{p}_r$ are the observed marginal proportions and $\boldsymbol{\pi}_r(\boldsymbol{\phi}))$ the theoretical moments up to order r evaluated under the model. These moments are marginal probabilities obtained by a linear transformation of the cell probabilities. Further, $\boldsymbol{C}_r$ can be viewed as the inverted covariance matrix conditional on and evaluated at an estimate of $\boldsymbol{\phi}$. In M_1 only univariate information is used, in M_2 univariate and bivariate information is used, and so forth up to M_I, a full information statistic equal to a Pearson Chi-square when ML estimation is used. The precision of the asymptotic approximation decreases with the magnitude of r. Maydeu-Olivares and Joe (2006) argued that most IRT models are identified from univariate and bivariate information, so M_2 is the preferred statistic for testing IRT models. Using ML estimation (or any other asymptotically normal estimator), M_r is asymptotically Chi-square-distributed with $s - q$ degrees of freedom. From simulation studies of a test of the 1PLM against the 2PLM, 3PLM, and multidimensional 1PL models, Maydue-Olivares and Montag (2013) concluded that, when used as a global test, the performance of M_2 is comparable to the performance of the LM test.

References

Aitchison, J. and Silvey, S. D. 1958. Maximum likelihood estimation of parameters subject to restraints. *Annals of Mathematical Statistics, 29,* 813–828.

Andersen, E. B. 1973. A goodness of for test for the Rasch model. *Psychometrika, 38,* 123–140.

Birch, M. W. 1964. A new proof of the Pearson-Fisher theorem. *Annals of Mathematical Statistics, 35,* 817–814.

Bishop, Y. M. M., Fienberg, S. E., and Holland, P. W. 1975. *Discrete Multivariate Analysis: Theory and Practice.* Cambridge, MA: MIT press.

Birnbaum, A. 1968. Some latent trait models. In F. M. Lord and M. R. Novick (Eds.), *Statistical Theories of Mental Test Scores.* Reading, MA: Addison-Wesley, pp. 395–479.

Buse, A. 1982. The likelihood ratio, Wald, and Lagrange multiplier tests: An expository note. *The American Statistician, 36,* 153–157.

Camilli, G. and Shepard, L. A. 1994. *Methods for Identifying Biased Test Items.* Thousand Oaks, CA: Sage.

Cressie, N. and Holland, P. W. 1983. Characterizing the manifest probabilities of latent trait models. *Psychometrika, 48,* 129–141.

Efron, B. 1977. Discussion on maximum likelihood from incomplete data via the EM algorithm (by A. Dempster, N. Liard, and D. Rubin). *Journal of the Royal Statistical Society, Series B, 39,* 29.

Fischer, G. H. 1983. Logistic latent trait models with linear constraints. *Psychometrika, 48,* 3–26.

Glas, C. A. W. 2006. Testing generalized Rasch models. In M. von Davier and C. H. Carstensen (Eds.), *Multivariate and Mixture Distribution Rasch Models: Extensions and Applications.* New York: Springer, pp. 37–56.

Glas, C. A. W. 1988. The derivation of some tests for the Rasch model from the multinomial distribution. *Psychometrika, 53,* 525–546.

Glas, C. A. W. 1999. Modification indices for the 2-pl and the nominal response model. *Psychometrika, 64,* 273–294.

Glas, C. A. W. (2010a). *MIRT: Multidimensional Item Response Theory.* (computer software). Enschede, the Netherlands: University of Twente. (http://www.utwente.nl/gw/omd/en/employees/employees/glas.doc/).

Glas, C. A. W. (2010b). Testing fit to IRT models for polytomously scored items. In M. L. Nering and R. Ostini (Eds.), *Handbook of Polytomous Item Response Models.* New York: Taylor and Francis Group, pp. 185–208.

Glas, C. A. W. and Suárez-Falćon, J. C. 2003. A comparison of item-fit statistics for the three-parameter logistic model. *Applied Psychological Measurement, 27,* 87–106.

Glas, C. A. W. and Verhelst, N. D. 1989. Extensions of the partial credit model. *Psychometrika, 54,* 635–659.

Glas, C. A. W. and Verhelst, N. D. 1995. Testing the Rasch model. In G. H. Fischer and I. W. Molenaar (Eds.), *Rasch Models: Foundations, Recent Developments and Applications.* New York: Springer, pp. 69–96.

Holland, P. W. and Wainer, H. (Eds.). 1993. *Differential Item Functioning.* Hillsdale, N.J., Erlbaum.

Jannarone, R. J. 1986. Conjunctive item response theory kernels. *Psychometrika, 51,* 357–373.

Kelderman, H. 1984. Loglinear RM tests. *Psychometrika, 49,* 223–245.

Kelderman, H. 1989. Item bias detection using loglinear IRT. *Psychometrika, 54,* 681–697.

Lehmann, E. L. 1986. *Testing Statistical Hypotheses* (2nd ed.). New York: Springer.

Lord, F. M. and Wingersky, M. S. 1984. Comparison of IRT true-score and equipercentile observed-score "equatings." *Applied Psychological Measurement, 8,* 453–461.

Louis, T. A. 1982. Finding the observed information matrix when using the EM algorithm. *Journal of the Royal Statistical Society, Series B, 44,* 226–233.

Martin-Löf, P. 1973. *Statistika Modeller. Anteckningar från seminarier Lasåret 1969–1970, utardeltade av Rolf Sunberg. Obetydligt ändrat nytryck, oktober 1973.* [Statistical Models. Lecture notes from the 1969–1970 seminar by Rolf Sunberg. Revised version, October, 1973]. Stockholm: Institutet för Försäkringsmatematik och Matematisk Statistik vid Stockholms Universitet.

Masters, G. N. 1982. A Rasch model for partial credit scoring. *Psychometrika, 47,* 149–174.

Maydeu-Olivares, A. and Joe, H. 2005. Limited- and full-information estimation and goodness-of-fit testing in 2(n) contingency tables: A unified framework. *Journal of the American Statistical Association, 100,* 1009–1020.

Maydeu-Olivares, A. and Joe, H. 2006. Limited information goodness-of-fit testing in multidimensional contingency tables. *Psychometrika, 71,* 713–732.

Maydeu-Olivares, A. and Montag, R. 2013. *Psychometrika, 78,* 116–133.

Mellenbergh, G. J. 1982. Contingency table models for assessing item bias. *Journal of Educational Statistics, 7,* 105–118.
Mellenbergh, G.J. 1995. Conceptual notes on models for discrete polytomous item responses. *Applied Psychological Measurement, 19,* 91–100.
Orlando, M. and Thissen, D. 2000. Likelihood-based item-fit indices for dichotomous item response theory models. *Applied Psychological Measurement, 24,* 50–64.
Pfanzagl, J. 1994. On item parameter estimation in certain latent trait models. In G. H. Fischer and D. Laming (Eds.), *Contributions to Mathematical Psychology, Psychometrics, and Methodology.* New York: Springer.
Ponocny, I. 2001. Nonparametric goodness-of-fit tests for the Rasch model. *Psychometrika, 66,* 437–460.
Rao, C. R. 1973. *Linear Statistical Inference and Its Applications.* New York: Wiley.
Samejima, F. 1969. Estimation of latent ability using a pattern of graded scores. *Psychometrika, Monograph Supplement, No. 17.*
Rasch, G. 1960. *Probabilistic Models for Some Intelligence and Attainment Tests.* Copenhagen: Danish Institute for Educational Research.
Suárez-Falćon, J. C. and Glas, C. A. W. 2003. Evaluation of global testing procedures for item fit to the Rasch model. *British Journal of Mathematical and Statistical Psychology, 56,* 127–143.
Teugels, J. L. 1990. Some representations of the multivariate Bernoulli and binomial distributions. *Journal of Multivariate Analysis, 32,* 256–268.
Thissen, D. 1991. *MULTILOG. Multiple, Categorical Item Analysis and Test Scoring Using Item Response Theory.* Chicago, IL: Scientific Software International, Inc.
Tutz, G. 1990. Sequential item response models with an ordered response. *British Journal of Mathematical and Statistical Psychology, 43,* 39–55.
van den Oord, E. J. C. G. 2005. Estimating Johnson curve population distributions in MULTILOG. *Applied Psychological Measurement, 29,* 45–64.
van den Wollenberg, A. L. 1982. Two new tests for the Rasch model. *Psychometrika, 47,* 123–140.
Verhelst, N. D. and Glas, C. A. W. 1995. The generalized one parameter model: OPLM. In G. H. Fischer and I. W. Molenaar (Eds.), *Rasch Models: Their Foundations, Recent Developments and Applications.* New York: Springer, pp. 213–237.
Verhelst, N. D., Glas, C. A. W., and de Vries, H. H. 1997. A steps model to analyze partial credit. In W. J. van der Linden and R.K.Hambleton (Eds.), *Handbook of Modern Item Response Theory.* New York: Springer, pp. 123–138.
Yen, W. M. 1984. Effects of local item dependence on the fit and equating performance of the three-parameter logistic model. *Applied Psychological Measurement, 8,* 125–145.
Yen, W.M. 1993. Scaling performance assessments: Strategies for managing local item dependence. *Journal of Educational Measurement, 30,* 187–213.

18

Information Criteria

Allan S. Cohen and Sun-Joo Cho

CONTENTS

18.1 Introduction

An item response theory (IRT) model needs to fit the data if the benefits of IRT and the model are to be realized. Selecting an appropriate model given the data is based, at least in part, on model-data fit. In general, a saturated model will fit the data at least as well if not better than a less saturated model. However, a model that is more complex than appropriate violates the principle of parsimony. That is, the simplest model should be selected that still provides a useful explanation of the data. In general, a more parsimonious model will have less chance of introducing inconsistencies, ambiguities, and redundancies into the explanation. The objective in model selection is to select a model that provides the best fit to the data but also has the capability for predicting future or different data.

In this chapter, we focus on criteria to decide on which of several candidate response models provide(s) the best fit to the data. The approach discussed here is based on the use of information indices derived from information theory, which essentially provide an estimate of the difference in information provided between a true and a candidate model (Burnham and Anderson, 2002). The objective is to choose the model (or models) that loses the least information and, therefore, leads to the best interpretation of the data.

In the following section, we first present a review of commonly used information criteria in the context of IRT. These criteria are separated into one of the two estimation methods typically employed in IRT, maximum-likelihood estimation (MLE), and Bayesian estimation. For each estimation method, we describe discrepancy indices and their estimators for the model-selection criteria being reviewed. In the next section, the use of these criteria is surveyed. Finally, the uses of the criteria in item response modeling are discussed.

18.2 Overview of Information Criteria

Standard significance testing is the most common approach for selecting the best fitting model. Unfortunately, most significance tests are sensitive to sample size such that parsimonious models may be rejected as nonfitting for large samples. In addition, standard significance testing for model fit, such as a likelihood-ratio test, only can be used for nested models. These limitations are not an issue when using information criteria to select the best fitting model.

Information criteria can be categorized into those based on the Kullback–Leibler distance (KLD) (Kullback and Leibler, 1951) and dimension-consistent criteria (Burnham and Anderson, 2002), where dimension means the same as the order of the model. Information criteria such as Akaike's information criterion (AIC; Akaike, 1973) and its extensions (Akaike, 1974, 1977) and Takeuchi's information criterion (TIC; Takeuchi, 1976) are based on the KLD. Although the AIC and its extensions are most often estimated with MLE methods, these indices have also been reformulated for use in a Bayesian estimation context (Akaike, 1977, 1978).

Dimension-consistent criteria have been developed to provide a consistent estimator of the order or dimension of the true (or full) model. For these indices, the probability of selecting the true model approaches 1 as sample size increases. The best known of these dimension-consistent criteria is the Bayesian information criterion (BIC; Schwartz, 1978).

For a more in-depth treatment of the derivation and use of information criteria for model selection, the reader is referred to Burnham and Anderson (2002), Claeskens and Hjort (2008), and Kuha (2004).

18.2.1 Frequentist Model Selection

Kullback–Leibler Information

The KLD (Kullback–Leibler, 1951; for a review, see Chapter 7) provides the basis for an information-theoretic approach to model selection. The KLD between two models provides a measure of the information lost, when a candidate model P (sometimes called the approximating model) is selected rather than the true model P^*. The KLD is generally interpreted as the distance from the candidate model P to the true model P^*, but this is a directional interpretation as the distance from the true model to the candidate model is not the same. The purpose of model selection based on the KLD, $I(P^*, P)$, then, is to find a candidate model that minimizes the loss of information relative to other candidate models that are being considered.

In the context of IRT, KLD is defined as

$$I(P^*, P) = \sum_{a=1}^{A} P^*\{U = a\} \ln\left(\frac{P^*\{U = a\}}{P\{U = a; \theta\}}\right), \tag{18.1}$$

where a is a response alternative ($a = 1, \ldots, A$), $P^*\{U = a\}$ is the true probability of observing $U = a$, and $P\{U = a; \theta\}$ is the probability of observing $U = a$ based on a candidate model having parameters θ.

Both P^* and parameter θ should be known in order to calculate $I(P^*, P)$. This requirement is not needed, however, when *relative* KLD is used. To derive a measure of *relative*

KLD, Equation 18.1 can be rewritten as

$$I(P^*, P) = \sum_{a=1}^{A} P^*\{U = a\} \ln(P^*\{U = a\}) - \sum_{a=1}^{A} P^*\{U = a\} \ln(P\{U = a; \theta\})$$
$$= E_{P^*}[\ln(P^*\{U = a\})] - E_{P^*}[\ln(P\{U = a; \theta\})]. \quad (18.2)$$

The first term, $E_{P^*}[\ln(P^*\{U = a\})]$ is unknown because it depends on an unknown true distribution. However, since it is the same for all candidate models, it can be cancelled out, leaving only the differences of $I(P^*, P)$ between the models. But then $I(P^*, P)$ can be estimated up to a constant C, where the second term, $E_{P^*}[\ln(P\{U = a; \theta\})]$, is computed as follows:

$$I(P^*, P) - C = -E_{P^*}[\ln(P\{U = a; \theta\})]. \quad (18.3)$$

$I(P^*, P) - C$ is called the *relative* KLD between P^* and P, and $\ln(P\{U = a; \theta\})$ is the quantity of interest when selecting the best model.

Three information criteria, commonly used as estimates of the *relative* KLD, are reviewed below: AIC (Akaike 1973), TIC (Takeuch, 1976), and the second-order information criterion (AIC_c; Sugiura, 1978).

Akaike's Information Criterion (AIC)

Akaike (1973) described a general information theoretic approach based on the relationship between the KLD and Fisher's maximum loglikelihood function ($L = P\{\theta; U\}$). First, denote θ_0 as the best estimate of θ for a candidate model P. Next, assume that the KLD is minimized at θ_0. Akaike showed that the MLE of θ can be used as a substitute for θ_0.

When two candidate models, P_h and $P_{h'}$, are compared, the following KLD difference can be used with MLEs $\widehat{\theta}_h^U$ and $\widehat{\theta}_{h'}^U$ estimated from the data U for P_h and $P_{h'}$, respectively:

$$I(P^*, P_{h'}) - P(P^*, P_h) = E_{P^*}[\ln(P^*\{U = a; \widehat{\theta}_{h'}^U\}) - \ln(P^*\{U = a; \widehat{\theta}_h^U\})]. \quad (18.4)$$

When P_h is nested within $P_{h'}$, $\ln(P^*\{U = a; \widehat{\theta}_{h'}^U\})$ can never be smaller than $\ln(P^*\{U = a; \widehat{\theta}_h^U\})$ (Kuha, 2004). This is because the same data U are used both for estimating θ and for judging the fit of the resulting model. Thus, the difference would never indicate a preference for the smaller model P_h.

To avoid this problem, assume that an MLE is obtained from a separate, independent sample of hypothetical data, V, from the same true model P^*. Denote this MLE as $\widehat{\theta}^V$. The expected KLD difference from $\widehat{\theta}^V$ as a target quantity T_A can then be described as

$$T_A = -2E_V[I(P^*, P_{h'}) - I(P^*, P_h)]$$
$$= -2E_V E_U[\ln(P^*\{U = a; \widehat{\theta}_{h'}^V\}) - \ln(P^*\{U = a; \widehat{\theta}_h^V\})]. \quad (18.5)$$

For a given candidate model P, a large-sample, approximately unbiased estimator of T_A for the "good" model (i.e., close to P^*) is

$$T_A = -2E_V[I(P^*, P)] \approx -2\ln(L) + \left(1 + \frac{N}{N_0}\right) K, \quad (18.6)$$

where K is a bias-adjustment term calculated as the number of parameters estimated in the candidate model P, N is the sample size of the data U, and N_0 is the sample size of the hypothetical data V. The loglikelihood $\ln(L) = \ln(P\{\theta; U\})$ can be interpreted as a measure of how well model P with θ^V predicts new data U. Given this interpretation, AIC is closely related to the idea of cross-validation (Kuha, 2004). With K as described, T_A is called a bias-corrected loglikelihood.

AIC is an approximation to T_A as follows:

$$\text{AIC} = -2\ln(L) + \left(1 + \frac{N}{N_0}\right) K = -2\ln(L) + 2K. \tag{18.7}$$

It is often assumed that N is equal to N_0 so that the quantity $1 + \frac{N}{N_0}$ is 2. To summarize the derivation of AIC, AIC is an *estimate* of the expected relative distance between the candidate model and the true model that generated the observed data. AIC is not a measure of the directed KLD.

AIC is a good estimate of T_A for large samples and the "good" model close to P^*. If the candidate models are poor (i.e., far from P^*), then TIC (described below) is an attractive alternative, if the sample size is very large.

Takeuchi's Information Criterion (TIC)

The TIC (Takeuchi, 1976) is a more general and asymptotically unbiased estimator of expected KLD. Unlike AIC, TIC can be used in cases where the candidate models are not close to the true model P^*. TIC is defined as

$$\text{TIC} = -2\ln(L) + 2 \cdot tr(Q(\theta)S(\theta)^{-1}), \tag{18.8}$$

where $tr(Q(\theta)S(\theta)^{-1})$ is a bias-adjustment term, with the $K \times K$ matrices $Q(\theta)$ and $S(\theta)$ involving first and second mixed partial derivatives of the loglikelihood function and where tr denotes the trace of a matrix.

One drawback to TIC is that it requires relatively large samples in order to estimate elements of the two $K \times K$ matrices for adjusting bias. Viewed in this way, AIC can be seen to be a special case of TIC, where $tr(Q(\theta)S(\theta)^{-1}) \approx K$.

The Second-Order Information Criterion

AIC can perform poorly when the ratio of estimated number of parameters, K, to sample size is large (Sugiura, 1978). Sugiura (1978) derived a second-order variant of AIC called AIC_c under Gaussian assumptions for linear models. AIC_c is defined as

$$\text{AIC}_c = \text{AIC} + \frac{2K(K+1)}{N-K-1}, \tag{18.9}$$

where $\frac{2K(K+1)}{N-K-1}$ is a bias adjustment term for sample size N. Burnham and Anderson (2002) recommend using AIC_c when the ratio N/K is small (say, less than 40).

To summarize, calculation of AIC, TIC, and AIC_c differs with respect to the bias-adjustment term, but these information criteria are all estimates of the relative KLD.

AIC, TIC, and AIC_c *Differences, Akaike Weights, and Evidence Ratios*

A useful practice is to compute AIC, TIC, and AIC_c over all candidate models (Burnham and Anderson, 2002). AIC, TIC, and AIC_c are relative indicators and are all dependent on sample size. Taking AIC as an example, the difference can be presented as follows:

$$\Delta_h = \text{AIC}_h - \text{AIC}_{\text{min}}, \tag{18.10}$$

where AIC_h is AIC for the hth candidate model and AIC_{min} is the minimum AIC among candidate models. The best model will be one that has $\Delta_h \equiv \Delta_{\text{min}} \equiv 0$. The ordering of the Δ_h from smallest to largest can be taken as an indication of how good each candidate model is as an approximation to expected KLD.

It also is possible to quantify the *plausibility* of each model being the best model using the model likelihood, $L(P_h; U)$.* $L(P_h; U)$ can be calculated as follows (Akaike, 1983):

$$L(P_h; U) \propto \exp(-0.5\Delta_h). \tag{18.11}$$

Akaike weights (Akaike, 1978), w_h, provide a basis for interpretation of the relative likelihood of a model. The weights are obtained by normalizing $L(P_h; U)$ given the set of R models:

$$w_h = \frac{\exp(-0.5\Delta_h)}{\sum_{r=1}^{R} \exp(-0.5\Delta_r)}. \tag{18.12}$$

The ratio of Akaike weights ($w_h/w_{h'}$) for each pair of models can be used as an indication of the relative fit of the models.

18.2.2 Bayesian Model Selection

For a general introduction to Bayesian approaches to model selection (see Chapter 19). Bayesian estimation of the parameters θ_h of model M_h is based on the posterior probability

$$P\{M_h \mid U\} = \frac{P\{U; M_h\} P\{M_h\}}{P\{U\}}, \tag{18.13}$$

where $P\{M_h\}$ is the prior probability of the model, $P\{U\}$ is the marginal probability of the data, and $P\{U; M_h\}$ the marginal probability of the data given the model calculated as

$$P\{U; M_h\} = \int P\{U; \theta_h, M_h\} P\{\theta_h; M_h\} d\theta_h. \tag{18.14}$$

The extent to which the data support M_h over a competing model $M_{h'}$ is measured by the posterior odds of M_h against $M_{h'}$:

$$\frac{P\{M_h; U\}}{P\{M_{h'}; U\}} = \frac{P\{M_h\}}{P\{M_{h'}\}} \cdot \frac{P\{U; M_h\}}{P\{U; M_{h'}\}}, \tag{18.15}$$

where $\frac{P\{M_h\}}{P\{M_{h'}\}}$ are the prior odds and $\frac{P\{U;M_h\}}{P\{U;M_{h'}\}}$ is the Bayes factor (BF), a measure of the evidence provided by the data in favor of M_h over $M_{h'}$. If there is no reason to favor either

* The likelihood of a model given the data should be distinguished from the likelihood of the parameters given both the data and model.

candidate model, equal prior probabilities are assigned to the two models. In this case, the posterior odds reduces to BF. Thus, a target quantity is

$$T_B = -2(\ln E[P\{U; M_h, \theta_h\}] - \ln E[P\{U; M_{h'}, \theta_{h'}\}]). \tag{18.16}$$

For given model M_h,

$$T_B = -2\ln E[P\{U; M_h, \theta_h\}] \approx -2\ln(L_h) + \ln(1 + \frac{N}{N_0})K_h \tag{18.17}$$

$$= -2\ln(L_h) + \ln(N)K_h, \tag{18.18}$$

where L_h is the likelihood of M_h, N_0 is the prior sample size, N is the sample size of the data U, and K_h is the number of parameters in M_h. The term, $\ln(1 + \frac{N}{N_0})$, is approximately $\ln(N)$, when N_0 is 1 and N is large.

BIC (Schwarz, 1978) is an approximation of a target quantity T_B. For a model M_h, BIC is often defined as

$$\text{BIC}_h = -2\ln(L_h) + \ln(N)K_h. \tag{18.19}$$

BIC_h implicitly assumes the prior $P\{\theta_h; M_h\}$ in Equation 18.14 to be

$$P\{\theta_h; M_h\} \sim N(\widehat{\theta}_h, \widehat{\Sigma}_h), \tag{18.20}$$

where $\widehat{\theta}_h$ is the MLE and $\widehat{\Sigma}_h$ is its estimated variance-covariance matrix. For this implicit prior, BIC_h is sometimes used as a non-Bayesian statistic. But when, in a fully Bayesian analysis, explicitly specified prior distributions are used, numerical methods such as Laplace approximations are sometimes more accurate than using BIC_h to calculate BF.

Deviance Information Criterion

Spiegelhalter et al. (2002) developed a Bayesian information measure of model complexity based on the concept of the excess of the true information relative to the estimated residual information. It is defined as

$$D\{\theta, \widehat{\theta}, U\} = -2P\{\theta; U\} + 2P\{\widehat{\theta}; U\}, \tag{18.21}$$

where $\widehat{\theta}$ is the estimated parameter vector. $D\{\theta, \widehat{\theta}, U\}$ can be thought of as the degree to which the candidate model overfits the data, as it provides an estimate of how much less the model with estimated parameters deviates from the data than the model with true parameters.

Spiegelhalter et al. (2002) proposed a Bayesian model complexity or effective number of parameters, K_D, using the posterior expectation of $D\{\theta, \widehat{\theta}, U\}$,

$$K_D \equiv \text{E}[D\{\theta, \widehat{\theta}, U\}; U] = \text{E}[-2P\{\theta; U\}; U] + 2P\{\widehat{\theta}; U\}. \tag{18.22}$$

The deviance information criterion (DIC) was suggested with K_D as

$$\text{DIC} \equiv \text{E}[-2P\{\theta; U\}; U] = -2P\{\widehat{\theta}; U\} + K_D. \tag{18.23}$$

From a Bayesian perspective, DIC is analogous to AIC.

18.2.3 Comparisons between AIC and BIC

AIC and BIC can be compared using several difference quantities.

Target Quantity

The expressions for the target quantities for AIC and BIC (T_A and T_B, respectively) are similar. The similarity is enhanced by the fact that the prior distribution assumed for θ in BIC (Equation 18.20) is the same as the sampling distribution of $\widehat{\theta}$ used to obtain AIC (Kuha, 2004).

Penalty Term

AIC and BIC have some similarity with respect to the penalty term, in spite of the different derivation of the two indices (Kuha, 2004). Quantity N_0 is used as both the sample size of the hypothetical data and the prior sample size in AIC and BIC, respectively. BIC can be interpreted as AIC with $N_0 = \frac{N}{\ln N - 1}$ and AIC can be interpreted as BIC with $N_0 = \frac{N}{e^{[2]}-1}$. Thus, AIC can be seen as corresponding to a BF when prior sample size N_0 is proportional to N.

Consistency and Efficiency

A model-selection method is consistent to the extent it leads to selection of the true model (Claeskens and Hjort, 2008). These authors differentiate weak from strong consistency. Weak consistency occurs when the selection method identifies the true model with a probability that approaches one in the limit. Strong consistency is present when the true model is identified with very high certainty.

According to the same authors, a model-selection method is efficient to the extent it leads to identification of a model that behaves like the theoretically best model, for example, when the selected model has almost the same mean squared error or expected prediction error as the theoretically best model. They also point out, however, that a model-selection method cannot be both consistent and efficient. It is for this reason that such information criterion methods as AIC and BIC sometimes yield different results; AIC is efficient and BIC is consistent.

Bozdogan (1987) proposed the consistent AIC (CAIC) criterion, in which the penalty function includes both the order of the model and sample size:

$$CAIC = -2\ln(L) + (\ln(N) + 1)K. \tag{18.24}$$

Claeskens and Hjort (2008) suggest that this index is essentially $BIC - K$.

Performance Comparisons Using Information Indices in Latent Variable Modeling

Both AIC and BIC enable a rank ordering of all candidate models for the purpose of selecting the best fitting model. The lowest AIC or BIC value is taken to indicate the best fitting model. BIC has been found to be somewhat more useful for model selection, in part because it is less sensitive to sample size than indices such as AIC (e.g., Nylund et al., 2007).

A comparison of model selection methods by Janssen and De Boeck (1999) suggested that AIC tends to select more saturated models in large samples. Kang and Cohen (2007) reported a simulation study with data generated for each of the 1-, 2-, and 3-parameter dichotomous IRT models. Their results indicated that selection for these models tended to be more accurate when the ability distribution matched test difficulty. AIC and BIC worked well, when data were generated for the 1- and 2-parameter models, but neither AIC, BIC, or DIC worked well with data for the 3-parameter model. In a subsequent simulation study, Kang et al. (2009) examined the consistency of information indices to detect the true model among polytomous IRT models. BIC was found to work better than several other information indices, including AIC. Li et al. (2009) used a similar simulation study approach with dichotomous mixture IRT models. In their case, BIC was also found to work best.

18.3 Use of Information Criteria in Item Response Modeling

Information criteria have been widely used to select the best among candidate IRT models, both with nested and non-nested models. In this section, we review some of these studies (see Table 18.1 for the full list), with the intent to illustrate some of the more common findings.

18.3.1 Nested Models

Even though information criteria have been used mainly for non-nested models, applications with for nested models do exist. Some of them include exploring the number of dimensions in multidimensional IRT models (MIRT) and selection of the most parsimonious model among constrained alternatives.

Exploring the Number of Dimensions

The number of dimensions in MIRT is a potentially important factor in determining the penalty term. Bolt and Johnson (2009) compared one-, two-, and three-dimensional versions of a multidimensional nominal response models (NRM) in a study of the effects of respondent's systematic tendencies to use different parts of the rating scale on a 68-item self-report measure of tobacco dependency. Their model included slope and intercepts parameter for each response category of an item on each dimension. Parameter estimates for each of the models were obtained using marginal MLE (MMLE), as implemented in the computer program Latent GOLD (Volume Three, Chapter 30; Vermunt and Magidson, 2004, 2006). The results were compared using three information criteria, AIC, BIC, and CAIC. The number of item parameters estimated was used penalty term, K. For identification, the sum of the intercepts and item slopes for each item and dimension were constrained to equal to zero. For the one-dimensional NRM, a total of 816 parameters were estimated (i.e., six slopes plus six intercepts for each of 68 items). The mean ability vector was constrained to zero. A three-dimensional NRM was chosen based on AIC, but a two-dimensional NRM was selected based on BIC and CAIC. For the two-dimensional model, 1224 parameters were estimated, that is, the 816 parameters from the one-dimensional model plus an additional 408 parameters (six categories for the second dimension for each of the 68 items).

Similarly, Janssen and De Boeck (1999) used MMLE to estimate item parameters with the computer program TESTMAP (McKinley, 1988, 1989, 1992) for one-, two-, and three-dimensional 2-parameter IRT models. Results based on AIC and CAIC were

TABLE 18.1
Applications of Information Indices for Model Selection

Papers	Estimation Methods	Models	Indices
		Nested Models	
Exploring the Number of Dimensions			
Bolt and Johnson (2009)	MMLE	1-, 2-, and 3-dimensional model	AIC, BIC, and CAIC
Janssen and De Boeck (1999)	MMLE	1-, 2-, and 3-dimensional model	AIC and CAIC
Rijmen and De Boeck (2005)	MMLE	1- and 2-dimensional model	BIC
Yao and Schwarz (2006)	MCMC	1-, 2-, and 3-dimensional model	AIC
A General Model with Constraints			
Abad et al. (2009)	MMLE	Multiple-choice model	AIC
Hickendorff et al. (2009)	MMLE	Explanatory IRT with covariates	BIC
Revuelta (2008)	MMLE	Logit linear IRT	AIC and BIC
May (2006)	MCMC	Multilevel IRT	BIC
Nested Models with Respect to Number of Item Parameters			
de Ayala (2009)	MMLE	1-,2-, and 3-PL	AIC and BIC
Revuelta (2008)	MCMC	Logit linear IRT	AIC and BIC
Modeling the Multilevel Data Structure in Multilevel IRT Models			
May (2006)	MCMC	Multilevel IRT	BIC
Wang and Jin (2009)	MMLE	MIRID and multilevel MIRID	AIC
		Non-Nested Models	
Different Families of Models			
Bolt and Lall (2003)	MCMC	MLTM, M2PL	Bayes factor
Patz et al. (2002)	MCMC	HRM,facet IRT	BIC
Goegebeur et al. (2008)	MMLE	Speeded IRT	AIC and BIC
Semmes et al. (2011)	REML	Models for speededness with fixed or random effects	AIC and BIC
De Boeck (2008)	Laplace	Rasch models with fixed or random effects	AIC and BIC
Klein Entink et al. (2009)	MCMC	Multivariate IRT for fixed or random effects	BF and DIC
Rijmen and De Boeck (2002)	MMLE	Random weight LLTM	BIC

compared to find the best fitting model, and again, the number of estimated item parameters was used for penalty K. Person ability parameters were fixed to identify the models. The two-dimensional model was best according to AIC while the one-dimensional model was best according to CAIC.

Alternative definitions of the penalty term have been used, though. Yao and Schwarz (2006), for example, fit one-, two-, and three-dimensional partial credit models using Markov chain Monte Carlo (MCMC) as implemented in the computer program BMIRT (Yao, 2003). AIC was used to find the best-fit model among one-, two-, and three-dimensional partial credit models. In this study, the sum of the number of item and ability parameters was used to compute penalty term K for the AIC penalty term.

General Model with Constraints

Imposing constraints on IRT models can sometimes be a useful way of determining which of the parameters in the model might explain the data best. Abad et al. (2009) investigated problems involved in the use of multiple-choice models, including large sample size requirements (e.g., sample sizes of 3000 or more), lack of uniqueness of parameter estimates, and the handling of omitted responses. Equality constraints on the item parameters in four polytomous models were imposed to detect differences in propensity for intentional omitting. More specifically, the following cases were compared: Constrained and unconstrained versions of a NRM (Bock, 1972), Samejima's multiple-choice model (Samejima, 1979), a restricted version of Samejima's multiple-choice model (Abad et al., 2009), and the multiple choice model (Thissen and Steinberg, 1984). Parameter estimates were obtained using MULTILOG (Thissen, 2003), and the comparisons were made using both a likelihood-ratio test and AIC. The number of unconstrained parameters was used to calculate K. The results produced by AIC indicated that the restricted version of Samejima's model had the best fit.

Other studies have used similar model constraints. Hickendorff et al. (2009) used BIC in a study of the use of covariates in building explanatory IRT models (Volume One, Chapter 33). Their constraints included different types of covariates, such as gender, parental background, parental education, and general mathematics level. The models were estimated using MMLE as implemented in the SAS NLMIXED proc (SAS Institute, 2002), and the number of persons was used as the sample size in the BIC penalty term. Their null model, without any covariates, can be considered as the constrained model of a full explanatory IRT model with all covariates to be tested.

Revuelta (2008) compared a NRM (Bock, 1972) with a version that constrained the item scale and difficulty parameters to be linear functions of the item parameters. AIC and BIC were used to compare the two models, with the parameters estimated using MMLE and the number of persons as penalty terms for BIC. Klein Entink et al. (2009) and May (2006) used multilevel IRT models to test differential item functioning (DIF). Their models had constraints on the item parameters, and BIC was used to select the best fitting model. The parameters were estimated using an MCMC algorithm.

Models Nested with Respect to the Number of Item Parameters

In addition to studies reported above, the following studies also used the number of item parameters in the penalty term. de Ayala (2009) used AIC and BIC to find the best fitting model among 1-, 2-, and 3-parameter logistic IRT models estimated using MMLE. Revuelta (2008) compared a 2-parameter logistic model with a 3-parameter logistic model using

AIC and BIC estimated using the computer program MULTILOG. Kang and Cohen (2007) investigated the performance of AIC and BIC for the selection of the true model for data sets generated with 1-, 2-, and 3-parameter models. The number of estimated parameters was used to define the penalty term K. Their results indicated that AIC and BIC worked well when the true model was either the 1- or 2-parameter model but neither worked well when the 3-parameter model was true.

Modeling the Multilevel Data Structure in Multilevel IRT Models

May (2006) tested items for DIF using multilevel IRT models. BIC was used to select the best fitting model in the context of Bayesian estimation using the MCMC algorithm implemented in WinBUGS (Spiegelhalter et al., 2003), using the number of persons in the penalty term for BIC. Wang and Jin (2009) compared a model with internal restrictions on item difficulty and its multilevel extension using AIC and based on MMLE results from SAS NLMIXED proc.

18.3.2 Non-Nested Models

Examples of the use of information criteria for non-nested models include comparisons across such different families of models as (1) compensatory and noncompensatory MIRT models, (2) models for specific problems such as test speededness, (3) models with fixed or random effects, and (4) models with random effects.

Different Families of Models

Bolt and Lall (2003) examined the use of MCMC estimation for compensatory and noncompensatory MIRT models. A multicomponent latent trait model (Whitely, 1980; Volume One, Chapter 14) is a noncompensatory model. This model was compared with a multidimensional 2PL logistic (M2PL) model (Reckase, 1985; Volume One, Chapter 12), a compensatory model, using BF. Results indicated better fit of the M2PL model to data from a test of English usage. Patz et al. (2002) compared a hierarchical rater model and a facet IRT model for rating items using BIC. The model parameters were estimated using MCMC. The sum of the number of items and population parameters of the person distribution was used for the penalty term. Goegebeur et al. (2008) used AIC and BIC to compare models designed to account for test speededness. MMLE was used to estimate model parameters. A standard 3PL model was compared with a gradual process change model using a likelihood-ratio test, AIC, and BIC.

In other studies, De Boeck (2008) used AIC and BIC to compare models with fixed or random effects with Laplace approximation and with restricted maximum likelihood (REML) (Semmes et al., 2011). Klein Entink et al. (2009) used BF to detect DIF and to compare multivariate IRT models with fixed or random effects. Rijmen and De Boeck (2002) used BIC to compare models with different kinds of random effects. The model parameters were estimated using MMLE as implemented in the computer program ConQuest (Wu et al., 2007).

18.4 Issues Using Information Criteria in IRT Models

In this section, several issues are discussed regarding the use of information criteria: (1) definition of model complexity terms in AIC and BIC, (2) comparisons of models with respect to different numbers of random effects, (3) comparisons of models with both fixed and random effects, (4) comparisons of models based on different integral approximations, (5) comparisons of models based on results using different estimation algorithms, (6) comparisons across models containing many sources of model complexity, (7) the use of information criteria with small samples, and (8) the use of information differences, weights, and evidence ratios.

Definition of Model Complexity Terms in AIC and BIC

As Skrondal and Rabe-Hesketh (2004) note, it is generally difficult to determine the number of parameters for the calculation of AIC and BIC for latent variable models, including IRT models. This is because the number of parameters can be taken as the number of model parameters (excluding the latent variables), the total number of model parameters or the number of realizations of the latent variables (Hodges and Sargent, 2001; Vaida and Blanchard, 2005). Inconsistent counts of number of parameters for use in calculating the model complexity terms can be found in several MCMC applications. For example, the number of parameters to be sampled in MCMC includes item parameters, person parameters (i.e., realizations of the latent variables as in MMLE), and population parameters of person parameters. Yao and Schwarz (2006) counted all parameters as the number of parameters while Cho and Cohen (2010) and Patz et al. (2002) did not count person parameters in calculating BIC.

It is also difficult to define the sample sizes in calculating BIC (Skrondal and Rabe-Hesketh, 2004). In IRT applications, the number of persons has been used for this purpose. In multilevel IRT, however, it is not clear whether the appropriate sample sizes are the number of persons, groups, or both. The number of persons has been used most often in applications (Bartolucci et al. 2011; Cho and Cohen, 2010; May, 2006).

Models with Different Numbers of Random Effects

When testing an IRT model with R random effects against a model with $R+1$ random effects, it is inappropriate to use a likelihood-ratio test (De Boeck, 2008; Rijmen and De Boeck, 2002). This holds even when the former is nested within the latter. The inappropriateness arises because the nested model is on the boundary of the parameter space of the alternative model. Although a likelihood-ratio test is no longer valid, information criteria may still be useful. The standard BIC performs well for models with a single random effect (Jiang and Rao, 2003). Jiang et al. (2008) described new conditions for consistency for the selection of mixed-effect models with more than one random effect. More research on this issue for IRT models is needed.

Comparisons of Models with Fixed and Random Effects

Model selection is often done between IRT models that include both fixed and random effects to determine which of these effects are required. In explanatory IRT models (Volume One, Chapter 33), for example, different kinds of covariates can be added to the model as fixed or random effects, including (1) item covariates that vary across items, (2) person covariates that vary across persons, and (3) person-by-item covariates that vary across persons and items. The issues are similar as in standard linear regression modeling. As another example, in multilevel IRT models, one needs to determine whether group-level random effects are necessary.

Approximate Estimation Methods and Information Criteria

As its name suggests, a quasi-likelihood is not an "exact" likelihood. Comparisons of models estimated via quasi-loglikelihood, such as the Laplace approximation, can be problematic, since quasi-likelihood is an approximation to the integrand.

Model Comparisons Across Different Estimation Methods

Different maximum-likelihood estimation methods exist. Their results are not automatically comparable, and the same may hold for information indices calculated on them. For example, the loglikelihood from a Laplace approximation is not an exact likelihood, whereas the marginal likelihood from MMLE is.

Multiple Sources of Model Complexity

When more than one source of model complexity is present, results from different information criteria will often differ. For example, AIC identified the correct number of classes in a mixture Rasch model using estimates from both rough and smooth person distribution, but BIC was only successful with smooth person distribution (Kreiner et al., 2006). Li et al. (2009) found that results for several different information indices, including AIC and BIC, were not consistent across different unidimensional mixture IRT models (e.g., 2-parameter mixture IRT model vs. 3-parameter mixture IRT model).

Use of Information Criteria for Small Sample Sizes

Alternative AIC definitions have been proposed to adjust for small sample sizes. One is the second order (i.e., small sample) approximation AIC_c (Hurvitz and Tsai, 1989; Sugiura, 1978). As noted above, use of it for small samples is recommended, when the ratio N/K is small (say < 40) (Burnham and Anderson, 2002). AIC_c was derived for the linear models. It is unknown if it can be applied to such nonlinear models as in IRT.

Use of Information Differences, Weights, and Evidence Ratios

Burnham and Anderson (2002, page 47) note that the use of information criteria for model selection does not necessarily lead to the single best model. Rather, multiple hypotheses should be formulated and sets of models should be constructed to test these hypotheses. This confirmatory approach is mandatory in science. In this multimodel approach, the candidate models can be ranked from best to worst, scaling them using Akaike weights and evidence ratios. For example, Janssen and De Boeck (1999) ranked several candidate models based on AIC values to interpret their results. However, in most IRT model applications, AIC and BIC have been used to find a single best model. In addition to looking at the lowest values of AIC or BIC, therefore, more emphasis on the use of Akaike weights and evidence ratios is recommended.

18.5 Conclusion

It is appropriate to provide a few cautionary notes for use of such information criteria as AIC and TIC for model selection. They are not statistical hypothesis tests but rather estimate the actual loss of information involved in the selection of a particular model. That is, their use is mainly exploratory and recommended for the early stages of inquiry. It is possible, however, to use them as confirmatory tools (Burnham and Anderson, 2002). In either case, however, the candidate models need to have a more substantive basis as well. In other words, it is not appropriate to simply generate multiple different models to see which one fits the best, if there is no additional justification for them.

In sum, it is useful, particularly for new or complex models, to provide additional evidence on the performance of information criteria in the specific substantive context(s) being studied (e.g., Cho and Cohen, 2010; Klein Entink et al., 2009; Kreiner et al., 2006; Suh and Bolt, 2010). Also, more general study on performance of information criteria in IRT is needed, particularly with respect to the issues described in the previous section.

References

Abad, F. J., Olea, J., and Ponsoda, V. 2009. The multiple-choice model: Some solutions for estimation of parameters in the presence of omitted responses. *Applied Psychological Measurement, 33,* 200–221.

Akaike, H. 1973. Information theory and an extension of the maximum likelihood principle. In B. N. Petrov and F. Csáki (Eds.), *Proceedings of the Second International Symposium on Information Theory.* Budapest: AkadémiaiKiadó.

Akaike, H. 1974. A new look at the statistical model identification. *IEEE Transactions on Automatic Control, 19,* 716–723.

Akaike, H. 1977. On entropy maximization principle. In P. R. Krishnaiah (Ed.), *Proceedings of the Symposium on Applications of Statistics.* Amsterdam: North-Holland, pp. 27–41.

Akaike, H. 1978. On newer statistical approaches to parameter estimation and structure determination. *Internal Federation of Automatic Control, 3,* 1877–1884.

Akaike, H. 1983. Information measures and model selection. *International Statistical Institute, 44,* 277–291.

Bartolucci, F., Pennoni, F., and Vittadini, G. 2011. Assessment of school performance through a multilevel latent Markov Rasch model. *Journal of Educational and Behavioral Statistics, 36,* 491–522.

Bock, R. D. 1972. Estimating item parameters and latent ability when responses are scored in two or more nominal categories. *Psychometrika, 37,* 29–51.

Bolt, D. M. and Johnson, T. R. 2009. Addressing score bias and differential item functioning due to individual differences in response style. *Applied Psychological Measurement, 33,* 335–352.

Bolt, D. M. and Lall, V. F. 2003. Estimation of compensatory and noncompensatory multidimensional item response models using Markov chain Monte Carlo. *Applied Psychological Measurement, 27,* 395–414.

Bozdogan, H. 1987. Model selection and Akaike's information criterion (AIC): The general theory and its analytical extensions. *Psychometrika, 52,* 345–370.

Burnham, K. P. and Anderson, D. R. 2002. *Model Selection and Multimodel Inference: A Practical Information-Theoretic Approach* (2nd ed.). New York: Springer.

Cho, S.-J. and Cohen, A. S. 2010. A multilevel mixture IRT model with an application to DIF. *Journal of Educational and Behavioral Statistics, 35,* 336–370.

Claeskens, G. and Hjort, N. L. 2008. *Model Selection and Model Averaging.* New York: Cambridge.

De Boeck, P. 2008. Random item IRT models. *Psychometrika, 73,* 533–559.

de Ayala, R. J. 2009. *The Theory and Practice of Item Response Theory.* New York: The Guilford Press.

De Boeck, P. and Wilson, M. 2004. *Explanatory Item Response Models.* New York: Springer.

Goegebeur, Y., De Boeck, P., Wollack, J. A., and Cohen, A. S. 2008. A speeded item response model with gradual process change. *Psychometrika, 73,* 65–87.

Hickendorff, M., Heiser, W. J., van Putten, C, M., and Verhelst, N. D. 2009. Solution strategies and achievement in complex arithmetic latent variable modeling of change. *Psychometrika, 74,* 331–350.

Hodges, I. and Sargent, D. 2001. Counting degrees of freedom in hierarchical and other richly-parameterised models. *Biometrika, 88,* 367–379.

Hurvitz, C. M. and Tsai, C.-L. 1995. Model selection for extended quasi-likelihood models in small samples. *Biometrika, 51,* 1077–1084.

Janssen, R. and De Boeck, P. 1999. Confirmatory analyses of componential test structure using multidimensional item response theory. *Multivariate Behavioral Research, 34,* 245–268.

Jiang, J. and Rao, J. S. 2003. Consistent procedures for mixed linear model selection. *Sankhyā, 65,* 23–42.

Jiang, J., Rao, J. S., Gu, Z., and Nguyen, T. 2008. Fence methods for mixed model selection. *The Annals of Statistics, 36,* 1669–1692.

Kang, T.-H. and Cohen, A. S. 2007. IRT model selection methods for dichotomous items. *Applied Psychological Measurement, 31*(4), 331–358.

Kang, T.-H., Cohen, A. S., and Sung, H.-J. 2009. IRT model selection methods for polytomous items. *Applied Psychological Measurement, 33* (7), 499–518.

Klein Entink, R. H., Fox, J.-P., and van der Linden, W. J. 2009. A multivariate multilevel approach to the modeling of accuracy and speed of test takers. *Psychometrika, 74,* 21–48.

Kreiner, S., Hansen, M., and Hansen, C. R. 2006. On local homogeneity and stochastically ordered mixed Rasch models. *Applied Psychological Measurement, 30,* 271–297.

Kuha, J. 2004. AIC and BIC: Comparisons of assumptions and performance. *Sociological Methods and Research, 33,* 188–228.

Kullback, S. and Leibler, R. A. 1951. On information and sufficiency. *Annals of Mathematical Statistics, 22,* 79–86.

Li, F., Cohen, A. S., Kim, S.-H., and Cho, S.-J. 2009. Model selection methods for dichotomous mixture IRT models. *Applied Psychological Measurement, 33,* 353–373.

May, H. 2006. A multilevel Bayesian item response theory method for scaling socioeconomic status in international studies of education. *Journal of Educational and Behavioral Statistics, 31,* 63–79.

McKinley, R. L. 1988. Assessing dimensionality using confirmatory multidimensional IRT. Paper presented at the annual meeting of the American Educational Research Association, New Orleans.

McKinley, R. L. 1989. Confirmatory analysis of test structure using multidimensional item response theory (Research Report No. 89–21). Princeton, NJ: Educational Testing Service.

McKinley, R. L. 1992. *TESTMAP Version 2.1 User's Guide.* Unpublished manuscript.

Nylund, K. L., Asparouhov, T., and Muthén, B. O. 2007. Deciding on the number of classes in latent class analysis and growth mixture modeling: A Monte Carlo simulation study. *Structural Equation Modeling, 14,* 535–569.

Patz, R. J., Junker, B. W., Johnson, M. S., and Mariano, L. T. 2002. The hierarchical rater model for rated test items and its application to large-scale educational assessment data. *Journal of Educational and Behavioral Statistics, 27,* 341–384.

Revuelta, J. 2008. The generalized logit-linear item response model for binary-designed items. *Psychometrika, 73,* 385–405.

Reckase, M. D. 1985. The difficulty of test items that measure more than one ability. *Applied Psychological Measurement, 9,* 401–412.

Rijmen, F. and De Boeck, P. 2002. The random weights linear logistic test model. *Applied Psychological Measurement, 26,* 271–285.

Rijmen, F. and De Boeck, P. 2005. A relation between a between-item multidimensional IRT model and the mixture Rasch model. *Psychometrika, 70,* 481–496.

Samejima, F. 1979. A new family of models for multiple-choice item. (sic) Research Report #79–4. University of Tennessee, Knoxville: Tennessee.

SAS Institute 2002. *SAS Online Doc (version 9).* Cary, NC: SAS Institute Inc.

Schwartz, G. 1978. Estimating the dimension of a model. *Annals of Statistics, 6,* 461–464.

Semmes, R., Davison, M. L., and Close, C. 2011. Modeling individual differences in numerical reasoning speed as a random effect of time limits. *Applied Psychological Measurement, 35,* 433–446.

Skrondal, A. and Rabe-Hesketh, S. 2004. *Generalized Latent Variable Modeling: Multilevel, Longitudinal and Structural Equation Models.* Boca Raton, FL: Chapman and Hall.

Spiegelhalter, D. J., Best, N. G., Carlin, B. P., and van der Linde, A. 2002. Bayesian measures of model complexity and fit. *Journal of the Royal Statistical Society, Series B, 64,* 583–639.

Spiegelhalter, D. J., Thomas, A., Best, N. G., and Lunn, D. 2003. *WinBUGS 1.4* User Manual* [Computer program]. Cambridge, UK: MRC Biostatistics Unit.

Sugiura, N. 1978. Further analysis of the data by Akaike's information criterion and the finite corrections. *Communications in Statistics, Theory and Methods, A7,* 13–26.

Suh, Y. and Bolt, D. M. 2010. Nested logit models for multiple-choice item response data. *Psychometrika, 75,* 454–473.

Takeuchi, K. 1976. Distribution of informational statistics and a criterion of model fitting. *Suri-Kagaku (Mathematical Sciences), 153,* 12–18 (in Japanese).

Thissen, D. 2003. *MULTILOG: Multiple Categorical Item Analysis and Test Scoring Using Item Response Theory* (version 7.03). Chicago: Scientific Software International.

Thissen, D. and Steinberg, L. 1984. A response model for multiple-choice items. *Psychometrika, 49,* 501–519.

Vaida, F. and Blanchard, S. 2005. Conditional Akaike information for mixed effects models. *Biometrika, 92,* 351–370.

Vermunt, J. K. and Magidson, J. 2004. *LatentGOLD 4.0. User's Guide.* Belmont, MA: Statistical Innovations.

Vermunt, J. K. and Magidson, J. 2006. Latent GOLD 4.0 and IRT modeling. Retrieved May 10, 2012, from http://www.statisticalinnovations.com/products/LGIRT.pdf.

Wang, W.-C. and Jin, K.-Y. 2010. Multilevel, two-parameter, and random-weights generalizations of a model with internal restrictions on item difficulty. *Applied Psychological Measurement, 34,* 46–65.

Whitely, S. E. 1980. Multicomponent latent trait models for ability tests. *Psychometrika, 45,* 479–494.

Wu, M. L., Adams, R. J., Wilson, M. R., and Haldane, S.A. 2007. *ACER ConQuest Version 2: Generalised Item Response Modelling Software.* Camberwell: Australian Council for Educational Research.

Yao, L. 2003. *BMIRT: Bayesian Multivariate Item Response Theory* [Computer software]. Monterey, CA: CTB/McGraw-Hill.

Yao, L. and Schwarz, R. D. 2006. A multidimensional partial-credit model with associated item and test statistics: An application to mixed-format tests. *Applied Psychological Measurement, 30,* 469–492.

19

Bayesian Model Fit and Model Comparison

Sandip Sinharay

CONTENTS

19.1 Introduction

According to the Standard 3.9 of the Standards for Educational and Psychological Testing (American Educational Research Association, American Psychological Association and National Council for Measurement in Education, 1999), evidence of model fit should be provided when an item response theory (IRT) model is used to make inferences from a test data set. Researchers, such as Yen (1981) and Wainer and Thissen (1987) have suggested that an incorrect choice of an IRT model can lead to biased estimates of examinee performance, unfair rankings of examinees, and wrongly equated scores. To avoid making an incorrect choice of an IRT model and to ensure the best possible model-data fit, it is essential to apply model-fit and model-comparison methods.

A model-fit method allows an investigator to assess the fit of the model to make sure that the model captures the important features of the data set. Serious misfit, that is, failure of the model to explain a number of aspects of the data that are of practical interest, should result in the replacement or extension of the model, if possible. Even if a model has been appointed as the final model for an application, it is important to assess its fit in order to be aware of its limitations before making any inferences. In contrast, in the application of a model-comparison method, an investigator will develop a set of plausible models, apply the method to compare the fit of the models to the data and then select the best fitting model for analyzing the data.

Model fit and model comparison are complementary in practice. Iterative fitting and fit assessment of models can be used to suggest a range of plausible models to serve as input to model comparison. Also, once the best fitting model or combinations of models is selected, the fit of that model must still be checked.

There has been a recent surge in the use of Bayesian estimation in IRT. Beguin and Glas (2001), Bradlow et al. (1999), Fox (2005, 2010), Fox and Glas (2001, 2003), Geerlings et al. (2011), Janssen et al. (2000), Henson et al. (2009), Klein Entink et al. (2009), Mariano et al. (2010), and Patz and Junker (1999a,b) are only some of the recent examples of application of Bayesian estimation, mostly using the Markov chain Monte Carlo (MCMC) algorithm to fit a wide variety of psychometric models. For details on the MCMC algorithm, see, for example, Gelman et al. (2003) or Junker et al. (see Chapter 15). Also, see several chapters in this book to learn more about recent developments in Bayesian IRT modeling. Naturally, research on Bayesian model-fit and model-comparison methods has also flourished in an attempt to keep pace with this advancement in Bayesian IRT modeling.

The next two sections include further details on Bayesian model-fit and model-comparison methods, respectively, in the context of IRT models. The application section includes examples of applications of some of the techniques to operational data sets. The final section provides conclusions and recommendation.

19.2 Bayesian Model-Fit Methods

A statistical model that is appropriate for a data set should be able to explain the key features in the data set adequately. In other words, if one uses the model to predict important features of the data, those predicted values should be close to the corresponding observed values. This is a self-consistency check. This intuitive idea is the basis of several model-fit methods, including traditional residual analysis for linear models. Checks of this kind can detect patterns in the data that indicate the inappropriateness of the fitted model. They can also identify individual observations that are not consistent with the model. These observations are called outliers, or influential points in linear models. Some popular Bayesian model-fit methods are as follows: (1) Bayesian residual analysis; (2) prior predictive checks; and (3) posterior predictive checks. These approaches are described below.

19.2.1 Bayesian Residual Analysis

In linear models, analysis of residuals is a popular and intuitive tool for diagnosing model misfit. See, for example, Sinharay et al. (2011) and Wells and Hambleton (see Chapter 20) for examples of frequentist residual analysis in the context of IRT models. Chaloner and Brant (1988) and Albert and Chib (1995) suggested a Bayesian analog to the frequentist residual analysis. Suppose U_p denotes the value of a variable for person p, $p = 1, 2, \ldots, P$. For example, U_p could be the score of the person on an item. Suppose further that $E(U_p|\omega) = \mu_p$, where ω denotes the vector of parameters in the model. The residual $\epsilon_p = U_p - \mu_p$ is a function of the model parameters through μ_p. In Bayesian residual analysis, U_p is considered outlying if the posterior distribution of the residual ϵ_p is located far from zero (Chaloner and Brant, 1988). Examination of residual plots based on posterior means of the residuals or randomly chosen posterior draws may help identify patterns that call the model assumptions into question. Sinharay and Almond (2007) applied Bayesian residual analysis to assess fit of cognitive diagnostic models (see, e.g., Wang and Chang, Chapter 16 in *Handbook of Item Response Theory, Volume Three: Applications,* for details on these models).

Albert and Chib (1995) suggested the Bayesian latent residual that is the difference between a latent response and its expected value. Fox (2004) and Geerlings et al. (2011) applied Bayesian latent residual analysis to hierarchical IRT models.

In an application of an IRT model to, for example, a 50-item test taken by 1000 examinees, there are 50,000 responses. It is not straightforward to make meaningful conclusions from Bayesian residual analyses of so many responses in a timely manner. It is conceptually possible to apply Bayesian residual analysis to data summaries such as item proportion-correct or item biserial correlations, but such applications have been lacking in the IRT literature.

19.2.2 Prior Predictive Checks

Let $\boldsymbol{U}$ denote the observed data. Let $\boldsymbol{U}^{\text{rep}}$ denote replicate data that one might observe if the procedure that generated $\boldsymbol{U}$ is replicated and let $\boldsymbol{\omega}$ denote all the parameters of the model. Box (1980) suggested assessing the fit of Bayesian models using the prior predictive distribution

$$p(\boldsymbol{U}^{\text{rep}}) = \int p(\boldsymbol{U}^{\text{rep}}|\boldsymbol{\omega})p(\boldsymbol{\omega})d\boldsymbol{\omega} \tag{19.1}$$

as a reference distribution for the observed data $\boldsymbol{U}$, where $p(\boldsymbol{U}^{\text{rep}}|\boldsymbol{\omega})$ denotes the likelihood of the replicate data $\boldsymbol{U}^{\text{rep}}$ and $p(\boldsymbol{\omega})$ denotes the prior distribution on the parameters. In practice, one defines diagnostic measures $D(\boldsymbol{U})$ and compares the observed value $D(\boldsymbol{U})$ to the reference distribution of $D(\boldsymbol{U}^{\text{rep}})$. Any significant difference between them indicates a misfit of the model.

A major drawback of the prior predictive approach is the important role played by the prior distribution in defining the reference distribution. When this approach indicates model misfit, either the likelihood or the prior distribution or both could be misspecified. The prior predictive distribution is undefined under improper prior distributions and can be quite sensitive to the prior distribution if vague prior distributions are used. Prior predictive checks were applied to IRT models by Zhang (2008).

19.2.3 Posterior Predictive Checks

In posterior predictive model checking (PPMC) method (Rubin, 1984), one uses as a reference distribution using the posterior distribution of the model parameters

$$p(\boldsymbol{\omega}|\boldsymbol{U}) \propto p(\boldsymbol{U}|\boldsymbol{\omega})p(\boldsymbol{\omega}). \tag{19.2}$$

The method involves assessing the fit of a model by examining whether the observed data appear extreme with respect to the *posterior predictive distribution* (PPD) of replicated data $\boldsymbol{U}^{\text{rep}}$, where the PPD is given by

$$p(\boldsymbol{U}^{\text{rep}}|\boldsymbol{U}) = \int p(\boldsymbol{U}^{\text{rep}}|\boldsymbol{\omega})p(\boldsymbol{\omega}|\boldsymbol{U})d\boldsymbol{\omega}. \tag{19.3}$$

In practice, *test quantities* or *discrepancy measures* $D(\boldsymbol{U}, \omega)$ are defined (Gelman et al. 1996) and the posterior distribution of $D(\boldsymbol{U}, \omega)$ is compared to the PPD of $D(\boldsymbol{U}^{\text{rep}}, \omega)$, with substantial differences between them indicating model misfit. A researcher may use $D(\boldsymbol{U}, \omega) = D(\boldsymbol{U})$, a discrepancy measure depending on the data only (which can also be called a *test statistic*). In that case, the PPMC method consists in comparing $D(\boldsymbol{U})$ to the PPD of $D(\boldsymbol{U}^{\text{rep}})$. The comparison of observed and replicated discrepancy measures is mostly performed using graphical plots. It can also be useful to examine a quantitative measure of lack of fit, a tail-area probability also known as the *posterior predictive p-value* (PPP value)

$$P\left(D(\boldsymbol{U}^{\text{rep}}, \omega) \geq D(\boldsymbol{U}, \omega) | \boldsymbol{U}\right) = \int\limits_{D(\boldsymbol{U}^{\text{rep}}, \omega) \geq D(\boldsymbol{U}, \omega)} p(\boldsymbol{U}^{\text{rep}} | \omega) p(\omega | \boldsymbol{U}) \mathrm{d}\boldsymbol{U}^{\text{rep}} \mathrm{d}\omega, \tag{19.4}$$

which can be a useful supplement to a graphical plot.

Because of the difficulty in dealing with Equations 19.3 or 19.4 analytically for all but simple problems, Rubin (1984) suggested simulating replicated (or *posterior predictive*) data sets from the PPD in applications of the PPMC method. One draws N simulations $\omega^1, \omega^2, \ldots, \omega^N$ from the posterior distribution $p(\omega|\boldsymbol{U})$ of ω (most likely using an MCMC algorithm) and then draws $\boldsymbol{U}^{\text{rep},n}$ from the distribution $p(\boldsymbol{U}|\omega^n)$ for $n = 1, 2, \ldots, N$. The process results in N draws from the joint posterior distribution $p(\boldsymbol{U}^{\text{rep}}, \omega|\boldsymbol{U})$. One then computes the *predictive discrepancies* $D(\boldsymbol{U}^{\text{rep},n}, \omega^n)$ and *realized discrepancies* $D(\boldsymbol{U}, \omega^n)$, $n = 1, 2, \ldots, N$. The values $D(\boldsymbol{U}^{\text{rep},n}, \omega^n)$ are actually draws from the PPD of $D(\boldsymbol{U}^{\text{rep}}, \omega)$. It is possible then to create a graphical plot of $D(\boldsymbol{U}^{\text{rep},n}, \omega^n)$ versus $D(\boldsymbol{U}, \omega^n)$, $n = 1, 2, \ldots, N$; points lying consistently above or below the 45° line indicate model misfit. The proportion of the N replications for which $D(\boldsymbol{U}^{\text{rep},n}, \omega^n)$ exceeds $D(\boldsymbol{U}, \omega^n)$ provides an estimate of the PPP value. Extreme PPP values (close to 0, or 1, or both, depending on the nature of the discrepancy measure) indicate model misfit. Figure 19.1 graphically describes the PPMC method.

Next, a description is provided of the PPMC method in the context of IRT models. Let U_{pi} denote the binary score of the pth individual for the ith item in an educational assessment. Suppose that the IRT model of interest is the two-parameter logistic (2PL) model, which implies that

$$P(U_{pi} = 1|\theta) = \frac{\exp(a_i(\theta - b_i))}{1 + \exp(a_i(\theta - b_i))} \tag{19.5}$$

with symbols having their usual meanings. In this context, if one treats the examinee proficiency θ as nuisance parameters, then ω is the collection of all item parameters. The

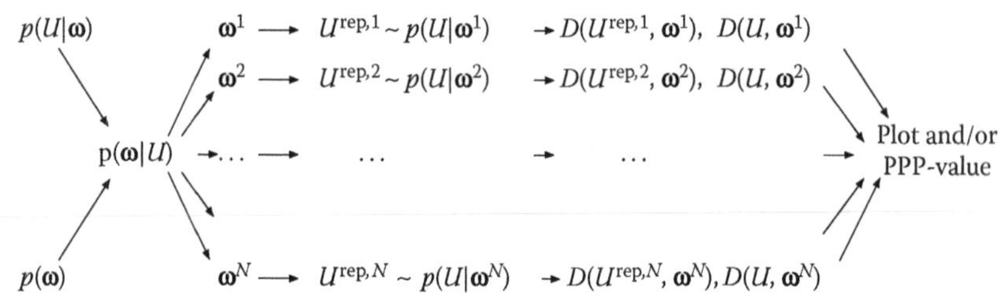

FIGURE 19.1
A graph describing the PPMC method.

marginal posterior distribution $p(\omega|U)$ is given by

$$p(\omega|U) \propto \left[\prod_p \int_\theta \left\{\prod_i [P(U_{pi}=1)]^{U_{pi}}[1-P(U_{pi}=1)]^{1-U_{pi}}\right\} p(\theta)\mathrm{d}\theta\right] p(\omega), \qquad (19.6)$$

where $p(\theta)$ is the prior distribution on the θ.

To assess the fit of the 2PL model to a data set, one has to repeat the following steps a large number of times:

1. Generate a draw of item parameters from the posterior distribution given by Equation 19.6 using an MCMC algorithm to fit the 2PL model. Draw proficiency parameters $\theta_1, \theta_2, \ldots, \theta_P$ from $p(\theta)$.
2. Draw a data set from the 2PL model, using the item parameters and proficiency parameters drawn in the above step. This is a posterior predictive data set.
3. Compute the values of the predictive and realized discrepancy measures from the above draws of parameters and data set.

One then creates plots and/or computes PPP values for the discrepancy measures.

The choice of discrepancy measures is crucial in the application of the PPMC method. While the flexibility of the method allows one to use any function of data and parameters as a discrepancy measure, some measures may be more useful than others. Ideally, discrepancy measures should be chosen to reflect aspects of the model that are relevant to the scientific purposes to which the inference will be applied, and to measure features of the data not directly addressed by the probability model (Gelman et al. 2003, p. 172).

For IRT models, a number of discrepancy measures may be of interest, depending on the context of the problem. These models assume unidimensionality, local independence, a specific shape of the response function, and, often, normality of the ability distribution, and each of these assumptions should be checked using suitable discrepancy measures. There does not seem to exist a single, omnibus discrepancy measure that can detect violations of all of these assumptions together.

The PPMC methods have been criticized for being conservative. The PPP values are not necessarily uniformly distributed when the fitted model is in fact correct, and there is some evidence that PPP-values under the correct model tend to be closer to 0.5 more often than would be expected under a uniform distribution (Bayarri and Berger, 2000; Sinharay and Stern, 2003; Sinharay et al. 2006). However, Hjort et al. (2006) suggested an approach to modify the PPP values so that the resulting p values are uniformly distributed when the fitted model is correct. In addition, researchers such as Beguin and Glas (2001), Fox and Glas (2003), Hoijtink (2001), Li et al. (2006), Levy and Svetina (2011), Levy et al. (2009), Glas and Meijer (2003), Sinharay (2005), Sinharay (2006a,b), Sinharay and Johnson (2008), Sinharay et al. (2006), Toribio and Albert (2011), and Zhu and Stone (2011) successfully applied the PPMC method to various types of IRT models such as unidimensional IRT models, multidimensional IRT models, multilevel IRT models, and the testlet model. Together, they employed a variety of discrepancy measures such as the observed score distribution, item proportion-correct, item-pair proportion correct, item-pair odds ratio, item-biserial correlation, item-fit Chi-type statistics, and person-fit statistics to assess different aspects of IRT model fit such as item fit, person fit, dimensionality, differential item functioning, and overall model fit.

19.3 Bayesian Model-Comparison Methods

19.3.1 Deviance Information Criterion

Spiegelhalter et al. (2002) suggested the deviance information criterion (DIC), which is a natural model-comparison method for comparing models fitted using an MCMC algorithm. The DIC for a model is defined as

$$\text{DIC} = \bar{D} + p_D, \tag{19.7}$$

where

$$p_D = \bar{D} - D(\bar{\omega}), \tag{19.8}$$

$\bar{\omega}$ is the posterior mean of the parameter vector ω,

$$D(\omega) = -2 \times \text{log-likelihood at } \omega, \tag{19.9}$$

and $\bar{D}$ is the posterior mean of $D(\omega)$. If an IRT model has been fitted to a data set using an MCMC algorithm, $\bar{D}$ is the arithmetic mean of the values of the likelihood for each draw from the posterior distribution. The DIC involves the term p_D, where p_D is a measure of the effective number of parameters, which penalizes models with more parameters. Also, DIC is approximately equivalent to Akaike's information criterion or AIC (Akaike, 1974) for models with negligible prior information. Note that AIC or its close relative Bayesian information criterion (BIC) are usually not employed as Bayesian model-comparison methods; these involve the computation of the maximum likelihood estimates of the model parameters. Among two competing models, the one with less DIC fits the data better. Applications of DIC to IRT models include Geerlings et al. (2011), Kang and Cohen (2007), Kang et al. (2009), Sheng and Wikle (2007), and Sinharay and Almond (2007). For further details on DIC and other information criteria, see Cohen and Cho (see Chapter 18).

19.3.2 Bayes Factors

Suppose that a model M with parameters ω has been fitted to the observed data $\boldsymbol{U}$. Let

$$p(\boldsymbol{U}|M) = \int p(\boldsymbol{U}|\omega, M)p(\omega|M)d\omega \tag{19.10}$$

denote the marginal density of the data $\boldsymbol{U}$ under model M. A popular Bayesian approach to compare two models M_1 and M_2 is to compute the Bayes factor (BF; Kass and Raftery, 1995)

$$\text{BF}^{12} = \frac{p(\boldsymbol{U}|M_1)}{p(\boldsymbol{U}|M_2)} \tag{19.11}$$

that compares the marginal densities of the observed data $\boldsymbol{U}$ under the two models.

The BF is also the ratio of posterior odds and prior odds, that is,

$$\text{BF}^{12} = \frac{p(M_1|\boldsymbol{U})}{p(M_2|\boldsymbol{U})} \Big/ \frac{p(M_1)}{p(M_2)}. \tag{19.12}$$

Kass and Raftery (1995) provided a comprehensive review of BF including information about their interpretation. They also noted that the BIC is a rough approximation to the logarithm of the BF. The marginal densities $p(\boldsymbol{U}|M)$ cannot be computed analytically if M is an IRT model. Different approaches exist for estimating the BF for models such as IRT models for which the computation of $p(\boldsymbol{U}|M)$ is not straightforward. Some of the more popular approaches are importance sampling methods (e.g., DiCiccio et al. 1997), Chib's method (Chib, 1995; Chib and Jeliazkov, 2001), reversible jump MCMC method (Green, 1995), and bridge sampling (Meng and Wong, 1996). However, there are few applications of the BF to IRT models except for research works such as Fox (2010), Klein Entink et al. (2009), and Toribio (2006).

19.3.3 Cross-Validation Likelihood and Partial BF

O'Hagan (1995) suggested partitioning the data $\boldsymbol{U}$ into a training set of observations $\boldsymbol{U}_T$ and a cross-validation set of observations $\boldsymbol{U}_{CV}$. The posterior distribution of the parameters is computed using only the training set $\boldsymbol{U}_T$. Let us denote this posterior distribution under model M as $p(\boldsymbol{\omega}|\boldsymbol{U}_T, M)$. The likelihood of the cross-validation observations $\boldsymbol{U}_{CV}$ under model M, also called cross-validation likelihood, is then computed using this updated posterior distribution as

$$p(\boldsymbol{U}_{CV}|M) = \int p(\boldsymbol{U}_{CV}|\boldsymbol{\omega}, \boldsymbol{U}_T, M)\, p(\boldsymbol{\omega}|\boldsymbol{U}_T, M) d\boldsymbol{\omega}, \tag{19.13}$$

where $p(\boldsymbol{U}_{CV}|\boldsymbol{\omega}, \boldsymbol{U}_T, M)$ is the conditional likelihood of $\boldsymbol{U}_{CV}$ given $\boldsymbol{U}_T$. One then computes the partial BF (O'Hagan, 1995)

$$\text{PBF}^{12} = \frac{p(\boldsymbol{U}_{CV}|M_1)}{p(\boldsymbol{U}_{CV}|M_2)} \tag{19.14}$$

that compares the cross-validation likelihoods under the two models.

Bolt et al. (2001), Bolt and Lall (2003), Kang and Cohen (2007), and Kang et al. (2009) applied the cross-validation likelihood and partial BFs in comparing mixture IRT models, multidimensional IRT models, unidimensional IRT models for dichotomous items, and unidimensional IRT models for polytomous items, respectively.

19.4 Application

Data were available from two forms of a test that measures school students' progress toward achieving the academic content standards adopted by a U.S. state in one subject area. The test results aim to describe what students should know and be able to do in each grade and subject tested. The test consists of 65 multiple choice (MC) items, each with four answer options. The two forms were taken by about 46,000 and 31,000 examinees, respectively. The Rasch model was operationally employed with a normal ability distribution to perform IRT true score equating of the score on the first form to the score on the second form of the test. The equating design is the nonequivalent groups with anchor test (NEAT) design with an internal anchor. The number of anchor items for the test was 30. The first step in the Bayesian analysis was the fitting of the Rasch model with a normal ability distribution using an MCMC algorithm to the data sets. The prior distributions on the item parameters were as follows:

$$\log(a) \sim \mathcal{N}(0, 10), b_i \sim \mathcal{N}(0, 10). \tag{19.15}$$

Ten thousand iterations of the MCMC algorithm were used after a burn-in of 2000 iterations. Every 10th iteration out of the 10,000 was retained, leading to a sample of size 1000 from the posterior distribution of the parameters. Appropriate convergence diagnostics ensured the convergence of the MCMC algorithm.

19.4.1 Fit to the Observed-Score Distribution

The top panel of Figure 19.2 shows the fit of the Rasch model to the marginal score distribution for one form of the test, where the marginal score distribution refers to the distribution of the raw scores and specifies the number of examinees who obtained the raw scores of

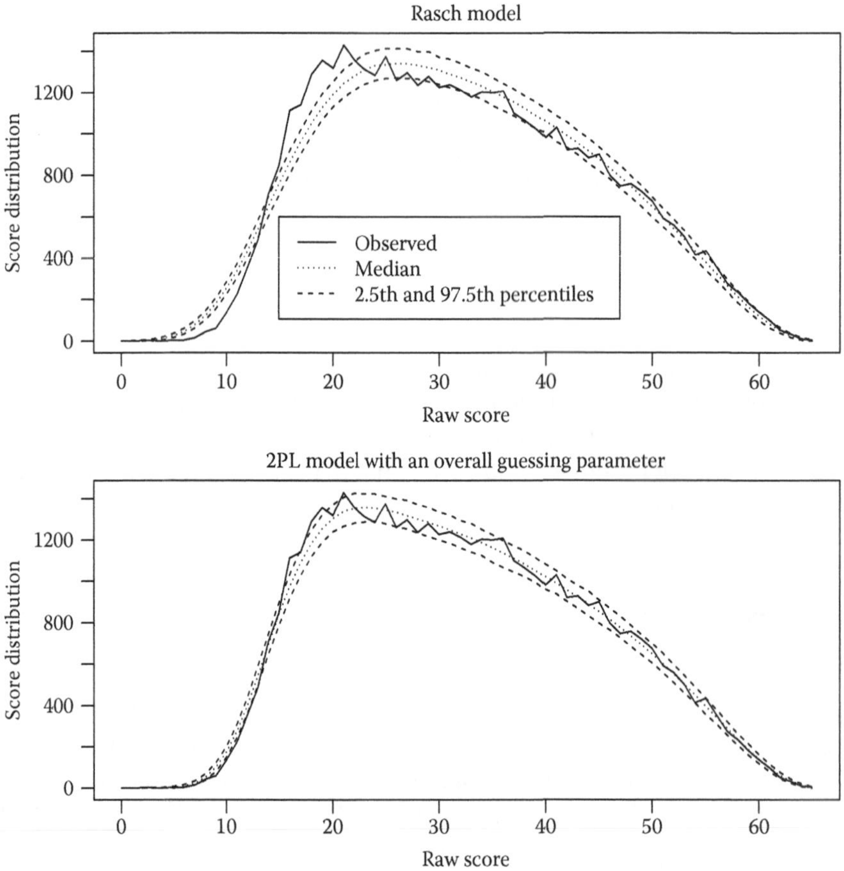

FIGURE 19.2
Fit of the Rasch model (top) and the 2PL model with an overall guessing parameter (bottom) to the marginal score distribution for one form of the state test.

$0, 1, 2, \ldots, 65$. The Rasch model is used for equating of this test and the marginal score distribution plays a key role in equating. Therefore, it was appropriate to examine the fit of the Rasch model to the marginal score distribution. The PPMC method has been applied to create Figure 19.2. The solid line joins the points indicating the observed number of examinees at each score point. The dotted line joins the medians of the posterior predictive distribution of the number of examinees at each score point. One thousand posterior predictive data sets were used to obtain the median. The two dashed lines join the 2.5th and 97.5th percentiles of the posterior predictive distribution of the number of examinees at each score point. At any score point, a misfit of the model is indicated if the solid line lies far from the dotted line and outside the 95% posterior predictive interval formed by the two dashed lines.

The figure shows strong evidence of misfit of the Rasch model to the marginal score distribution. The observed score distribution appears skewed to the left. All but a couple of the observed values for scores below 23 lie outside the 95% posterior predictive interval. Thus, Figure 19.2 suggests that the Rasch model does not adequately fit the observed score distribution of the data.

It is a common belief that examinees, especially low-scoring ones, often randomly guess the answer in MC tests (see, e.g., Birnbaum, 1968, pp. 303–305). If examinees guess answers, then the Rasch model that does not allow guessing would not fit the data well, especially at the low end of the score distribution. In addition, the Rasch model did not adequately predict the biserial correlations of the data set (results not shown). The misfit occurred because the Rasch model forces the slope parameter to be the same over all the items. To examine if allowance of guessing and varying slope parameters lead to a better model-data fit, the 2PL model with an overall guessing parameter was fitted to the data. For this model, the probability of a correct response to Item i of an examinee with ability θ is given by

$$c + (1 - c)\frac{\exp[a_i(\theta - b_i)]}{1 + \exp[a_i(\theta - b_i)]}, \tag{19.16}$$

where c denotes the overall guessing parameter. The prior distributions on the item parameters were as follows:

$$\log(a_i) \sim \mathcal{N}(0, 10), b_i \sim \mathcal{N}(0, 10), \log\left(\frac{c}{1 - c}\right) \sim \mathcal{N}(-1.4, 1). \tag{19.17}$$

For the data set, the posterior mean of c was 0.18, which is substantially larger than 0 that represents no guessing, but somewhat smaller than 0.25 that is the probability of a correct answer by random guessing on a 4-option MC item.

The bottom panel of Figure 19.2 shows the fit of the 2PL model with an overall guessing parameter to the observed-score distribution. The fit is substantially better in comparison to that of the 1PL model. The observed marginal score distribution is within the 95% posterior predictive interval for almost whole of the score range except for a few low score points. Thus, the 2PL model with an overall guessing parameter seems to fit the marginal score distribution better than the Rasch model.

19.4.2 Item Fit

An item-fit analysis using the PPMC method (Sinharay, 2006a) indicated that the Rasch model fits very few items in any of the two data sets and the 2PL model with an overall guessing parameter performs much better. Figure 19.3 shows item fit plots (Sinharay,

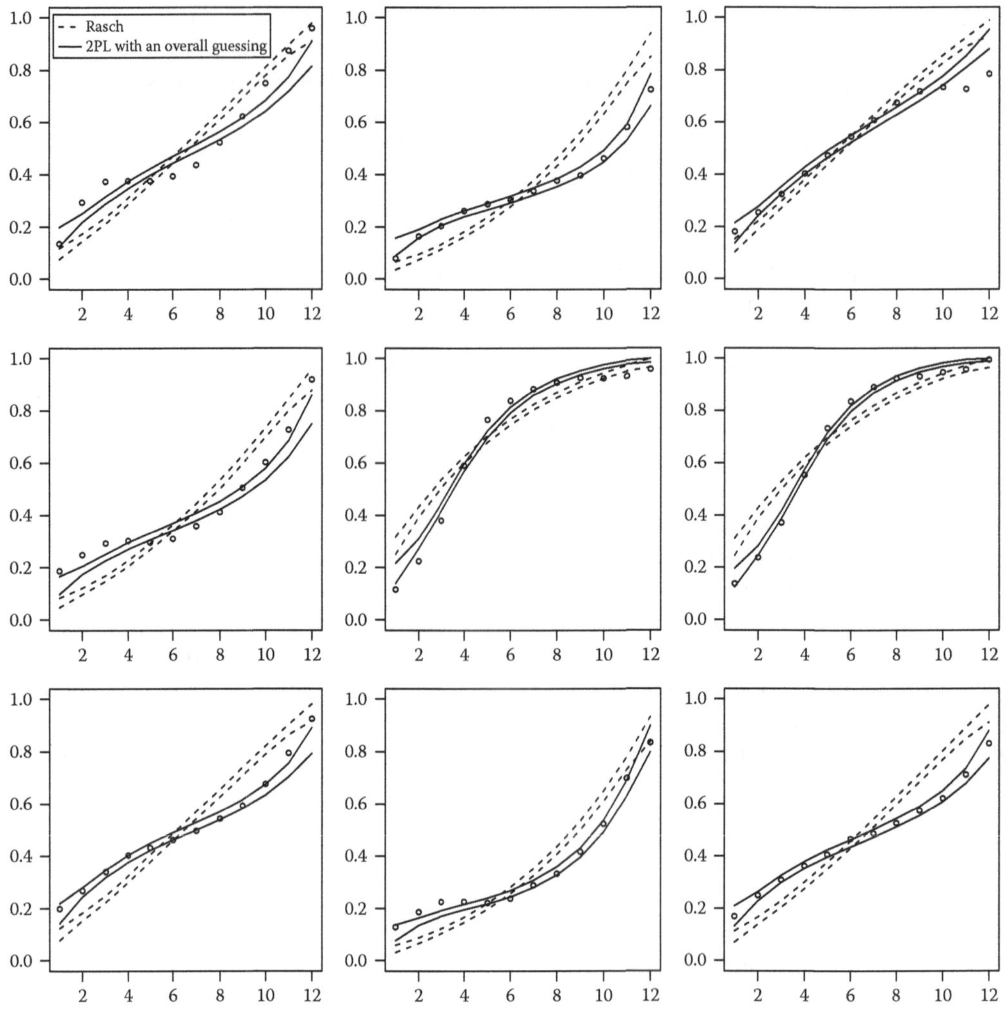

FIGURE 19.3
Item fit for the two models for one form of the state test.

2006a) for nine items for which the misfit of the Rasch model appeared most severe for one form of the test. To create Figure 19.3, the examinees were divided into 12 groups based on their observed raw scores on the test. Examinees with raw scores 0–10 belonged to Group 1, those with raw scores 11–15 belonged to Group 2, · · · , those with raw scores 61–65 belonged to Group 12. Each panel in Figure 19.3 shows, for one item, the observed proportion correct score on that item for each score group (using a hollow circle), the 2.5th and 97.5th percentiles of the PPD of the proportion correct score under the Rasch model on that item for each score group (using dashed lines), and the corresponding 2.5th and 97.5th percentiles of the PPD of the proportion correct score under the 2PL model with an overall guessing parameter (using solid lines). For any item, too many hollow circles lying outside the interval formed by two dashed lines would indicate a misfit of the Rasch model to that item and too many hollow circles lying outside the interval formed by two solid lines would indicate a misfit of the 2PL model with an overall guessing parameter.

Figure 19.3 (and similar plots for other items not shown here) points to severe misfit of the Rasch model to the data set. For all the items shown in the figure, the observed proportions-correct are mostly well outside the corresponding 95% posterior predictive intervals under the Rasch model. The fit of the 2PL model with an overall guessing parameter, while not perfect, is much better than that of the Rasch model for all the items; the 95% posterior predictive intervals are much closer to the corresponding observed proportions-correct under this model.

19.4.3 Model Comparison Using DIC

The DICs were computed for comparing the Rasch model and the 2PL model with an overall guessing parameter for the two forms of the state test. To compute the DIC under a model, the marginal maximum likelihood was computed under the model for each draw of item parameters from the the MCMC algorithm and also for the posterior mean of the item parameters. The DIC values for Form 1 were approximately 3,637,036 and 3,598,659, respectively, for the Rasch model and the 2PL model with an overall guessing parameter. The DIC values for Form 2 were approximately 2,448,521 and 2,416,809, respectively, for the Rasch model and the 2PL model with an overall guessing parameter. Thus, for each form, the 2PL model with an overall guessing parameter has smaller value of DIC, indicating a better fit of the model compared to the Rasch model.

19.4.4 Assessment of Practical Significance of Misfit

Figures 19.2 and 19.3 point to severe misfit of the operationally used Rasch model to the state test data and to comparatively better fit of the 2PL model with an overall guessing parameter. However, it is the difference between the two models practically significant? In other words, does the 2PL model with an overall guessing parameter lead to a practically different conclusion?

Figure 19.4 shows the impact of the choice of the IRT model (1PL vs. 2PL model with an overall guessing) on equating. The figure shows for each score point the differences

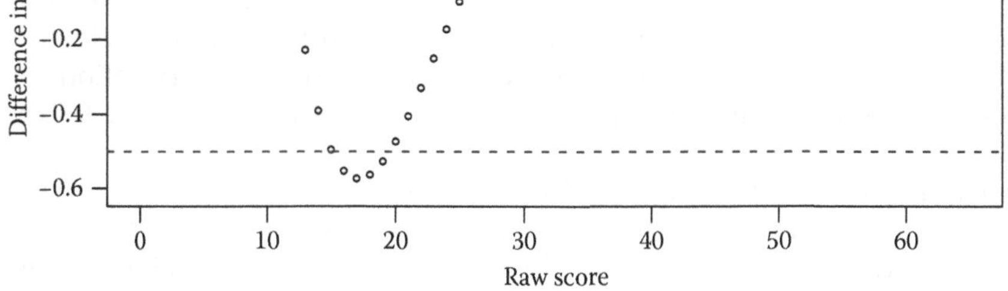

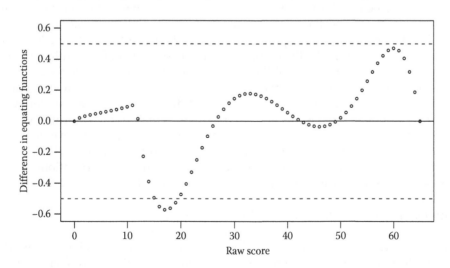

FIGURE 19.4
The difference in equating functions from the two models.

between the IRT true score equating functions obtained using the 1PL model and that obtained using the 2PL model with an overall guessing parameter. The IRT true score equating function of the Form 1 score to the Form 2 score was obtained under each of these models using the Stocking-Lord method (see, e.g., Kolen and Brennan, 2004). The posterior means of the item parameters were used in the computations. In interpreting this plot, the difference that matters or DTM criterion suggested by Dorans and Feigenbaum (1994) will be used as a difference in equating conversions that is practically large. For raw score conversions, the DTM is 0.5 according to Dorans and Feigenbaum (1994). There are two horizontal and dashed lines at −0.5 and 0.5 denoting the DTM criterion. The differences between the two equating functions exceed the DTM only for a few score points around 17. The figure shows that the reported scores of several examinees might change if the 2PL model with an overall guessing parameter is chosen instead of the operationally used 1PL model. The total percentage of examinees who obtained the scores at which the difference between the two equating functions is larger than the DTM is about 11%. Hence, it seems that the choice of the 2PL model with an overall guessing parameter over the 1PL model would lead to a practically significant difference for some examinees for this test; however, even this practically significant difference is not huge.

19.5 Conclusions

The literature review presented in this paper indicates that there is an abundance of Bayesian model-fit and model-comparison methods. Fox (2010), Sahu (2002), and Toribio (2006) also provide detailed discussions on the topic. Several of these methods were dependent on an MCMC algorithm. It is important to make sure, before the application of any model-fit or model-comparison method, that the MCMC algorithm has converged.

Additional methods that could be applied to IRT models and were not covered in this chapter include cross-validatory predictive checks (Marshall and Spiegelhalter, 2003), partial posterior predictive checks (Bayarri and Berger, 2000), Bayesian modification indices (Fox and Glas, 2004), the Bayesian goodness-of-fit test (Johnson, 2004), posterior BFs (Aitkin, 1991), pseudo-BF and expected predictive deviance (e.g., Sahu, 2002). An idea closely related to that of model fit and model comparison is that of examining the sensitivity of the inferences to the choice of the prior distribution and the likelihood. Any single model will usually underestimate the uncertainty in the inferences drawn. Other models could have fit the data equally well yet yielded different inferences. Model averaging (e.g., Carlin and Louis, 2008) is one approach to take account of this information. Another approach is to perform a *sensitivity analysis* that involves the fitting of several probability models to the same data set, altering either the prior distribution, the likelihood, or both, and studying how the primary inferences from the data change. Applications of sensitivity analysis can be found in Gelman et al. (2003), and in the context of IRT models, in Sheng (2010).

An important question that investigator of Bayesian IRT models faces, given so many choices of model-fit and model-comparison methods, is "Which of these methods are appropriate to my Bayesian IRT model?" The answer is specific to each problem. In applications of IRT models, one is often restricted to choose a specific IRT model. For example, several K-12 tests have to use the Rasch model by contract. In such cases, model comparison is not relevant and model-fit analysis is usually restricted to assessing item fit and removing the misfitting items. If the investigator is not restricted to a specific model and has a few possible models to choose the final model from, he/she should apply at least one model fit method and at least one model comparison method. Ideally, one fits an "initial" model, probably the simplest of the possible models, to the data and then iterates between the following steps:

1. Apply a model-fit method.
2. If misfit is found, fit an alternative model that, according to the knowledge of the investigator, is most likely to overcome that misfit.
3. Apply a model-comparison method; if the alternative model fits the data better than does the initial model and provides results that are practically different from those from the initial model, then go to Step 1; otherwise, discard the alternative model from the set of possible models and go to Step 2.

The iteration stops when one runs out of the possible models or finds one model with no misfit. If the final model does not fit the data adequately, one should extend the model, or, if that is not possible, use the model, but be aware of its practical limitations.

The most popular Bayesian model-fit and model-comparison methods in the context of IRT models are arguably the PPMC method and the DIC. Both of these have been applied by several researchers in the context of several types of models. Both of these methods can be implemented with little extra effort if one has already implemented an MCMC algorithm to fit the model of interest. Also, DIC is routinely computed by the WinBUGS software (Lunn et al. 2000) that is often used to fit Bayesian models using the MCMC algorithm. It is straightforward to generate posterior predictive data sets and employ the PPMC method using WinBUGS. For further details on BUGS, see Johnson (Chapter 21 in *Handbook of Item Response Theory, Volume Three: Applications*). In addition, as Muthen and Muthen (Chapter 28 in *Handbook of Item Response Theory, Volume Three: Applications*) and Asparouhov and Muthen (2010) show, Mplus can be used to apply the PPMC method to latent variable models. Therefore, it should be possible to use Mplus to apply the PPMC method to IRT models.

The issue of evaluating practical consequences of model misfit has been given little attention in IRT model fit and model comparison (Hambleton and Han, 2005; Sinharay, 2005), whether frequentist or Bayesian. A model can be demonstrably wrong but can still work for some purposes (Gelman et al. 2003, p. 176). Besides, it is possible that discrepancies between the test data and predictions from a model are of no practical consequences, but a statistical test indicates misfit (van der Linden and Hambleton, 1997, p. 16). Therefore, finding an extreme p-value and thus rejecting a model should not be the end of an analysis (Gelman et al. 2003, p. 176). The investigator should determine whether the misfit of the model has substantial practical consequences for the particular problem at hand and whether the better fit of an alternative model is of substantial practical consequences. The state test example in this paper includes an examination of whether the observed misfit of the Rasch model and the better fit of the 2PL model with an overall guessing parameter have any practical consequences. More such examples can be found in the Bayesian context in Sinharay (2005) and in the frequentist context in Hambleton and Han (2005) and Sinharay et al. (2011). Determination of the practical consequences of model misfit mostly involves additional statistical analysis.

Acknowledgments

The author thanks Wim van der Linden, Daniel Bolt, Jianbin Fu, Hanneke Geerlings, Shelby J. Haberman, Roy Levy, Sherwin Toribio, Peter van Rijn, Matthias von Davier, and Xiaowen Zhu for their helpful comments on an earlier draft. Most of the chapter was written when the author was an employee of Educational Testing Service, Princeton, New Jersey. The final version of this chapter was submitted in 2012. Therefore, references to some recent research are missing.

References

Aitkin, M. 1991. Posterior Bayes factors. *Journal of the Royal Statistical Society, Series B, 53*, 111–142.

Akaike, H. 1974. A new look at the statistical model identification. *IEEE Transactions on Automatic Control, 19*, 716–723.

Albert, J. H. and Chib, S. 1995. Bayesian residual analysis for binary response regression models. *Biometrika, 82*, 747–759.

American Educational Research Association, American Psychological Association and National Council for Measurement in Education. 1999. *Standards for Educational and Psychological Testing*. Washington DC: American Educational Research Association.

Asparouhov, T. and Muthen, B. 2010. *Bayesian Analysis Using mplus: Technical Implementation*. (Technical Report).

Bayarri, M. J. and Berger, J. O. 2000. P-values for composite null models. *Journal of the American Statistical Association, 95*, 1127–1142.

Beguin, A. A. and Glas, C. A. W. 2001. MCMC estimation and some fit analysis of multidimensional IRT models. *Psychometrika, 66*, 471–488.

Birnbaum, A. 1968. Some latent trait models and their use in inferring an examinee's ability. In F. M. Lord and M. R. Novick (Eds.), *Statistical Theories of Mental Test Scores*. Reading, MA: Addison-Wesley. (pp. 397–479).

Bolt, D. M., Cohen, A. S., and Wollack, J. A. 2001. A mixture item response model for multiple-choice data. *Journal of Educational and Behavioral Statistics, 26*, 381–409.

Bolt, D. M. and Lall, V. 2003. Estimation of compensatory and noncompensatory multidimensional item response models using Markov chain Monte Carlo. *Applied Psychological Measurement, 27*, 395–414.

Box, G. E. P. 1980. Sampling and Bayes' inference in scientific modelling and robustness. *Journal of the Royal Statistical Society, Series A, 143*, 383–430.

Bradlow, E. T., Wainer, H., and Wang, X. 1999. A Bayesian random effects model for testlets. *Psychometrika, 64*, 153–168.

Carlin, B. P. and Louis, T. A. 2008. *Bayesian Methods for Data Analysis*. Boca Raton, FL: Chapman and Hall.

Chaloner, K. and Brant, R. 1988. A Bayesian approach to outlier detection and residual analysis. *Biometrika, 75*, 651–659.

Chib, S. 1995. Marginal likelihood from the Gibbs output. *Journal of the American Statistical Association, 90*, 1313–1321.

Chib, S. and Jeliazkov, I. 2001. Marginal likelihood from the Metropolis-Hastings output. *Journal of the American Statistical Association, 96*, 270–281.

DiCiccio, J., T, Kass, R. E., Raftery, A., and Wasserman, L. 1997. Computing Bayes factors by combining simulation and asymptotic approximations. *Journal of the American Statistical Association, 92*, 903–915.

Dorans, N. J. and Feigenbaum, M. D. 1994. *Equating Issues Engendered by Changes to the SAT and PSAT/NMSQT* (ETS Research Memorandum No. 94–10). Princeton, NJ: ETS.

Fox, J. P. 2004. Multilevel IRT model assessment. In L. A. van der Ark, M. A. Croon, and K. Sijtsma (Eds.), *New Developments in Categorical Data Analysis for the Social and Behavioral Sciences*. London, UK: Lawrence Erlbaum (pp. 227–252).

Fox, J. P. 2005. Randomized item response theory models. *Journal of Educational and Behavioral Statistics, 30*, 189–212.

Fox, J. P. 2010. *Bayesian Item Response Modeling: Theory and Applications*. New York, NY: Springer.

Fox, J. P. and Glas, C. A. W. 2001. Bayesian estimation of a multilevel IRT model using Gibbs sampling. *Psychometrika, 66*, 269–286.

Fox, J. P. and Glas, C. A. W. 2003. Bayesian modeling of measurement error in predictor variables. *Psychometrika, 68*, 169–191.

Fox, J. P. and Glas, C. A. W. 2004. Bayesian modification indices for IRT models. *Statistica Neerlandica, 59*, 95–106.

Geerlings, H., Glas, C. A. W., and van der Linden, W. J. 2011. Modeling rule-based item generation. *Psychometrika, 76*, 337–359.

Gelman, A., Carlin, J. B., Stern, H. S., and Rubin, D. B. 2003. *Bayesian Data Analysis*. New York, NY: Chapman and Hall.

Gelman, A., Meng, X., and Stern, H. S. 1996. Posterior predictive assessment of model fitness via realized discrepancies. *Statistica Sinica, 6*, 733–807.

Glas, C. A. W. and Meijer, R. R. 2003. A Bayesian approach to person fit analysis in item response theory models. *Applied Psychological Measurement, 27*(3), 217–233.

Green, P. J. 1995. Reversible jump Markov chain Monte Carlo computation and Bayesian model determination. *Biometrika, 82*, 711–732.

Hambleton, R. K. and Han, N. 2005. Assessing the fit of IRT models to educational and psychological test data: A five step plan and several graphical displays. In W. R. Lenderking and D. Revicki (Eds.), *Advances in Health Outcomes Research Methods, Measurement, Statistical Analysis, and Clinical Applications*. Washington, DC: Degnon Associates (pp. 57–78).

Henson, R., Templin, J., and Willse, J. 2009. Defining a family of cognitive diagnosis models using log-linear models with latent variables. *Psychometrika, 74*, 191–210.

Hjort, N. L., Dahl, F. A., and Steinbakk, G. 2006. Post-processing posterior predictive *p*-values. *Journal of the American Statistical Association, 101*, 1157–1174.

Hoijtink, H. 2001. Conditional independence and differential item functioning in the two-parameter logistic model. In A. Boomsma, M. A. J. van Duijn, and T. A. B. Snijders (Eds.), *Essays in Item Response Theory*. New York, Springer. (pp. 109–130).

Janssen, R., Tuerlinckx, F., Meulders, M., and De Boeck, P. 2000. A hierarchical IRT model for criterion-referenced measurement. *Journal of Educational and Behavioral Statistics*, 25, 285–306.

Johnson, V. 2004. A Bayesian χ^2 test for goodness-of-fit. *Annals of Statistics, 32*, 2361–2384.

Kang, T. and Cohen, A. S. 2007. IRT model selection methods for dichotomous items. *Applied Psychological Measurement, 31*, 331–358.

Kang, T., Cohen, A. S., and Sung, H. 2009. Model selection indices for polytomous items. *Applied Psychological Measurement, 33*, 499–518.

Kass, R. E. and Raftery, A. E. 1995. Bayes factors. *Journal of the American Statistical Association, 90*, 773–795.

Klein Entink, R. H., Fox, J. P., and van der Linden, W. J. 2009. A multivariate multilevel approach to the modeling of accuracy and speed of test takers. *Psychometrika, 74*, 21–48.

Kolen, M. J. and Brennan, R. L. 2004. *Test Equating, Scaling, and Linking* (2nd ed.). New York, NY: Springer.

Levy, R., Mislevy, R. J., and Sinharay, S. 2009. Posterior predictive model checking for multidimensionality in item response theory. *Applied Psychological Measurement, 33*, 519–537.

Levy, R. and Svetina, D. 2011. A generalized dimensionality discrepancy measure for dimensionality assessment in multidimensional item response theory. *British Journal of Mathematical and Statistical Psychology, 64*, 208–232.

Li, Y., Bolt, D. M., and Fu, J. 2006. A comparison of alternative models for testlets. *Applied Psychological Measurement, 30*, 3–21.

Lunn, D. J., Thomas, A., Best, N., and Spiegelhalter, D. 2000. WinBUGS—a Bayesian modelling framework: Concepts, structure, and extensibility. *Statistics and Computing, 10*, 325–337.

Mariano, L. T., McCafferty, D. F., and Lockwood, J. R. 2010. A model for teacher effects from longitudinal data without assuming vertical scaling. *Journal of Educational and Behavioral Statistics, 35*, 253–279.

Marshall, E. C. and Spiegelhalter, D. J. 2003. Approximate cross-validatory predictive checks in disease mapping models. *Statistics in Medicine, 22*, 1649–1660.

Meng, X. and Wong, W. H. 1996. Simulating ratios of normalizing constants via a simple identity: a theoretical exploration. *Statistica Sinica, 6*, 831–860.

O'Hagan, A. 1995. Fractional Bayes factors for model comparison (with discussion). *Journal of the Royal Statistical Society, Series B, 57*, 99–138.

Patz, R. and Junker, B. 1999a. A straightforward approach to Markov chain Monte Carlo methods for item response models. *Journal of Educational and Behavioral Statistics, 24*, 146–178.

Patz, R. and Junker, B. 1999b. Applications and extensions of MCMC in IRT: Multiple item types, missing data, and rated responses. *Journal of Educational and Behavioral Statistics, 24*, 342–366.

Rubin, D. B. 1984. Bayesianly justifiable and relevant frequency calculations for the applied statistician. *Annals of Statistics, 12*, 1151–1172.

Sahu, S. 2002. Bayesian estimation and model choice in item response models. *Journal of Statistical Computation and Simulation, 72*, 217–232.

Sheng, Y. 2010. A sensitivity analysis of Gibbs sampling for 3PNO IRT models: Effects of prior specifications on parameter estimates. *Behaviormetrika, 37*, 87–110.

Sheng, Y. and Wikle, C. K. 2007. Comparing multi-unidimensional and unidimensional item response theory models. *Educational and Psychological Measurement, 67*, 899–919.

Sinharay, S. 2005. Assessing fit of unidimensional item response theory models using a Bayesian approach. *Journal of Educational Measurement, 42*, 375–394.

Sinharay, S. 2006a. Bayesian item fit analysis for unidimensional item response theory models. *British Journal of Mathematical and Statistical Psychology, 59*, 429–449.

Sinharay, S. 2006b. Model diagnostics for Bayesian networks. *Journal of Educational and Behavioral Statistics, 31*, 11–33.

Sinharay, S. and Almond, R. G. 2007. Assessing fit of cognitive diagnostic models: A case study. *Educational and Psycholgical Measurement, 67*, 239–257.

Sinharay, S., Haberman, S. J., and Jia, H. 2011. *Fit of Item Response Theory Models: A Survey of Data From Several Operatioanal Tests* (ETS Research Report No. RR-11-29). Princeton, NJ: ETS.

Sinharay, S. and Johnson, M. S. 2008. Use of item models in a large-scale admissions test: A case study. *International Journal of Testing*, 209–236.

Sinharay, S., Johnson, M. S., and Stern, H. S. 2006. Posterior predictive assessment of item response theory models. *Applied Psychological Measurement, 30*, 298–321.

Sinharay, S. and Stern, H. S. 2003. Posterior predictive model checking in hierarchical models. *Journal of Statistical Planning and Inference, 111*, 209–221.

Spiegelhalter, D. J., Best, N. G., Carlin, B. P., and Van Der Linde, A. 2002. Bayesian measures of model complexity and fit. *Journal of the Royal Statistical Society, Series B*, 583–640.

Toribio, S. G. 2006. Bayesian Model Checking Strategies for Dichotomous Item Response Theory Models. Unpublished doctoral dissertation, Bowling Green State University.

Toribio, S. G. and Albert, J. H. 2011. Discrepancy measures for item fit analysis in item response theory. *Journal of Statistical Computation and Simulation, 81*, 1345–1360.

van der Linden, W. J. and Hambleton, R. K. 1997. *Handbook of Modern Item Response Theory*. New York, NY: Springer.

Wainer, H. and Thissen, D. 1987. Estimating ability with the wrong model. *Journal of Educational Statistics, 12*, 339–368.

Yen, W. 1981. Using simulation results to choose a latent trait model. *Applied Psycholgoical Measurement, 5*, 245–262.

Zhang, S. 2008. Prior Predictive Checking of Item Response Theory (IRT) Models. Unpublished doctoral dissertation, University of Iowa.

Zhu, X. and Stone, C. 2011. Assessing fit of unidimensional graded response models using Bayesian methods. *Journal of Educational Measurement, 48*, 81–97.

20

Model Fit with Residual Analyses

Craig S. Wells and Ronald K. Hambleton

CONTENTS

20.1 Introduction

Item response theory (IRT) is a powerful scaling technique that uses a mathematical model to depict the probability of a correct response given item and person parameters. The advantages and attractive features of IRT are based on the invariance property in which the item parameter values retain the same values regardless of the ability distribution and the person parameters (Embretson and Reise, 2000; Hambleton et al., 1991). However, the invariance property only holds when the assumptions of the specific IRT model, such as undimensionality, local independence, and monotonicity 1 (Volume One, Chapter 2) are satisfied. Model fit, which is the focus of this chapter, indicates that these assumptions adequately match or represent the examinee response data.

Applying an IRT model that does not fit the data may lead to a loss of the all-important invariance property (Bejar, 1983; Bolt, 2002; Hambleton et al., 1991; Swaminathan et al., 2007; Wells and Keller, 2010). Therefore, model fit plays an important role in the valid application of IRT models in developing and maintaining a stable score scale. As a result, evaluating the fit of an IRT model is an important step in any situation where an IRT model is being applied. The purpose of this chapter is to describe how the analysis of residuals can help in assessing model fit. In doing so, our focus will be entirely on the evaluation of model fit for items. However, there are available methods too that examine person fit (see, e.g., Meijer and Sijtsma, 2001).

20.2 Raw and Standardized Residuals

Model fit at the item level is often assessed by comparing the observed proportions of examinees responding to a particular category to predictions based on the model being applied. The set of observed proportions is used as a proxy for the response function that is not constrained to follow any particular shape. If the model fits the data, then we expect the observed proportions for an item to be close to the model predictions given sampling error. The difference between the observed proportions and model-based predictions are referred to as *raw residuals* and provide the basis for evaluating model fit via a summary statistic (e.g., Yen's Q_1) (Yen, 1981) or graphical displays.

Although the steps used in obtaining the observed proportions can vary, a common procedure implemented in practice is as follows. First, the IRT model (e.g., 3-parameter logistic (3PL) model) is selected and the item and ability parameters are estimated. Second, the ability scale is divided into intervals in which the examinees are placed based on their ability estimates. The width of each interval and the number of intervals are dictated by the sample size and the variability in the sample with respect to the ability estimates, that is, more intervals can be used for larger sample sizes presuming the variability in the ability estimates is sufficiently large. Ideally, it is desirable to have as many intervals as possible so that we have many estimates of the response functions. In addition, we would like to have narrow intervals so that the examinees in each subgroup are homogeneous with respect to ability. However, having too many intervals with a narrow width can lead to unstable results due to small sample sizes per subgroup. For data sets with very large sample sizes (e.g., >10,000) and sufficient variability with respect to the ability estimates, many narrow-width intervals (e.g., 15–30) can be used. However, for smaller sample sizes (e.g., <1,000), one may be forced to use as few as 8–10 intervals. Third, once the examinees have been placed into their respective intervals based upon ability parameter estimates, the observed proportions, denoted p_{ij}, for item i in subgroup (i.e., interval) j can be computed by dividing the number of examinees who answered the item correctly by the number of examinees in the subgroup.

Once the observed proportions for each subgroup are computed for an item, the residuals for item i and subgroup j, denoted r_{ij}, are determined as follows: $r_{ij} = p_{ij} - P_{ij}$, where P_{ij} represents the model-based prediction for item i and subgroup j. There are a few ways of computing P_{ij}. For example, it can be based on the probability of a correct response at the midpoint of the interval. Another method is to use the average of the probabilities for all examinees within the respective subgroup. A software program by Liang et al. (2009) provides the flexibility to carry out residual and standardized residual (SR) analyses with many variations (e.g., number of intervals, equal or unequal widths, and positioning of the data within the score intervals) and observe the plots.

One of the challenges in interpreting model fit using raw residuals is that their magnitude is partly influenced by sampling error (i.e., we expect the raw residuals to fluctuate more for subgroups with smaller sample sizes). To address these limitations, the residuals are often standardized by dividing the raw residuals by their respective standard error. The SR for item i and subgroup j is computed as follows:

$$SR_{ij} = \frac{p_{ij} - P_{ij}}{\sqrt{\dfrac{P_{ij}(1 - P_{ij})}{N_j}}}, \tag{20.1}$$

TABLE 20.1

Example of Computing Observed Proportions, Raw Residuals, and Standardized Residuals for a Dichotomously Scored Item

Group	Ability Interval	Frequency	Frequency Correct	p_{ij}	P_{ij}	Raw Residual (r_{ij})	Standardized Residual (SR_{ij})
1	< −2.32	277	58	0.209	0.222	−0.013	−0.521
2	−2.32 to −1.96	193	43	0.223	0.230	−0.007	−0.231
3	−1.96 to −1.61	296	67	0.226	0.243	−0.017	−0.682
4	−1.61 to −1.25	447	120	0.268	0.267	0.001	0.048
5	−1.25 to −0.89	725	231	0.319	0.307	0.012	0.701
6	−0.89 to −0.54	1001	373	0.373	0.372	0.001	0.065
7	−0.54 to −0.18	1283	569	0.443	0.465	−0.022	−1.580
8	−0.18 to 0.18	1346	776	0.577	0.581	−0.004	−0.297
9	0.18 to 0.54	1340	931	0.695	0.702	−0.007	−0.560
10	0.54 to 0.89	1135	925	0.815	0.806	0.009	0.767
11	0.89 to 1.25	792	709	0.895	0.883	0.012	1.051
12	1.25 to 1.61	509	478	0.939	0.933	0.006	0.541
13	1.61 to 1.96	291	276	0.948	0.962	−0.014	−1.249
14	1.96 to 2.32	176	173	0.983	0.979	0.004	0.370
15	>2.32	189	188	0.995	0.989	0.006	0.791

where N_j represents the sample size for subgroup j. When the model fits the data, we expect the SRs to be relatively small and randomly distributed around zero along the ability scale.

As an example, suppose we have 40 dichotomously scored items and we want to evaluate the 3PL model for the data. First, the item and ability parameters are estimated using a software package, such as BILOG-MG (Zimowski et al., 2003). Second, examinees are placed into one of, say, 15 equally spaced intervals based on their ability parameter estimate. For this example, the boundaries that define the intervals along the ability scale range from −2.32 to 2.32 in increments of 0.36 (i.e., −2.32, −1.96, −1.61, . . . ,2.32). The number of intervals and the width of each intervals are selected so that the intervals would be narrow enough to produce homogeneous subgroups of ability scores but wide enough so that the size of each subgroup is sufficient to produce stable statistics. Table 20.1 provides the number of examinees in each interval or subgroup. Third, the observed proportions for item i are computed by dividing the number of examinees who answered the item correctly within a respective subgroup by the sample size for that subgroup. For example, of the 447 examinees in subgroup 4, 120 answered the item correctly, which produced an observed proportion correct of 0.268 (Table 20.1). Fourth, the model-based predictions for each subgroup on item i are determined using the item-parameter estimates and the ability parameter value at the midpoint of the interval (as already noted, one could also consider using the average of the model-based probabilities for each subgroup or the probability associated with the average of the ability parameter estimates for each subgroup). Fifth, the raw residual is computed by taking the difference between the observed proportion correct and the model-based prediction for each subgroup. For example, in subgroup 4, the raw residual is $0.268–0.267 = 0.001$ (Table 20.1). The raw residuals are standardized by dividing them by their respective standard error. For example, in subgroup 4, the standard error equals $\sqrt{0.267(1-0.267)/447} = 0.021$; therefore, the SR is $0.001/0.021 = 0.048$.

Since the SRs are small and there is no apparent pattern of positive and negative residuals indicating systematic misfit, we can conclude that model provides reasonable fit for this item.

It is interesting to note that the SRs for an item can be combined to produce a summary statistic to assess model fit; for example, the sum of the squared standardized residuals (SR^2) is typically treated as an approximately Chi-square distributed statistic, $\sum_j SR_{ij}^2 \approx \chi_i^2$. Although it may seem appealing to use a summary statistic to evaluate model fit via a statistical hypothesis test, there are several drawbacks to this approach. First, the actual degrees of freedom for the test statistic is unclear, making it difficult to know which Chi-square distribution to use (Orlando and Thissen, 2000; Stone and Zhang, 2003). Second, grouping examinees into the intervals based on ability parameter estimates, which contain error, influences the distribution of the test statistic (Orlando and Thissen, 2000; Stone, 2000; Stone and Hansen, 2000; Stone and Zhang, 2003). Third, even if the test statistic values were distributed as a Chi-square with known degrees of freedom, rejecting the null hypothesis only allows one to conclude that the model does not fit perfectly. In fact, with very large sample sizes, nearly, every item would be statistically significant since no parametric model will be able to represent the underlying response function perfectly (Wainer and Thissen, 1987). In other words, since mathematical models, such as parametric IRT models, are a simplification of reality, they cannot be reasonably expected to represent reality entirely. Therefore, every model will misfit to a certain degree. However, even a misfitting model can be useful. As a result, the goal of evaluating model fit is to determine whether the model provides a reasonable approximation to the data so that it is still useful for its intended purpose. Graphically displaying the residual information, which is described elsewhere, will help accomplish this goal.

20.3 Graphical Techniques for Assessing Model Fit

Graphical techniques are useful for portraying evidence of model fit in a convenient and informative display. Although residuals provide useful information for evaluating item-level model fit, they are difficult to interpret in tabular form, especially when analyzing many items. Using graphs provides the information in a visually appealing way such that a practitioner can easily and quickly judge if the model provides reasonable fit. One of the simplest methods of displaying model-fit information graphically is to plot the observed proportions along with the estimated response functions. Figure 20.1a provides an example of a response function from the 3PL model based on real data from a multiple-choice test with 10,000 examinees. The y-axis represents the probability of a correct response and the x-axis represents the ability scale. The solid line represents the model-based estimates and the points about the line represent the observed proportion correct for each subgroup. Given that the points are close to the response function and they appear to be randomly distributed about it indicates that the model provides reasonable fit. The raw residuals can be used to provide further information regarding the difference between the observed proportions and the response functions (Figure 20.1b). The y-axis represents the raw residual and the x-axis represents the ability scale. Again, the points in the raw residual plot are close to and randomly distributed around zero, and both these features support adequate model fit.

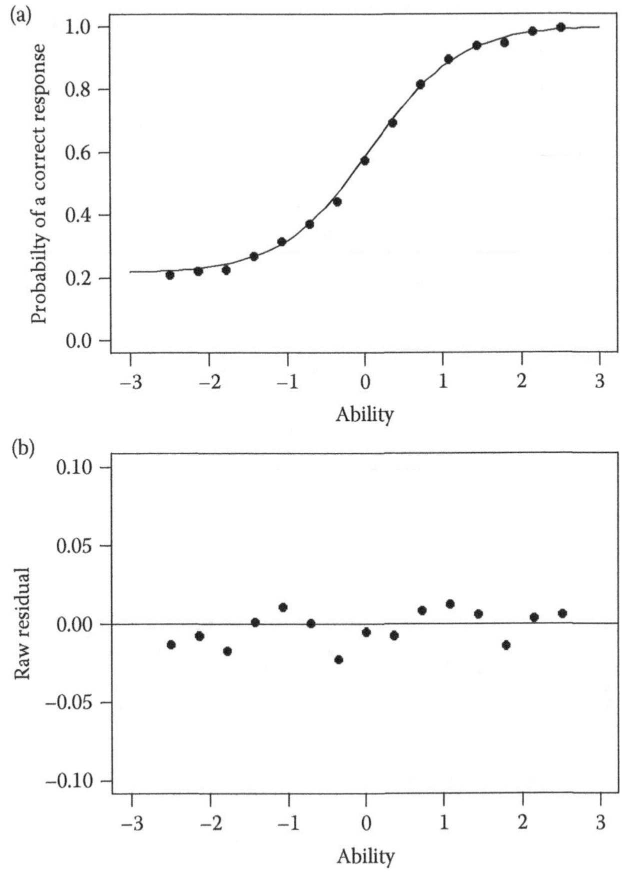

FIGURE 20.1
Example of a raw residual plot for an item fit by the 3PLM ($a = 1.75$, $b = 0.02$, $c = 0.21$). (a) Response function and observed proportions. (b) Raw residuals.

The graphs can be improved by incorporating standard errors to address the effect of sampling error. For example, we can insert confidence intervals on the functions to help interpret model fit. Figure 20.2a provides the response functions and observed proportions, but also includes vertical lines to represent the confidence interval for each subgroup's estimate. The widths of the intervals are based on the standard error shown in the denominator of Equation 20.1. Typically, the intervals are two or three standard errors above and below the estimates provided by the response functions. However, because of the very large sample sizes, we used three standard deviations. The width of the intervals are influenced by two factors, as was mentioned earlier—sample size and the conditional probability of a correct response. As the sample size in a subgroup increases, the width of the intervals tend to decrease (because the standard error is smaller). The intervals also tend to be smaller as the conditional probabilities approach 1 (or 0), controlling for sample size. When most of the observed proportions are within the confidence intervals, evidence supporting reasonable model fit is available.

It is also informative to plot the SRs. Figure 20.2b illustrates a plot of the SRs for each of the subgroups. The x-axis represents the ability scale and the y-axis represents the SR.

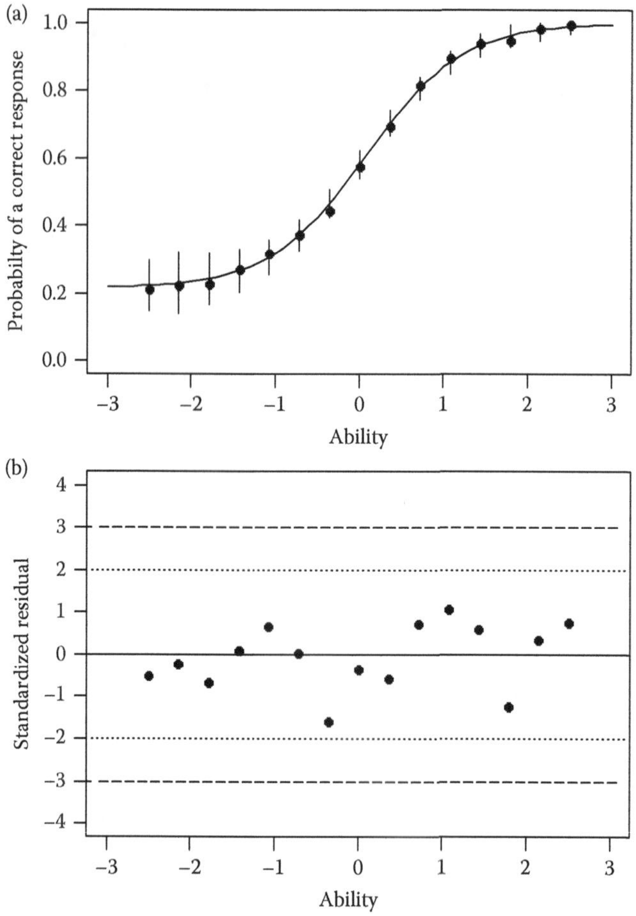

FIGURE 20.2
Example of standardized residuals and confidence intervals. (a) Response function with confidence intervals. (b) Standardized residuals.

The dotted and dashed lines at two and three SRs, respectively, help visually identify parts of the actual response function that the model may not accurately represent. For models that provide reasonable fit, the SRs are expected to be randomly distributed around 0 and within two to three standard errors, say. It is interesting to note the effect of the subgroup sample size on the SRs. For example, the raw residuals for the third and seventh subgroups are similar (−0.017 and −0.022, as reported in Table 20.1), and yet the SR for the seventh subgroup is much larger (−1.58) compared to the third subgroup's SR (−0.68). This occurs in part because the sample size for the seventh subgroup ($n = 1283$) is much larger than the sample size for the third subgroup ($n = 296$) which influences the standard error.

To understand how the residuals will vary across different models, we fit the 3PL, 2PL, and 1PL models to the same set of real data from a multiple-choice test. Figure 20.3a–c provide the fit plots for the three respective models. For the 3PL model (Figure 20.3a), the observed proportions are within the confidence intervals, with end points that are three standard errors above and below the response function. None of the SRs exceed two units. For the 2PL model, where the lower asymptote is set equal to zero, the SRs are slightly

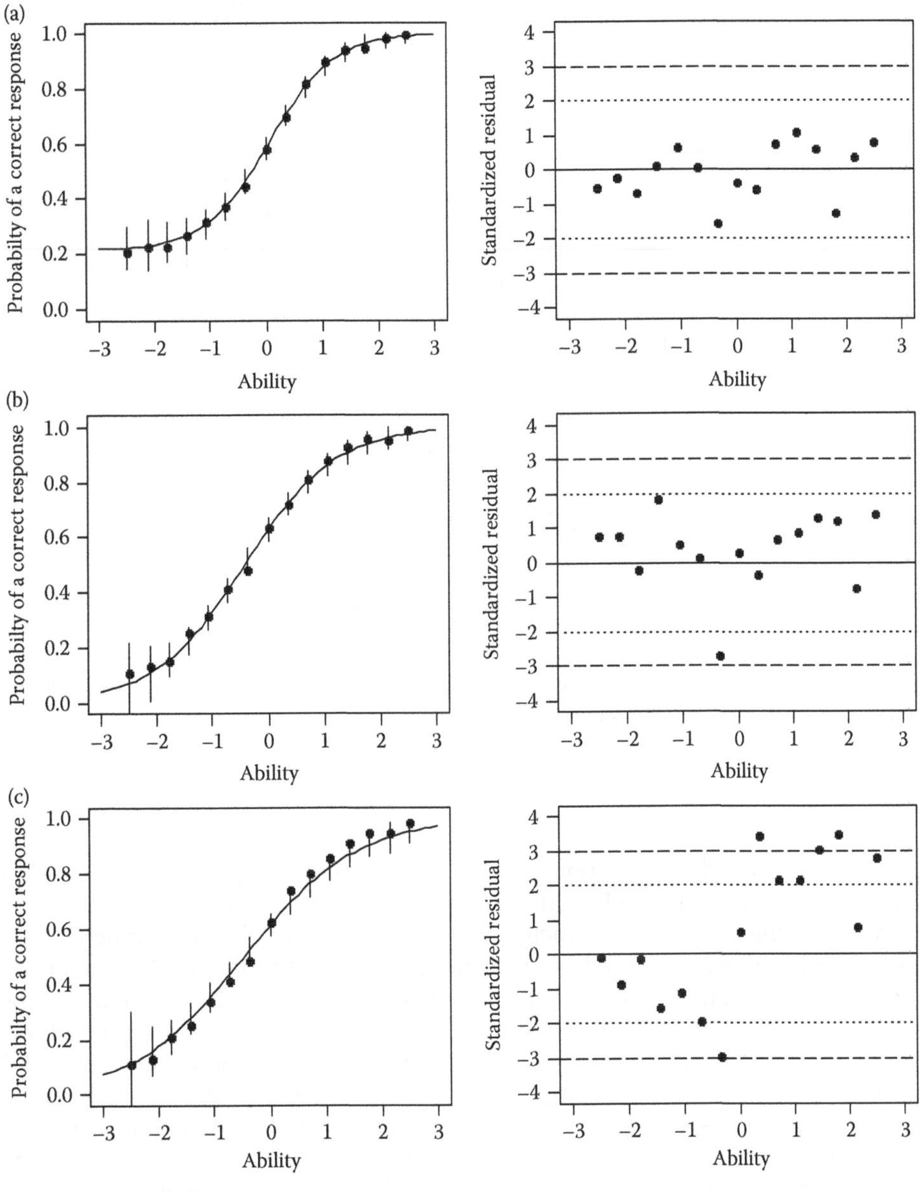

FIGURE 20.3
Comparison of (a) 3PL, (b) 2PL, and (c) 1PL model fit to the same item.

larger than for the 3PL model. However, the observed proportions are still within the confidence intervals, suggesting that the model appears to provide reasonable fit. For the 1PL model, where the discrimination parameters for all items are expected to be equal and the lower asymptote is set equal to zero, the observed proportions for subgroups 9, 12, and 13 are outside the confidence intervals. The SRs are also much larger compared to the other two models. In addition, the residuals do not appear to be randomly distributed (notice the distinct pattern of the SRs), with negative values at the lower portion of the ability scale

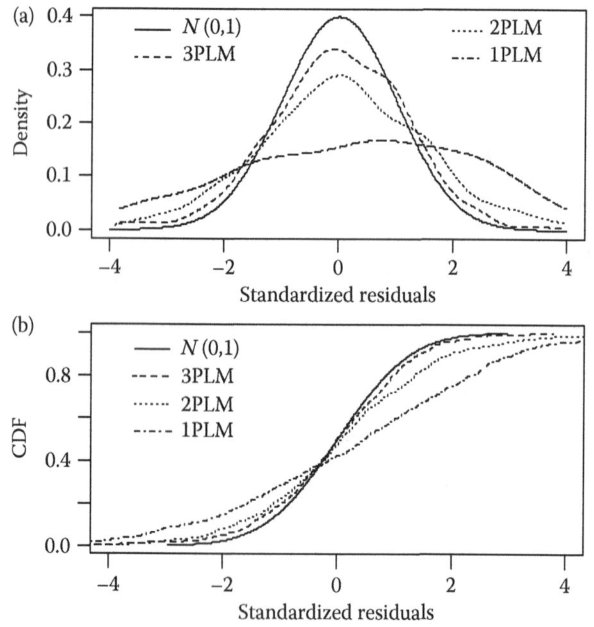

FIGURE 20.4
Distribution of standardized residuals. (a) Distribution of standardized residuals from fitting the 3PL, 2PL, and 1PL models. (b) Cdf of standardized residuals.

and positive values at the upper portion of it. When systematic patterns are spotted in SR plots, usually a better fitting IRT model can be found. From the response-function plot, it is apparent that the 1PLM overpredicted the probabilities for ability parameter estimates less than zero but underpredicted the probabilities for ability parameter estimates above zero. In this case, the large and systematically distributed residuals indicated that the 1PL model did not provide reasonable fit to the data. A 2PL or 3PL model would allow item discrimination to vary across the items to better match the data and improve model fit.

Once the residuals have been computed at the item level, it is possible to examine the distribution of residuals over all items. Figure 20.4a provides the observed distribution of the standardized residuals for the 3PL, 2PL, and 1PL models. The x-axis represents the SRs and the y-axis represents their relative frequencies. Although the SRs are theoretically not normally distributed for the same reasons as discussed for the assumption of the Chi-square distribution above, the standard normal curve is typically used as a reference point to judge how well the models fit the data based on the distribution of SRs. It is apparent from the graph that the residuals for the 1PLM across the items were much larger than expected given the standard normal distribution (i.e., there were more residuals in the tails of the distribution relative to the standard normal). The residuals for the 3PL model were more reasonably distributed, although there were still slightly more extreme values compared to the normal distribution. However, such slight misfit is to be expected since no parametric model fits perfectly and the normal distribution holds only approximately. The 2PL model provided much better fit over all of the items compared to the 1PL model but its fit was slightly worse than for the 3PL.

The distribution of residuals can be further explored by graphing the empirical cumulative distribution function (cdf) for each model. Figure 20.4b provides the cdfs for the 3PL, 2PL, and 1PL model as well as the standard normal distribution. The cdf for the 3PL model is reasonably close to the cdf for the standard normal. However, the cdf for the 1PL model is much flattering indicating that there were more residuals in the extreme of the distribution. Overall, it is apparent that the 3PL model did provide the best fit for these data.

20.4 Using Residuals to Evaluate Model Fit for Polytomous IRT Models

The same techniques previously described can be extended to assess the fit of polytomous IRT models, such as Samejima's (1969; Volume One, Chapter 6) graded-response model (GRM). The steps for computing the residuals are essentially the same as those used for dichotomous models. First, the model is selected and the item and ability parameters are estimated using a computer program. Second, the ability scale is divided into intervals and the examinees are placed into the intervals based on their ability parameter estimate. Third, the observed proportions for each score category are computed by dividing the number of examinees within a particular interval by the number of examinees who responded to the specific score category. Figure 20.5 provides an example of the observed proportions for a five score category item in which Samejima's GRM was used to model the data. The solid line in each plot provides the model-based estimate for responding to the respective score category and the points represent the proportion of examinees who responded to that particular score category. Similar to assessing fit for dichotomous models, if the polytomous model fits reasonably well, then we expect the observed proportions to be relatively close to and randomly distributed around the score category functions. For the data shown in Figure 20.5, Samejima's GRM appears to provide adequate fit due to the small and randomly distributed residuals. One could also plot the raw residuals for each score category function (SCF) as well as the SRs to help interpret model fit. The raw and SRs plots for each score category function would be very similar to those shown in Figures 20.1b and 20.2b.

For Samejima's graded-response model (GRM) where the score category functions (SCFs) are based on cumulative score category functions (CSCFs) (Volume One, Chapter 6), it is also possible to plot the observed proportions with the CSCFs instead of the SCFs (see, e.g., Liang et al., 2009). Figure 20.6 provides an example of the CSCFs and observed proportions for the same item shown in Figure 20.5. The solid lines represent the CSCFs and the points represent the observed proportions for the respective CSCFs. When analyzing CSCFs, the observed proportions are computed differently than when using SCFs. For Samejima's GRM, instead of computing the proportion of examines who responded to a particular category, the proportions are based on the examinees who responded at or above category k, where k ranges from 0 to K (and K equals the maximum score on the test item). For example, the first CSCF in Figure 20.6 (the curve on the left) represents the probability of responding to the score category or higher.

Although examining fit using the CSCFs or SCFs will produce identical results, one advantage of examining the CSCFs is that it is possible to observe the nature of the misfit. For example, one version of Samejima's GRM (the homogeneous model case) assumes that the slopes of the CSCFs are equal across score categories within an item. If the data

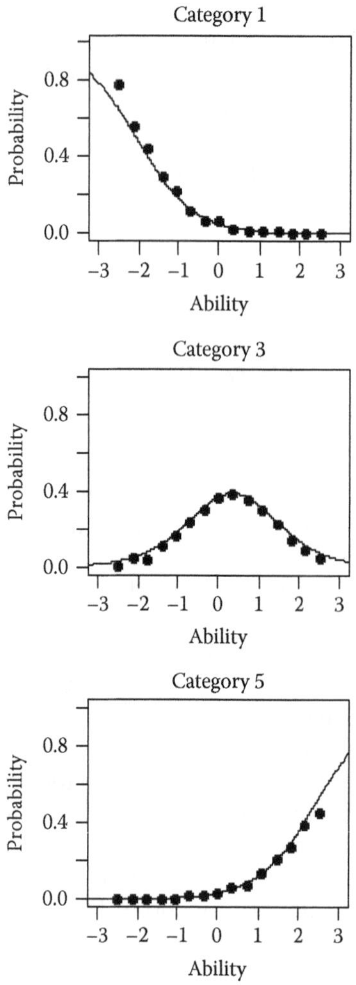

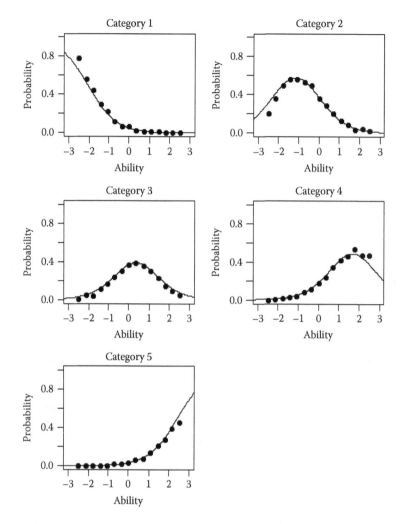

FIGURE 20.5
Example of model fit plot based on score category functions for Samejima's GRM ($a = 1.39$, $b_1 = -2.05$, $b_2 = -0.23$, $b_3 = 0.91$, $b_4 = 2.39$).

do not suggest equal slopes, however, then the residuals may reveal a systematic pattern. Figure 20.7 illustrates a misfitting item in which the best fitting slopes for the CSCFs are not equal for a specific item. Confidence intervals using three standard errors have been included to help interpret model fit. The standard errors used to determine the confidence intervals are computed in the same fashion as shown in the denominator of Equation 20.1. It is apparent from the large residuals that the GRM did not fit the data well. Several observed proportions for the first and last CSCFs are well beyond three standard errors from the model-based estimates. In addition, the residuals are systematically distributed above or below the CSCFs indicating the equal slope assumption is questionable.

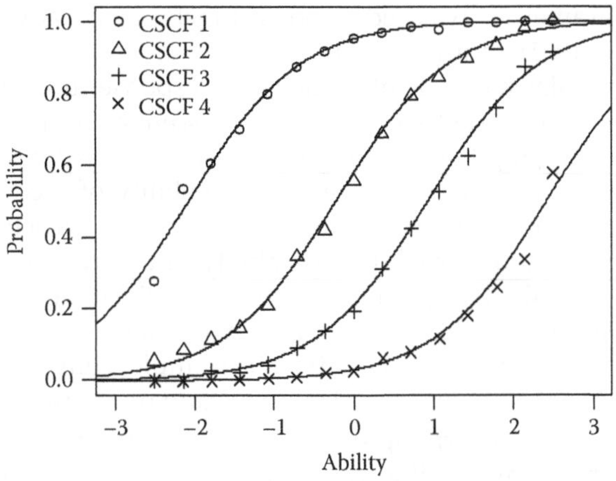

FIGURE 20.6
Example of model fit plot based on cumulative score category functions for Samejima's GRM ($a = 1.39$, $b_1 = -2.05$, $b_2 = -0.23$, $b_3 = 0.91$, $b_4 = 2.39$).

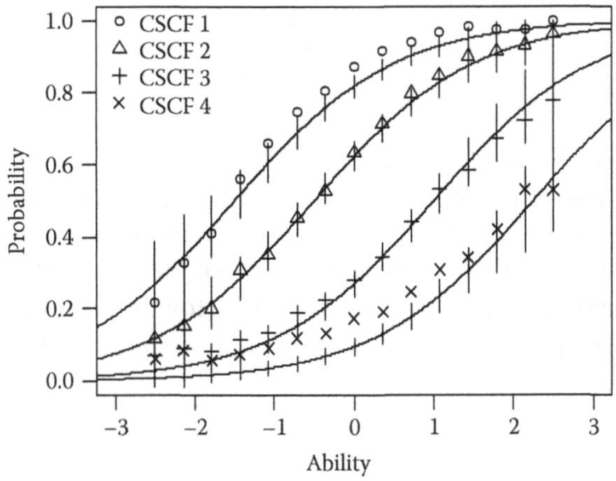

FIGURE 20.7
Example of Samejima's GRM with misfit due to unequal slopes across cumulative score category functions (CSCF) 1 and 4.

20.5 Using Nonparametric IRT to Assess Model Fit

Nonparametric IRT in the form of kernel smoothing (Ramsay, 1991; Volume One, Chapter 20) has been shown to offer a promising tool for evaluating the fit of parametric models (Douglas and Cohen, 2001; Liang and Wells, 2009; Liang et al., 2014; Lord, 1970; Wells and Bolt, 2008). Essentially, model fit is assessed by comparing estimates of nonparametric response functions against the functions for the parametric model of interest. Generally speaking, a nonparametric approach has fewer restrictions imposed on the shape of the response function; therefore, if the parametrically based response function differs from the nonparametric response function, then the parametric model may be incorrect. Lord (1970) first introduced this method with the purpose of determining if empirically estimated response functions followed a specific logistic form. Since Lord (1970), the methods for obtaining a nonparametrically based response function have advanced (Douglas, 1997; Eubank, 1988; Hardle, 1991; Ramsay, 1991).

One of the more popular methods of modeling a response function nonparametrically, used by Ramsay (1991), is kernel smoothing. The essential principle underlying kernel smoothing is local averaging. Local averaging is useful in response function estimation because, for a dichotomous item response, the probability of a correct response can be portrayed as an average response for examinees close to a particular ability parameter value. Kernel-smoothed estimates of the probability of a correct response for item i are obtained for a set of Q evaluation points, denoted x_q, defined along the ability parameter scale (e.g., $x_1 = -3.00$, $x_2 = -2.88, \ldots, x_{50} = 2.88$, $x_{51} = 3.00$, where $Q = 51$). An estimate of ability is provided for each examinee p by, for example, converting the rest score (i.e., total score excluding the item being modeled) to a standard score. We obtain the standard score to represent $\hat{\theta}_p$ by transforming each examinee's empirical percentile of the total score distribution to the corresponding value from the ability distribution (Ramsay, 1991). For example, if the ability distribution follows the standard normal distribution, then an examinee with a rest score at the 5th percentile would receive an estimate of -1.645.

The kernel-smoothed response function for item i and evaluation point q is estimated as follows:

$$\hat{P}_{non,iq} = \sum_{j=1}^{N} w_{pq} U_{ip}, \tag{20.2}$$

where w_{pq} represents the weight assigned to each examinee at each evaluation point; N indicates the number of examinees; and U_{ip} is the response by examinee p on item i (i.e., 0 or 1). Weights are assigned to each examinee with respect to each evaluation point; therefore, each examinee receives Q weights. The weight for examinee p increases as the evaluation point moves closer to $\hat{\theta}_p$ and decreases toward zero as the evaluation point moves further away from $\hat{\theta}_p$. The weight for examinee p at evaluation point x_q may be calculated as

$$w_{pq} = \frac{K(\hat{\theta}_p - x_q/h)}{\sum_{p=1}^{N} K(\hat{\theta}_j - x_q/h)}, \tag{20.3}$$

where x_q refers to evaluation point q and $\hat{\theta}_p$ is the ability estimate for examinee p. Two other important components of the formula are h and K. The value h is referred to as the bandwidth or smoothing parameter because it controls the amount of bias and inaccuracy in the estimated response function. As h decreases, the amount of bias is reduced but the inaccuracy is increased (i.e., less smoothness). The opposite occurs when h is increased. Parameter h is often set equal to $1.1^* N^{-.2}$ so as to produce a smoothed function with small bias. K is referred to as the kernel-smoothing function. K is always greater than or equal to 0 and approaches zero as $\hat{\theta}_p$ moves away from a particular evaluation point, x_q. Two commonly used kernel functions are the Gaussian $[K(u) = \exp(-u^2)]$ and uniform $[K(u) = 1$ if $|u| \leq 1$, else 0$]$. Given the previous information, it is apparent that the further an examinee is from the evaluation point, the less weight that examinee has in determining $\hat{P}_{non,iq}$, especially compared to an examinee who has a $\hat{\theta}$ equal to the evaluation point of interest since the density for a Gaussian, for instance, is largest at $u = 0$.

The nonparametric approach based on kernel smoothing is similar to using the observed proportions previously described, in that both the nonparametric IRT model and the observed proportions are intended to be proxies to the underlying response function. In fact, Douglas (1997) showed that for medium-length tests and medium sample sizes, the underlying curve can be consistently estimated using nonparametric response functions under a set of weak assumptions. However, an advantage of using nonparametric IRT to represent the underlying response function is that the kernel-smoothed estimates are not influenced by the parametric model in comparison to the previous methods where examinees are placed into intervals based their ability parameter estimate. The same advantage does not hold for grouping examinees into intervals based on the respective IRT model's ability parameter estimate. For example, the observed proportions in each interval for the 3PLM, 2PLM, and 1PLM shown in Figure 20.3 vary across the models, even though the intervals are the same. Furthermore, it is apparent that for the 1PLM and 2PLM, the observed proportions are closer to zero for extreme negative ability parameter estimates (i.e., near the lower asymptote) compared to the 3PLM. As a result, the model misfit may not seem as severe as it should be for the 1PLM and 2PLM since both models fix the lower asymptote to zero. This is an unattractive feature because the observed data are not accurately representing the underlying response function. The nonparametric approach does not suffer this fate since the rest scores are used to represent the ability parameter estimate.

The solid line shown in Figure 20.8 represents the kernel-smoothed response function for a multiple-choice item from a large-scale assessment. The vertical lines represent the confidence intervals based on three standard errors for each smoothed estimate. As standard error for item i and evaluation point q, we suggest computing

$$\hat{P}_{non,iq} \pm 3 * \sqrt{\sum_{p=1}^{N} w_{pq}^2 \hat{P}(\theta_p)[1 - \hat{P}(\theta_p)]}, \tag{20.4}$$

where $\hat{P}(\theta_p)$ represents the probability of examinee p correctly answering item i based on the smoothed response function.

Once the nonparametric response function is obtained, the next step is to estimate the function for the respective parametric model. There are several ways of obtaining the parametric curve that may be compared to the nonparametric response function. One method

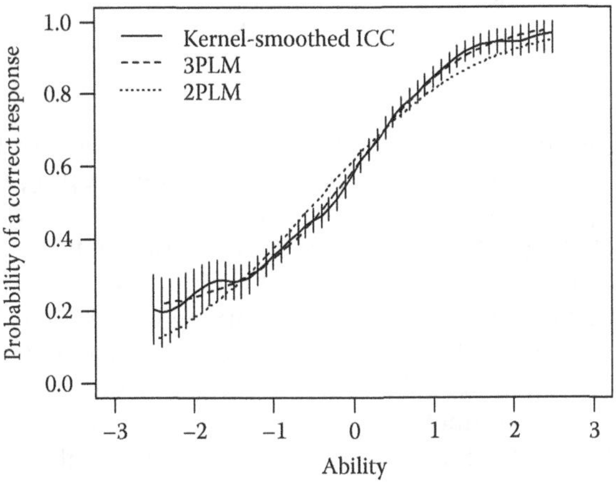

FIGURE 20.8
Comparison of kernel-smoothed response function to the same data fit by the 3PLM and 2PLM.

is to estimate the parametric curve using, for example, marginal maximum-likelihood estimation (MMLE). It is important to note that for this method to be appropriate, the ability distribution must be defined in the same manner for both the nonparametric and parametric response functions (Douglas and Cohen, 2001). For example, if the standard normal distribution is used in MMLE, then ability estimates used to obtain the nonparametric curve must follow the same distribution. A second method, illustrated by Douglas and Cohen (2001) for the 2PL model, finds the item parameter estimates by regressing the logit $\hat{P}_{non,iq}$ onto the evaluation points. A third method that has been shown to be effective via simulation studies (Liang and Wells, 2009; Liang et al., 2014; Wells and Bolt, 2008) is an analytic solution that implements maximum likelihood estimation (MLE) developed by Bolt (2002). The MLE procedure is essentially the same as estimating the item parameters when using real data and known θs (i.e., maximizing the log-likelihood) except for two primary differences. First, the observed proportions are replaced by probabilities from the nonparametric response function. Second, a weight associated with each quadrature point is incorporated into the log-likelihood. The weight, which is based on the density that represents the θ-distribution in the population (e.g., standard normal), is normalized (i.e., $\sum_{q=1}^{Q} w(x_q) = 1$) and determines the relative importance of the probabilities matched to the nonparametric response function probabilities at the respective quadrature point.

The dashed and dotted lines shown in Figure 20.8 represent the best fitting response functions for the 3PLM and 2PLM, respectively. It appears that the 3PLM provides reasonable fit since the model-based estimates are very close to the nonparametric response function. However, the 2PLM response function diverges from the nonparametric curve beyond the confidence intervals in several places throughout the ability parameter scale indicating the model does not provide adequate fit.

Douglas and Cohen (2001) proposed a statistic to summarize the difference between the nonparametric and parametric response functions referred to as the root integrated squared error (RISE). A discrete approximation of the RISE statistic for a dichotomously scored item is computed as follows:

$$\text{RISE}_i = \sqrt{\frac{\sum_{q=1}^{Q}(\hat{P}_{iq} - \hat{P}_{non,iq})^2}{Q}}, \quad (20.5)$$

where $\hat{P}_q$ and $\hat{P}_{non,iq}$ represent points on the response functions for the model-based and nonparametric methods, respectively. A parametric bootstrapping method was adopted to approximate the significance level for each item. The bootstrapping procedure entails simulating M data sets (e.g., $M = 500$) under the condition that the parametric model fits and sampling N θ's from the standard normal distribution. The statistic RISE is calculated for each simulated data set, producing a sampling distribution for RISE. The p value associated with the observed RISE statistic for each item, based on the original data, is determined by the proportion of RISEs from the simulated data that were greater than the observed RISE value. The parametric bootstrapping procedure and *RISE* have exhibited controlled (or slightly deflated) Type-I error rates and adequate power to detect model misfit (Liang and Wells, 2009; Liang et al., 2014; Wells and Bolt, 2008). However, items that are identified as misfiting should be examined graphically to determine whether the misfit is non-negligible.

20.6 Evaluating Model Fit at the Test Score Level

To date, we examined techniques for evaluating model fit at the item level. It is possible, and also informative, to evaluate model fit at the test score level. For example, the nonparametric approach can be extended to examine model fit at the test score level by comparing test characteristic curves (TCCs) based on the nonparametric and parametric response functions. The TCCs are simply determined by summing the response probabilities across items for each evaluation point. Figure 20.9 provides the TCCs based on the kernel-smoothed response functions (solid line), 3PLM (dashed line), and 2PLM (dotted line) for a 45-item, multiple-choice test. The TCCs for the 3PLM and 2PLM are based on the best fitting response functions given the kernel-smoothed estimates. In this example, the TCC based on the 3PLM is very similar to the nonparametric TCC indicating that at the test score level the 3PLM provides a reasonable approximation to the expected true scores. However, the TCC based on the 2PLM diverges from the nonparametric TCC for much of the ability parameter scale. It is clear in this example that the 3PLM provides better fit to the data compared to the 2PLM.

In addition to comparing the TCCs, we can examine the residuals of the observed score (cumulative) probability functions to determine how well the model predicts the observed raw score distribution using the item and ability parameter estimates (Ferrando and Lorenzo, 2001; Hambleton and Traub, 1973). There are basically two ways to obtain the model-predicted raw score distribution. The first method implements the Lord and Wingersky (1984) recursive formula. For dichotomously scored items, the model-based raw score distribution given its ability parameter is a compound binomial distribution. For polytomously scored items, the model-predicted raw score distribution is a compound multinomial distribution. In both cases, the results need to be integrated over the ability parameter if the interest is in the marginal observed-score distribution for a given population. Details for implementing the Lord–Wingersky method both for dichotomous and polytomous items can be found in Kolen and Brennan (1995, pp. 181–183), (see Chapter 6), and Wang et al. (2000).

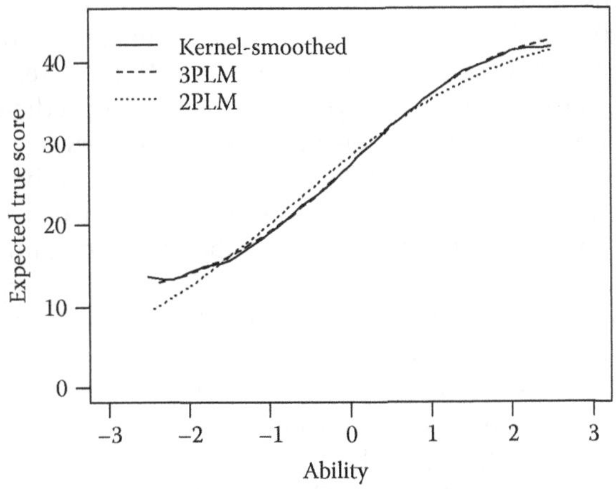

FIGURE 20.9
Comparison of TCCs based on kernel-smoothed response functions and the 3PLM and 2PLM.

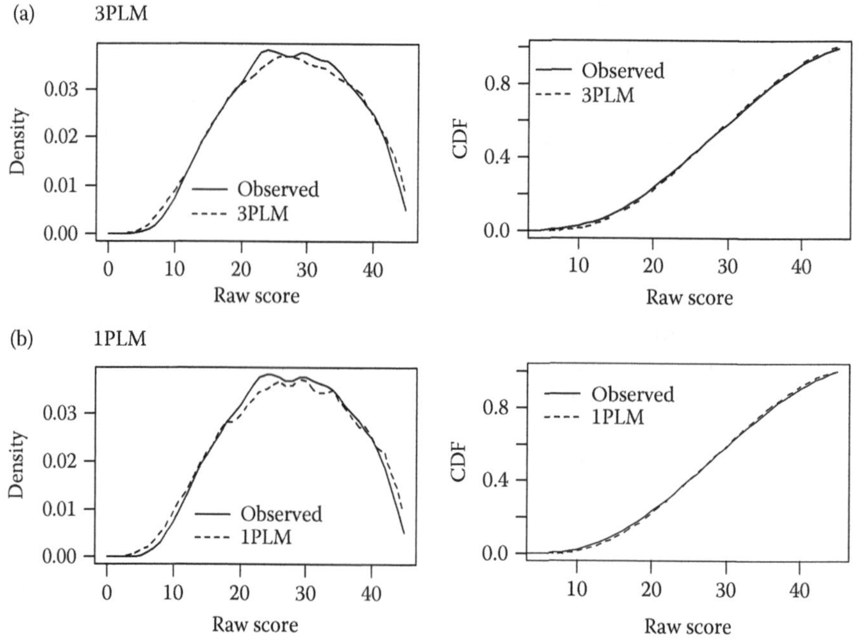

FIGURE 20.10
Comparison of model-predicted and observed-score distribution for the (a) 3PLM and (b) 1PLM.

The second method, which treats the distribution of the ability estimates for the examinees in the data set as the distribution of their true abilities, uses a Monte Carlo simulation technique to create raw scores using the item and ability parameter estimates from the respective model (Hambleton and Han, 2005). Essentially, the item and ability parameter estimates are treated as true values and used to simulate dichotomous and/or polytomous item responses. The model-predicted score distribution is simply based on the raw scores from the simulated data. The plots in the first column in Figure 20.10 show the predicted- and observed-score distributions when fitting the 3PLM and 1PLM to a 45-item multiple-choice test with 10,000 examinees. The y-axis represents the density of the distribution and the x-axis represents the raw score. The solid line in both plots represents the observed-scored distribution and the dashed line represents the model-predicted distribution based on the respective model. Since the model-predicted and observed-score distributions are similar, it appears that both models are able to reasonably predict the observed score distribution. The observed and predicted raw score distributions can also be compared using the cumulative distribution function (cdf). The plots in the second column of Figure 20.10 show the cdfs for the observed and predicted distributions. The solid line represents the observed data and the dashed line represents the model-predicted distribution for the respective model. Consistent with the plots shown in the first column of Figure 20.10, the model-predicted and observed cdfs are very similar indicating that the models appear to predict the observed-score distribution reasonably well.

It is also possible, if desired, to test the hypothesis that the distributions are identical using the Kolmogorov–Smirnov (KS) test. The KS test compares the cdfs for two distributions and the sample statistic, denoted D, is based on the maximum difference in the cdfs for any raw score value. In this example, the D statistic for the 3PLM was 0.015 with a p

value of 0.24. For the 1PLM, the D statistic was 0.017 with a p value of 0.15. Although the hypothesis test can be useful, it is possible to obtain a significant result and yet the model produces a reasonable approximation to the observed data. Therefore, it is still important to examine the graphical displays to determine if the size of the residuals (i.e., difference in the distributions) is non-negligible. Furthermore, the KS test can be used in a descriptive manner by comparing several attempts to predict the raw score distribution, for example, across several models.

20.7 Summary and Conclusions

Model fit plays an important role in the valid application of IRT models. The purpose of the present chapter was to describe how to use residuals to assess model fit. Raw residuals were defined as the difference between the observed proportions correct minus the model-based probabilities. To aid interpretability, the residuals are standardized by dividing them by their standard error. Although the SRs can be combined to produce a summary statistic, this is often unhelpful since a large statistic does not necessarily imply problematic misfit. Therefore, graphical displays of the residuals are emphasized since they provide more useful information about the type and magnitude of the misfit. Furthermore, we described how to compare the fit at the test score level as well as comparing score distributions.

Although the techniques described in this chapter are effective in assessing model fit, it is important to note that there is room for further research on the use of residuals to assess model fit. For example, Haberman and Sinharay (2013) recently developed a method of computing residuals that have the attractive feature of being, in theory, asymptotically normally distributed. The advantage of their approach is that it is theoretically more appropriate to use their residuals as a summary statistic for hypothesis testing compared to combining the SRs described in this chapter. However, simulation research has shown that the summary statistic based on their residuals has inflated Type I error rates for moderate to long test lengths. In addition, since no parametric model will fit real data, testing a null hypothesis without any specific alternative is often meaningless. Nonetheless, the Haberman and Sinharay approach can still be used graphically as illustrated in this chapter.

It is helpful to have readily available software to create residual information and plots illustrated in this chapter. Although IRT software such as *BILOG-MG* (Volume Three, Chapter 22) and *MULTILOG* can produce plots of response functions with observed proportions, the information provided is limited. The software package *ResidPlots-2* (Liang et al., 2009) can be used to obtain much of the information described in this chapter. *ResidPlots-2* uses item and person parameters that have been estimated from one of the commercial programs (e.g., *BILOG-MG*) with the raw data to assess model fit using raw and standardized residuals, observed and model-based proficiency distributions, and summary statistics. The software produces several useful graphical displays to report the information in an informative manner.

When assessing model fit using residuals, especially via graphical inspection, a challenge one faces is how to report the results. For test developers working in an operational testing program, it may be possible to simply provide the graphs in an Appendix as part of a technical report. However, that is not possible when reporting model fit results in a journal article where space is limited. In such a case, it may be prudent to report only the worst fitting items and/or the distribution of residuals for several models to illustrate that the final model was the most appropriate option.

References

Bejar, I. I. 1983. Introduction to item response models and their assumptions. In R. K. Hambleton (Ed.), *Applications of Item Response Theory* (pp. 1–23). Vancouver, B.C.: Educational Research Institute of British Columbia.

Bolt, D. M. 2002. A Monte Carlo comparison of parametric and nonparametric polytomous DIF detection methods. *Applied Measurement in Education, 15,* 113–141.

Douglas, J. 1997. Joint consistency of nonparametric item characteristic curve and ability estimation. *Psychometrika, 62,* 7–28.

Douglas, J. and Cohen, A. S. 2001. Nonparametric item response function estimation for assessing parametric model fit. *Applied Psychological Measurement, 25,* 234–243.

Embretson, S. E. and Reise, S. P. 2000. *Item Response Theory for Psychologists.* Mahwah, NJ: Lawrence Erlbaum Associates, Publishers.

Eubank, R. L. 1988. *Spline Smoothing and Nonparametric Regression.* New York: Marcel Dekker.

Ferrando, P. J. and Lorenzo, U. 2001. Checking the appropriateness of item response theory models by predicting the distribution of observed scores: The program EP-Fit. *Educational and Psychological Measurement, 61,* 895–902.

Haberman, S. J. and Sinharay, S. 2013. Assessing item fit for unidimensional item response theory models using residuals from estimated item response functions. *Psychometrika, 78,* 417–440.

Hambleton, R. K. and Han, N. 2005. Assessing the fit of IRT models to educational and psychological test data: A five-step plan and several graphical displays. In R. R. Lenderking and D. A. Revicki (Eds.), *Advancing Health Outcomes Research Methods and Clinical Applications.* McLean, VA: Degnon Associates, pp. 57–77.

Hambleton, R. K., Swaminathan, H., and Rogers, H. J. 1991. *Fundamentals of Item Response Theory.* Newbury Park, CA: SAGE Publications, Inc.

Hambleton, R. K. and Traub, R. E. 1973. Analysis of empirical data using two logistic latent trait models. *British Journal of Mathematical and Statistical Psychology, 26,* 195–211.

Hardle, W. 1991. *Smoothing Techniques with Implementation in S.* New York: Springer-Verlag.

Kolen, M. J. and Brennan, R. L. 1995. *Test Equating: Methods and Practices.* New York, NY: Springer-Verlag.

Liang, T., Han, K. T., and Hambleton, R. K. 2009. ResidPlots-2: Computer software for IRT graphical residual analyses. *Applied Psychological Measurement, 33(5),* 411–412.

Liang, T. and Wells, C. S. 2009. A model fit statistic for the generalized partial credit model. *Educational and Psychological Measurement, 69,* 913–928.

Liang, T., Wells, C. S., and Hambleton, R. K. 2014. An assessment of the nonparametric approach for evaluating the fit of item response models. *Journal of Educational Measurement, 51,* 1–17.

Lord, F. M. 1970. Item characteristic curves estimated without knowledge of their mathematical form – a confrontation of Birnbaum's logistic model. *Psychometrika, 35,* 43–50.

Lord, F. M. and Wingersky, M. S. 1984. Comparison of IRT true-score and equipercentile observed-score equating. *Applied Psychological Measurement, 8,* 452–461.

Meijer, R. R. and Sijtsma, K. 2001. Methodology review: Evaluating person fit. *Applied Psychological Measurement, 25,* 107–135.

Orlando, M. and Thissen, D. 2000. Likelihood-based item-fit indices for dichotomous item response theory models. *Applied Psychological Measurement, 24*(1), 50–64.

Ramsay, J. O. 1991. Kernel smoothing approaches to nonparametric item characteristic curve estimation. *Psychometrika, 60,* 323–339.

Samejima, F. 1969. Estimation of latent trait ability using a response pattern of graded scores. *Psychometika Monograph,* No. 17.

Stone, C. A. 2000. Monte Carlo based null distribution for an alternative goodness-of-fit statistic in IRT models. *Journal of Educational Measurement, 37,* 58–75.

Stone, C. A. and Hansen, M. A. 2000. The effect of errors in estimating ability on goodness-of-fit tests for IRT models. *Educational and Psychological Measurement, 60,* 974–991.

Stone, C., A. and Zhang, B. 2003. Assessing goodness of fit of item response theory models: A comparison of traditional and alternative procedures. *Journal of Educational Measurement, 40,* 331–352.

Swaminathan, H., Hambleton, R. K., and Rogers, H. J. 2007. Assessing the fit of item response theory models. In C. R. Rao and S. Sinharay (Eds.), *Handbook of Statistics 26: Psychometrics*. Amsterdam: North-Holland, pp. 683–713.

Wainer, H. and Thissen, D. 1987. Estimating ability with the wrong model. *Journal of Educational and Behavioral Statistics, 12,* 339–368.

Wang, T., Kolen, M. W., and Harris, D. J. 2000. Psychometric properties of scale scores and performance levels for performance assessments using polytomous IRT. *Journal of Educational Measurement, 37,* 141–163.

Wells, C. S. and Bolt, D. M. 2008. Investigation of a nonparametric procedure for assessing goodness-of-fit in item response theory. *Applied Measurement in Education, 21*(1), 22–40.

Wells, C. S. and Keller, L. A. 2010. Effect of model misfit on parameter invariance. Paper presented at the meeting of the National Council on Measurement in Education, Denver, CO.

Yen, W. M. 1981. Using simulation results to choose a latent trait model. *Applied Psychological Meas*urement, *5,* 245–262.

Zimowski, M. F., Muraki, E., Mislevy, R. J., and Bock, R. D. 2003. *BILOG-MG 3.0* [computer software]. Lincolnwood, IL: Scientific Software International.

Index

C

D

E

F

G

N

R

S

T

U